STUDENT SOLUTIONS MANUAL

REVISED PRINTING

to accompany

CALCULUS for Biology and Medicine

Second Edition

Claudia Neuhauser
University of Minnesota

Upper Saddle River, NJ 07458

Editor-in-Chief: Sally Yagan
Acquisitions Editor: Adam Jaworski
Project Manager: Dawn Murrin
Executive Managing Editor: Kathleen Schiaparelli
Senior Managing Editor: Nicole M. Jackson
Assistant Managing Editor: Karen Bosch Petrov
Production Editor: Jessica Barna
Supplement Cover Manager: Paul Gourhan
Supplement Cover Designer: Christopher Kossa
Manufacturing Buyer: Ilene Kahn
Manufacturing Manager: Alexis Heydt-Long

Pearson Prentice Hall
Pearson Education, Inc.
Upper Saddle River, NJ 07458

Printed in the United States of America

10 9 8 7 6 5 4 3 2

ISBN 0-13-199672-X

Pearson Education Ltd., *London*
Pearson Education Australia Pty. Ltd., *Sydney*
Pearson Education Singapore, Pte. Ltd.
Pearson Education North Asia Ltd., *Hong Kong*
Pearson Education Canada, Inc., *Toronto*
Pearson Educación de Mexico, S.A. de C.V.
Pearson Education—Japan, *Tokyo*
Pearson Education Malaysia, Pte. Ltd.

Contents

1 Preview and Review

1.1 Preliminaries

Prob. 1.

(a) Walking 3 units to the right and to the left we get the numbers 2 and -4, respectively.

(b) $|x-(-1)| = 3$, so $x+1 = \pm 3$, yielding, $x = 2$ or $x = -4$.

Prob 3.

(a) $2x - 4 = \pm 6$, so either $2x = 10$ giving the solution $x = 5$ or $2x = -2$ and the other solution is $x = -1$.

(b) $x - 3 = \pm 2$, from which we see that $x = 5$ or $x = 1$.

(c) $2x + 3 = \pm 5$, so $2x = 2$ giving $x = 1$ or $2x = -8$ and $x = -4$.

(d) The equation has no solutions, since the absolute value of a number cannot be negative, and so could never be equal to -7.

Prob. 5.

(a) We can change the absolute value for two inequalities

$$-4 \leq 5x - 2 \leq 4$$

which then can be solved to

$$-4 + 2 \leq 5x \leq 4 + 2$$

and dividing by 5 we get:

$$-\frac{2}{5} \leq x \leq \frac{6}{5}$$

(b) It will change to the inequalities: $1 - 3x < -8$ or $1 - 3x > 8$. Solving the first we get: $9 < 3x$ or $x > 3$, and solving the second we get $-3x > 7$, diving by -3 and reverting the side of the inequality we get $x < -\frac{7}{3}$.

(c) The first inequality will be $7x + 4 \geq 3$, that solves to $x \geq -\frac{1}{7}$. Solving the second equation we have $7x \leq -7$ or $x \leq -1$.

(d) This will produce the two inequalities: $-7 < 6 - 5x < 7$, and subtracting 6 on both sides we get: $-13 < -5x < 1$, dividing by -5 and reverting the inquality sign, we have $\frac{13}{5} > x > -\frac{1}{5}$, or in thenatural order of the real line $-\frac{1}{5} < x < \frac{13}{5}$.

Prob. 7. We use the point-slope formula to get: $y - y_0 = m(x - x_0)$, and in this case: $y - 4 = -\frac{1}{3}(x - 2)$, so multiplying both sides by 3, we get: $3(y-4) = -(x-2)$, or $3y - 12 = -x + 2$, and finally writing it into standard form: $x + 3y - 14 = 0$.

Prob. 9. $(X_0, y_0) = (0, -2)$ $m = -3$
$y - y_0 = m(x - x_0)$
$y - (-2) = -3(x - 0)$
$y + 2 = -3x$
Here $3x + y + 2 = 0$ is Standard Form $A = 3$, $B = 1$, $C = 2$.

Prob. 11. $(x, y_1) = (-2, -3)$
$(x_2, y_2) = (1, 4)$
First find slope of one
$m = \frac{y_2 - y_1}{x_2 - x_1} = \frac{(4)-(-3)}{(1)-(-2)} = \frac{4+3}{1+2}$
$m = \frac{7}{3}$
Second, get Standard Form with point-slope method:
$y - y_0 = m(x - x_0)$
Substitute into (x_0, y_0) either (x_1, y_1) or (x_2, y_2) with the same result
$y - 4 = \frac{7}{3}(x - 1)$
$3(y - 4) = 7(x - 1)$
$3y - 12 = 7x - 7$
$3y - 12 = 7x - 7$
$3y - 7x - 12 + 7 = 0$
$-7x + 3y - 5 = 0$ is Standard Form
or $7x - 3y + 5 = 0$.

Prob. 13. $(x_1, y_1) = (0, 4)$
$(x_2, y_2) = (3, 0)$
$m = \frac{y_2 - y_1}{x_2 - x_1} = \frac{(0)-(4)}{(3)-(0)} = -\frac{4}{3}$
$y - y_0 = m(x - x_0)$
$y - 4 = -\frac{4}{3}(x - 0)$
$3(y - 4) = -4x$
$3y - 12 = -4x$
$4x + 3y - 12 = 0$ is Standard Form.

Prob. 15. Horizontal lines are always $y = k$.
A horizontal line through $(3, \frac{3}{2})$ is $y = \frac{3}{2}$ Standard Form is $2y - 3 = 0$

Prob. 17. Vertical lines are always $x = h$. A vertical line through $(-1, \frac{7}{2})$ is $x = -1$. Standard form is $x + 1 = 0$.

Prob. 19. $m = 3$ y-intercept $(0, 2)$. Use slope-intercept form $y = mx + b$.
Here $y = 3x + 2$. Standard form is $3x - y + 2 = 0$.

Prob. 21. $m = \frac{1}{2}$ y-intercept $(0, 2)$
$y = mx + b$
$y = \frac{1}{2}x + 2$ is slope-intercept form
$x - 2y + 4 = 0$ is standard form.

Prob. 23. $m = -2$ x-intercept $(1, 0)$
$y - y_0 = m(x - x_0)$
$y - 0 = -2(x - 1)$
$y = -2x + 2$ slope-intercept form
$2x + y - 2 = 0$ standard form

Prob. 25. $m = -\frac{1}{4}$ x-intercept $(3, 0)$
$y - y_0 = m(x - x_0)$
$y - 0 = -\frac{1}{4}(x - 3)$
$y = -\frac{1}{4}x + \frac{3}{4}$ slope-intercept form and $x + 4y - 3 = 0$ is standard form.

Prob. 27. Line through $(2, -3)$ parallel to $x + 2y - 4 = 0$
$x + 2y - 4 = 0$ is Standard form change to slope-intercept form to find slope of the given line
$x + 2y - 4 = 0$
$2y - 4 = -x$
$\frac{2y}{2} = -\frac{x+4}{2}$
$y = -\frac{1}{2}x + 2$, slope is $m = -\frac{1}{2}$. Now
$m = -\frac{1}{2}$, point is $(2, -3)$
$y - y_0 = m(x - x_0)$
$y - (-3) = -\frac{(}{x} - 2)$
$y + 3 = -\frac{1}{2}(x - 2)$
$y + 3 = -\frac{1}{2}x + 1$
$2(y + 3) = (-\frac{1}{2}x + 1)2$
$2y + 6 = -x + 2$
$x + 2y + 4 = 0$ is Standard form of the line parallel to $x + 2y - 4 = 0$ passing through $(2, -3)$.

Prob. 29. Line passing through $(-1,-1)$ parallel to line passing through $(0,1)$ and $(3,0)$.

First find slope $m = \frac{y_2-y_1}{x_2-x_1}$
$m = \frac{(0)-(1)}{(3)-(0)} = -\frac{1}{3}$
Second use point-slope form
$y - y_0 = m(x - x_0)$
$y - (-1) = -\frac{1}{3}(x - (-1))$
$y + 1 = \frac{1}{3}(x+1)(-1)$
$-3(y+1) = x + 1$
$-(3y+3) = x + 1$
or $x + 3y + 4 = 0$ standard form.

Prob. 31. Line through $(1,4)$ perpendicular to $2y - 5x + 7 = 0$

$2y - 5x + 7 = 0$
$2y = 5x - 7$
$\frac{2y}{2} = \frac{5x-7}{2}$
$y = \frac{5}{2}x - \frac{7}{2}$, $m = \frac{5}{2}$
$m_\perp = -\frac{1}{m}$ (recall $m_1 \cdot m_2 = -1$)
$m_\perp = \frac{-1}{(\frac{5}{2})} = -\frac{2}{5}$
use pt-slope form with $(1,4)$
$y - y_0 = m_\perp(x - x_0)$
$y - 4 = -\frac{2}{5}(x - 1)$
$5(y-4) = -(x-1)$
$5y - 20 = -2x + 2$
$2x - 2 + 5y - 20 = 0$
$2x + 5y - 22 = 0$

Prob. 33. Line through $(5,-1)$ perpendicular to line passing through $(-2,1)$ and $(1,-2)$

$m = \frac{y_2-y_1}{x_2-x_1} = \frac{(-2)-(1)}{(1)-(-2)} = -\frac{3}{3}$
$m = -1$
$m_\perp = -\frac{1}{m} = \frac{-1}{-1} = 1$
use pt-slope form with $(5,-1)$
$y - y_0 = m_\perp(x - x_0)$
$y - (-1) = 1(x - 5)$
$y + 1 = x - 5$
$y = x - 6$

Prob. 35. So the line is horizontal through $(4,2)$, and the equation is $y = 2$, in standard form the equation is $y - 2 = 0$.

Prob. 37. The line is vertical with equation $x = -1$, so in standard form the equation is $x + 1 = 0$.

Prob. 39. The line is vertical with equation $x = -1$, so in standard form the equation is $x + 1 = 0$.

Prob. 41. The line is horizontal with equation $y = 3$, so in standard form the equation is $y - 3 = 0$.

Prob. 43. $y = 30.5x$ is in the form of $y = mx$ meaning the two quantities x and y are linearly related. Being that y is proportional to x, then m is the constant of proportionality.

(a) To use this relationship, recall that 1 foot = 30.5cm. Therefore let x be the variable for feet, and y be the variable for centimeters.

(b) Convert into centimeters.

(i) $y = 30.5x$, $x = 6ft$
$y = 30.5(6) = 183$cm

(ii) $y = 30.5x$, $x = 3ft$ 2in. Note that 2 inches $= \frac{2}{12}$ feet $\approx 0,167t$
Let $x = 3.167$ feet (approx.)
$y = 30.5(3.167) = 96.58$cm

(iii) $y = 30.5x$ $x = 1ft$ 7ins.
Note $7ins = \frac{7}{12}$ feet $\approx 0.583ft$.
Let $x = 1.583ft$ (approx.)
$y = 30.5(1.583) = 48.28$cm.

(c) Convert into feet

(i) $y = 30.5x$, $y = 173$cm
$173 = 30.5x$
$x = 5.67$ feet

(ii) $y = 30.5x$, $y = 75$cm
$75 = 30.5x$
$x = 2.459$ feet

(iii) $y = 48$cm
$y = 30.5x$
$48 = 30.5x$
$x = 1.574$ feet

Prob. 45. Distance = rate · time (Recall $y = m \cdot x$)
time = 15 mins = $\frac{1}{4}$ hour = 0,25 hrs
distance = 10 mi
Constant of proportionality is miles per hour or "mph"
distance = speed · time
10 mi = speed · 0,25 hrs
$\frac{10mi}{0{,}25hrs}$ = speed (the constant of proportionality)
speed = 40 $\frac{miles}{hour}$ (or "mph")

Prob. 47. 1 foot = 0.305 meters so 3.279 ft = 1 meter, now converting $1m^2 = (1m)(1m) = (3.279ft)(3.279ft) = 10.75ft^2$ (approx.)

Prob. 49. 1 liter = 33.81 ounces

(a) $y = mx$, $m = \frac{1liter}{33.81ounces}$
$y(liters) = m \cdot x(ounces)$

(b) $x = 12ounces$
$y = \frac{1}{33.8}(12) = \frac{1liter}{33.81ounces} \cdot (12ounces)$
$y = 0.355liters$

Prob. 51. $1mile = 1.609Kilometers$

(a) $\left(5\frac{miles}{hour}\right)\left(\frac{1.609Kilometers}{1mile}\right)$
$Speed = 88.5\frac{Kilometers}{hour}$

(b) $\left(130\frac{Kilometers}{hour}\right)\left(\frac{mile}{1.609Kilometers}\right)$
$Speed = 80.8\frac{miles}{hour}$

Prob. 53.

(a) Denoting the two scales by K and C, they relate y equation

$$K = mC + b$$

and the slope of the line m is equal to 1 because of the conversion factor, so

$$K = C + b$$

and substituting the two points $(-273.15, 0)$ we get $b = 273.15$ so

$$K = C + 273.15$$

(b) Nitrogen $\to 77.4K$ and Oxygen $\to 90.2K$

So the boiling points will be given in Celsius by the equation

$$C = K - 273.15$$

so then are $-195.75°$ for Nitrogen and $-182.95°$ for Oxygen.

$$F = \frac{9}{5}{}^{0}C + 32$$

Nitrogen: $F = \frac{9}{5}(-195.75) + 32 = -320.35$

Oxygen: $F = \frac{9}{5}(-182.95) + 32 = -297.31$

and the Nitrogen will end up being distilled first since it has the lower boiling point.

Prob. 55. $r^2 = (x - x_0)^2 + (y - y_0)^2$
$r = 3, \quad (x_0, y_0) = (-1, 4)$
$9 = (x - (-1))^2 + (y - 4)^2$
$9 = (x + 1)^2 + (y - 4)^2$

Prob. 57. $r^2 = (x - x_0)^2 + (y - y_0)^2$

(a) $r = 3, \quad (x_0, y_0) = (2, 5)$

$3^2 = (x - 2)^2 + (y - 5)^2$

$9 = (x - 2)^2 + (y - 5)^2$

(b) where does the circle intersect the y-axis?

When $x = 0$ the circle is on the y-axis.

$9 = (x - 2)^2 + (y - 5)^2$

$9 = (0 - 2)^2 + (y - 5)^2$

$9 = 4 + (y - 5)^2$

$5 = (y - 5)^2$

$\sqrt{5} = y - 5$

$5 \pm \sqrt{5} = y$

(c) Does the circle intersect the x-axis? If $y = 0$ the circle would be on the x-axis.

$9 = (x - 2)^2 + (y - 5)^2$

$9 = (x-2)^2 + (0-5)^2$
$9 = (x-2)^2 + 25$
$-16 = (x-2)^2$ cannot be solved hence does not intersect x-axis.

Prob. 59. Find center and radius.
$(x-2)^2 + y^2 = 16$
$(x-2)^2 + (y-0)^2 = 4^2$
center $(x_0, y_0) = (2, 0)$, $\quad r = 4$

Prob. 61. $0 = x^2 + y^2 - 4x + 2y - 11$
$0 = (x^2 - 4x + 4) + (y^2 + 2y + 1) - 11 - 5$
$0 = (x-2)^2 + (y+1)^2 - 16$
$16 = (x-2)^2 + (y+1)^2$
$16 = (x-2)^2 + (y+1)^2$
$r = 4, \quad (x_0, y_0) = (2, -1)$

Prob. 63.

(a) Convert to radian measure

$$(75^0)(\frac{\pi}{180^0}) = \frac{5}{12}\pi$$

(b) Convert to degree measure

$$(\frac{17}{12}\pi)(\frac{180^0}{\pi}) = \frac{17.180^0}{12} = 255^0$$

Prob. 65.

(a) Evaluate without a calculator

$$\sin(\frac{-5}{4}\pi)$$

$$\sin(\frac{-5}{4}\pi) = \frac{\sqrt{2}}{2}$$

(b)

$$\cos(\frac{5}{6}\pi)$$

$$\cos(\frac{5}{6}\pi) = \frac{-\sqrt{3}}{2}$$

(c)

$$\tan(\frac{\pi}{3}) = \frac{\sqrt{3}}{1}$$
$$\tan(\frac{\pi}{3}) = \sqrt{3}$$

Prob. 67.

(a) Find values of $\alpha \in [0, 2\pi)$

$$\sin\alpha = -\frac{1}{2}\sqrt{3}$$

$$\pi + \frac{\pi}{3} = \frac{4}{3}\pi$$

and

$$2\pi - \frac{\pi}{3} - \frac{5}{3}\pi$$
$$\alpha = \frac{4}{3}\pi \text{ and } \frac{5}{3}\pi$$

(b) Find values of $\alpha \in [0, 2\pi)$

$$\tan\alpha = \sqrt{3}$$

$\frac{\pi}{3}$ and $\pi + \frac{\pi}{3}$

$$\alpha = \frac{\pi}{3} \text{ and } \frac{4}{3}\pi$$

Prob. 69.

$$\begin{aligned}
\sin^2\theta + \cos^2\theta &= 1 \\
\frac{\sin^2\theta + \cos^2\theta}{\cos^2\theta} &= \frac{1}{\cos^2\theta} && \text{Divide both sides by } \cos^2\theta \\
\frac{\sin^2\theta}{\cos^2\theta} + \frac{\cos^2\theta}{\cos^2\theta} &= \frac{1}{\cos^2\theta} && \text{Distribute}
\end{aligned}$$

$$\left(\text{Recall} \quad \frac{\sin\theta}{\cos\theta} = \tan\theta, \quad \text{and} \quad \frac{1}{\cos\theta} = \sec\theta\right)$$

$$\tan^2\theta + 1 = \sec^2\theta \quad \text{make substitution}$$

Prob. 71. If $\sin\theta \neq 0$:

$$\begin{aligned} 2\cos\theta\sin\theta &= \sin\theta, \quad [0, 2\pi) \\ \frac{2\cos\theta\sin\theta}{\sin\theta} &= \frac{\sin\theta}{\sin\theta} \\ 2\cos\theta &= 1 \\ \cos\theta &= \frac{1}{2} \end{aligned}$$

$$\theta = \frac{\pi}{3} \text{ and } 2\pi - \frac{\pi}{3}$$

Hence

$$\theta = \frac{\pi}{3} \text{ and } \frac{5}{3}\pi$$

If $\sin\theta = 0$, the first equality holds, and hence $\theta = 0$ and π are additional solutions.

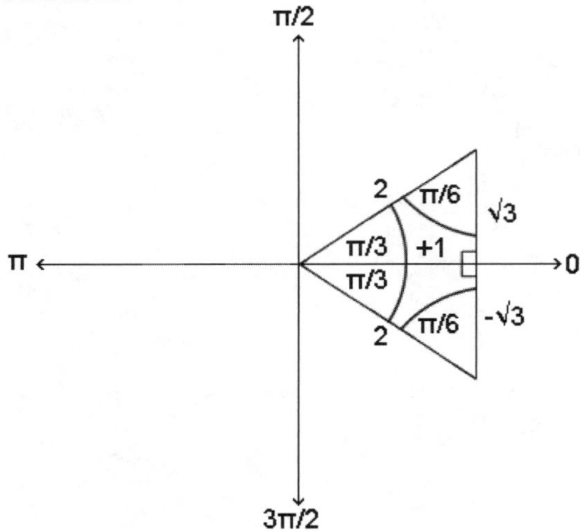

Prob. 73.

(a) $4^3 4^{-2/3}$ Recall

$$\begin{aligned} a^r a^s &= a^{r+s} \\ 4^{3+(-2/3)} &= 4^{7/3} \end{aligned}$$

(b)

$$\begin{aligned} \frac{3^2 3^{1/2}}{3^{-1/2}} &= \frac{3^{5/2}}{3^{-1/2}} = 3^{5/2} \cdot 3^{1/2} \\ &= 3^{6/2} = 3^3 = 27 \end{aligned}$$

(c)

$$\begin{aligned}\frac{5^k 5^{2k-1}}{5^{1-k}} &= 5^k \cdot 5^{2k-1} \cdot 5^{k-1} \\ 5^{k+2k-1+k-1} &= 5^{4k-2}\end{aligned}$$

Prob. 75.

(a)

$$\begin{aligned}\log_4 x &= -2 \\ x &= 4^{-2} = \frac{1}{4^2} = \frac{1}{16}\end{aligned}$$

(b)

$$\begin{aligned}\log_{1/3} x &= -3 \\ x &= (\frac{1}{3})^{-3} = \frac{1}{(\frac{1}{3})^3} = \frac{1}{(\frac{1}{27})} = 27\end{aligned}$$

(c)

$$\begin{aligned}\log_{10} x &= -2 \\ x &= 10^{-2} = \frac{1}{10^2} = \frac{1}{100}\end{aligned}$$

Prob. 77.

(a)

$$\begin{aligned}\log_{1/2} 32 &= x \\ 32 &= (\frac{1}{2})^x = \frac{1^x}{2^x} = \frac{1}{2^x} = \frac{1}{32} = \frac{1}{2^5} \\ 2^x &= 2^{-5} \\ x &= -5\end{aligned}$$

(b)

$$\begin{aligned}\log_{1/3} 81 &= x \\ 81 &= (\frac{1}{3})^x = \frac{1^x}{3^x} \\ 3^x &= \frac{1}{81} = \frac{1}{3^4} = 3^{-4} \\ x &= -4\end{aligned}$$

(c)

$$\begin{aligned}\log_{10} 0.001 &= x \\ 10^x &= 0.001 = \frac{1}{1000} = \frac{1}{10^3} \\ 10^{-3} &= 10^x \\ x &= -3\end{aligned}$$

Prob. 79.

(a) $-\ln(\frac{1}{3}) = \ln(\frac{1}{3})^{-1} = \ln 3$

(b) $\log_4(x^2 - 4) = \log_4[(x-2)(x+2)] = \log_4(x-2) + \log_4(x+2)$

(c) $\log_2 4^{(3x-1)} = (3x-1)\log_2(2^2) = (3x-1)(2 \cdot \log_2 2) = (3x-1)(2 \cdot 1) = 6x - 2$

Prob. 81.

(a)

$$\begin{aligned}e^{3x-1} &= 2 \\ \ln e^{(3x-1)} &= \ln 2 \\ (3x-1)(\ln e) &= \ln 2 \qquad \text{Recall } \ln e = 1 \\ 3x - 1 &= \ln 2 \\ 3x &= 1 + \ln 2 \\ x &= \frac{1 + \ln 2}{3}\end{aligned}$$

(b)

$$\begin{aligned}e^{-2x} &= 10 \\ \ln e^{(-2x)} &= \ln 10 \\ (-2x)(\ln e) &= \ln 10 \\ -2x &= \ln 10 \\ x &= \frac{\ln 10}{-2} = -\frac{1}{2} \cdot \ln 10 \\ \text{or } x &= \ln 10^{-1/2} = \ln(\frac{1}{\sqrt{10}})\end{aligned}$$

(c)

$$\begin{aligned} e^{x^2-1} &= 10 \\ \ln e^{(x^2-1)} &= \ln 10 \\ (x^2-1)(\ln e) &= \ln 10 \qquad (\ln e = 1) \\ x^2 - 1 &= \ln 10 \\ x^2 &= 1 + \ln 10 \\ x &= \pm\sqrt{1+\ln 10} \end{aligned}$$

Prob. 83.

(a)

$$\begin{aligned} \ln(x-3) &= 5 \\ e^{\ln(x-3)} &= e^5, \qquad \text{Recall } e^{\ln x} = x \\ x - 3 &= e^5 \\ x &= e^5 + 3 \end{aligned}$$

(b)

$$\begin{aligned} \ln(x+2) + \ln(x-2) &= 1 \\ \ln(x+2)(x-2) &= 1 \\ e^{\ln[(x+2)(x-2)]} &= e \\ (x+2)(x-2) &= e \\ x^2 - 4 &= e \\ x^2 &= 4 + e \\ x &= \sqrt{4+e} \end{aligned}$$

(c)

$$\begin{aligned} \log_3 x^2 - \log_3 2x &= 2 \\ \log_3 \frac{x^2}{2x} &= 2 \\ \frac{x^2}{2x} &= 3^2 \\ \frac{x}{2} &= 9 \\ x &= 18 \end{aligned}$$

Prob. 85. $3 - 2i - (-2 + 5i) = 3 - 2i + 2 - 5i = 5 - 7i$

Prob. 87. $(4 - 2i) + (9 + 4i) = 4 + 9 - 2i + 4i = 13 + 2i$

Prob. 89. $3(5 + 3i) = 15 + 9i$

Prob. 91. $(6 - i)(6 + i) = 36 - (i)^2 = 36 - (-1) = 37$

Prob. 93.

$$\begin{aligned} z &= 3 - 2i \\ \text{conjugate } \overline{z} &= 3 + 2i \end{aligned}$$

Prob. 95. $\overline{z + v} = \overline{3 - 2i} + \overline{3 + 5i} = 3 + 2i + 3 - 5i = 6 - 3i$

Prob. 97.

$$\overline{vw} = \overline{(3 + 5i)(1 - i)} = \overline{(8 + 2i} = 8 - 2i$$

Prob. 99. Let

$$\begin{aligned} z &= a + bi \\ z + \overline{z} &= a + bi + \overline{a + bi} \\ &= a + bi + a - bi = 2a \end{aligned}$$

and

$$\begin{aligned} z - \overline{z} &= (a + bi) - \overline{(a + bi)} \\ &= a + bi - (a - bi) \\ &= a + bi - a + bi = 2bi \end{aligned}$$

Prob. 101. $2x^2 - 3x + 2 = 0$. Here $a = 2$, $b = -3$, $c = 2$
Recall

$$\begin{aligned} x_{1,2} &= \frac{-b \pm \sqrt{b^2 - 4ac}}{2a} \\ x_{1,2} &= \frac{-(-3) \pm \sqrt{(-3)^2 - 4(2)(2)}}{2(2)} \\ x_{1,2} &= \frac{3 \pm \sqrt{9 - 16}}{4} \\ x_{1,2} &= \frac{3 \pm \sqrt{-7}}{4} = \frac{3 \pm \sqrt{7i^2}}{4} \end{aligned}$$

Recall $i^2 = -1$, and $\sqrt{ab} = \sqrt{a} \cdot \sqrt{b}$

$$x_{1,2} = \frac{3 \pm \sqrt{7} \cdot \sqrt{i^2}}{4} = \frac{3 \pm \sqrt{7}i}{4}$$

$$x_1 = \frac{3 + \sqrt{7}i}{4} \quad \text{and} \quad x_2 = \frac{3 - \sqrt{7}i}{4}$$

Prob. 103. Here $a = -1, b = 1$, and $c = 2$

$$x_{1,2} = \frac{-b \pm \sqrt{b^2 - 4ac}}{2a}$$

$$x_{1,2} = \frac{-(-1) \pm \sqrt{(1)^2 - 4(-1)(2)}}{2(-1)}$$

$$x_{1,2} = \frac{-1 \pm \sqrt{1+8}}{-2}$$

$$x_{1,2} = \frac{-1 \pm \sqrt{9}}{-2} = -\frac{1 \pm 3}{-2}$$

$$x_1 = \frac{-1+3}{-2} = \frac{2}{-2} = -1$$

$$x_2 = \frac{-1-3}{-2} = \frac{-4}{-2} = 2$$

Prob. 105. $4x^2 - 3x + 1 = 0$

$$a = 4, \quad b = -3, \quad c = 1$$

$$x_{1,2} = \frac{-b \pm \sqrt{b^2 - 4ac}}{2a}$$

$$x_{1,2} = \frac{-(-3) \pm \sqrt{(-2)^2 - 4(4)(1)}}{2(4)}$$

$$x_{1,2} = \frac{3 \pm \sqrt{9 - 16}}{8}$$

$$x_{1,2} = \frac{3 \pm \sqrt{-7}}{8} = \frac{3 \pm \sqrt{7i^2}}{8} \qquad \text{Note } i^2 = -1$$

$$x_{1,2} = \frac{3 \pm \sqrt{7}i}{8}$$

Prob. 107. $3x^2 - 4x - 7 = 0$. Compute the discriminant, $b^2 - 4ac$

$$a = 3, \quad b = -4, \quad c = -7$$

$$(-4)^2 - 4(3)(-7) = 16 + 84 = 100 > 0$$

Hence, two real solutions: $x_{1,2} = \frac{4 \pm \sqrt{100}}{(2)(3)}$. Hence, $x_1 = \frac{7}{3}$ and $x_2 = -1$.

Prob. 109. $-x^2 + 2x - 1 = 0$. Compute the discriminant, $b^2 - 4ac$

$$a = -1, \quad b = 2, \quad c = -1$$

$$(2)^2 - 4(-1)(-1)$$

$$4 - 4 = 0$$

Two identical real solutions, that is, only one real solution: $x_{1,2} = \frac{-2 \pm \sqrt{0}}{-2}$. Hence, $x_1 = x_2 = 1$.

Prob. 111. $3x^2 - 5x + 6 = 0$

$$a = 4, \quad b = -1, \quad c = 1$$

Compute the discriminant

$$b^2 - 4ac$$

$$(-5)^2 - 4(3)(6)$$

$$25 - 72 = -47 < 0$$

Hence, two complex solutions which are conjugates of each other: $x_{1,2} = \frac{5 \pm \sqrt{-47}}{6}$. Hence, $x_1 = \frac{5}{6} + i\frac{\sqrt{47}}{6}$ and $x_2 = \frac{5}{6} - i\frac{\sqrt{47}}{6}$.

Prob. 113. Let $z = a + bi$. Conjugate $\overline{z} = a - bi$

$$\overline{(\overline{z})} = \overline{(a - bi)} = a + bi = z.$$

Hence $z = \overline{(\overline{z})}$.

Prob. 115. Let

$$\begin{aligned} z &= a + bi \\ w &= c + di \end{aligned}$$

Does $\overline{zw} = \overline{z} \cdot \overline{w}$?

For Left Hand Side

$$\begin{aligned}\overline{z \cdot w} &= \overline{(a+bi)(c+di)} = \overline{(ac-bd)+(ad+bc)i} \\ \overline{z \cdot w} &= (ac-bd)-(ad+bc)i\end{aligned}$$

For Right Hand Side

$$\begin{aligned}\overline{z} \cdot \overline{w} &= \overline{(a+bi)} \cdot \overline{(c+di)} \\ &= (a-bi) \cdot (c-di) \\ &= ac - adi - bci + bdi^2 \\ &= ac - (ad+bc)i - bd \\ \overline{z} \cdot \overline{w} &= (ac-bd)-(ad+bc)i\end{aligned}$$

Hence, Left Hand Side = Right Hand Side.
So $\overline{zw} = \overline{z} \cdot \overline{w}$ is true.

1.2 Elementary Functions

Prob. 1. $f(x) = x^2$, $x \in \mathbb{R}$. Range $[0, +\infty)$

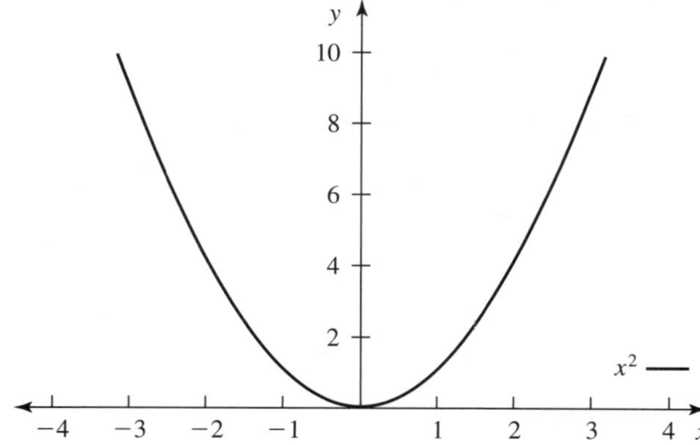

Prob. 3. $f(x) = x^2$, $-1 < x \leq 0$. Range $[0, 1)$

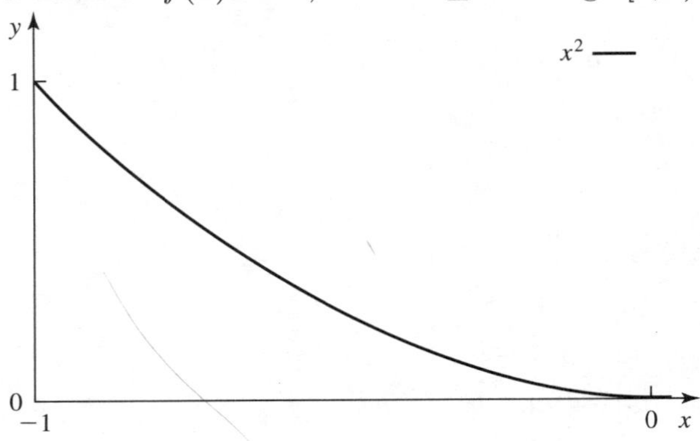

Prob. 5.

(a) Factor $\frac{x^2-1}{x-1} = \frac{(x+1)(x-1)}{x-1} = x + 1$ (if $x \neq 1$)

$$f(x) = \frac{x^2 - 1}{x - 1}, \quad x \neq 1$$

(b) No the functions are not equal, they have different domains, f is not defined at the point 1.

Prob. 7. The function is odd, since $f(-x) = -f(x)$.

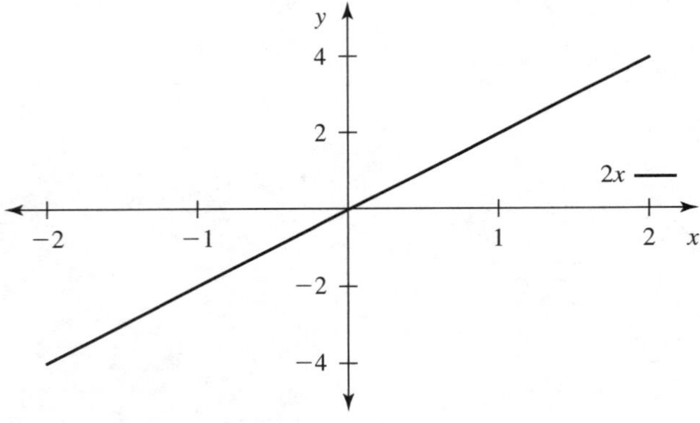

Test for "Even";

$$\begin{aligned} \text{Does } f(x) &= f(x)? \\ \text{Ask if ; } 2x &= 2(-x)? \\ \text{Does } 2x &= -2x? \quad \text{no, not "even"} \end{aligned}$$

Test for "Odd";

$$\begin{aligned} \text{Does } f(x) &= -f(-x) \\ \text{Ask if ; } 2x &= -2(-x)? \\ \text{Does } 2x &= 2x? \quad \text{yes, hence "odd"} \end{aligned}$$

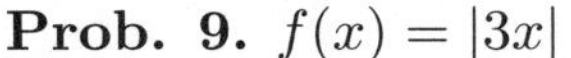

Prob. 9. $f(x) = |3x|$

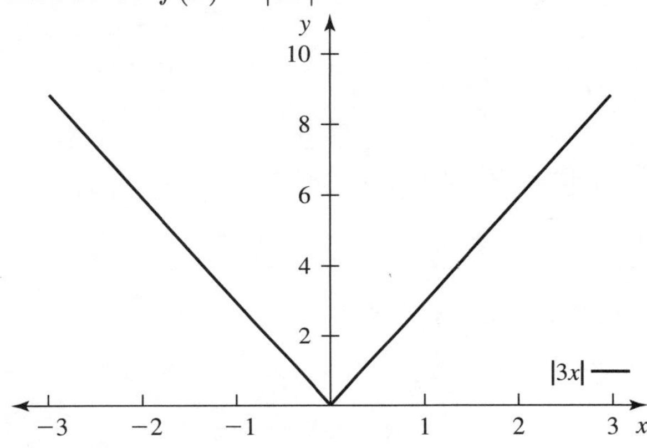

Test for "Even";

$$\begin{aligned} \text{Does } f(x) &= f(-x)? \\ \text{Ask if ; } |3x| &= |3(-x)|? \\ \text{Does } |3x| &= |-3x|? \quad \text{yes, hence "even"} \end{aligned}$$

Test for "Odd";

$$\begin{aligned} \text{Does } f(x) &= -f(-x)? \\ \text{Ask if ; } |3x| &= -|3(-x)|? \\ \text{Does } |3x| &= -|3x|? \quad \text{no, not "odd"} \end{aligned}$$

Prob. 11.

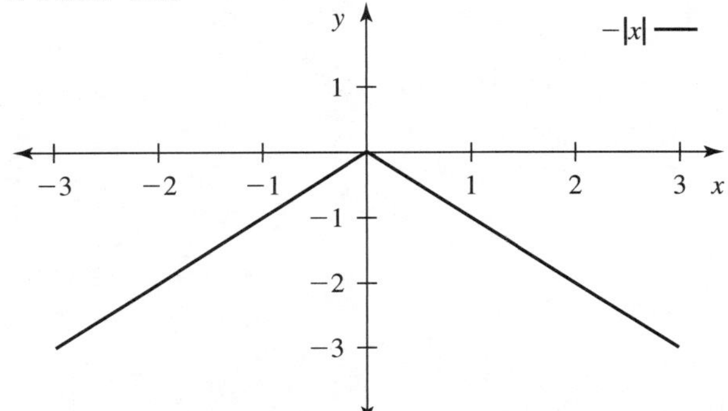

Test for "Even";

$$\begin{aligned} \text{Does } f(x) &= f(-x)? \\ \text{Does } -|x| &= -|-x|? \quad \text{yes, hence "even"} \end{aligned}$$

Prob. 13.

(a) $(f \circ g)(x) = f[g(x)]$

$$\begin{aligned} f(x) &= x^2 \\ f(g(x)) &= (3+x)^2 \\ (f \circ g)(x) &= (3+x)^2 \end{aligned}$$

(b) $(g \circ f)(x) = g[f(x)]$

$$\begin{aligned} g(f(x)) &= 3+(x^2) \\ (g \circ f)(x) &= 3+x^2 \end{aligned}$$

Prob. 15.

(a)

$$\begin{aligned} (f \circ g)x &= f(g(x)) \\ f(g(x)) &= f(2x) = 1-(2x)^2 \\ f(g(x)) &= 1-4x^2 \end{aligned}$$

Domain $x \geq 0$

(b)

$$\begin{aligned} (g \circ f)(x) &= g(f(x)) \\ g(x) &= 2x \\ g(f(x)) &= 2(f(x)) = 2(1-x^2) \\ g(f(x)) &= 2-2x^2 \end{aligned}$$

Domain $x \in \mathbb{R}$

Prob. 17.

$$\begin{aligned} f(x) &= 3x^2 \quad x \geq 3 \\ g(x) &= \sqrt{x} \quad x \geq 0 \end{aligned}$$

$$\begin{aligned}(f \circ g)(x) &= f(g(x)) \\ &= f(\sqrt{x}) = 3(\sqrt{x})^2 \\ (f \circ g)(x) &= f(g(x)) = 3x\end{aligned}$$

Domain $x \geq 9$

Prob. 19.

$$\begin{aligned}f(x) &= x^2 \quad x \geq 0 \\ g(x) &= \sqrt{x} \quad x \geq 0\end{aligned}$$

Note $f \circ g \neq g \circ f$ in general show $f \circ g = g \circ f$

$$\begin{aligned}f \circ g &= f(g(x)) = f(\sqrt{x}) = (\sqrt{x})^2 \\ f \circ g &= x, x \geq 0\end{aligned}$$

and

$$\begin{aligned}g \circ f &= g(f(x)) = g(x^2) = \sqrt{x^2} \\ g \circ f &= x, x \geq 0\end{aligned}$$

Prob. 21.

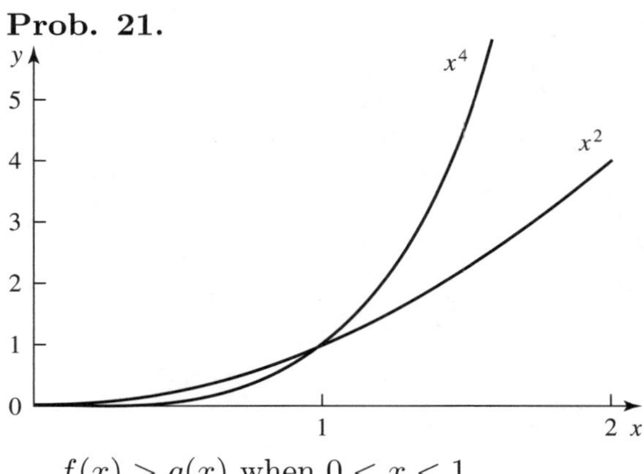

$f(x) > g(x)$ when $0 < x < 1$
$f(x) < g(x)$ when $x > 1$

Prob. 23. $y = x^n,\ x \geq 0$

$y_1 = x^1$
$y_2 = x^2$
$y_3 = x^3$
$y_4 = x^4$

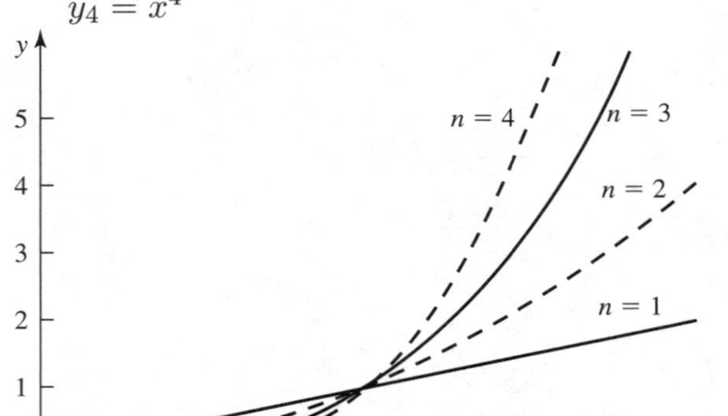

Prob. 25.

(a) $f(x) = x^2$

$g(x) = x^3 \qquad x \geq 0$

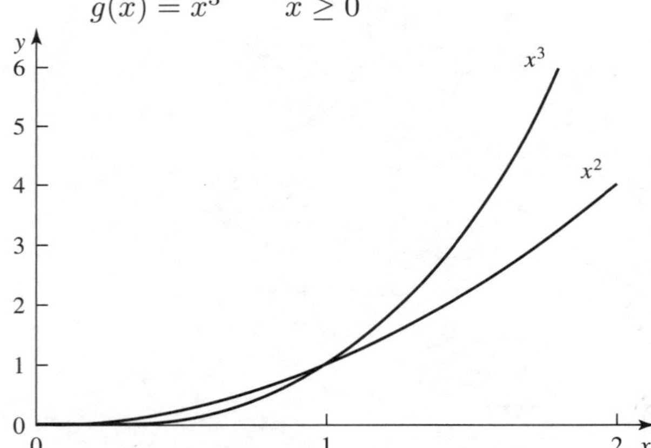

(b) We can start with the inequalities $0 \leq x \leq 1$, and multiply both sides by x^2 which is positive, so it will not change the direction of the inequalities that become $0 \leq x^3 \leq x^2$

(c) In the same way, we start with $x \geq 1$, and multiply both sides by the positive value x^2, which then becomes $x^3 \geq x^2$

Prob. 27.

(a) Show $y = x^2, x \in \mathbb{R}$ is even.

$f(x) = x^2, f(-x) = (-x)^2$

Does $x^2 = (-x)^2$?

$x^2 = x^2$ hence even.

(b) Show $y = x^3$, $x \in \mathbb{R}$ is odd.

$f(x) = x^3, -f(-x) = -(-x)^3$

Does $f(x) = -f(-x)$?

$x^3 = -(-x^3) = -1 \cdot (-1 \cdot x^3)$

$x^3 = x^3$ hence odd.

Prob. 29. $A + B \rightarrow AB$, $a = [A] = 3$, $b = [B] = 4$

(a) $R(x) = k(a - x)(b - x)$

$R = k[A] \cdot [B]$

$9 = k \cdot 1$

$k = 9$

(b) $R(x) = 9(3 - x)(4 - x)$

$0 = 9(3 - x)(4 - x)$. To find domain set $R(x) = 0$

$0 = (3 - x)(4 - x)$. Is quadratic

$0 = 12 - 7x + x^2$

$0 = x^2 - 7x + 12$

$x = \frac{-(-7) \pm \sqrt{(-7)^2 - 4(1)(12)}}{2(1)}$

$x = \frac{7 \pm \sqrt{49-48}}{2} = \frac{7 \pm \sqrt{1}}{2} = \frac{7 \pm 1}{2} = \begin{cases} \frac{8}{2} = 4 \\ \frac{6}{2} = 3 \end{cases}$

$R(x) = 9(3 - x)(4 - x)$

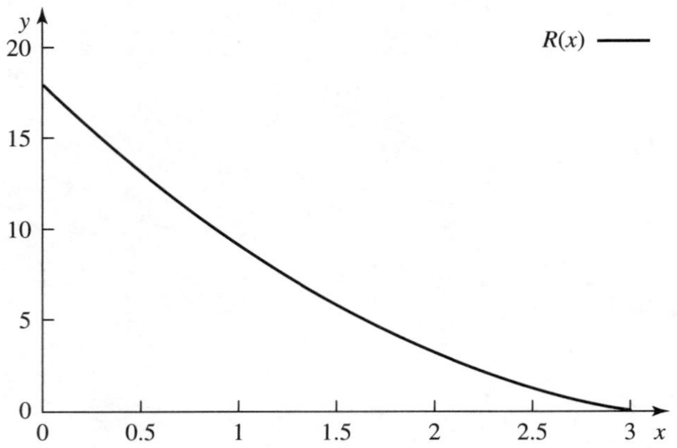

Prob. 31. Constant speed of beetle $= 1 \frac{\text{meter}}{\text{hour}}$

$$(\text{rate})(\text{times}) = \text{distance}.$$

In one hour

$$\left(1\ \frac{\text{meter}}{\text{hour}}\right)(1 \text{ hour}) = 1 \text{ meter}.$$

In two hours

$$\left(1\ \frac{\text{meter}}{\text{hour}}\right)(2 \text{ hours}) = 2 \text{ meters}.$$

In three hours

$$\left(1\ \frac{\text{meter}}{\text{hour}}\right)(3 \text{ hours}) = 3 \text{ meters}.$$

Now distance = rate · times,

Let distance $= y$

rate = slope $= m$

time $= x$

This is a polynomial of degree 1.

Prob. 33. $f(x) = \frac{1}{1-x}$

Domain $(-\infty, 1) \cup (1, +\infty)$

Range $(-\infty, 0) \cup (0, +\infty)$

Prob. 35. $f(x) = \frac{x-2}{x^2-9}$

Domain $(-\infty, -3) \cup (-3, +3) \cup (3, +\infty)$

Range $(-\infty, +\infty)$

Prob. 37.

$$y = \frac{1}{x} \quad y = \frac{1}{x^2}, \quad x > 0$$

The curves intersect at $x = 1$, when $\frac{1}{x^2} = \frac{1}{x}$

$$\frac{1}{x} > \frac{1}{x^2}, \text{ if } x > 1 \quad \text{and} \quad \frac{1}{x} < \frac{1}{x^2}, \text{ if } 0 < x < 1$$

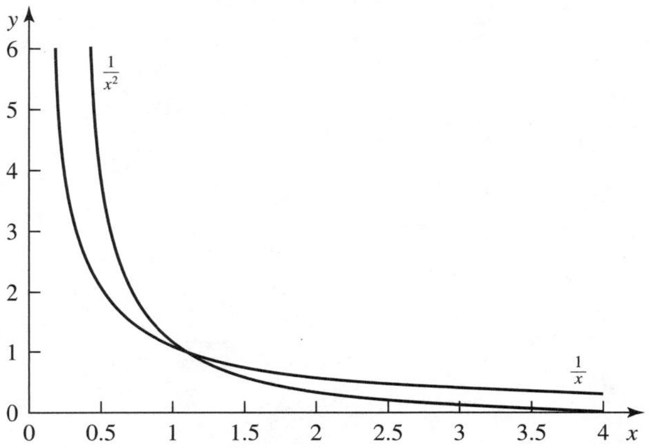

Prob. 39. (a)

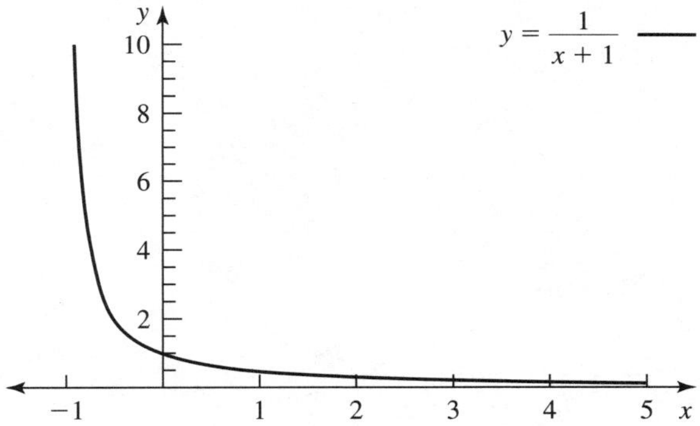

(b) range of $f(x)$ is $(0, \infty)$, (c) $x = -\frac{1}{2}$,
(d) exactly one.

Prob. 41. (a)

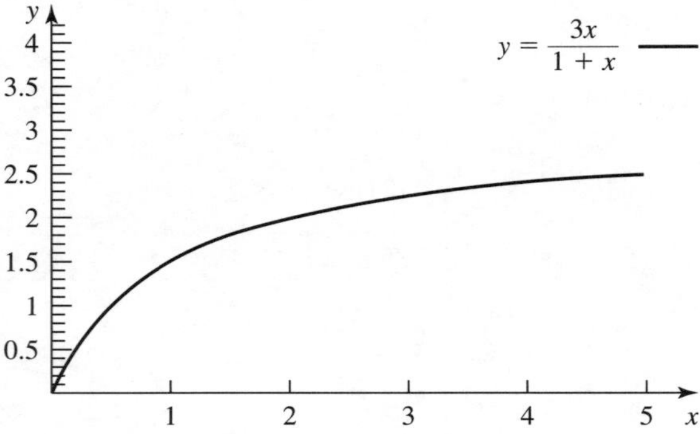

(b) range of $f(x)$ is $[0,3)$, (c) $x = 2$,
(d) exactly one, $x = \frac{a}{3-a}$.

Prob. 43. 83.3%, 4.76%

Prob. 45. (a)

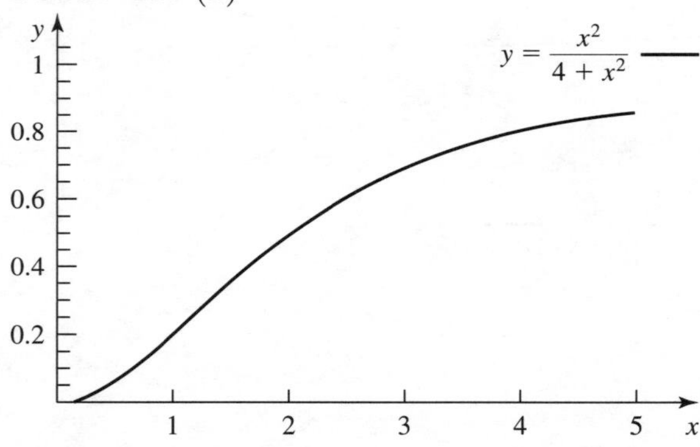

(b) range of $f(x)$ is $[0,1)$, (c) $f(x)$ approaches 1.

Prob. 47.

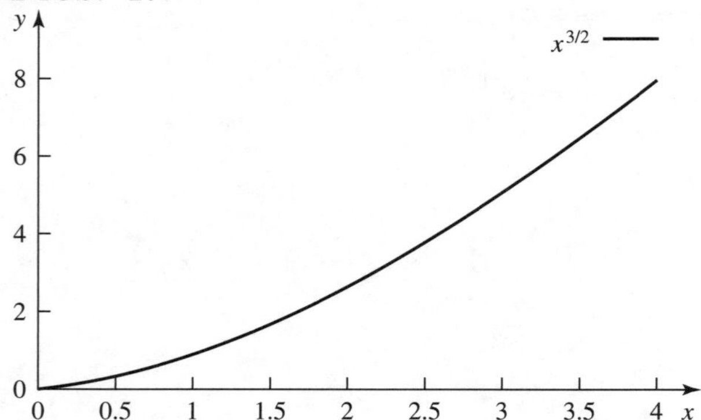

Prob. 49.

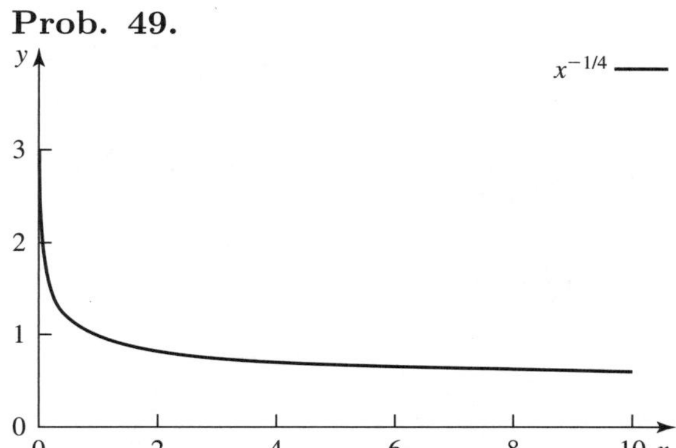

Prob. 51. (a)

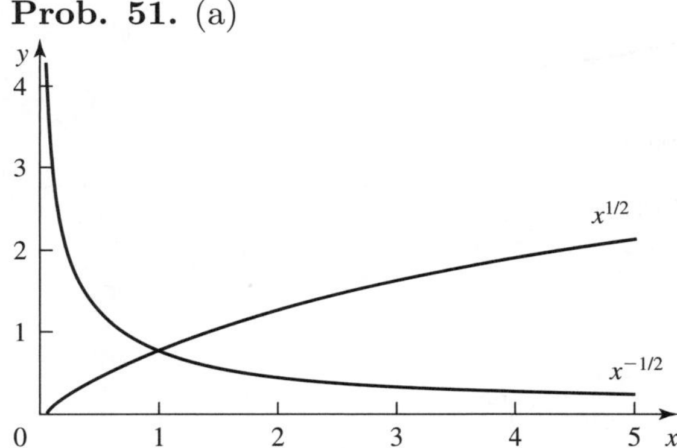

Prob. 53. increases

Prob. 55. increases

Prob. 57. (a) 1,2,4,8,16 (b)

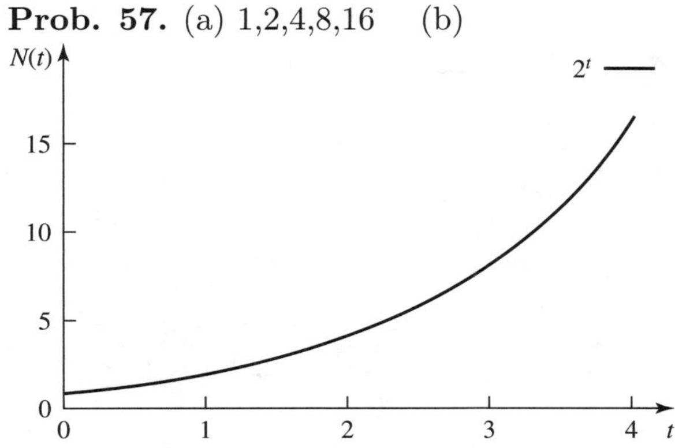

Prob. 59. $20\exp\left[-\frac{\ln 2}{5730}2000\right]$

Prob. 61. $\lambda = \frac{\ln 2}{7\,\text{days}}$

Prob. 63. (a) $W(t) = (300\,\text{gr})\exp\left[-\frac{\ln 2}{140\,\text{days}}t\right]$
(b) $t = \frac{\ln 5}{\ln 2}140\,\text{days} \approx 325\,\text{days}$ (c)

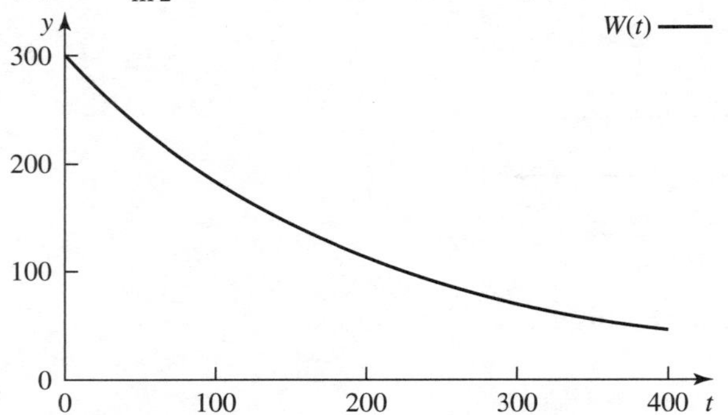

Prob. 65. $\frac{W(t)}{W(0)} = \exp\left[-\frac{\ln 2}{5730}15{,}000\right] \approx 16.3\%$

Prob. 67.

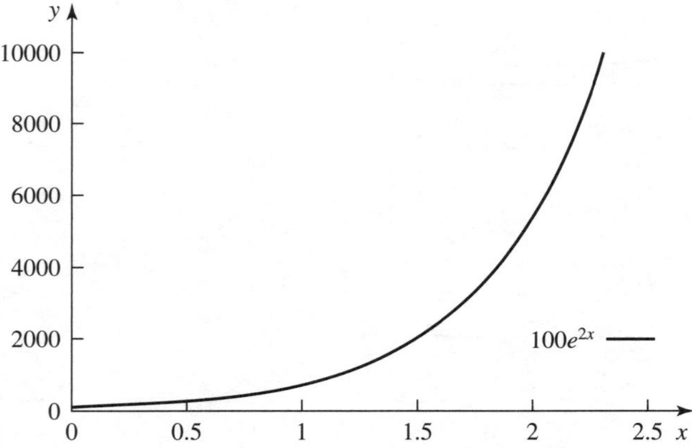

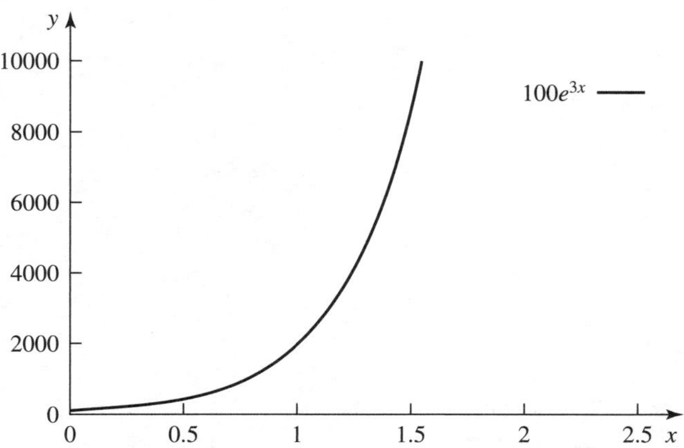

(a) The population with $r = 3$ grows faster.

(b) $\frac{N(t+1)}{N(t)} = \frac{N_0 e^{r(t+1)}}{N_0 e^{rt}} = \frac{250}{200}$, simplifying the fraction we get $e^r = 1.25$, or $r = \ln 1.25$.

Prob. 69. (a) yes (b) no (c) yes (d) yes (e) no (f) yes

Prob. 71. (a) $f^{-1}(x) = \sqrt{x-1}$, $x \geq 1$ (b)

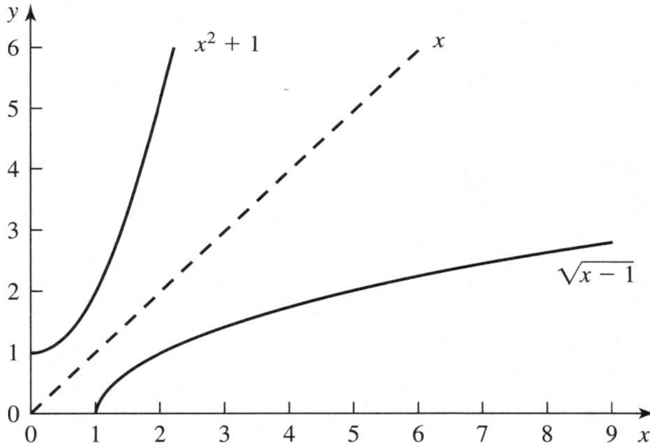

Prob. 73. $f^{-1}(x) = \sqrt[3]{\frac{1}{x}}$, $x > 0$

Prob. 75. $f^{-1}(x) = \log_3 x$, $x > 0$

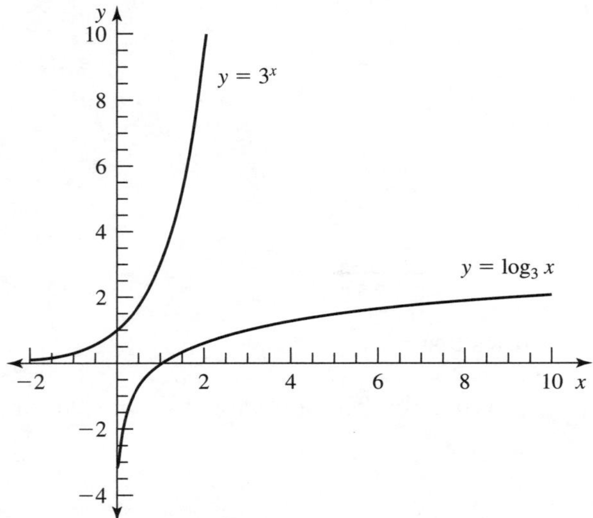

Prob. 77. $f^{-1}(x) = \log_{1/4} x$, $x > 0$

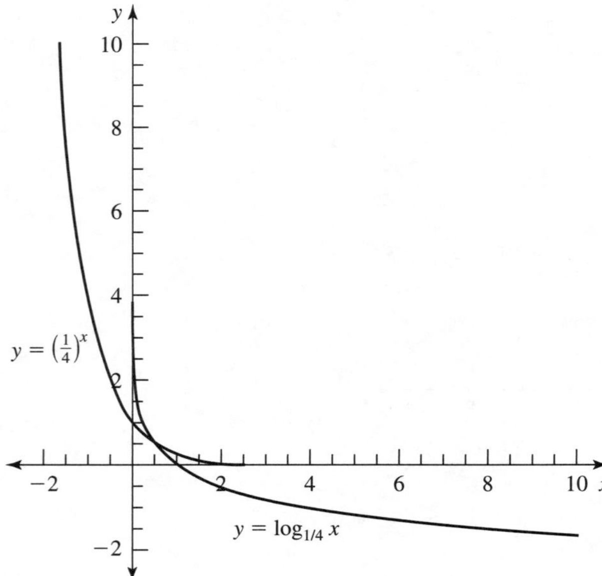

Prob. 79. $f^{-1}(x) = \log_2(x)$, $x \geq 1$

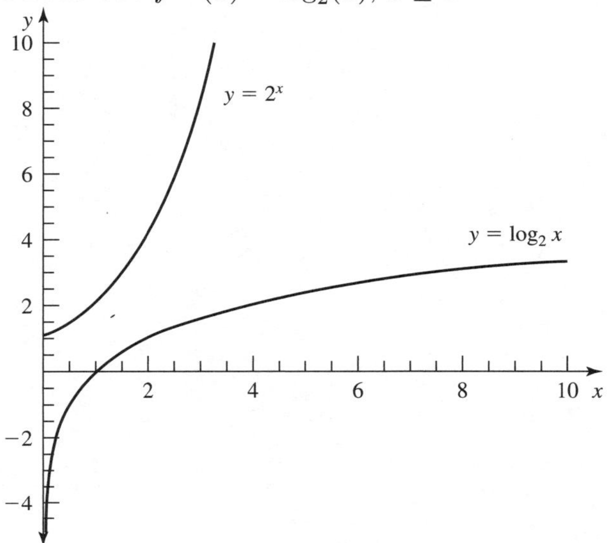

Prob. 81. (a) x^5, (b) x^4, (c) x^{-5}, (d) x^{-4}, (e) x^{-3}, (f) x^2

Prob. 83. (a) $5\ln x$, (b) $6\ln x$, (c) $\ln(x-1)$, (d) $-4\ln x$

Prob. 85. (a) $e^{x\ln 3}$, (b) $e^{(x^2-1)\ln 4}$, (c) $e^{-(x+1)\ln 2}$, (d) $e^{(-4x+1)\ln 3}$

Prob. 87. $\mu = \ln 2$

Prob. 89. $K = -\frac{3}{4}\ln\left(1 - \frac{4}{3} \cdot \frac{47}{300}\right)$

Prob. 91. Same period; $2\sin x$ has twice the amplitude of $\sin x$.

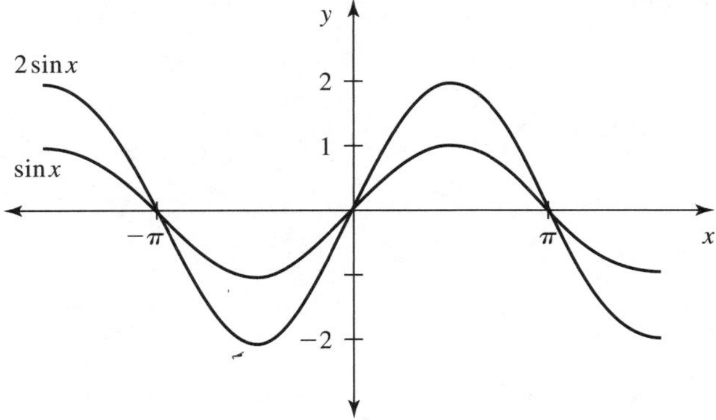

Prob. 93. Same period; $2\cos x$ has twice the amplitude of $\cos x$.

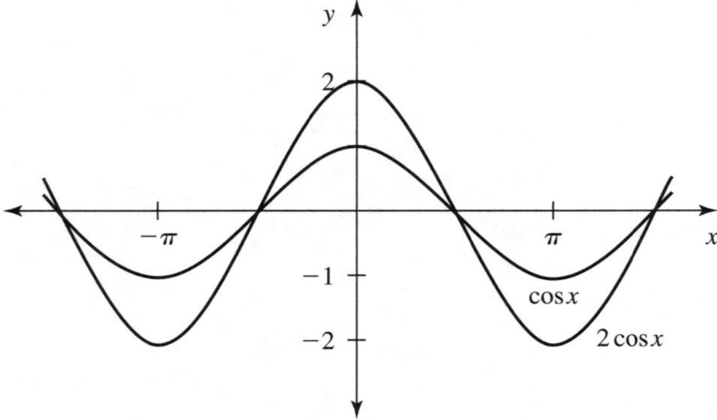

Prob. 95. Same period; $y = 2\tan x$ is stretched by a factor of 2.

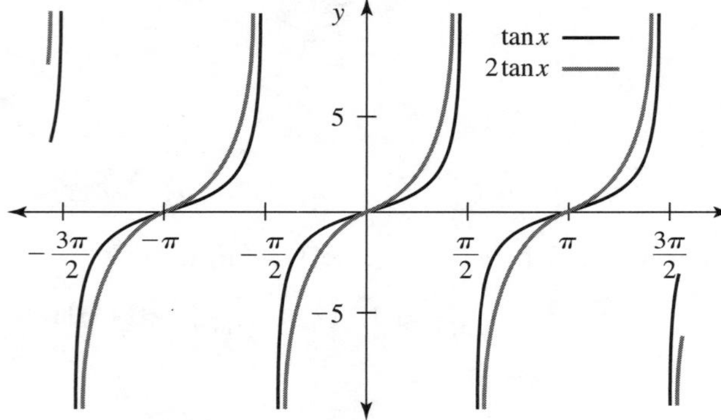

Prob. 97. Amplitude: 3; period: $\frac{\pi}{2}$

Prob. 99. Amplitude: 4; period: 1

Prob. 101. Amplitude: 4; period: 8π

Prob. 103. Amplitude: 3; period: 10

1.3 Graphing

Prob. 1.

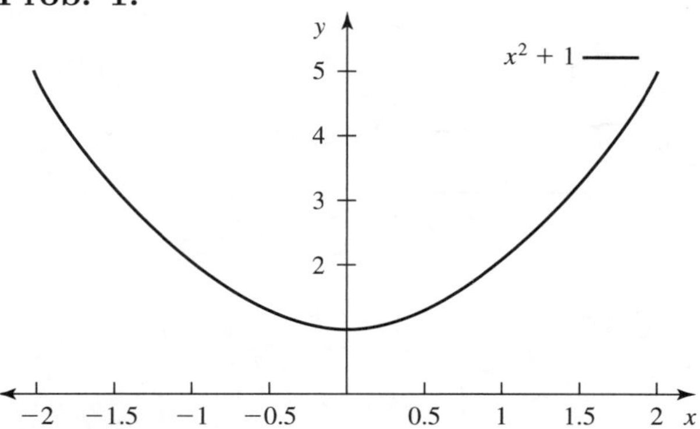

Prob. 3.

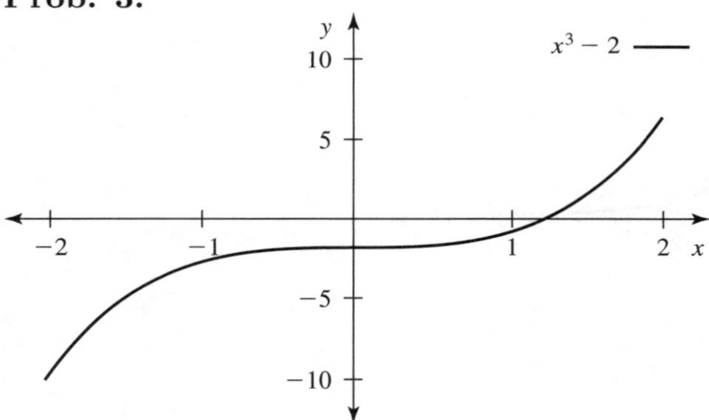

Prob. 5.

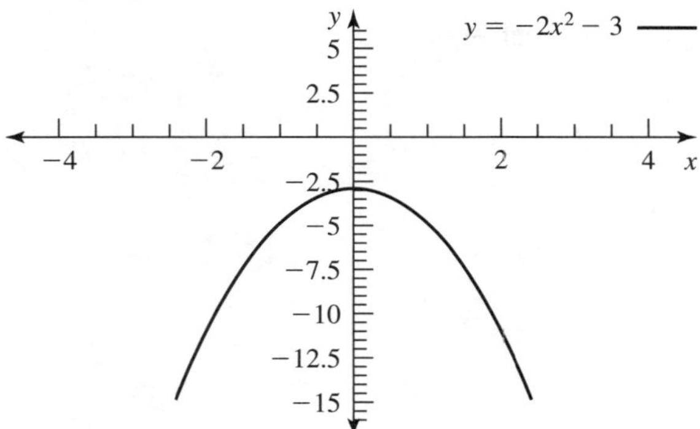

Prob. 7.

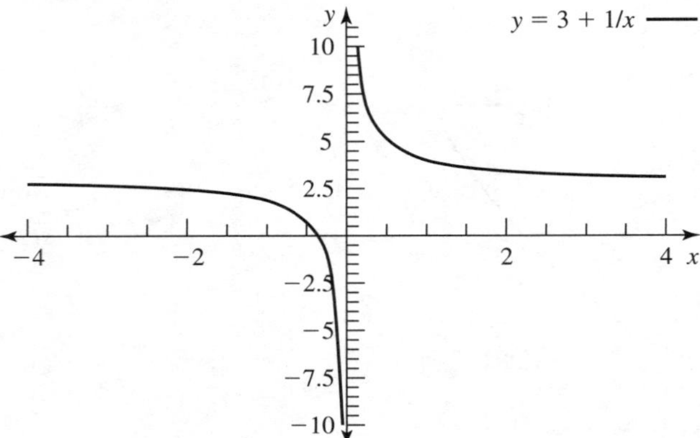

Prob. 9.

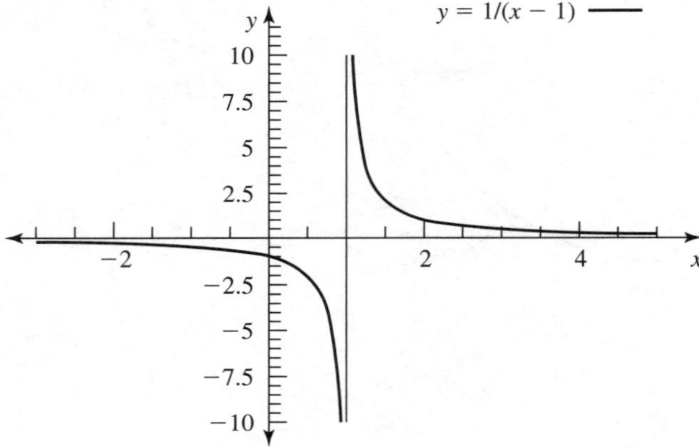

Prob. 11.

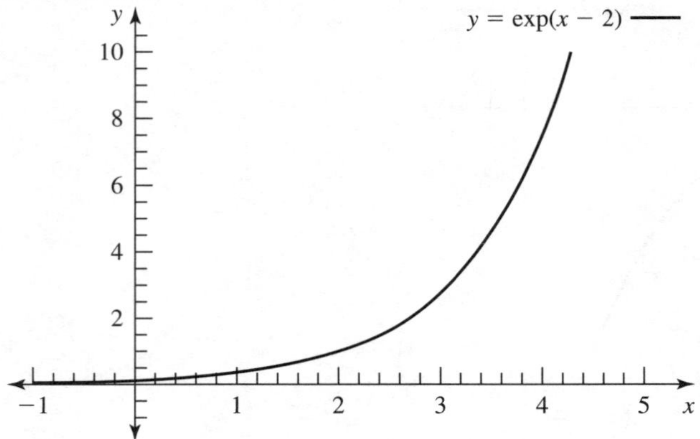

Prob. 13.

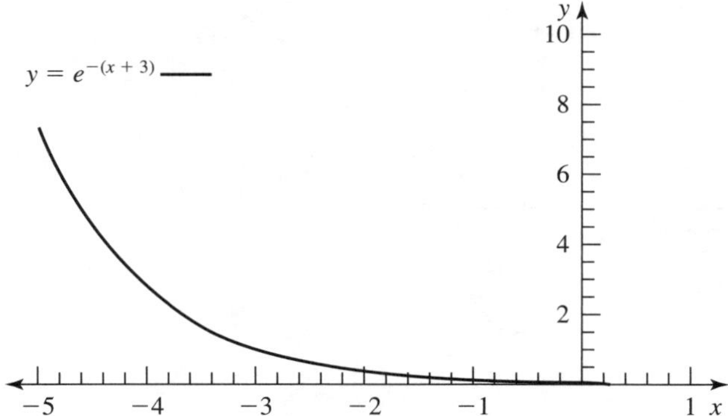

Prob. 15.

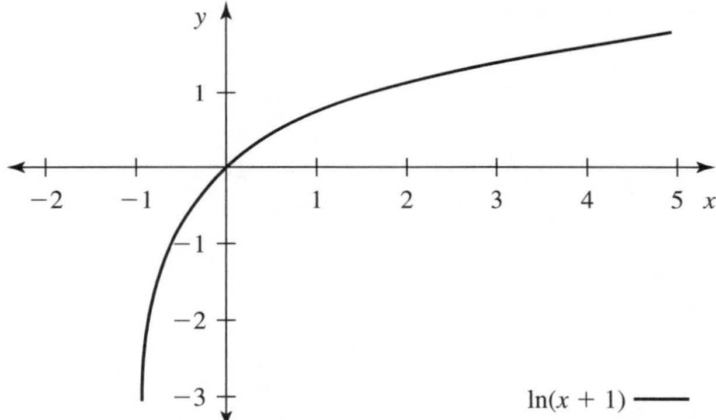

Prob. 17.

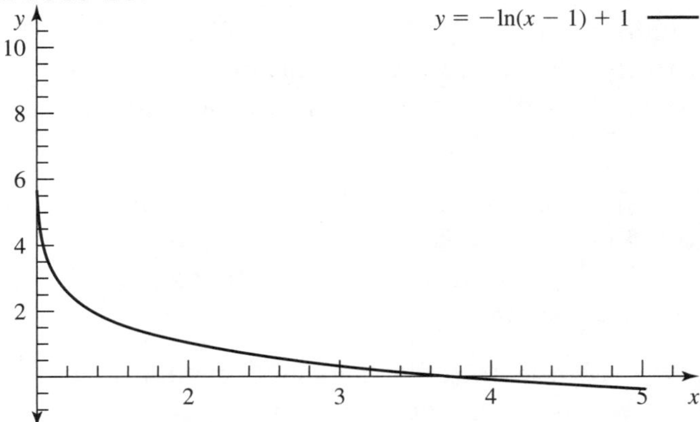

Prob. 19.

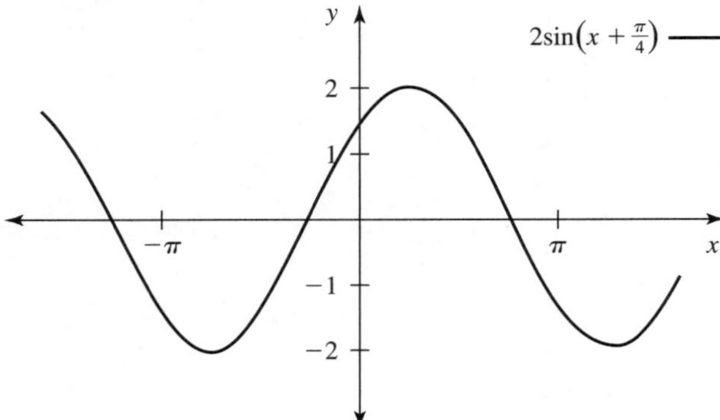

Prob. 21.

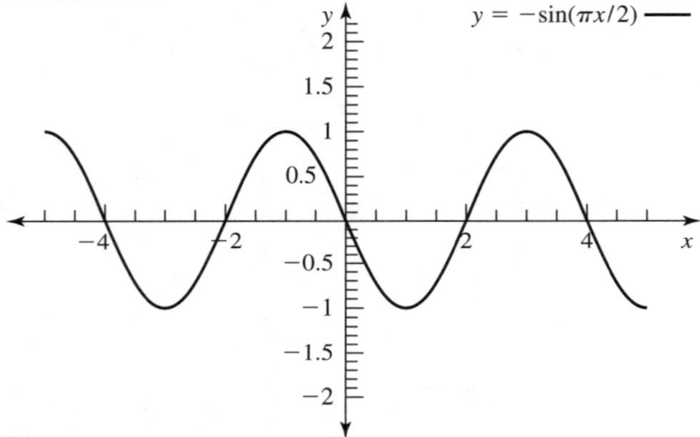

Prob. 23. (a) Shift two units down.
(b) Shift $y = x^2$ one unit to the right and then one unit up.
(c) Shift $y = x^2$ two units to the left, stretch by a factor of 2, and reflect about the x-axis.

Prob. 25. (a) 1. Reflect $\frac{1}{x}$ about the x-axis. 2. Shift up one unit.
(b) 1. Shift $\frac{1}{x}$ one unit to the right. 2. Reflect about the x-axis.
(c) 1. Write $y = \frac{x+1-1}{x-1} = 1 - \frac{1}{x+1}$. Shift $y = \frac{1}{x}$ one unit to the left. 2. Reflect about x-axis. 3. Shift up one unit.

Prob. 27. (a) Stretch $y = e^x$ by a factor of 2, then shift one unit down.
(b) Reflect $y = e^x$ about the y-axis, then reflect about the x-axis.
(c) Shift $y = e^x$ two units to the right, then shift one unit up.

Prob. 29. (a) Shift $y = \ln x$ one unit to the right.
(b) Reflect $y = \ln x$ about the x-axis, then shift up one unit.

(c) Shift $y = \ln x$ three units to the left, then down one unit.

Prob. 31. (a) Reflect $y = \sin x$ about the x-axis, then one unit up.
(b) Shift $y = \sin x$ by $\pi/4$ units to the right.
(c) Shift $y = \sin x$ by $\pi/3$ units to the left, then reflect about the x-axis.

Prob. 33. To locate the point, we have to compute the log of each of the numbers, which are:

$$\begin{aligned}
\log 0.0002 &= \log 2 \times 10^{-4} = \log 2 + \log 10^{-4} = -4 + \log 2 \approx -3.7 \\
\log 0.02 &= \log 2 \times 10^{-2} = -2 + \log 2 \approx -1.7 \\
\log 1 &= 0 \\
\log 5 &\approx 0.7 \\
\log 50 &= \log 5 \times 10 = \log 5 + \log 10 = 1 + \log 5 \approx 1.7 \\
\log 100 &= \log 10^2 = 2 \log 10 = 2 \\
\log 1000 &= \log 10^3 = 3 \log 10 = 3 \\
\log 8000 &= \log 8 \times 10^3 = \log 8 + \log 10^3 = 3 + log8 \approx 3.9 \\
\log 20000 &= \log 2 \times 10^4 = \log 2 + \log 10^4 = 4 + log2 \approx 4.3
\end{aligned}$$

Prob. 35. (b) No, (c) No

Prob. 37. four

Prob. 39. one, three

Prob. 41. six to seven

Prob. 43. $y = 5 \times (0.58)^x$

Prob. 45. $y = 3^{1/3} \times (3^{-1/3})^x$

Prob. 47. $\log y = \log 3 - 2x$

Prob. 49. $\log y = \log 2 - (1.2)(\log e)x$

Prob. 51. $\log y = \log 5 + (4 \log 2)x$

Prob. 53. $\log y = \log 4 + (2 \log 3)x$

Prob. 55. $y = 2x^{-(\log 2)/\log 5}$

Prob. 57. $y = \frac{1}{8}x^2$

Prob. 59. $\log y = \log 2 + 5 \log x$

Prob. 61. $\log y = 6 \log x$

Prob. 63. $\log y = -2 \log x$

Prob. 65. $\log y = \log 4 - 3 \log x$

Prob. 67. $\log y = \log 3 + 1.7 \log x$, log-log transformation

Prob. 69. $\log N(t) = \log 130 + (1.2t) \log 2$, log-linear transformation

Prob. 71. $\log R(t) = \log 3.6 + 1.2 \log t$, log-log transformation

Prob. 73. $y = 1.8x^{0.2}$

Prob. 75. $y = 4 \times 10^x$

Prob. 77. $y = (5.7)x^{2.1}$

Prob. 79. $\log_2 y = x$

Prob. 81. $\log_2 y = -x$

Prob. 83. (a) $\log N = \log 2 + 3t \log e$
(b) slope: $3 \log e \approx 1.303$

Prob. 85. $\log S = \log C + z \log A$, z = slope of straight line

Prob. 87. $v_{\max}$ =horizontal line intercept, $\frac{v_{\max}}{K_m}$ =vertical line intercept

Prob. 89. (a) $\log y = \log 1.162 + 0.933 \log x$

Prob. 91. (a) $\alpha = -\ln 0.91/\text{m}$,
(b) 10%,
(c) 1 m: 90%, 2 m: 81%, 3 m: 72.9%,
(e) slope $= \log 0.9 = -\alpha / \ln 10$,
(f) $z = -\frac{1}{\alpha} \ln(0.01) = \frac{\ln(0.01)}{\ln(0.9)}$,
(g) Clear lake: small α; milky lake: large α

Prob. 93. $y = (100)(10^{1/3})^x$

Prob. 95. $y = (2^{1/3})(2^{2/3})^x$

Prob. 97. $y = \log x$

Prob. 99.

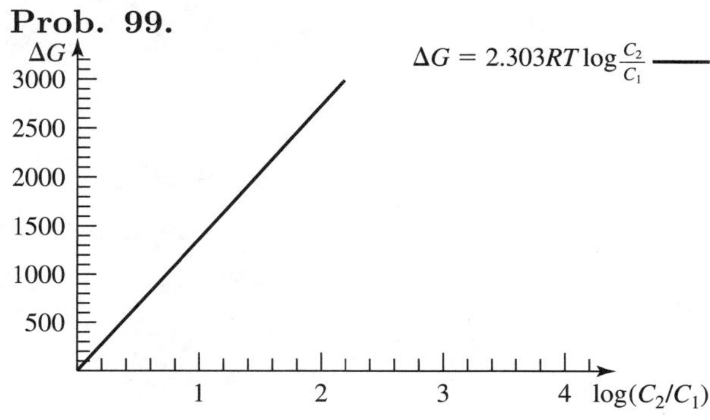

Prob. 101.

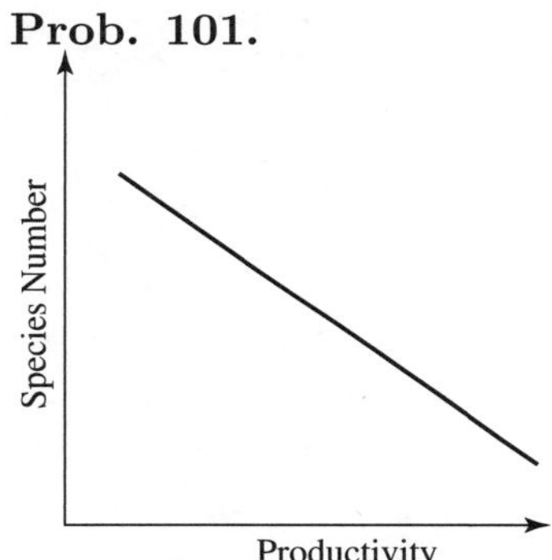

Prob. 103.

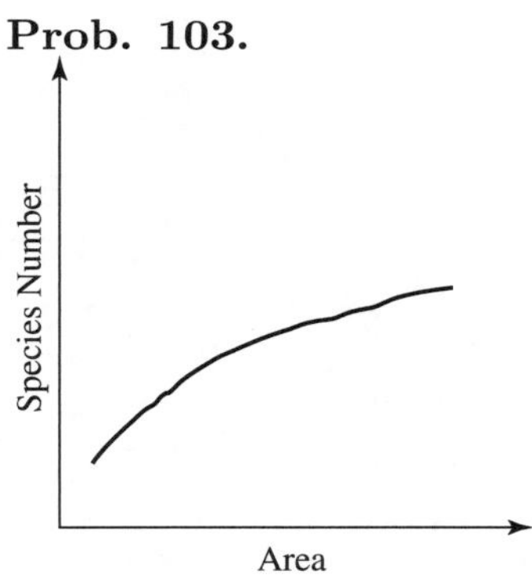

Prob. 105.

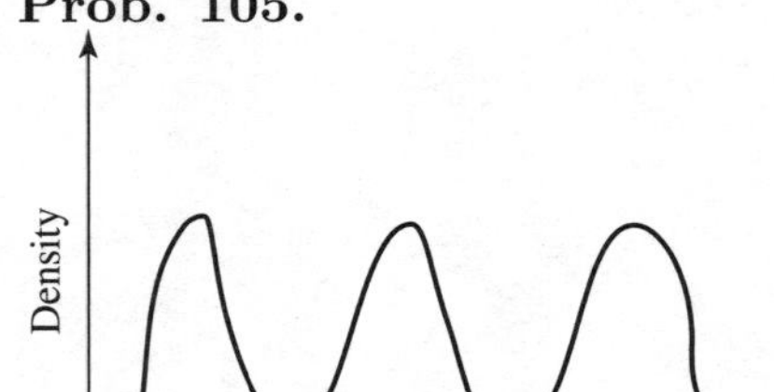

Prob. 107.

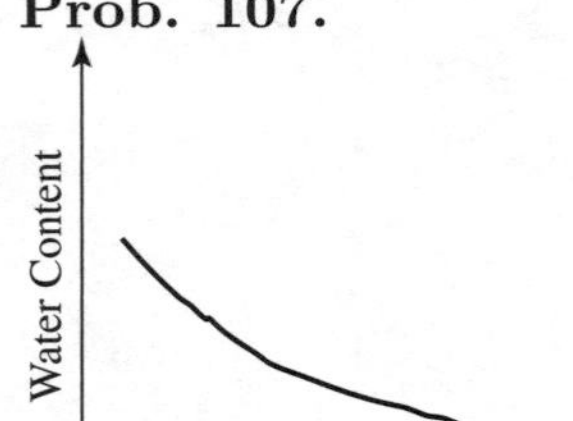

Prob. 109.

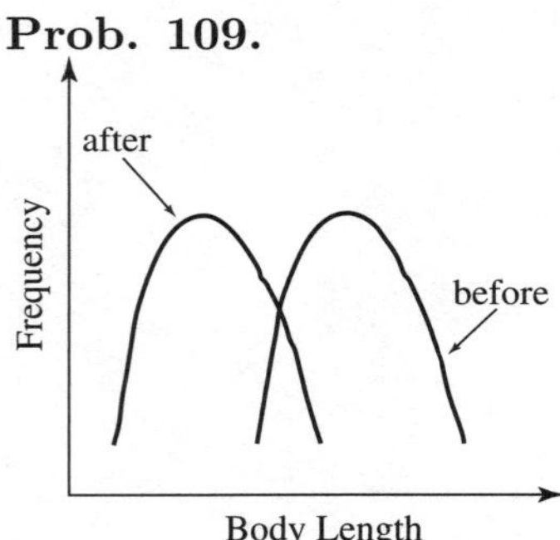

Prob. 111.

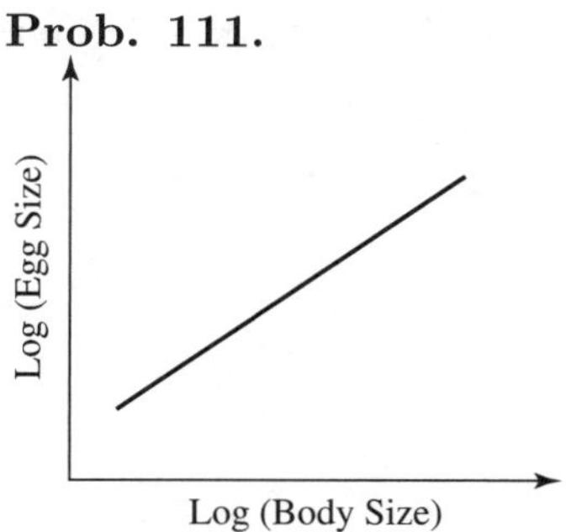

Prob. 113.

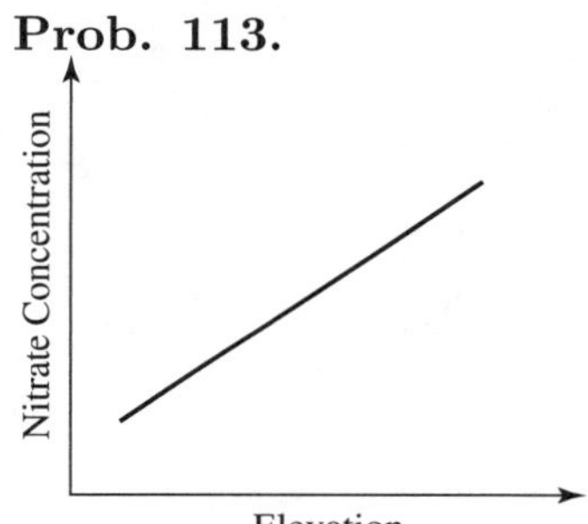

Prob. 115.

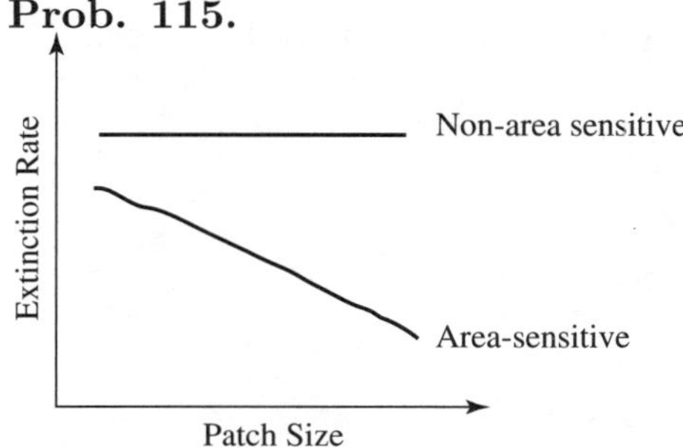

1.5 Review Problems

Prob. 1. (a) $10^4, 1.1 \times 10^4, 1.22 \times 10^4, 1.35 \times 10^4, 1.49 \times 10^4$
(b) $t = 10 \ln 10 \approx 23.0$

Prob. 3. (b) $R(x) = -4kx^3 + 4k(a+b)x^2 - kb(4a-b)x + kab^2$, polynomial of degree 3

(c) $R(x) = (0.3)(5-x)(6-2x)^2$, $0 \leq x \leq 3$

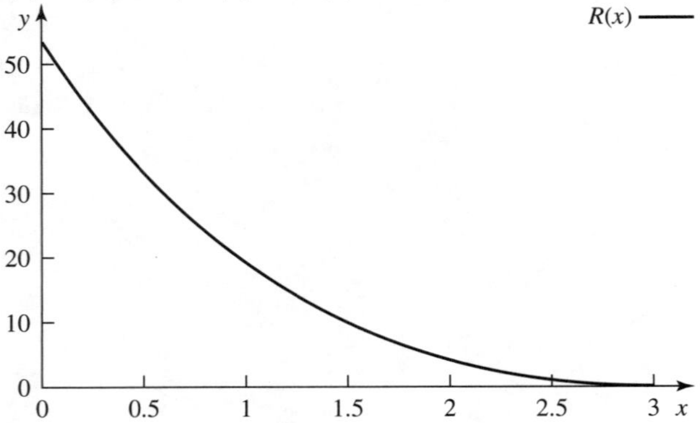

Prob. 5. (a) $L(t) = 0.69, 2.40, 4.62, 6.91$, $E(t) = 1.72, 2.20 \times 10^4, 2.69 \times 10^{43}, 1.97 \times 10^{434}$

(b) 20.93 years, 3.09 ft

(c) $10^{536,000,000}$ years, 21.98 ft

(d) $L = 19.92$ ft, $E = 10^{195 \times 10^6}$ ft

Prob. 7. $T = \frac{\ln 2}{\ln(1+\frac{9}{100})}$, T goes to infinity as q gets closer to 0.

Prob. 9. (a)

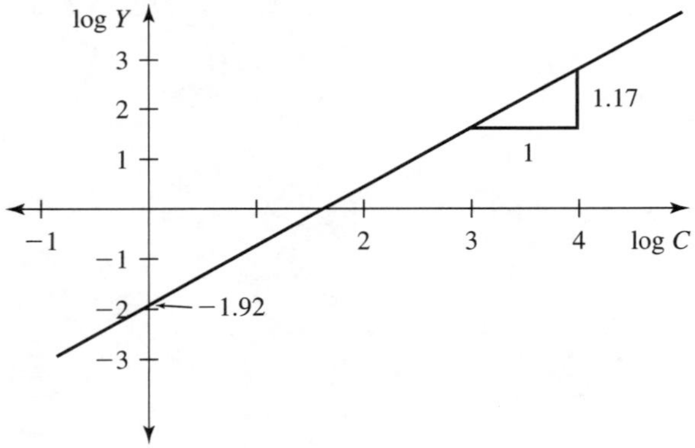

(b) $Y = C^{1.17}10^{-1.92}$

(c) $Y_p = 2.25Y_c$

(d) 8.5%

Prob.11. (a) 400 days per year
(b) $y = 4320 - 180x$
(c) 376×10^6 to 563×10^6 years ago

Prob. 13. (a) males: $S(t) = \exp[-(0.019t)^{3.41}]$; females: $S(t) = \exp[-(0.022t)^{3.24}]$
(b) males: 47.27 days; females: 40.59 days
(c) males should live longer

Prob. 15. (a) $x = k$, $v = \frac{a}{2}$
(b) $x_{0.9} = 81x_{0.1}$

Prob. 17.

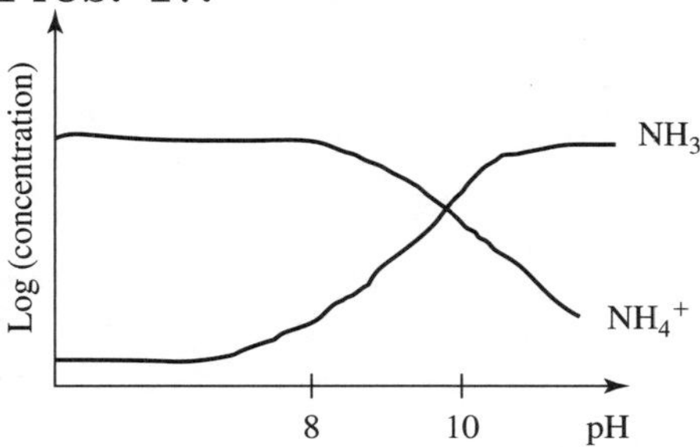

Prob. 19. $g(s) = \frac{v_{\max}}{S_k} S$

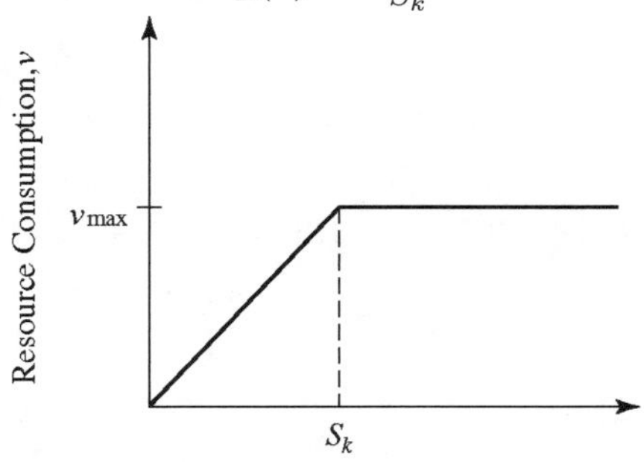

Prob. 21. (a) $\alpha = -\frac{\ln(0.01)}{18 \text{ m}} \approx 0.25\frac{1}{\text{m}}$
(b) 4.87 m

2 Discrete Time Models, Sequences Difference Equations

2.1 Exponential Growth and Decay

Prob. 1.

Table for the function $f(n) = 1/(n+2)$

n	0	1	2	3	4	5
$1/n+2$	1/2	1/3	1/4	1/5	1/6	1/7

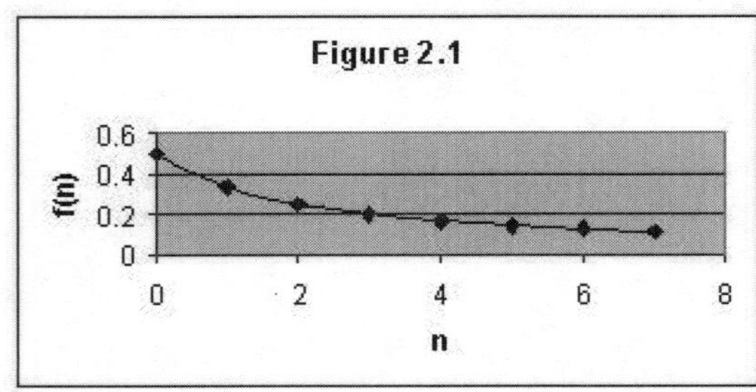

Plot for the function $f(n) = 1/n+2$

Prob. 3.

Table for the function $f(n) = 1/(n+1)^2$

n	0	1	2	3	4	5
$1/(1+n)^2$	1	1/4	1/9	1/16	1/25	1/36

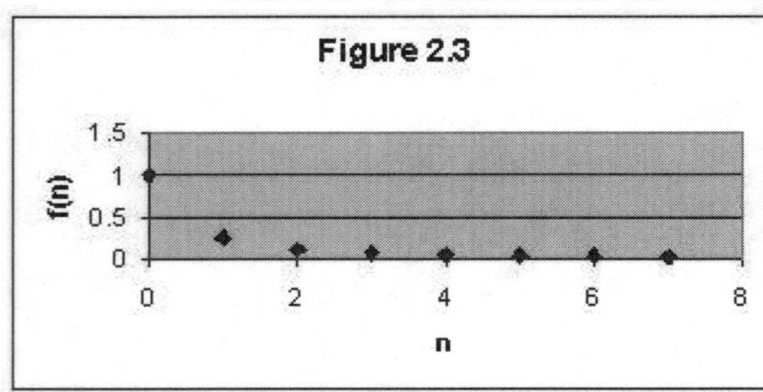

Plot for the function $f(n) = 1/(n+1)^2$

Prob. 5.

Table for the function $f(n) = n^2 - 1$

n	0	1	2	3	4	5
$n^2 - 1$	-1	0	3	8	15	24

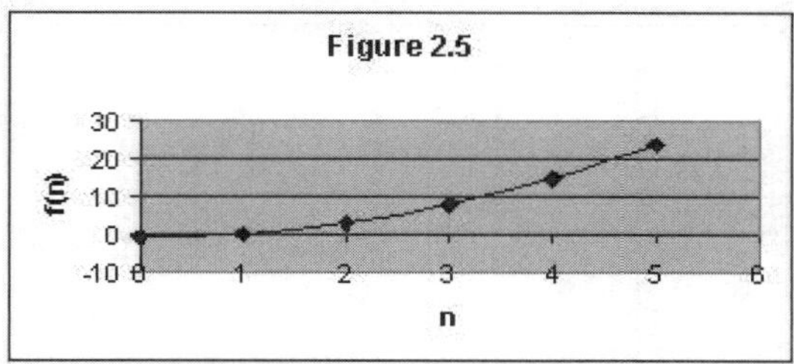

Plot for the function $f(n) = n^2 - 1$

Prob. 7.

Table for the function $f(n) = (n+1)^2$

n	0	1	2	3	4	5
$(n+1)^2$	1	4	9	16	25	36

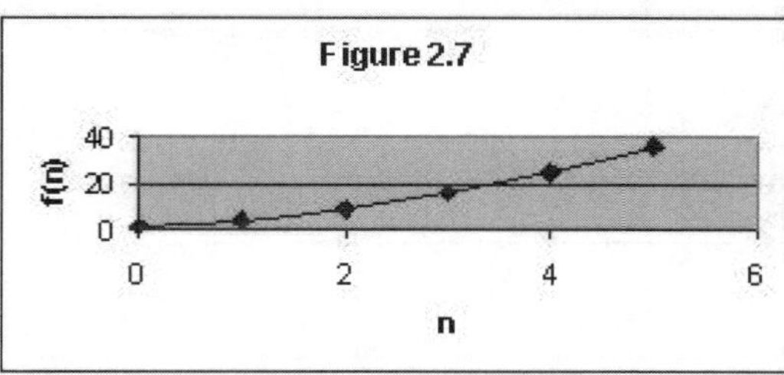

Plot for the function $f(n) = (n+1)^2$

Prob. 9.

Table for the function $f(n) = e^{\sqrt{n}}$

n	0	1	2	3	4	5
$e^{\sqrt{n}}$	1	e	$e^{\sqrt{2}}$	$e^{\sqrt{3}}$	e^2	$e^{\sqrt{5}}$

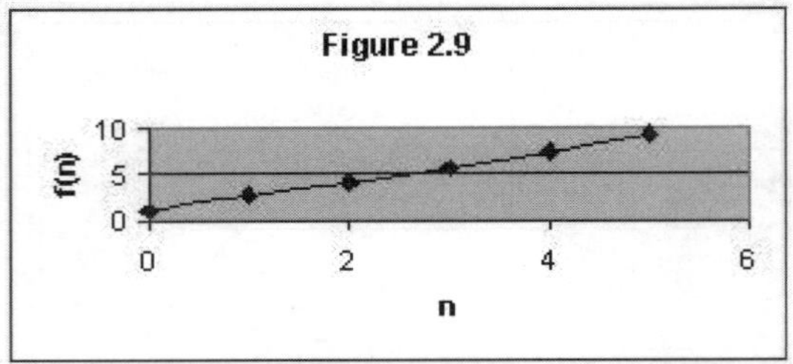

Plot for the function $f(n) = e^{\sqrt{n}}$

Prob. 11.

Table for the function $f(n) = (1/3)^n$

n	0	1	2	3	4	5
$(1/3)^n$	1	1/3	$(1/3)^2$	$(1/3)^3$	$(1/3)^4$	$(1/3)^5$

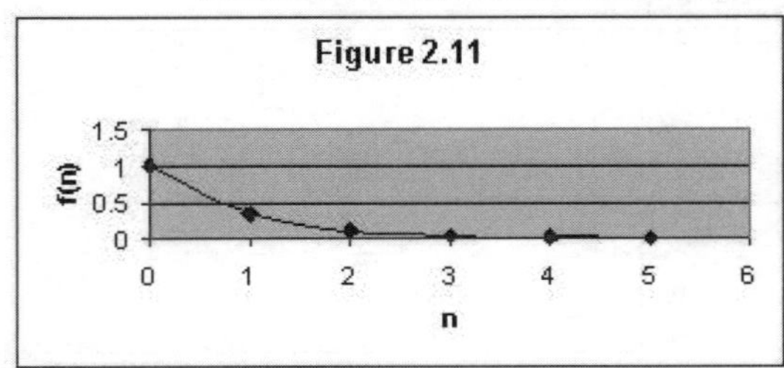

Plot for the function $f(n) = (1/3)^n$

Prob. 13. At time 0, there is one bacterium. After 1 hour, the bacterium split in to two, so there are two bacteria at time 1. 1 hour later, each of the bacteria splits again, resulting in 4 bacteria and so on.

Population Size table for function $N(t) = 2^t$

t	0	1	2	3	4	5
$N(t)$	1	2	4	8	16	32

Prob. 15. At time 0, there is one bacterium. After 23 min, the bacterium split in to two, so there are two bacteria at time 23 min. 23 minutes later,

each of the bacteria splits again, resulting in 4 bacteria and so on.
One unit of time equals 23 minutes.
From the equation N(t) = 2^t
We get, $128 = 2^t; 2^7 = 2^t$
Therefore, $t = 7$
Time taken in the production of 128 bacteria is $23 * 7 = 161 = 2$ hours 41 minutes.

Prob. 17. At time 0, there are 3 bacteria. After 10 min, each bacteria split in to two, so there are 6 bacteria at time 10 min. 10 minutes later, each of the bacteria splits again, resulting in 12 bacteria and so on.
One unit of time equals 10 minutes.
From the equation $N(t) = (N_o) * 2^t$
We get, $96 = 3.2^t; 2^5 = 2^t$
Therefore, $t = 5$
Total time taken in the production of 96 bacteria s $10 * 5 = 50 minutes$

Prob. 19. Let N(t) be the population size at time t, where t is measured as unit time, like in the above example (unit time equals 23). If initially there is one bacterium and each bacteria split in to two in a unit time, then from the table shown below we will get:

Population Size table for function $N(t) = 2^t$

t	0	1	2	3	4	5
$N(t)$	1	2	4	8	16	32

$N(t) = 2^t, t = 0, 1, 2, \ldots$ Base 2 reflects that the population doubles every unit time. Now, consider that we had 40 bacteria in the beginning. The new table now will be:

Population Size table for the new function $N(t)$

t	0	1	2	3	4	5
$N(t)$	40	80	160	320	640	1280

We see that the initial population size appears to be a multiplicative factor in front of the term 2^t. So we can now write the new equation as: $N(t) = 40 * 2^t, t = 0, 1, 2, \ldots$

Prob. 21. As already been discussed in the previous examples we know that the base 2 indicates that the population size doubles every unit time. For representing a population that triples in size every unit time, we will modify our previous equation as:

$N(t) = 3^t, t = 0, 1, 2\ldots$ Now, consider that we had 20 bacteria in the beginning. The new table now will be:

Population Size table for the this function $N(t)$

t	0	1	2	3	4	5
$N(t)$	20	60	180	540	1620	4860

We see that the initial population size appears to be a multiplicative factor in front of the term 3^t. So we can now write the new equation as: $N(t) = 20 * 3^t, t = 0, 1, 2, \ldots$

Prob. 23. As already been discussed in the previous examples we know that the base 2 indicates that the population size doubles every unit time. For representing a population, which triples in size every unit time, we will modify our previous equation as:
$N(t) = 4^t, t = 0, 1, 2, \ldots$

Now, consider that we had 72 bacteria in the beginning. The new table now will be:

Population Size table for the this function $N(t)$

t	0	1	2	3	4	5
$N(t)$	5	20	80	320	1280	5120

We see that the initial population size appears to be a multiplicative factor in front of the term 4^t. So we can now write the new equation as: $N(t) = 5 * 4^t, t = 0, 1, 2, \ldots$

Prob. 25. While constructing the tables for the population that doubles in size every unit time, we simply doubled the population size from one time step to another time step. This is equivalent to of computing the population size at time t+1 based on the population size at time t.

Take an example:

Lets take $N(0) = 1$ at $t = 0$

From our previous experience if a simple population which doubles itself starting from one bacterium we know
N(1) = 2 which can also be written as
N(1) = 2 N (0) = 2
Consecutively for the population time in the next period we can write
N(2) = 2 N(1) = 4
N(3) = 2N(2) = 8
and so on.

If we denote N(t) as N_t, We thus have two equivalent ways of describing this population. for $t = 0, 1, 2, \ldots$

$N_t = 2^t$ is equivalent to $N_{t+1} = 2N_t$ with $N_0 = 1$

If now as given in the problem $N_0 = 20$

So the above given generalized equation will remain same with initial condition modified, which would be:

$N_{\ t+1} = 2N_t$ with $N_0 = 20$.

Prob. 27. While constructing the tables for the population that triples in size every unit time, we simply tripled the population size from one time step to another time step. This is equivalent to of computing the population size at time t+1 based on the population size at time t.

Take an example:

Lets take $N(0) = 1$ at$t = 0$

From our previous experience if a simple population which triples itself starting from one bacterium we know

N(1) = 3 which can also be written as

N(1) = 3 N (0) = 3

Consecutively for the population time in the next period we can write

N(2) = 3 N(1) = 9

N(3) = 3 N(2) = 27

and so on.

If we denote N(t) as N_t, We thus have two equivalent ways of describing this population. for $t = 0, 1, 2, \ldots$

$N_t = 3^t$ is equivalent to $N_{t+1} = 3N_t$ with $N_0 = 1$

If now as given in the problem $N_0 = 10$

So the above given generalized equation will remain same with initial condition modified, which would be:

$N_{\ t+1} = 3N_t$ with $N_0 = 10$.

Prob. 29. While constructing the tables for the population that quadruples in size every unit time, we simply quadrupled the population size from one time step to another time step. This is equivalent to of computing the population size at time t+1 based on the population size at time t.

Take an example:

Lets take $N(0) = 1$ at $t = 0$

From our previous experience if a simple population which quadruples itself starting from one bacterium we know

N(1) = 4 which can also be written as

N(1) = 4 N (0) = 4

Consecutively for the population time in the next period we can write
N(2) = 4 N(1) = 16
N(3) = 4 N(2) = 64
and so on.
If we denote N(t) as N_t, We thus have two equivalent ways of describing this population. for $t = 0, 1, 2, \ldots$

$N_t = 4^t$ is equivalent to $N_{t+1} = 4N_t$ with $N_0 = 1$

If now as given in the problem $N_0 = 30$
So the above given generalized equation will remain same with initial condition modified, which would be:
$N_{\ t+1} = 4N_t$ with $N_0 = 30$.

Prob. 31.

Plot of the functions $f(x) = a^x$ and $N_t = R^t$ for
$(1)a = R = 2, (2)a = R = 3, (3)a = R = 1/2, (4)a = R = 1/3$

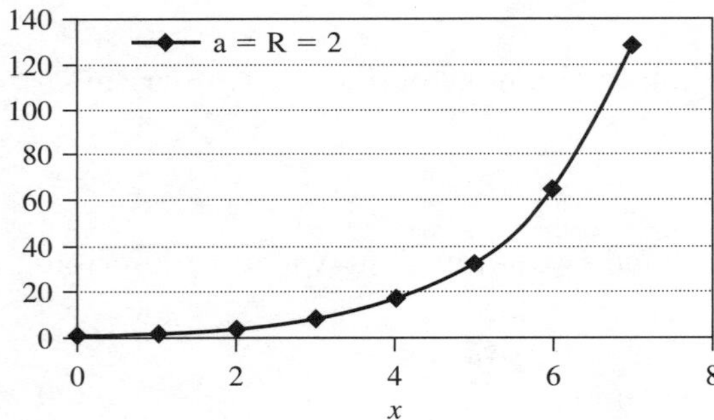

Prob. 33.

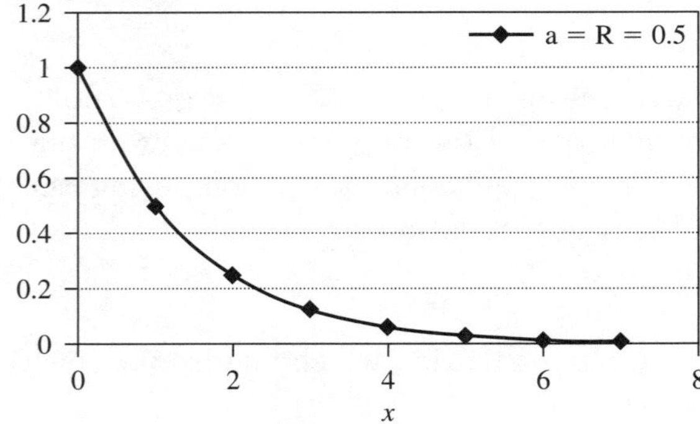

Prob. 35. $N_{t+1} = 2N_t$ with $N_0 = 3 fort = 0, 1, 2, \ldots 5$

Table for the Population Size

t	0	1	2	3	4	5
N_t	3	6	12	24	48	96

Prob. 37. $N_{t+1} = 3N_t$ with $N_0 = 2 fort = 0, 1, 2, \ldots 5$

Table for the Population Size

t	0	1	2	3	4	5
N_t	2	6	18	54	162	486

Prob. 39. $N_{t+1} = 5N_t$ with $N_0 = 1 fort = 0, 1, 2, \ldots 5$

Table for the Population Size

t	0	1	2	3	4	5
N_t	1	5	25	125	625	3125

Prob. 41. $N_{t+1} = 1/2N_t$ with $N_0 = 1024 fort = 0, 1, 2, \ldots 5$

Table for the Population Size

t	0	1	2	3	4	5
N_t	1024	512	256	128	64	32

Prob. 43. $N_{t+1} = 1/3N_t$ with $N_0 = 729 fort = 0, 1, 2, \ldots 5$

Table for the Population Size

t	0	1	2	3	4	5
N_t	729	243	81	27	9	3

Prob. 45. $N_{t+1} = 1/5N_t$ with $N_0 = 31250 fort = 0, 1, 2, \ldots 5$

Table for the Population Size

t	0	1	2	3	4	5
N_t	31250	6250	1250	250	50	10

Prob. 47. $N_{t+1} = 2N_t$ with $N_0 = 15$ N_t as a function of t would be $N_t = 15 * 2^t$

Prob. 49. $N_{\ t+1} = 3N_t$ with $N_0 = 12$ N_t as a function of t would be $N_t = 12 * 3^t$

Prob. 51. $N_{\ t+1} = 4N_t$ with $N_0 = 24$ N_t as a function of t would be $N_t = 24 * 4^t$

Prob. 53. $N_{t+1} = 1/2N_t$ with $N_0 = 5000$ N_t as a function of t would be $N_t = 5000 * (1/2)^t$

Prob. 55. $N_{\ t+1} = 1/3N_t$ with $N_0 = 8000$ N_t as a function of t would be $N_t = 8000 * (1/3)^t$

Prob. 57. $N_{\ t+1} = 1/5N_t$ with $N_0 = 1200$ N_t as a function of t would be $N_t = 1200 * (1/5)^t$

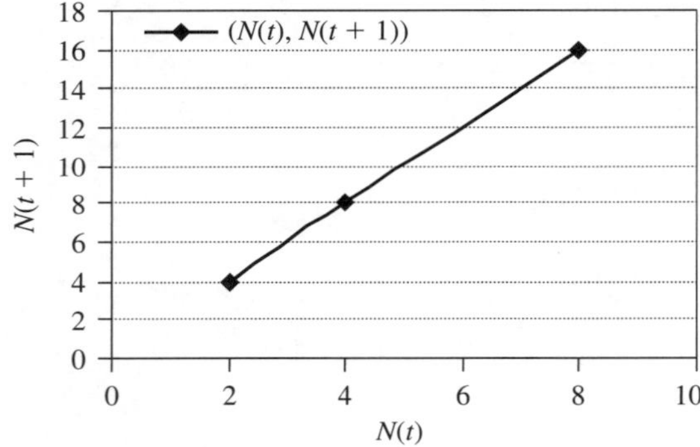

Prob. 59.

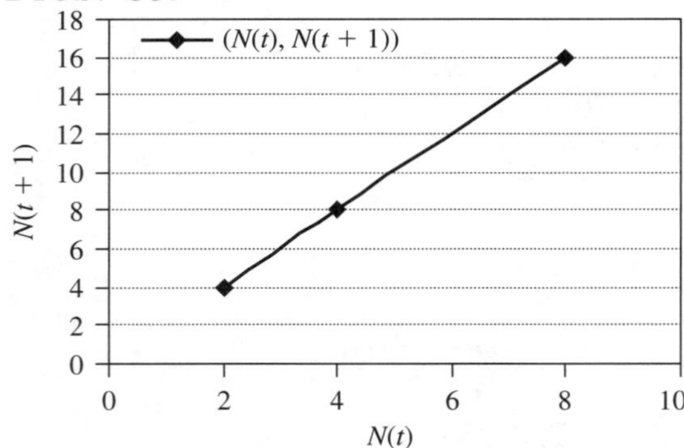

Plot of $(N_t, N_{\ t+1})$ for $R = 2, N_0 = 2$

Prob. 61.

Plot of (N_t, N_{t+1}) for $R = 3, N_0 = 1$

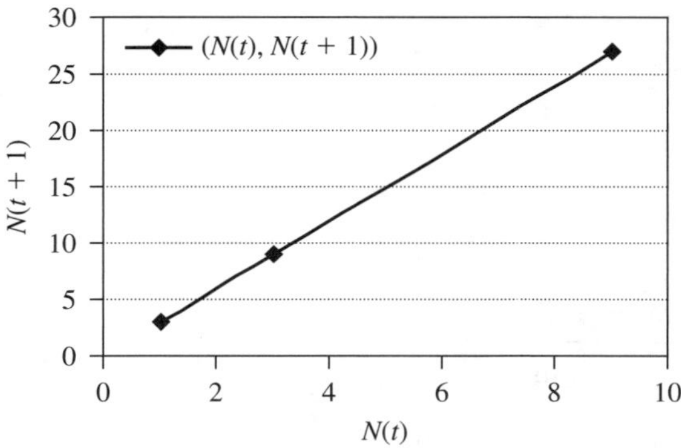

Prob. 63.

Plot of (N_t, N_{t+1}) for $R = 1/2, N_0 = 16$

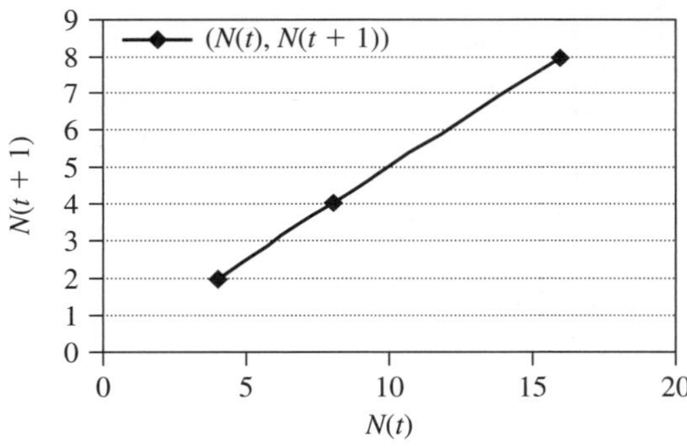

Prob. 65.

Plot of (N_t, N_{t+1}) for $R = 1/3, N_0 = 81$

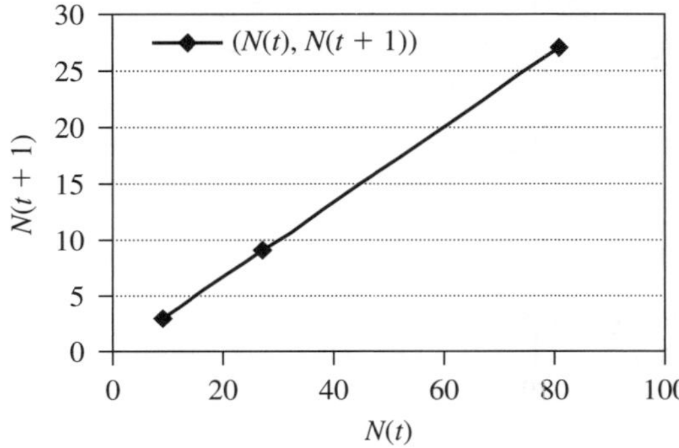

Prob. 67.

Plot of $(N_t, N_t/N_{\ t+1})$ for $R = 2, N_0 = 2$

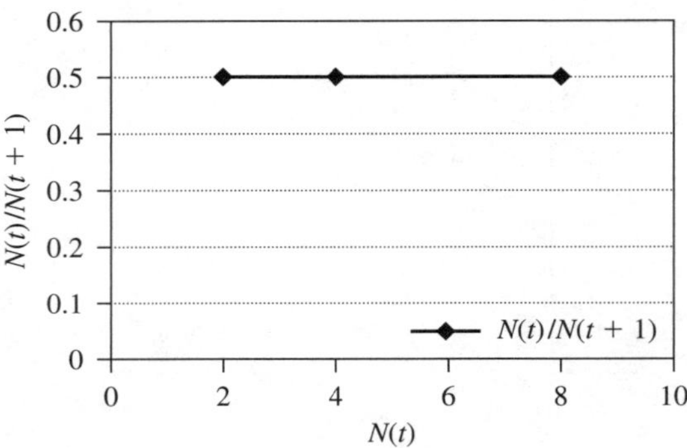

Prob. 69.

Plot of $(N_t, N_t/N_{\ t+1})$ for $R = 3, N_0 = 2$

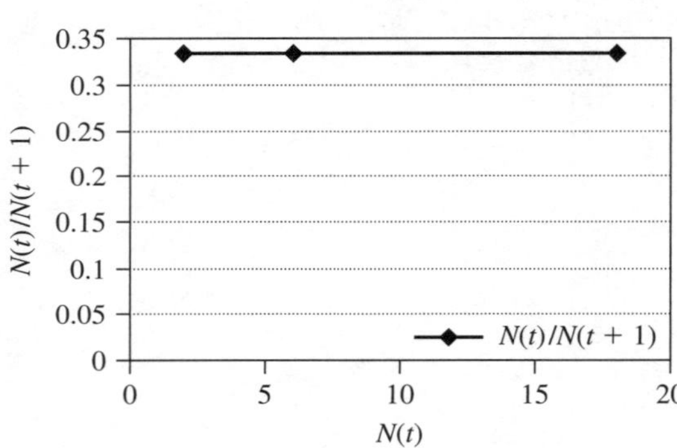

Prob. 71.

Plot of $(N_t, N_t/N_{\ t+1})$ for $R = 1/2, N_0 = 16$

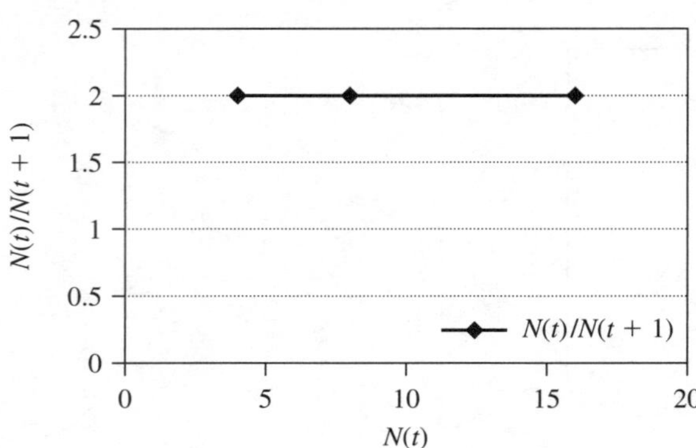

Prob. 73.

Plot of $(N_t, N_t/N_{t+1})$ for $R = 1/3, N_0 = 27$

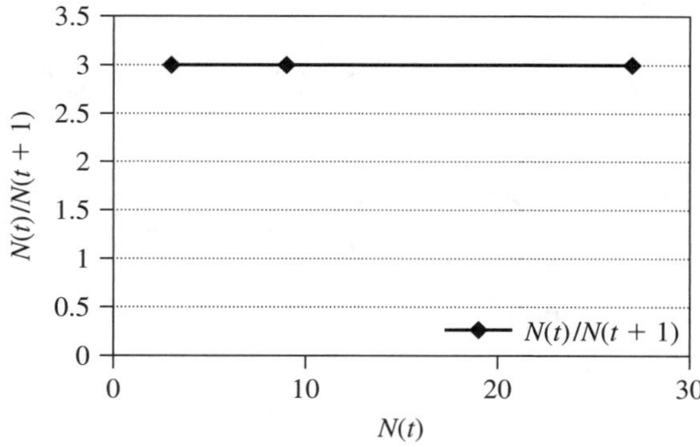

Prob. 75. Lets consider each case separately.

Case 1. As given in the Section 2.1, we know that if parent-offspring ratio is constant, parents produce same number of offsprings regardless of the current population density. But as the nesting sites is a limiting factor in population growth, if all the nesting sites are occupied there would not be an exponential population growth.

Case 2. This is a most suitable case as the nesting sites are much more than the population. Assuming that all other conditions are favorable, population growth would be exponential.

Case 3. As a hurricane had killed a large number of birds, it is very probable that the nesting conditions and other important conditions necessary for survival are not suitable. There is a very high probability that the growth rate would not be exponential though such a situation can not totally be ruled out.

Prob. 77. A statement like this can only be true when the conditions are same and favorable throughout the time span. As discussed in Problem 76, problems like pollution, population of other inhabitants, invasion of other species, deforestation and big disasters like floods, hurricanes can be reasons of any instability in the population growth.

2.2 Sequences

Prob. 1. The sequence a_n = n. n = 0, 1, 2, . . . , 5. takes on value 0, 1, 2, 3, 4, 5.

Prob. 3. The sequence $a_n = 1/(n+2)$. n = 0, 1, 2...., 5. takes on value 0.5, 0.3333, 0.25, 0.20, 0.1667, 0.1428.

Prob. 5. The sequence $a_n = (-1)^n n$. n = 0, 1, 2,..., 5. takes on value 0, -1, 2, -3, 4, -5.

Prob. 7. The sequence $a_n = n^2/(n+1)$. n = 0, 1, 2,..., 5. takes on value 0, 0.5, 1.33, 2.25, 3.33, 4.16.

Prob. 9. Looking at the question, we can guess the sequence to be as follows $a_n = n+1, n = 0, 1, 2, 3, 4$ Thus for n = 5, 6, 7, 8 a_n = 6, 7, 8, 9

Prob. 11. Looking at the question, we can guess the sequence to be as follows

$$a_n = 1/(n+1)^2, n = 0, 1, 2, 3, 4$$

Thus for $n = 5, 6, 7, 8$ $a_n = 1/36, 1/49, 1/64, 1/81$

Prob. 13. Looking at the question, we can guess the sequence to be as follows

$$a_n = (n+1)/(n+2), n = 0, 1, 2, 3, 4$$

Thus for $n = 5, 6, 7, 8$ $a_n = 6/7, 7/8, 8/9, 9/10$

Prob. 15. Looking at the question, we can guess the sequence to be as follows

$$a_n = \sqrt{(}(n+1) + e^(n+1)), n = 0, 1, 2, 3, 4$$

Thus for $n = 5, 6, 7, 8$ $a_n = \sqrt{(}6+e^6),\ \sqrt{(}7+e^7),\ \sqrt{(}8+e^8),\ \sqrt{(}9+e^9)$

Prob. 17. Looking at the sequence, we can guess the next terms, namely, 0, 1, 2, 3, 4 and so on. We thus find a_n = n for n= 0, 1, 2, ...

Prob. 19. Looking at the sequence, we can guess the next terms, namely, 1, 2, 4, 8, 16 and so on. We thus find $a_n = 2^n$ for n= 0, 1, 2,...

Prob. 21. Looking at the sequence, we can guess the next terms, namely, 1, 1/3, 1/9, 1/27, 1/81 and so on. We thus find $a_n = 1/3^n$ for n= 0, 1, 2,...

Prob. 23. Looking at the sequence, we can guess the next terms, namely, -1, 2, -3, 4, -5 and so on. We thus find $a_n = (-1)^{(n+1)}(n+1)$ for n= 0, 1, 2,...

Prob. 25. Looking at the sequence, we can guess the next terms, namely, -1/2, 1/3, -1/4, 1/5, -1/6 and so on. We thus find $a_n = (-1)^{(n+1)}/(n+2)$ for n= 0, 1, 2,...

Prob. 27. Looking at the sequence, we can guess the next terms, namely,

$$\sin(\pi), \sin(2\pi), \sin(3\pi), \sin(4\pi), \sin(5\pi)$$

and so on. We thus find $a_n = \sin((n+1)\pi)$ for n= 0, 1, 2,...

Prob. 29. Plugging successive values of n into a_n, we find that a_n is the sequence

$$1/2, 1/3, 1/4, 1/5, 1/6...$$

and guess that the terms will approach 0 as n tends to infinity. This is indeed the case, and we can show that

$$\lim_{n\to\infty} (1/(n+2)) = 0$$

Since the limiting value is a unique number, we say that the limit exists.

Prob. 31. Plugging successive values of n into a_n, we find that a_n is the sequence

$$0, 1/2, 2/3, 3/4, 4/5...$$

and guess that the terms will approach 1 as n tends to infinity. This is indeed the case, and we can show that

$$\lim_{n\to\infty} n/(n+1) = 1$$

Since the limiting value is a unique number, we say that the limit exists.

Prob. 33. Plugging successive values of n into a_n, we find that a_n is the sequence

$$1, 1/2, 1/5, 1/10, 1/17...$$

and guess that the terms will approach 0 as n tends to infinity. This is indeed the case, and we can show that

$$\lim_{n\to\infty} 1/(n^2+1) = 0$$

Since the limiting value is a unique number, we say that the limit exists.

Prob. 35. Plugging successive values of n into a_n, we find that a_n is the sequence

$$1, -1/2, 1/3, -1/4, 1/5...$$

and guess that the terms will approach 0 as n tends to infinity. This is indeed the case, and we can show that

$$\lim_{n\to\infty} (-1)^n/(n+1) = 0$$

Since the limiting value is a unique number, we say that the limit exists.

Prob. 37. Plugging successive values of n into a_n, we find that a_n is the sequence

$$1, 1/\sqrt{2}, 1/\sqrt{3}, 1/2, 1/\sqrt{5}...$$

and guess that the terms will approach 0 as n tends to infinity. This is indeed the case, and we can show that

$$\lim_{n\to\infty}(1/\sqrt{n+1}) = 0$$

Since the limiting value is a unique number, we say that the limit exists.

Prob. 39. Plugging successive values of n into a_n, we find that a_n is the sequence

$$1, 1/2, 1/3, 1/4, 1/5...$$

and guess that the terms will approach 0 as n tends to infinity.

This is indeed the case, and we can show that

$$\lim_{n\to\infty} 1/(n+1) = 0$$

Since the limiting value is a unique number, we say that the limit exists.

Prob. 41. Plugging successive values of n into a_n, we find that a_n is the sequence 0, 1/2, 4/3, 9/4, 16/5 . . . and guess that the terms will approach infinity as n tends to infinity. This is indeed the case, and we can show that

$$\lim_{n\to\infty} n^2/(n+1) = \infty$$

Since the limiting value is not a real number, we say that the limit does not exist.

Prob. 43. Plugging successive values of n into a_n, we find that a_n is the sequence

$$0, 1, \sqrt{2}, \sqrt{3}, 2...$$

and guess that the terms will approach infinity as n tends to infinity. This is indeed the case, and we can show that

$$\lim_{n\to\infty}\sqrt{n} = \infty$$

Since the limiting value is not a real number, we say that the limit does not exist.

Prob. 45. Plugging successive values of n into a_n, we find that a_n is the sequence

$$1, 2, 4, 8, 16...$$

and guess that the terms will approach infinity as n tends to infinity. This is indeed the case, and we can show that

$$\lim_{n\to\infty} 2^n = \infty$$

Since the limiting value is not a real number, we say that the limit does not exist.

Prob. 47. Plugging successive values of n into a_n, we find that a_n is the sequence $1, 3, 9, 27, 81...$ and guess that the terms will approach infinity as n tends to infinity. This is indeed the case, and we can show that

$$\lim_{n\to\infty} 3^n = \infty$$

Since the limiting value is not a real number, we say that the limit does not exist.

Prob. 49. Plugging successive values of n into a_n, we find that a_n is the sequence 0, 1, 1/2, 1/3, 1/4 . . and guess that the terms will approach 0 as n tends to infinity. This is indeed the case, and we can show that

$$\lim_{n\to\infty} 1/n = 0$$

Now We need to find an integer N so that $|1/n - 0| < 0.01$ whenever $n > N$.

Solving the inequality $|1/n - 0| < 0.01$ for n positive, we find $|1/n| < 0.01$ or $n > 1/0.01 = 100$

The smallest value for N we can choose is N = 100, which is the largest integer less than or equal to 1/0.01. Successive values for $n > 100$ give us confidence that we are on the right track but dont prove that our choice is correct:

$$a_{(101)} = 1/101 = 0.0099, a_{(102)} = 1/102 = 0.0098,$$

and so on. To see that our choice for N works, we need to show that $n > 100$ implies $|1/n| < 0.01$. Now, $n > 100$ implies $1/n \leq 1/101 = 0.0099$. Since $n > 0$, we have that $|1/n - 0| < 0.01$ whenever $n > 100$

Prob. 51. Plugging successive values of n into a_n, we find that a_n is the sequence 0, 1, 1/4, 1/9, 1/16 . . . and guess that the terms will approach 0 as n tends to infinity. This is indeed the case, and we can show that

$$\lim_{n\to\infty} (1/n)^2 = 0$$

Now We need to find an integer N so that $|1/n^2 - 0| < 0.01$ whenever $n > N$

Solving the inequality $|1/n^2 - 0| < 0.01$ for n positive, we find $|1/n^2| < 0.01$ or $n > 1/\sqrt{0.01} = 10$

The smallest value for N we can choose is N = 10, which is the largest integer less than or equal to $1/\sqrt{0.01}$. Successive values for $n > 10$ give us confidence that we are on the right track but dont prove that our choice is correct:

$a_{11} = 1/121 = 0.00826$, $a_{12} = 1/144 = 0.00694$, and so on.

To see that our choice for N works, we need to show that $n > 10$ implies $|1/n^2| < 0.01$. Now, $n > 10$ implies $1/n^2 \leq 1/121 = 0.00826$. Since $n > 0$, we have that

$|1/n^2 - 0| < 0.01$ whenever $n > 10$

Prob. 53. Plugging successive values of n into a_n, we find that a_n is the sequence 0, 1, $1/\sqrt{2}$, $1/\sqrt{3}$, 1/2... and guess that the terms will approach 0 as n tends to infinity. This is indeed the case, and we can show that

$$\lim_{n\to\infty} 1/\sqrt{n} = 0$$

Now We need to find an integer N so that $|1/\sqrt{n} - 0| < 0.1$ whenever $n > N$.

Solving the inequality $|1/\sqrt{n} - 0| < 0.1$ for n positive, we find $|1/\sqrt{n}| < 0.1$ or $n > (1/0.1)^2 = 100$

The smallest value for N we can choose is N = 100, which is the largest integer less than or equal to $(1/0.1)^2$. Successive values for $n > 100$ give us confidence that we are on the right track but do not prove that our choice is correct:

$$a_{101} = 1/\sqrt{101} = 0.0995, a_{102} = 1/\sqrt{102} = 0.0990,$$

and so on. To see that our choice for N works, we need to show that $n > 100$ implies $|1/\sqrt{n}| < 0.1$. Now, $n > 100$ implies $1/\sqrt{n} \leq 1/\sqrt{101} = 0.0995$. Since $n > 0$, we have that

$$|1/\sqrt{n} - 0| < 0.1$$

whenever $n > 100$

Prob. 55. Plugging successive values of n into a_n, we find that a_n is the sequence 0, -1, 1/2, -1/3, 1/4 . . . and guess that the terms will approach 0 as n tends to infinity. This is indeed the case, and we can show that

$$\lim_{n\to\infty} (-1)^n/n = 0$$

Now We need to find an integer N so that

$|(-1)^n/n - 0| < 0.01$ whenever $n > N$

Solving the inequality

$|(-1)^n/n - 0| < 0.01$ for n positive, we find

$|(-1)^n/n| < 0.01$ or $n > 1/0.01 = 100$

The smallest value for N we can choose is N = 100, which is the largest integer less than or equal to 1/0.01. Successive values for $n > 100$ give us confidence that we are on the right track but dont prove that our choice is correct:

$a_{101} = 1/101 = 0.0099$, a_{102} =1/102 = 0.0098, and so on.

To see that our choice for N works, we need to show that $n > 100$ implies $|(-1)^n/n| < 0.01$. Now, $n > 100$ implies $1/n \leq 1/101 = 0.0099$. Since $n > 0$, we have that

$|(-1)^n/n - 0| < 0.01$ whenever $n > 100$

Prob. 57. Plugging successive values of n into a_n, we find that a_n is the sequence 0, -1, 1/4, -1/9, 1/16 . . . and guess that the terms will approach 0 as n tends to infinity. This is indeed the case, and we can show that

$$\lim_{n\to\infty} (-1)^n/n^2 = 0$$

Now We need to find an integer N so that $|(-1)^n/n^2 - 0| < 0.001$ whenever $n > N$

Solving the inequality $|(-1)^n/n^2 - 0| < 0.001$ for n positive, we find $|(-1)^n/n^2| < 0.001$ or $n > 1/\sqrt{0.001} = 31.6227$

The smallest value for N we can choose is N = 31, which is the largest integer less than or equal to 1/0.001. Successive values for $n > 31$ give us confidence that we are on the right track but dont prove that our choice is correct:

$a_{32} = 1/32^2 = .0009765$, $a_{33} = 1/33^2 = 0.0009183$, and so on.

To see that our choice for N works, we need to show that $n > 31$ implies $|(-1)^n/n^2| < 0.001$. Now, $n > 31$ implies $1/n^2 \leq 1/32^2 = 0.00099765$. Since $n > 0$, we have that

$|(-1)^n/n^2 - 0| < 0.001$ whenever $n > 31$

Prob. 59. Plugging successive values of n into a_n, we find that a_n is the sequence 0, 1/2, 2/3, 3/4, 4/5 . . . and guess that the terms will approach 1 as n tends to infinity. This is indeed the case, and we can show that

$$\lim_{n\to\infty} n/(1+n) = 1$$

Now We need to find an integer N so that
$|n/(1+n) - 1| < 0.01$ whenever $n > N$
Solving the inequality
$|n/(1+n) - 1| < 0.01$ for n positive, we find
$|n/(1+n) - 1| < 0.01$ or $n > (1/0.01 - 1) = 99$

The smallest value for N we can choose is N = 99, which is the largest integer less than or equal to (1/0.01 -1). Successive values for $n > 99$ give us confidence that we are on the right track but do not prove that our choice is correct:

$a_{100} = 100/101 = 0.990099$, a_{101} =101/102 = 0.990091, and so on.

To see that our choice for N works, we need to show that $n > 99$ implies $|n/(n+1) - 1| < 0.01$. Now, $n > 99$ implies $1 > n/(n+1) \geq 100/101 =$ 0.990099. This implies $0 > \frac{n}{n+1} - 1 \geq -\frac{1}{100} = -0.01$. Since $n > 0$, we have that

$|n/(n+1) - 1| < 0.01$ whenever $n > 99$

Prob. 61. Plugging successive values of n into a_n, we find that a_n is the sequence 0, 2, 3/2, 4/3, 5/4 . . . and guess that the terms will approach 1 as n tends to infinity. This is indeed the case, and we can show that

$$\lim_{n\to\infty} (1+n)/n = 1$$

Now We need to find an integer N so that
$|(1+n)/n - 1| < 0.01$ whenever $n > N$
Solving the inequality
$|(1+n)/n - 1| < 0.01$ for n positive, we find
$|(1+n)/n - 1| < 0.01$ or $n > 1/0.01 = 100$

The smallest value for N we can choose is N = 100, which is the largest integer less than or equal to 1/0.01. Successive values for $n > 100$ give us confidence that we are on the right track but do not prove that our choice is correct:

$a_{101} = 102/101 = 1.0099$, a_{102} =103/102 = 1.0098, and so on.

To see that our choice for N works, we need to show that $n > 100$ implies $|(n+1)/n - 1| < 0.01$. Now, $n > 100$ implies $1 < (n+1)/n \leq 102/101 =$ 1.0099. This implies $0 < \frac{n+1}{n} - 1 \leq \frac{1}{101} < 0.01$. Since $n > 0$, we have that

$|(n+1)/n - 1| < 0.01$ whenever $n > 100$

Prob. 63. Plugging successive values of n into a_n, we find that a_n is the sequence 0, 1/2, 4/5, 9/10, 16/17 . . . and guess that the terms will approach 1 as n tends to infinity. This is indeed the case, and we can show that

$$\lim_{n\to\infty} n^2/(n^2+1) = 1$$

Now We need to find an integer N so that

$|n^2/(n^2+1) - 1| < 0.01$ whenever $n > N$

Solving the inequality

$|n^2/(n^2+1) - 1| < 0.01$ for n positive, we find

$|n^2/(n^2+1) - 1| < 0.01$ or $n > \sqrt{(1/0.01 - 1)} = 9.94987$

The smallest value for N we can choose is N = 9, which is the largest integer less than or equal to $\sqrt{1/0.01 - 1}$. Successive values for $n > 9$ give us confidence that we are on the right track but do not prove that our choice is correct:

$a_{10} = 100/101 = 0.990099$, a_{11} =121/122 = 0.99180, and so on.

To see that our choice for N works, we need to show that $n > 9$ implies $|n^2/(n^2+1) - 1| < 0.01$. Now, $n > 9$ implies $1 > n^2/(n^2+1) = 100/101 =$ 0.99002. Since $n > 0$, we have that

$|n^2/(n^2+1) - 1| < 0.01$ whenever $n > 9$

Prob. 65. We have to show that for every $\epsilon > 0$ we can find N so that

$|1/n - 0| < \epsilon$ whenever $n > N$

To find a candidate for N, we solve the inequality $|1/n| < \epsilon$. Since $1/n > 0$, we can drop the absolute values signs and find

$1/n < \epsilon$ or $n > 1/\epsilon$

Lets choose N to be the largest integer less than or equal to $1/\epsilon$. If $n > N$, then $n \geq N+1$, which is equivalent to $1/n \leq 1/(N+1)$. Since N is the largest integer less than or equal to $1/\epsilon$, it follows that $1/n \leq 1/(N+1) < \epsilon \leq 1/N$ for $n > N$. This, together with $n > 0$, shows that if N is the largest integer less than or equal to $1/\epsilon$, then $|1/n - 0| < \epsilon$ whenever $n > N$

Prob. 67. We have to show that for every $\epsilon > 0$ we can find N so that

$|1/n^2 - 0| < \epsilon$ whenever $n > N$

To find a candidate for N, we solve the inequality $|1/n^2| < \epsilon$. Since $1/n^2 > 0$, we can drop the absolute values signs and find

$1/n^2 < \epsilon$ or $n > (1/\sqrt{\epsilon})$

Lets choose N to be the largest integer less than or equal to $(1/\sqrt{\epsilon})$.

If $n > N$, then $n \geq N+1$, which is equivalent to $1/n \leq 1/(N+1)$.

Since N is the largest integer less than or equal to $(1/\sqrt{\epsilon})$, it follows that $1/n^2 \leq 1/(N+1)^2 < \epsilon \leq 1/(N)^2$ for $n > N$.

This, together with $n > 0$, shows that if N is the largest integer less than or equal to $1/\sqrt{\epsilon}$, then

$|1/n^2 - 0| < \epsilon$ whenever $n > N$

Prob. 69. We have to show that for every $\epsilon > 0$ we can find N so that
$|(n+1)/n - 1| < \epsilon$ whenever $n > N$
To find a candidate for N, we solve the inequality $|(n+1)/n - 1| < \epsilon$.
Since $|(n+1)/n - 1| > 0$, On solving the equation we will get,
$(n + 1 - n)/n = 1/n$
Now we can drop the absolute values signs and find
$1/n < \epsilon$ or $n > 1/\epsilon$
Lets choose N to be the largest integer less than or equal to $1/\epsilon$.
If $n > N$, then $n \geq N + 1$, which is equivalent to $1/n \leq 1/(N + 1)$.

Since N is the largest integer less than or equal to $1/\epsilon$, it follows that $1/n \leq 1/(N + 1) < \epsilon \leq 1/N$ for $n > N$.

This, together with $n > 0$, shows that if N is the largest integer less than or equal to $1/\epsilon$, then
$|(n+1)/n - 1| < \epsilon$ whenever $n > N$

Prob. 71. We break $1/n^2$ in to multiple of two terms, namely $1/n.1/n$. Since $\lim_{n\to\infty} 1/n$ exists, and is equal to 0. We find,

$$\lim_{n\to\infty}(1/n + 1/n^2) = \lim_{n\to\infty}(1/n) + (\lim_{n\to\infty} 1/n)(\lim_{n\to\infty} 1/n) = 0 + 0 = 0$$

Prob. 73. We break $(n + 1)/n$ in to sum of two terms, namely $1 + 1/n$. Since $\lim_{n\to\infty} 1/n$ and $\lim_{n\to\infty} 1$ exist, and are equal to 0 and 1 respectively. We find,
$\lim_{n\to\infty}(n + 1/n) = \lim_{n\to\infty}(1/n) + \lim_{n\to\infty} 1 = 0 + 1 = 1$

Prob. 75. We break $(n^2+1)/n^2$ in to sum of two terms, namely $1+1/n.1/n$. Since $\lim_{n\to\infty} 1/n$ and $\lim_{n\to\infty} 1$ exist, and are equal to 0 and 1 respectively. We find,

$$\lim_{n\to\infty}(n^2 + 1)/n^2 = \lim_{n\to\infty}(1) + (\lim_{n\to\infty} 1/n)(\lim_{n\to\infty} 1/n) = 1 + 0 = 1$$

Prob. 77. We break denominator of $(n + 1)/(n^2 - 1)$ into two terms and will write the terms as $(n + 1)/[(n + 1)(n - 1)]$. Since $\lim_{n\to\infty} 1/n$ exists, and are equal to 0. We find,

$$\begin{aligned}
\lim_{n\to\infty}(n+1)/(n^2-1) &= \lim_{n\to\infty}(n+1)/(n-1)(n+1) \\
&= \lim_{n\to\infty} 1/(n-1) \\
&= \lim_{n\to\infty}(1/n)/(1-1/n) \\
&= \lim_{n\to\infty}(1/n)/(\lim_{n\to\infty}(1-1/n)) \\
&= 0/(1-0) = 0
\end{aligned}$$

Prob. 79. Since $\lim_{n\to\infty}(1/3)^n$ and $\lim_{n\to\infty}(1/2)^n$ exist, and are equal to 0. We find,

$$\lim_{n\to\infty}[(1/3)^n + (1/2)^n] = 0 + 0 = 0$$

Prob. 81. The expression $(n + 2^{-n}/n)$ can also be written as $(1 + (1/2)^n.1/n)$. Since $\lim_{n\to\infty} 1$, $\lim_{n\to\infty} 1/n$ and $\lim_{n\to\infty}(1/2)^n$ exist, and are equal to 1, 0 and 0 respectively. We find,

$$\begin{aligned}\lim_{n\to\infty}(1 + 2^{-n}/n) &= \lim_{n\to\infty}(1 + 1/2^n.1/n)\\ &= \lim_{n\to\infty}(1) + (\lim_{n\to\infty}(1/2)^n)/(\lim_{n\to\infty}(1/n))\\ &= 1 + 0.0 = 1\end{aligned}$$

Prob. 83. By repeatedly applying the recursion to the equation,
$a_{n+1} = 2.a_n, a_0 = 1$ we find
$a_1 = 2.a_0 = 2.1 = 2$
$a_2 = 2.a_1 = 2.2 = 4$
$a_3 = 2.a_2 = 2.4 = 8$
$a_4 = 2.a_3 = 2.8 = 16$
$a_5 = 2.a_4 = 2.16 = 32$

Prob. 85. By repeatedly applying the recursion to the equation,
$a_{n+1} = 3.a_n - 2, a_0 = 1$
we find
$a_1 = 3.a_0 - 2 = 3.1 - 2 = 1$
$a_2 = 3.a_1 - 2 = 3.1 - 2 = 1$
$a_3 = 3.a_2 - 2 = 3.1 - 2 = 1$
$a_4 = 3.a_3 - 2 = 3.1 - 2 = 1$
$a_5 = 3.a_4 - 2 = 3.1 - 2 = 1$

Prob. 87. By repeatedly applying the recursion to the equation,
$a_{n+1} = 4 - 2.a_n, a_0 = 5$ we find
$a_1 = 4 - 2.a_0 = 4 - 2.5 = -6$
$a_2 = 4 - 2.a_1 = 4 - 2. - 6 = 16$
$a_3 = 4 - 2.a_2 = 4 - 2.16 = -28$
$a_4 = 4 - 2.a_3 = 4 - 2. - 28 = 60$
$a_5 = 4 - 2.a_4 = 4 - 2.60 = -116$

Prob. 89. By repeatedly applying the recursion to the equation,
$a_{n+1} = a_n/(1 + a_n), a_0 = 1$ we find
$a_1 = a_0/(1 + a_0) = 1/(1 + 1) = 0.5$

$a_2 = a_1/(1 + a_1) = 0.5/(1 + 0.5) = 0.3333$
$a_3 = a_2/(1 + a_2) = 0.3333/(1 + 0.33333) = 0.25$
$a_4 = a_3/(1 + a_3) = 0.25/(1 + 0.25) = 0.20$
$a_5 = a_4/(1 + a_4) = 0.20/(1 + 0.20) = 0.1667$

Prob. 91. By repeatedly applying the recursion to the equation,
$a_{n+1} = a_n + (1/a_n), a_0 = 1$ we find
$a_1 = a_0 + (1/a_0) = 1 + (1/1) = 2$
$a_2 = a_1 + (1/a_1) = 2 + (1/2) = 2.5$
$a_3 = a_2 + (1/a_2) = 2.5 + (1/2.5) = 2.9$
$a_4 = a_3 + (1/a_3) = 2.9 + (1/2.9) = 3.2448$
$a_5 = a_4 + (1/a_4) = 3.2448 + (1/3.2448) = 3.5530$

Prob. 93. if $a_{n+1} = f(a_n)$, then this means that if $a_0 = a$ and a is a fixed point, then $a_1 = f(a_0) = f(a) = a, a_2 = f(a_1) = f(a) = a$, and so on.

That is, a fixed point satisfies the equation

$a = f(a)$

We have the recursion $a_{n+1} = 1/2.a_n + 2$. Fixed points for recursion thus satisfy

$a = 1/2.a + 2$

Solving this for a, we find fixed point, $a = 4$.

Prob. 95. if $a_{n+1} = f(a_n)$, then this means that if $a_0 = a$ and a is a fixed point, then $a_1 = f(a_0) = f(a) = a, a_2 = f(a_1) = f(a) = a$, and so on.

That is, a fixed point satisfies the equation

$a = f(a)$

We have the recursion $a_{n+1} = 2/5.a_n - 9/5$. Fixed points for recursion thus satisfy

$a = 2/5.a - 9/5$

Solving this for a, we find fixed point, $a = -3$.

Prob. 97. if $a_{n+1} = f(a_n)$, then this means that if $a_0 = a$ and a is a fixed point, then $a_1 = f(a_0) = f(a) = a, a_2 = f(a_1) = f(a) = a$, and so on.

That is, a fixed point satisfies the equation

$a = f(a)$

We have the recursion $a_{n+1} = 4/a_n$. Fixed points for recursion thus satisfy

$a = 4/a$

Solving this for a, we find fixed points, $a = -2 and 2$.

Prob. 99. If $a_{n+1} = f(a_n)$, then this means that if $a_0 = a$ and a is a fixed point, then $a_1 = f(a_0) = f(a) = a, a_2 = f(a_1) = f(a) = a$, and so on.

That is, a fixed point satisfies the equation

$a = f(a)$

We have the recursion $a_{n+1} = 2/(a_n + 2)$. Fixed points for recursion thus satisfy

$a = 2/(a + 2)$

Solving this for a, we find fixed points, $a = (-1 - \sqrt{3}), (-1 + \sqrt{3})$.

Prob. 101. If $a_{n+1} = f(a_n)$, then this means that if $a_0 = a$ and a is a fixed point, then $a_1 = f(a_0) = f(a) = a, a_2 = f(a_1) = f(a) = a$, and so on. That is, a fixed point satisfies the equation $a = f(a)$ We have the recursion $a_{n+1} = \sqrt{5a_n}$. Fixed points for recursion thus satisfy $a = \sqrt{5a}$ Solving this for a, we find fixed point, $a = 5$. Another fixed point is $a = 0$.

Prob. 103. Since the problem tells us that the limit exists, we do not have to worry about existence. The problem that remains is to identify the limit.

To do this, we compute the fixed points. We solve

$$a = 1/2.(a + 5)$$

which has one solution, namely $a = 5$. When $a_0 = 1$, then $a_n > 1$ for all $n = 1, 2, 3, \cdots$, so we conclude

$$\lim_{n \to \infty} a_n = 5$$

Using a calculator, we can find successive values of a_n, which we collect in the following table (accurate to two decimals). The tabulated values suggest that the limit is indeed 5.

n	a_n
0	1
1	3
2	4
3	4.5
4	4.75
5	4.875
6	4.9375

Prob. 105. Since the problem tells us that the limit exists, we do not have to worry about existence. The problem that remains is to identify the limit.

To do this, we compute the fixed points. We solve

$$a = \sqrt{2a}$$

which has two solutions, namely a = 2 and a = 0. When $a_0 = 1$, then $a_n > 1$ for all n = 1, 2, 3, ...

So we conclude that $\lim_{n\to\infty} a_n = 2$.

Prob. 107. Since the problem tells us that the limit exists, we do not have to worry about existence. The problem that remains is to identify the limit.

To do this, we compute the fixed points. We solve

$$a = 2a(1 - a)$$

which has two solutions, namely a = 0 and 0.5. When $a_0 = 0.1$, then $a_n > 0.1$ for all $n = 1, 2, 3, \cdots$, so we conclude.

$$\lim_{n\to\infty} a_n = 0.5$$

Using a calculator, we can find successive values of a_n, which we collect in the following table (accurate to two decimals). The tabulated values suggest that the limit is indeed 0.5.

n	a_n
0	0.1
1	0.18
2	0.2952
3	0.41611
4	0.48592
5	0.4996
6	0.49999

Prob. 109. Since the problem tells us that the limit exists, we do not have to worry about existence. The problem that remains is to identify the limit.

To do this, we compute the fixed points. We solve $a = 1/2(a + 4/a)$ which has two solutions, namely $a = -2$ and 2. When $a_0 = 1$, then $a_n > 1$ for all $n = 1, 2, 3, \cdots$, so we conclude.

$$\lim_{n\to\infty} a_n = 2$$

Using a calculator, we can find successive values of a_n, which we collect in the following table (accurate to two decimals). The tabulated values suggest that the limit is indeed 2.

n	a_n
0	1
1	2.5
2	2.05
3	2.0006
4	2.00000
5	2.0000
6	2.0000

2.3 More Population Models

Prob. 1. The recursion to the Beverton-Holt recruitment curve is

$$N_{t+1} = \frac{2N_t}{1 + \frac{1}{15}N_t}$$

Graph of $\frac{N_t}{N_{t+1}}$ as a function of N_t

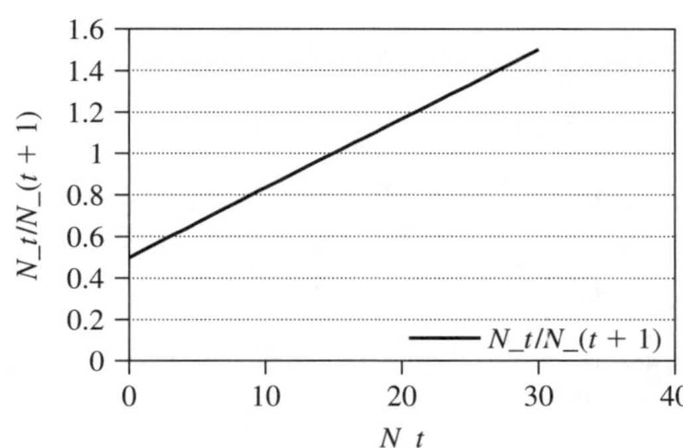

Prob. 3. The recursion to the Beverton-Holt recruitment curve is

$$N_{t+1} = \frac{1.5N_t}{1 + \frac{0.5}{40}N_t}$$

Graph of $\frac{N_t}{N_{t+1}}$ as a function of N_t

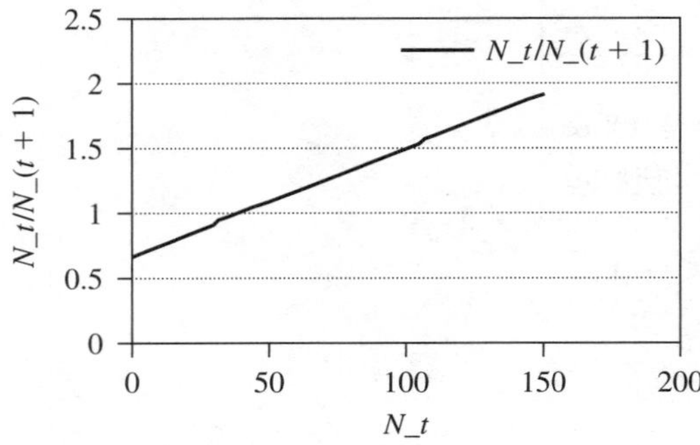

Prob. 5. The recursion to the Beverton-Holt recruitment curve is

$$N_{t+1} = \frac{1.5N_t}{1 + \frac{0.5}{40}N_t}$$

Graph of $\frac{N_t}{N_{t+1}}$ as a function of N_t

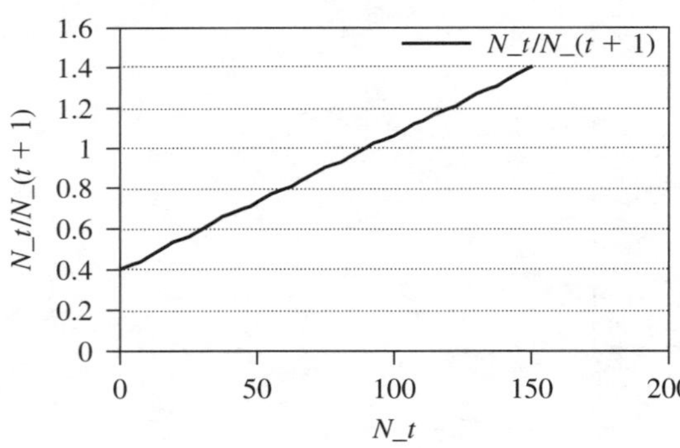

Prob. 7. The population growth equation described by the Beverton-Holt recruitment curve with growth parameter R and carrying capacity K is given by:

$$N_{t+1} = \frac{RN_t}{1 + \frac{R-1}{K}N_t}$$

The equation:

$$N_{t+1} = \frac{2N_t}{1 + \frac{1}{20}N_t}$$

can be written as:

$$N_{t+1} = \frac{2N_t}{1 + \frac{2-1}{20} N_t}$$

Comparing this equation with the equation of population growth by the Beverton-Holt recruitment curve we get:

$$R = 2, K = 20$$

Prob. 9. The population growth equation described by the Beverton-Holt recruitment curve with growth parameter R and carrying capacity K is given by:

$$N_{t+1} = \frac{RN_t}{1 + \frac{R-1}{K} N_t}$$

The equation:

$$N_{t+1} = \frac{1.5N_t}{1 + \frac{0.5}{30} N_t}$$

can be written as:

$$N_{t+1} = \frac{1.5N_t}{1 + \frac{1.5-1}{30} N_t}$$

Comparing this equation with the equation of population growth by the Beverton-Holt recruitment curve we get:

$$R = 1.5, K = 30$$

Prob. 11. The population growth equation described by the Beverton-Holt recruitment curve with growth parameter R and carrying capacity K is given by:

$$N_{t+1} = \frac{RN_t}{1 + \frac{R-1}{K} N_t}$$

The equation:

$$N_{t+1} = \frac{4N_t}{1 + \frac{1}{150} N_t}$$

can be written as:

$$N_{t+1} = \frac{4N_t}{1 + \frac{4-1}{150*3} N_t} = N_{t+1} = \frac{4N_t}{1 + \frac{4-1}{450} N_t}$$

Comparing this equation with the equation of population growth by the Beverton-Holt recruitment curve we get:

$$R = 4, K = 450$$

Prob. 13. We have the equation:

$$N_{t+1} = \frac{4N_t}{1 + \frac{1}{30}N_t}$$

We can compute the fixed points of the equation, Solving:

$$N = \frac{4N_t}{1 + \frac{1}{30}N}$$

for N, we immediately find $N = 0$. If $N \neq 0$, we divide both sides by N,

$$1 = \frac{4}{1 + \frac{1}{30}N}$$

and solve for N,

$$1 + \frac{1}{30}N = 4, \frac{1}{30}N = 3$$

from which we obtain

$$N = \frac{3}{\frac{1}{30}}, N = 30 * 3 = 90$$

We thus have two fixed points when $R > 1$: the fixed point $N = 0$, which we call trivial since it corresponds to the absence of the population, and the fixed point $N = 90$, which we call nontrivial since it corresponds to a positive population size.

Prob. 15. We have the equation:

$$N_{t+1} = \frac{2N_t}{1 + \frac{1}{30}N_t}$$

We can compute the fixed points of the equation, Solving:

$$N = \frac{2N_t}{1 + \frac{1}{30}N}$$

for N, we immediately find $N = 0$. If $N \neq 0$, we divide both sides by N,

$$1 = \frac{2}{1 + \frac{1}{30}N}$$

and solve for N,

$$1 + \frac{1}{30}N = 2, \frac{1}{30}N = 1$$

from which we obtain

$$N = \frac{1}{\frac{1}{30}}, N = 30 * 1 = 30$$

We thus have two fixed points when $R > 1$: the fixed point $N = 0$, which we call trivial since it corresponds to the absence of the population, and the fixed point $N = 30$, which we call nontrivial since it corresponds to a positive population size.

Prob. 17. We have the equation:

$$N_{t+1} = \frac{3N_t}{1 + \frac{1}{30}N_t}$$

We can compute the fixed points of the equation, Solving:

$$N = \frac{3N_t}{1 + \frac{1}{30}N}$$

for N, we immediately find $N = 0$. If $N \neq 0$, we divide both sides by N,

$$1 = \frac{3}{1 + \frac{1}{30}N}$$

and solve for N,

$$1 + \frac{1}{30}N = 3, \frac{1}{30}N = 2$$

from which we obtain

$$N = \frac{2}{\frac{1}{30}}, N = 30 * 2 = 60$$

We thus have two fixed points when $R > 1$: the fixed point $N = 0$, which we call trivial since it corresponds to the absence of the population, and the fixed point $N = 60$, which we call nontrivial since it corresponds to a positive population size.

Prob. 19. The population growth equation described by the Beverton-Holt recruitment curve with growth parameter R and carrying capacity K is given by:

$$N_{t+1} = \frac{RN_t}{1 + \frac{R-1}{K}N_t}$$

The population sizes for $t = 1, 2, 3, 4, 5$ and $\lim_{t\to\infty} N_t$ for the given values of R, K and N_0 are:

$$N_1 = \frac{2N_0}{1 + \frac{2-1}{10}N_0} = \frac{2 * 2}{1 + \frac{1}{10}2} = \frac{4}{\frac{6}{5}} = \frac{10}{3}$$

$$N_2 = \frac{2N_1}{1 + \frac{2-1}{10}N_1} = \frac{2 * \frac{10}{3}}{1 + \frac{1}{10}\frac{10}{3}} = \frac{\frac{20}{3}}{\frac{4}{3}} = 5$$

$$N_3 = \frac{2N_2}{1 + \frac{2-1}{10}N_2} = \frac{2 * 5}{1 + \frac{1}{10}5} = \frac{10}{\frac{3}{2}} = \frac{20}{3}$$

$$N_4 = \frac{2N_3}{1 + \frac{2-1}{10}N_3} = \frac{2 * \frac{20}{3}}{1 + \frac{1}{10}\frac{20}{3}} = \frac{\frac{40}{3}}{\frac{5}{3}} = 8$$

$$N_5 = \frac{2N_4}{1 + \frac{2-1}{10}N_4} = \frac{2 * 8}{1 + \frac{1}{10}8} = \frac{16}{\frac{9}{5}} = \frac{80}{9}$$

and $\lim\limits_{t\to\infty} N_t = 10$.

Prob. 21. The population growth equation described by the Beverton-Holt recruitment curve with growth parameter R and carrying capacity K is given by:

$$N_{t+1} = \frac{RN_t}{1 + \frac{R-1}{K}N_t}$$

The population sizes for $t = 1, 2, 3, 4, 5$ and $\lim\limits_{t\to\infty} N_t$ for the given values of R, K and N_0 are:

$$N_1 = \frac{3N_0}{1 + \frac{3-1}{15}N_0} = \frac{3 * 1}{1 + \frac{2}{15}1} = \frac{3}{\frac{17}{15}} = \frac{45}{17}$$

$$N_2 = \frac{3N_1}{1 + \frac{3-1}{15}N_1} = \frac{3 * \frac{45}{17}}{1 + \frac{2}{15}\frac{45}{17}} = \frac{\frac{135}{17}}{\frac{23}{17}} = \frac{135}{23}$$

$$N_3 = \frac{3N_2}{1 + \frac{3-1}{15}N_2} = \frac{3 * \frac{135}{23}}{1 + \frac{2}{15}\frac{135}{23}} = \frac{\frac{405}{23}}{\frac{41}{23}} = \frac{405}{41}$$

$$N_4 = \frac{3N_3}{1+\frac{3-1}{15}N_3} = \frac{3*\frac{405}{41}}{1+\frac{2}{15}\frac{405}{41}} = \frac{\frac{1215}{23}}{\frac{95}{23}} = \frac{243}{19}$$

$$N_5 = \frac{3N_4}{1+\frac{3-1}{15}N_4} = \frac{3*\frac{243}{19}}{1+\frac{2}{15}\frac{243}{19}} = \frac{\frac{729}{19}}{\frac{257}{95}} = \frac{3645}{257}$$

and $\lim_{t\to\infty} N_t = 15$.

Prob. 23. The population growth equation described by the Beverton-Holt recruitment curve with growth parameter R and carrying capacity K is given by:

$$N_{t+1} = \frac{RN_t}{1+\frac{R-1}{K}N_t}$$

The population sizes for $t = 1, 2, 3, 4, 5$ and $\lim_{t\to\infty} N_t$ for the given values of R, K and N_0 are:

$$N_1 = \frac{4N_0}{1+\frac{4-1}{40}N_0} = \frac{4*3}{1+\frac{3}{40}3} = \frac{12}{\frac{49}{40}} = \frac{480}{49}$$

$$N_2 = \frac{4N_1}{1+\frac{4-1}{40}N_1} = \frac{4*\frac{480}{49}}{1+\frac{3}{40}\frac{480}{49}} = \frac{\frac{1920}{49}}{\frac{85}{49}} = \frac{1920}{85}$$

$$N_3 = \frac{4N_2}{1+\frac{4-1}{40}N_2} = \frac{4*\frac{1920}{85}}{1+\frac{3}{40}\frac{1920}{85}} = \frac{\frac{7680}{85}}{\frac{229}{85}} = \frac{7680}{229}$$

$$N_4 = \frac{4N_3}{1+\frac{4-1}{40}N_3} = \frac{4*\frac{7680}{229}}{1+\frac{3}{40}\frac{7680}{229}} = \frac{\frac{30720}{229}}{\frac{805}{229}} = \frac{30720}{805}$$

$$N_5 = \frac{4N_4}{1+\frac{4-1}{40}N_4} = \frac{4*\frac{30720}{805}}{1+\frac{3}{40}\frac{30720}{805}} = \frac{\frac{122880}{805}}{\frac{3109}{805}} = \frac{122880}{3109}$$

and $\lim_{t\to\infty} N_t = 40$.

Prob. 25. The discrete logistic equation with parameter R and K is given by:

$$N_{t+1} = N_t[1 + R(1 - \frac{N_t}{K})]$$

Substituting the values of given R and K, we can write the equation as:

$$N_{t+1} = N_t[1 + (1 - \frac{N_t}{10})]$$

In the canonical form this equation can be written as:

$$x_{t+1} = rx_t(1 - x_t) = (R + 1)x_t(1 - x_t)$$

Where:

$$r = R + 1, x_t = \frac{R}{K(1 + R)}N_t, x_{t+1} = \frac{R}{K(1 + R)}N_{t+1}$$

Substituting the values of R and K in the canonical form of logistic equation, we get:

$$r = 2, x_t = \frac{1}{10(1 + 1)}N_t$$

$$r = 2, x_t = \frac{1}{20}N_t$$

Prob. 27. The discrete logistic equation with parameter R and K is given by:

$$N_{t+1} = N_t[1 + R(1 - \frac{N_t}{K})]$$

Substituting the values of given R and K, we can write the equation as:

$$N_{t+1} = N_t[1 + 2(1 - \frac{N_t}{15})]$$

In the canonical form this equation can be written as:

$$x_{t+1} = rx_t(1 - x_t) = (R + 1)x_t(1 - x_t)$$

Where:

$$r = R + 1, x_t = \frac{R}{K(1 + R)}N_t, x_{t+1} = \frac{R}{K(1 + R)}N_{t+1}$$

Substituting the values of R and K in the canonical form of logistic equation, we get:

$$r = 3, x_t = \frac{2}{15(1 + 2)}N_t$$

$$r = 3, x_t = \frac{2}{45}N_t$$

Prob. 29. The discrete logistic equation with parameter R and K is given by:

$$N_{t+1} = N_t[1 + R(1 - \frac{N_t}{K})]$$

Substituting the values of given R and K, we can write the equation as:

$$N_{t+1} = N_t[1 + 2.5(1 - \frac{N_t}{30})]$$

In the canonical form this equation can be written as:

$$x_{t+1} = rx_t(1 - x_t) = (R + 1)x_t(1 - x_t)$$

Where:

$$r = R + 1, x_t = \frac{R}{K(1+R)}N_t, x_{t+1} = \frac{R}{K(1+R)}N_{t+1}$$

Substituting the values of R and K in the canonical form of logistic equation, we get:

$$r = 3.5, x_t = \frac{2.5}{30(1+2.5)}N_t$$

$$r = 3.5, x_t = \frac{0.5}{21}N_t$$

Prob. 31.

(a) The original variable N_t has units (or dimension) "number of individuals"; the parameter K has the same units. By dividing N_t by K, the units cancel and we can say that: $x_t = \frac{N_t}{K}$ is *dimensionless.*

(b) The variable M_t has units (or dimension) "1000 number of individuals"; the parameter L has the same units. By dividing M_t by L, the units cancel and we can say that: $y_t = \frac{M_t}{L}$ is *dimensionless.*

(c) The variable M_t has units (or dimension) "1000 number of individuals" while the variable N_t has units (or dimension) "number of individuals". On dividing M_t by N_t, the common units, number of individuals cancel and we get:

$$\frac{M_t}{N_t} = \frac{1}{1000}$$

Similarly, The variable L has units (or dimension) "1000 number of individuals" while the variable K has units (or dimension) "number of individuals". On dividing L by K, the common units, number of individuals cancel and we get:

$$\frac{L}{K} = \frac{1}{1000}$$

(d) from part (c), we know that:

$$\frac{M_t}{N_t} = \frac{1}{1000}, \frac{L}{K} = \frac{1}{1000}$$

Substituting the values of N_t and K from the given question:

$$\frac{M_t}{20,000} = \frac{1}{1000}, \frac{L}{5000} = \frac{1}{1000}$$

$$M_t = 20, L = 5$$

(e) We know that:

$$y_t = \frac{M_t}{L}, x_t = \frac{N_t}{K}$$

In the above given expression, substituting the values of M_t, N_t, L and K, we will get:

$$y_t = \frac{20}{5}, x_t = \frac{20,000}{5000}$$

So:

$$y_t = 4, x_t = 4$$

From the values obtained above, it can be deduced that:

$$y_t = x_t$$

Prob. 33. The unit or dimension of T is characteristic time and the unit or dimension of t is time elapsed since the beginning of experiment. z is the obtained by dividing t by T, i.e time elapsed by characteristic time. As both (t and T) have the same units, they will cancel and we will get: $z = \frac{t}{T}$ is *dimensionless*. If t = 120 minutes and L = 20 minutes, $z = \frac{t}{T} = \frac{120}{20} = 6$; also z is dimensionless as t and T has the same units which will cancel. If t and T, both will have the units of hours instead of minutes, on dividing t by T, the units will get cancelled and same results will be obtained for z.

Prob. 35. The discrete logistic equation is given by:

$$x_{t+1} = rx_t(1 - x_t)$$

for $t = 0, 1, 2, 3, 4, \ldots 20$

$$x_0 = 0.2$$
$$x_1 = rx_0(1 - x_0) = 2 * 0.2(1 - 0.2) = 0.32$$
$$x_2 = rx_1(1 - x_1) = 2 * 0.32(1 - 0.32) = 0.4352$$
$$x_3 = rx_2(1 - x_2) = 2 * 0.4352(1 - 0.4352) = 0.4916$$
$$x_4 = rx_3(1 - x_3) = 2 * 0.4916(1 - 0.4916) = 0.4998$$
$$x_5 = rx_4(1 - x_4) = 2 * 0.4998(1 - 0.4998) = 0.4999$$
$$x_6 = rx_5(1 - x_5) = 2 * 0.4999(1 - 0.4999) = 0.4999$$
$$x_7 = rx_6(1 - x_6) = 2 * 0.4999(1 - 0.4999) = 0.4999$$
$$x_8 = rx_7(1 - x_7) = 2 * 0.4999(1 - 0.4999) = 0.4999$$
$$x_9 = rx_8(1 - x_8) = 2 * 0.4999(1 - 0.4999) = 0.4999$$
$$x_{10} = rx_9(1 - x_9) = 2 * 0.4999(1 - 0.4999) = 0.4999$$
$$x_{11} = rx_{10}(1 - x_{10}) = 2 * 0.4999(1 - 0.4999) = 0.4999$$
$$x_{12} = rx_{11}(1 - x_{11}) = 2 * 0.4999(1 - 0.4999) = 0.4999$$
$$x_{13} = rx_{12}(1 - x_{12}) = 2 * 0.4999(1 - 0.4999) = 0.4999$$
$$x_{14} = rx_{13}(1 - x_{13}) = 2 * 0.4999(1 - 0.4999) = 0.4999$$
$$x_{15} = rx_{14}(1 - x_{14}) = 2 * 0.4999(1 - 0.4999) = 0.4999$$
$$x_{16} = rx_{15}(1 - x_{15}) = 2 * 0.4999(1 - 0.4999) = 0.4999$$
$$x_{17} = rx_{16}(1 - x_{16}) = 2 * 0.4999(1 - 0.4999) = 0.4999$$
$$x_{18} = rx_{17}(1 - x_{17}) = 2 * 0.4999(1 - 0.4999) = 0.4999$$
$$x_{19} = rx_{18}(1 - x_{18}) = 2 * 0.4999(1 - 0.4999) = 0.4999$$
$$x_{20} = rx_{19}(1 - x_{19}) = 2 * 0.4999(1 - 0.4999) = 0.4999$$

The Graph of x_t as a function of t is given by:

Graph of x_t as a function of t with $r = 2, x_0 = 0.2$

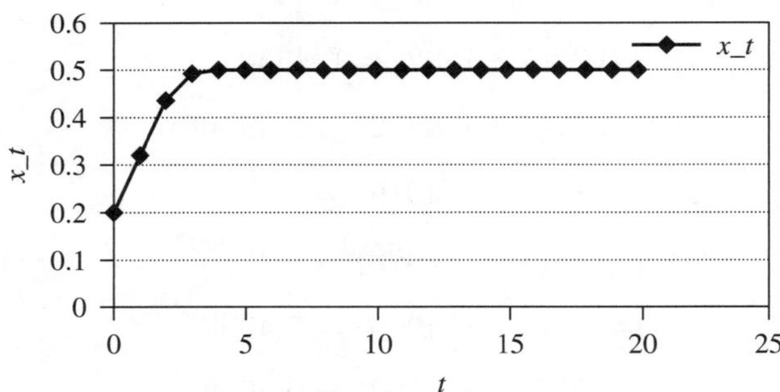

Prob. 37. The discrete logistic equation is given by:

$$x_{t+1} = rx_t(1 - x_t)$$

for $t = 0, 1, 2, 3, 4, \ldots 20$

$$x_0 = 0.9$$

$$x_1 = rx_0(1 - x_0) = 2 * 0.9(1 - 0.9) = 0.18$$

$$x_2 = rx_1(1 - x_1) = 2 * 0.18(1 - 0.18) = 0.2952$$

$$x_3 = rx_2(1 - x_2) = 2 * 0.2952(1 - 0.2952) = 0.4161$$

$$x_4 = rx_3(1 - x_3) = 2 * 0.4161(1 - 0.4161) = 0.4859$$

$$x_5 = rx_4(1 - x_4) = 2 * 0.4859(1 - 0.4859) = 0.4996$$

$$x_6 = rx_5(1 - x_5) = 2 * 0.4996(1 - 0.4996) = 0.5000$$

$$x_7 = rx_6(1 - x_6) = 2 * 0.5000(1 - 0.5000) = 0.5000$$

$$x_8 = rx_7(1 - x_7) = 2 * 0.5000(1 - 0.5000) = 0.5000$$

$$x_9 = rx_8(1 - x_8) = 2 * 0.5000(1 - 0.5000) = 0.5000$$

$$x_{10} = rx_9(1 - x_9) = 2 * 0.5000(1 - 0.5000) = 0.5000$$

$$x_{11} = rx_{10}(1 - x_{10}) = 2 * 0.5000(1 - 0.5000) = 0.5000$$

$$x_{12} = rx_{11}(1 - x_{11}) = 2 * 0.5000(1 - 0.5000) = 0.5000$$

$$x_{13} = rx_{12}(1 - x_{12}) = 2 * 0.5000(1 - 0.5000) = 0.5000$$

$$x_{14} = rx_{13}(1 - x_{13}) = 2 * 0.5000(1 - 0.5000) = 0.5000$$

$$x_{15} = rx_{14}(1 - x_{14}) = 2 * 0.5000(1 - 0.5000) = 0.5000$$

$$x_{16} = rx_{15}(1 - x_{15}) = 2 * 0.5000(1 - 0.5000) = 0.5000$$
$$x_{17} = rx_{16}(1 - x_{16}) = 2 * 0.5000(1 - 0.5000) = 0.5000$$
$$x_{18} = rx_{17}(1 - x_{17}) = 2 * 0.5000(1 - 0.5000) = 0.5000$$
$$x_{19} = rx_{18}(1 - x_{18}) = 2 * 0.5000(1 - 0.5000) = 0.5000$$
$$x_{20} = rx_{19}(1 - x_{19}) = 2 * 0.5000(1 - 0.5000) = 0.5000$$

The Graph of x_t as a function of t is given by:

Graph of x_t as a function of t with $r = 2, x_0 = 0.9$

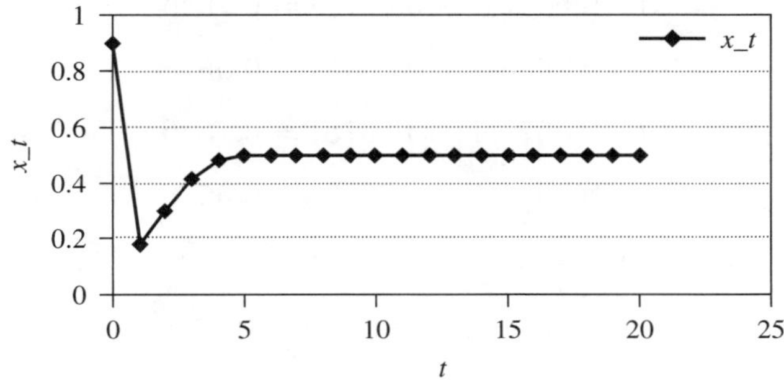

Prob. 39. The discrete logistic equation is given by:

$$x_{t+1} = rx_t(1 - x_t)$$

for $t = 0, 1, 2, 3, 4....20$

$$x_0 = 0.5$$
$$x_1 = rx_0(1 - x_0) = 3.1 * 0.5(1 - 0.5) = 0.775$$
$$x_2 = rx_1(1 - x_1) = 3.1 * 0.775(1 - 0.775) = 0.5406$$
$$x_3 = rx_2(1 - x_2) = 3.1 * 0.5406(1 - 0.5406) = 0.7699$$
$$x_4 = rx_3(1 - x_3) = 3.1 * 0.7699(1 - 0.7699) = 0.5491$$
$$x_5 = rx_4(1 - x_4) = 3.1 * 0.5491(1 - 0.5491) = 0.7675$$
$$x_6 = rx_5(1 - x_5) = 3.1 * 0.7675(1 - 0.7675) = 0.5531$$
$$x_7 = rx_6(1 - x_6) = 3.1 * 0.5531(1 - 0.5531) = 0.7662$$
$$x_8 = rx_7(1 - x_7) = 3.1 * 0.7662(1 - 0.7662) = 0.5552$$
$$x_9 = rx_8(1 - x_8) = 3.1 * 0.5552(1 - 0.5552) = 0.7655$$

$$x_{10} = rx_9(1 - x_9) = 3.1 * 0.7655(1 - 0.7655) = 0.5565$$
$$x_{11} = rx_{10}(1 - x_{10}) = 3.1 * 0.5565(1 - 0.5565) = 0.7651$$
$$x_{12} = rx_{11}(1 - x_{11}) = 3.1 * 0.7651(1 - 0.7651) = 0.5572$$
$$x_{13} = rx_{12}(1 - x_{12}) = 3.1 * 0.5572(1 - 0.5572) = 0.7648$$
$$x_{14} = rx_{13}(1 - x_{13}) = 3.1 * 0.7648(1 - 0.7648) = 0.5575$$
$$x_{15} = rx_{14}(1 - x_{14}) = 3.1 * 0.5575(1 - 0.5575) = 0.7647$$
$$x_{16} = rx_{15}(1 - x_{15}) = 3.1 * 0.7647(1 - 0.7647) = 0.5577$$
$$x_{17} = rx_{16}(1 - x_{16}) = 3.1 * 0.5577(1 - 0.5577) = 0.7646$$
$$x_{18} = rx_{17}(1 - x_{17}) = 3.1 * 0.7646(1 - 0.7646) = 0.5578$$
$$x_{19} = rx_{18}(1 - x_{18}) = 3.1 * 0.5578(1 - 0.5578) = 0.7646$$
$$x_{20} = rx_{19}(1 - x_{19}) = 3.1 * 0.7646(1 - 0.7646) = 0.5578$$

The Graph of x_t as a function of t is given by:

Graph of x_t as a function of t with $r = 3.1, x_0 = 0.5$

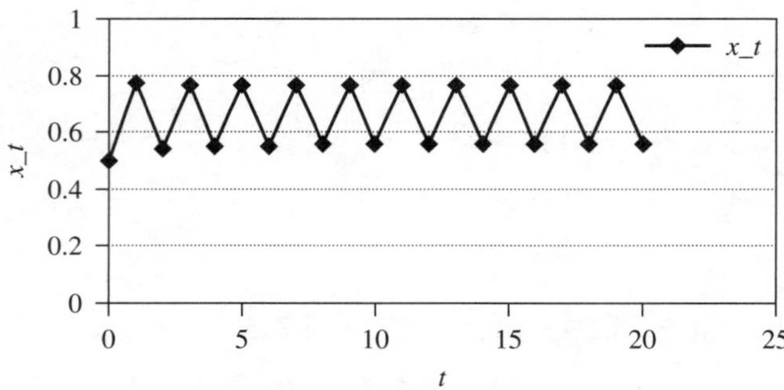

Prob. 41. The discrete logistic equation is given by:

$$x_{t+1} = rx_t(1 - x_t)$$

for $t = 0, 1, 2, 3, 4....20$

$$x_0 = 0.9$$
$$x_1 = rx_0(1 - x_0) = 3.1 * 0.9(1 - 0.9) = 0.279$$
$$x_2 = rx_1(1 - x_1) = 3.1 * 0.279(1 - 0.279) = 0.6236$$
$$x_3 = rx_2(1 - x_2) = 3.1 * 0.6236(1 - 0.6236) = 0.7276$$

$$x_4 = rx_3(1 - x_3) = 3.1 * 0.7276(1 - 0.7276) = 0.6143$$
$$x_5 = rx_4(1 - x_4) = 3.1 * 0.6143(1 - 0.6143) = 0.7345$$
$$x_6 = rx_5(1 - x_5) = 3.1 * 0.7345(1 - 0.7345) = 0.6045$$
$$x_7 = rx_6(1 - x_6) = 3.1 * 0.6045(1 - 0.6045) = 0.7412$$
$$x_8 = rx_7(1 - x_7) = 3.1 * 0.7412(1 - 0.7412) = 0.5947$$
$$x_9 = rx_8(1 - x_8) = 3.1 * 0.5947(1 - 0.5947) = 0.7472$$
$$x_{10} = rx_9(1 - x_9) = 3.1 * 0.7472(1 - 0.7472) = 0.5856$$
$$x_{11} = rx_{10}(1 - x_{10}) = 3.1 * 0.5856(1 - 0.5856) = 0.7522$$
$$x_{12} = rx_{11}(1 - x_{11}) = 3.1 * 0.7522(1 - 0.7522) = 0.5778$$
$$x_{13} = rx_{12}(1 - x_{12}) = 3.1 * 0.5778(1 - 0.5778) = 0.7562$$
$$x_{14} = rx_{13}(1 - x_{13}) = 3.1 * 0.7562(1 - 0.7562) = 0.5714$$
$$x_{15} = rx_{14}(1 - x_{14}) = 3.1 * 0.5714(1 - 0.5714) = 0.7592$$
$$x_{16} = rx_{15}(1 - x_{15}) = 3.1 * 0.7592(1 - 0.7592) = 0.5668$$
$$x_{17} = rx_{16}(1 - x_{16}) = 3.1 * 0.5668(1 - 0.5668) = 0.7612$$
$$x_{18} = rx_{17}(1 - x_{17}) = 3.1 * 0.7612(1 - 0.7612) = 0.5634$$
$$x_{19} = rx_{18}(1 - x_{18}) = 3.1 * 0.5634(1 - 0.5634) = 0.7625$$
$$x_{20} = rx_{19}(1 - x_{19}) = 3.1 * 0.7625(1 - 0.7625) = 0.5614$$

The Graph of x_t as a function of t is given by:

Graph of x_t as a function of t with $r = 3.1, x_0 = 0.9$

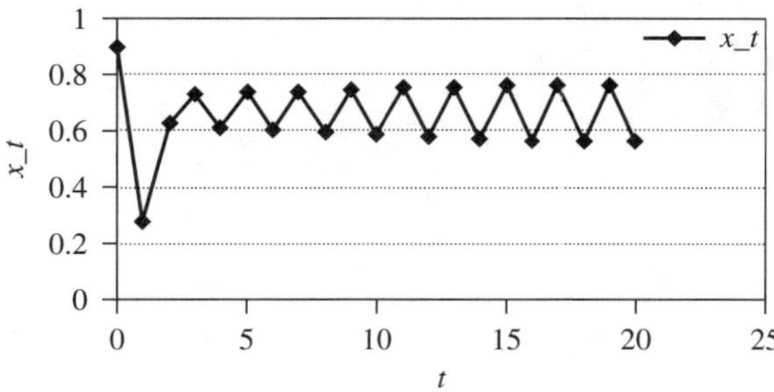

Prob. 43. The discrete logistic equation is given by $x_{t+1} = 3.8x_t(1 - x_t)$. The table lists the values for x_t for $t = 0, 1, 2, 3, \ldots, 20$

Time	x_t
0	0.5
1	0.95
2	0.1805
3	0.562095
4	0.935348
5	0.229794
6	0.672557
7	0.936851
8	0.518819
9	0.948654
10	0.185096
11	0.573174
12	0.929653
13	0.248514
14	0.709668
15	0.782949
16	0.645771
17	0.869254
18	0.431877
19	0.932365
20	0.23963

The graph of x_t with $r = 3.8$ and $x_0 = 0.5$ is below:

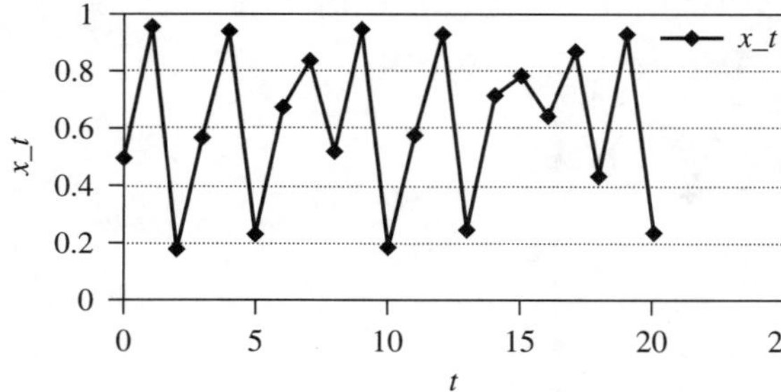

Prob. 45. The discrete logistic equation is given by $x_{t+1} = 3.8x_t(1 - x_t)$. The table lists the values for x_t for $t = 0, 1, 2, 3, \ldots, 20$

Time	x_t
0	0.9
1	0.342
2	0.855137
3	0.470736
4	0.946746
5	0.191589
6	0.588555
7	0.9202
8	0.27904
9	0.764472
10	0.684208
11	0.821056
12	0.558307
13	0.937081
14	0.224049
15	0.660634
16	0.851947
17	0.479305
18	0.948373
19	0.186056
20	0.575468

The graph of x_t with $r = 3.8$ and $x_0 = 0.9$ is below:

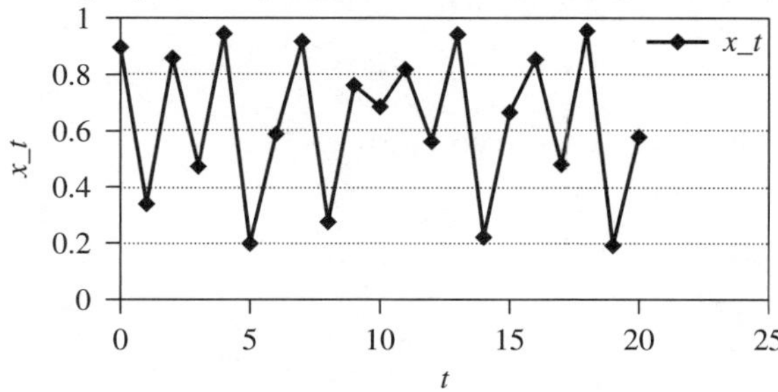

Prob. 47. The graph for the Ricker's curve in the $N_t - N_{t+1}$ plane is given below: The Graph of x_t as a function of t is given by:

Graph of N_{t+1} as a function of N_t

The points of intersection of this graph with the line $N_{t+1} = N_t$ are 0 and 10.

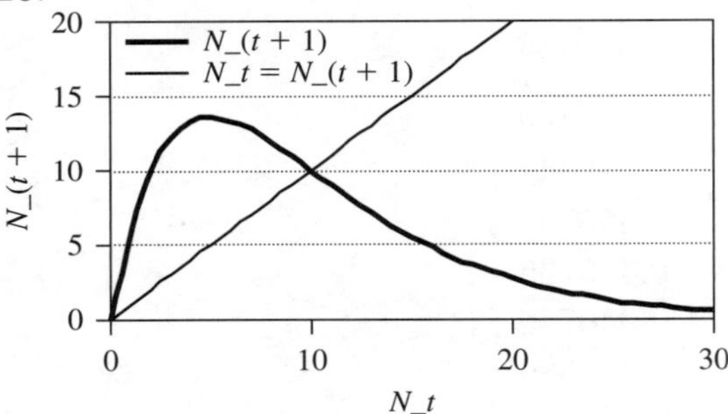

Prob. 49. The graph for the Ricker's curve in the $N_t - N_{t+1}$ plane is given below: The Graph of x_t as a function of t is given by:

Graph of N_{t+1} as a function of N_t

The points of intersection of this graph with the line $N_{t+1} = N_t$ are 0 and 12.

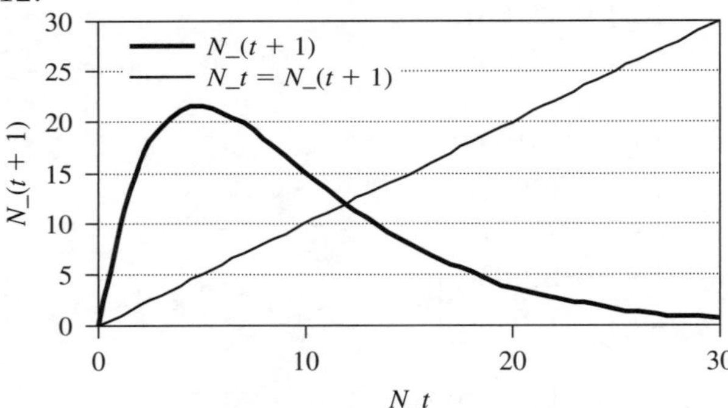

Prob. 51.

(a) We know the behavior of the Ricker's curve is given by:

$$N_{t+1} = N_t exp[R(1 - \frac{N_t}{K})]$$

N_t for $t = 1, 2, 3, 4, \ldots 20$ are:

$$N_1 = N_0 exp[1(1 - \frac{N_0}{20})] = 10.585$$

$$N_2 = N_1 exp[1(1 - \frac{N_1}{20})] = 16.94865$$

$$N_3 = N_2 exp[1(1 - \frac{N_2}{20})] = 19.74214$$

$$N_4 = N_3 exp[1(1 - \frac{N_3}{20})] = 19.99832$$

$$N_5 = N_0 exp[1(1 - \frac{N_4}{20})] = 20$$

$$N_6 = N_5 exp[1(1 - \frac{N_5}{20})] = 20$$

$$N_7 = N_6 exp[1(1 - \frac{N_6}{20})] = 20$$

$$N_8 = N_7 exp[1(1 - \frac{N_7}{20})] = 20$$

$$N_9 = N_8 exp[1(1 - \frac{N_8}{20})] = 20$$

$$N_{10} = N_9 exp[1(1 - \frac{N_9}{20})] = 20$$

$$N_{11} = N_{10} exp[1(1 - \frac{N_{10}}{20})] = 20$$

$$N_{12} = N_{11} exp[1(1 - \frac{N_{11}}{20})] = 20$$

$$N_{13} = N_{12} exp[1(1 - \frac{N_{12}}{20})] = 20$$

$$N_{14} = N_{13} exp[1(1 - \frac{N_{13}}{20})] = 20$$

$$N_{15} = N_{14} exp[1(1 - \frac{N_{14}}{20})] = 20$$

$$N_{16} = N_{15} exp[1(1 - \frac{N_{15}}{20})] = 20$$

$$N_{17} = N_{16} exp[1(1 - \frac{N_{16}}{20})] = 20$$

$$N_{18} = N_{17} exp[1(1 - \frac{N_{17}}{20})] = 20$$

$$N_{19} = N_{18} exp[1(1 - \frac{N_{18}}{20})] = 20$$

$$N_{20} = N_{19} exp[1(1 - \frac{N_{19}}{20})] = 20$$

The Graph of N_t as a function of t, for the given values of R = 1, K = 20, $N_0 = 5$ is given below,

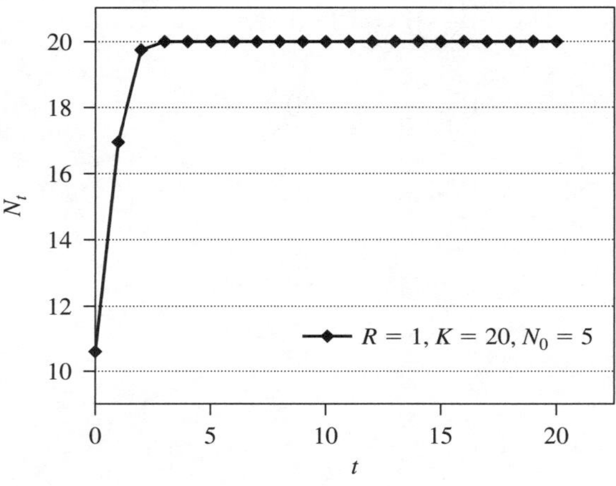

Graph of N_{t+1} as a function of t

(b) We know the behavior of the Ricker's curve is given by:

$$N_{t+1} = N_t exp[R(1 - \frac{N_t}{K})]$$

N_t for $t = 1, 2, 3, 4, \ldots 20$ are:

$$N_1 = N_0 exp[1(1 - \frac{N_0}{20})] = 16.48721$$

$$N_2 = N_1 exp[1(1 - \frac{N_1}{20})] = 19.653$$

$$N_3 = N_2 exp[1(1 - \frac{N_2}{20})] = 19.99695$$

$$N_4 = N_3 exp[1(1 - \frac{N_3}{20})] = 20$$

$$N_5 = N_0 exp[1(1 - \frac{N_4}{20})] = 20$$

$$N_6 = N_5 exp[1(1 - \frac{N_5}{20})] = 20$$

$$N_7 = N_6 exp[1(1 - \frac{N_6}{20})] = 20$$

$$N_8 = N_7 exp[1(1 - \frac{N_7}{20})] = 20$$

$$N_9 = N_8 exp[1(1 - \frac{N_8}{20})] = 20$$

$$N_{10} = N_9 exp[1(1 - \frac{N_9}{20})] = 20$$

$$N_{11} = N_{10} exp[1(1 - \frac{N_{10}}{20})] = 20$$

$$N_{12} = N_{11} exp[1(1 - \frac{N_{11}}{20})] = 20$$

$$N_{13} = N_{12} exp[1(1 - \frac{N_{12}}{20})] = 20$$

$$N_{14} = N_{13} exp[1(1 - \frac{N_{13}}{20})] = 20$$

$$N_{15} = N_{14} exp[1(1 - \frac{N_{14}}{20})] = 20$$

$$N_{16} = N_{15} exp[1(1 - \frac{N_{15}}{20})] = 20$$

$$N_{17} = N_{16} exp[1(1 - \frac{N_{16}}{20})] = 20$$

$$N_{18} = N_{17} exp[1(1 - \frac{N_{17}}{20})] = 20$$

$$N_{19} = N_{18} exp[1(1 - \frac{N_{18}}{20})] = 20$$

$$N_{20} = N_{19} exp[1(1 - \frac{N_{19}}{20})] = 20$$

The Graph of N_t as a function of t, for the given values of R = 1, K = 20, N_0 = 10 is given below,

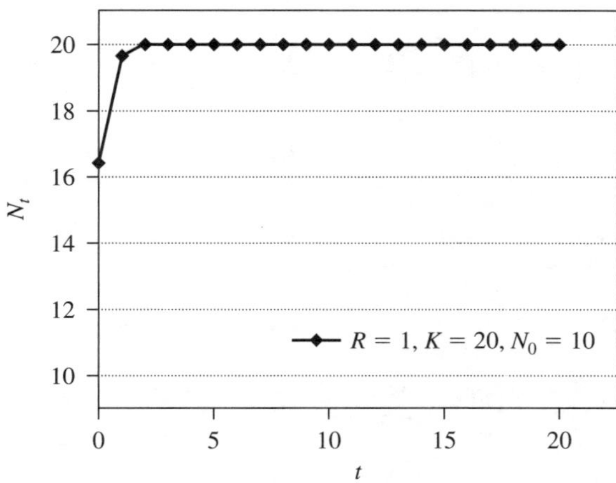

Graph of N_{t+1} as a function of t

(c) We know the behavior of the Ricker's curve is given by:

$$N_{t+1} = N_t exp[R(1 - \frac{N_t}{K})]$$

N_t for $t = 1, 2, 3, 4 \ldots 20$ are:

$$N_1 = N_0 exp[1(1 - \frac{N_0}{20})] = 20$$

$$N_2 = N_1 exp[1(1 - \frac{N_1}{20})] = 20$$

$$N_3 = N_2 exp[1(1 - \frac{N_2}{20})] = 20$$

$$N_4 = N_3 exp[1(1 - \frac{N_3}{20})] = 20$$

$$N_5 = N_0 exp[1(1 - \frac{N_4}{20})] = 20$$

$$N_6 = N_5 exp[1(1 - \frac{N_5}{20})] = 20$$

$$N_7 = N_6 exp[1(1 - \frac{N_6}{20})] = 20$$

$$N_8 = N_7 exp[1(1 - \frac{N_7}{20})] = 20$$

$$N_9 = N_8 exp[1(1 - \frac{N_8}{20})] = 20$$

$$N_{10} = N_9 exp[1(1 - \frac{N_9}{20})] = 20$$

$$N_{11} = N_{10} exp[1(1 - \frac{N_{10}}{20})] = 20$$

$$N_{12} = N_{11} exp[1(1 - \frac{N_{11}}{20})] = 20$$

$$N_{13} = N_{12} exp[1(1 - \frac{N_{12}}{20})] = 20$$

$$N_{14} = N_{13} exp[1(1 - \frac{N_{13}}{20})] = 20$$

$$N_{15} = N_{14} exp[1(1 - \frac{N_{14}}{20})] = 20$$

$$N_{16} = N_{15} exp[1(1 - \frac{N_{15}}{20})] = 20$$

$$N_{17} = N_{16}exp[1(1 - \frac{N_{16}}{20})] = 20$$

$$N_{18} = N_{17}exp[1(1 - \frac{N_{17}}{20})] = 20$$

$$N_{19} = N_{18}exp[1(1 - \frac{N_{18}}{20})] = 20$$

$$N_{20} = N_{19}exp[1(1 - \frac{N_{19}}{20})] = 20$$

The Graph of N_t as a function of t, for the given values of R = 1, K = 20, N_0 = 20 is given below,

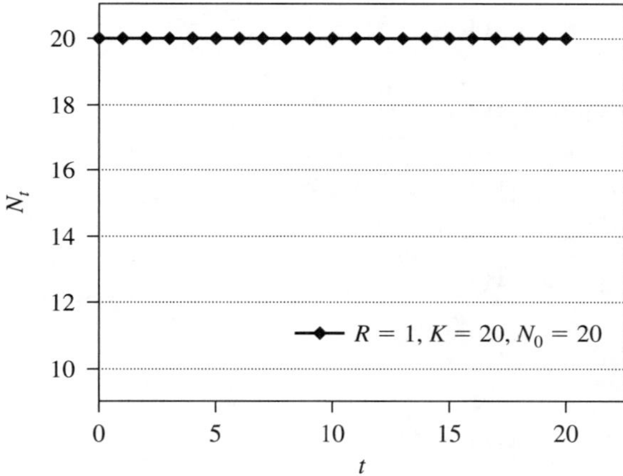

Graph of N_{t+1} as a function of t

(d) We know the behavior of the Ricker's curve is given by:

$$N_{t+1} = N_t exp[R(1 - \frac{N_t}{K})]$$

N_t for $t = 1, 2, 3, 4, \ldots 20$ are:

$$N_1 = N_0 exp[1(1 - \frac{N_0}{20})] = 0$$

$$N_2 = N_1 exp[1(1 - \frac{N_1}{20})] = 0$$

$$N_3 = N_2 exp[1(1 - \frac{N_2}{20})] = 0$$

$$N_4 = N_3 exp[1(1 - \frac{N_3}{20})] = 0$$

$$N_5 = N_0 exp[1(1 - \frac{N_4}{20})] = 0$$

$$N_6 = N_5 exp[1(1 - \frac{N_5}{20})] = 0$$

$$N_7 = N_6 exp[1(1 - \frac{N_6}{20})] = 0$$

$$N_8 = N_7 exp[1(1 - \frac{N_7}{20})] = 0$$

$$N_9 = N_8 exp[1(1 - \frac{N_8}{20})] = 0$$

$$N_{10} = N_9 exp[1(1 - \frac{N_9}{20})] = 0$$

$$N_{11} = N_{10} exp[1(1 - \frac{N_{10}}{20})] = 0$$

$$N_{12} = N_{11} exp[1(1 - \frac{N_{11}}{20})] = 0$$

$$N_{13} = N_{12} exp[1(1 - \frac{N_{12}}{20})] = 0$$

$$N_{14} = N_{13} exp[1(1 - \frac{N_{13}}{20})] = 0$$

$$N_{15} = N_{14} exp[1(1 - \frac{N_{14}}{20})] = 0$$

$$N_{16} = N_{15} exp[1(1 - \frac{N_{15}}{20})] = 0$$

$$N_{17} = N_{16} exp[1(1 - \frac{N_{16}}{20})] = 0$$

$$N_{18} = N_{17} exp[1(1 - \frac{N_{17}}{20})] = 0$$

$$N_{19} = N_{18} exp[1(1 - \frac{N_{18}}{20})] = 0$$

$$N_{20} = N_{19} exp[1(1 - \frac{N_{19}}{20})] = 0$$

The Graph of N_t as a function of t, for the given values of R = 1, K = 20, $N_0 = 0$ is given below,

Graph of N_{t+1} as a function of t

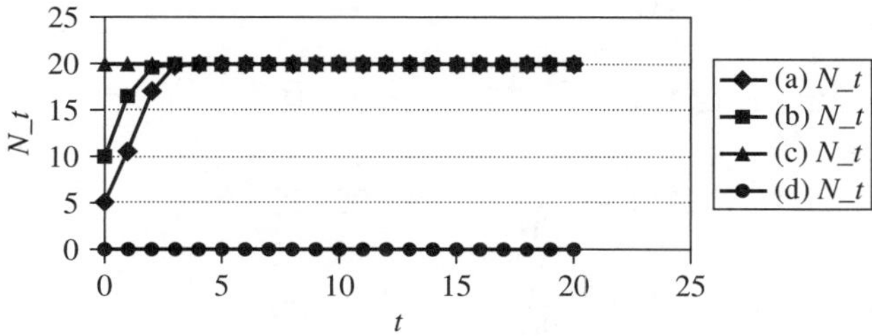

Prob. 53.

(a) We know the behavior of the Ricker's curve is given by:

$$N_{t+1} = N_t exp[R(1 - \frac{N_t}{K})]$$

N_t for $t = 1, 2, 3, 4, \ldots 20$ are:

$$N_1 = N_0 exp[2.1(1 - \frac{N_0}{20})] = 24.15371$$

$$N_2 = N_1 exp[2.1(1 - \frac{N_1}{20})] = 15.61604$$

$$N_3 = N_2 exp[2.1(1 - \frac{N_2}{20})] = 24.74478$$

$$N_4 = N_3 exp[2.1(1 - \frac{N_3}{20})] = 15.03548$$

$$N_5 = N_0 exp[2.1(1 - \frac{N_4}{20})] = 25.32235$$

$$N_6 = N_5 exp[2.1(1 - \frac{N_5}{20})] = 14.48105$$

$$N_7 = N_6 exp[2.1(1 - \frac{N_6}{20})] = 25.85052$$

$$N_8 = N_7 exp[2.1(1 - \frac{N_7}{20})] = 13.98557$$

$$N_9 = N_8 exp[2.1(1 - \frac{N_8}{20})] = 26.29927$$

$$N_{10} = N_9 exp[2.1(1 - \frac{N_9}{20})] = 13.57348$$

$$N_{11} = N_{10} exp[2.1(1 - \frac{N_{10}}{20})] = 27.65302$$

$$N_{12} = N_{11} exp[2.1(1 - \frac{N_{11}}{20})] = 13.25448$$

$$N_{13} = N_{12} exp[2.1(1 - \frac{N_{12}}{20})] = 26.91316$$

$$N_{14} = N_{13} exp[2.1(1 - \frac{N_{13}}{20})] = 13.02322$$

$$N_{15} = N_{14} exp[2.1(1 - \frac{N_{14}}{20})] = 27.09396$$

$$N_{16} = N_{15} exp[2.1(1 - \frac{N_{15}}{20})] = 12.86451$$

$$N_{17} = N_{16} exp[2.1(1 - \frac{N_{16}}{20})] = 27.21311$$

$$N_{18} = N_{17} exp[2.1(1 - \frac{N_{17}}{20})] = 12.76009$$

$$N_{19} = N_{18} exp[2.1(1 - \frac{N_{18}}{20})] = 27.2898$$

$$N_{20} = N_{19} exp[2.1(1 - \frac{N_{19}}{20})] = 12.8892$$

The Graph of N_t as a function of t, for the given values of R = 2.8, K = 20, $N_0 = 5$ is given below,

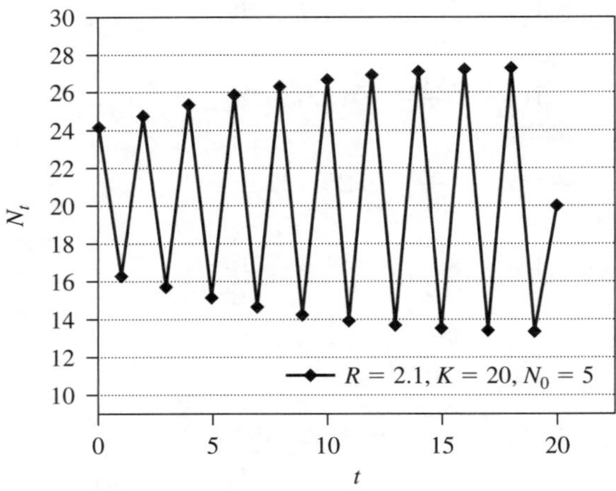

Graph of N_{t+1} as a function of t

(b) We know the behavior of the Ricker's curve is given by:

$$N_{t+1} = N_t exp[R(1 - \frac{N_t}{K})]$$

N_t for $t = 1, 2, 3, 4, \ldots 20$ are:

$$N_1 = N_0 exp[2.1(1 - \frac{N_0}{20})] = 28.57651$$

$$N_2 = N_1 exp[2.1(1 - \frac{N_1}{20})] = 11.61214$$

$$N_3 = N_2 exp[2.1(1 - \frac{N_2}{20})] = 28.01602$$

$$N_4 = N_3 exp[2.1(1 - \frac{N_3}{20})] = 12.07449$$

$$N_5 = N_0 exp[2.1(1 - \frac{N_4}{20})] = 27.75105$$

$$N_6 = N_5 exp[2.1(1 - \frac{N_5}{20})] = 12.29772$$

$$N_7 = N_6 exp[2.1(1 - \frac{N_6}{20})] = 27.60932$$

$$N_8 = N_7 exp[2.1(1 - \frac{N_7}{20})] = 12.41835$$

$$N_9 = N_8 exp[2.1(1 - \frac{N_8}{20})] = 27.52924$$

$$N_{10} = N_9 exp[2.1(1 - \frac{N_9}{20})] = 12.48688$$

$$N_{11} = N_{10} exp[2.1(1 - \frac{N_{10}}{20})] = 27.48268$$

$$N_{12} = N_{11} exp[2.1(1 - \frac{N_{11}}{20})] = 12.52685$$

$$N_{13} = N_{12} exp[2.1(1 - \frac{N_{12}}{20})] = 27.45518$$

$$N_{14} = N_{13} exp[2.1(1 - \frac{N_{13}}{20})] = 12.5505$$

$$N_{15} = N_{14} exp[2.1(1 - \frac{N_{14}}{20})] = 27.4388$$

$$N_{16} = N_{15} exp[2.1(1 - \frac{N_{15}}{20})] = 12.56461$$

$$N_{17} = N_{16}exp[2.1(1 - \frac{N_{16}}{20})] = 27.42898$$

$$N_{18} = N_{17}exp[2.1(1 - \frac{N_{17}}{20})] = 12.57307$$

$$N_{19} = N_{18}exp[2.1(1 - \frac{N_{18}}{20})] = 27.42308$$

$$N_{20} = N_{19}exp[2.1(1 - \frac{N_{19}}{20})] = 12.57816$$

The Graph of N_t as a function of t, for the given values of R = 2.1, K = 20, $N_0 = 10$ is given below,

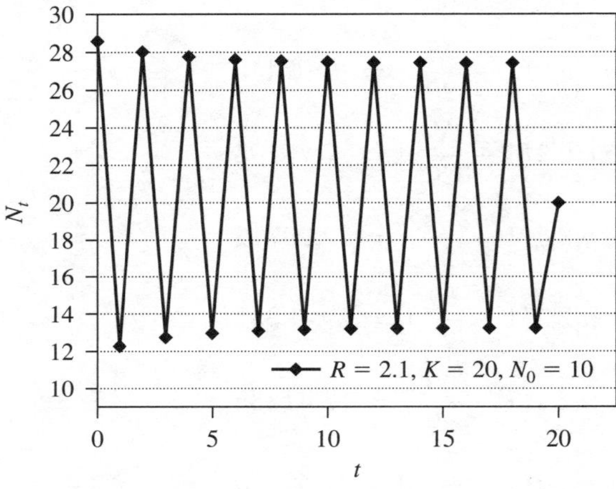

Graph of N_{t+1} as a function of t

(c) We know the behavior of the Ricker's curve is given by:

$$N_{t+1} = N_t exp[R(1 - \frac{N_t}{K})]$$

N_t for $t = 1, 2, 3, 4, \ldots 20$ are:

$$N_1 = N_0 exp[2.1(1 - \frac{N_0}{20})] = 20$$

$$N_2 = N_1 exp[2.1(1 - \frac{N_1}{20})] = 20$$

$$N_3 = N_2 exp[2.1(1 - \frac{N_2}{20})] = 20$$

$$N_4 = N_3 exp[2.1(1 - \frac{N_3}{20})] = 20$$

$$N_5 = N_0 exp[2.1(1 - \frac{N_4}{20})] = 20$$

$$N_6 = N_5 exp[2.1(1 - \frac{N_5}{20})] = 20$$

$$N_7 = N_6 exp[2.1(1 - \frac{N_6}{20})] = 20$$

$$N_8 = N_7 exp[2.1(1 - \frac{N_7}{20})] = 20$$

$$N_9 = N_8 exp[2.1(1 - \frac{N_8}{20})] = 20$$

$$N_{10} = N_9 exp[2.1(1 - \frac{N_9}{20})] = 20$$

$$N_{11} = N_{10} exp[2.1(1 - \frac{N_{10}}{20})] = 20$$

$$N_{12} = N_{11} exp[2.1(1 - \frac{N_{11}}{20})] = 20$$

$$N_{13} = N_{12} exp[2.1(1 - \frac{N_{12}}{20})] = 20$$

$$N_{14} = N_{13} exp[2.1(1 - \frac{N_{13}}{20})] = 20$$

$$N_{15} = N_{14} exp[2.1(1 - \frac{N_{14}}{20})] = 20$$

$$N_{16} = N_{15} exp[2.1(1 - \frac{N_{15}}{20})] = 20$$

$$N_{17} = N_{16} exp[2.1(1 - \frac{N_{16}}{20})] = 20$$

$$N_{18} = N_{17} exp[2.1(1 - \frac{N_{17}}{20})] = 20$$

$$N_{19} = N_{18} exp[2.1(1 - \frac{N_{18}}{20})] = 20$$

$$N_{20} = N_{19} exp[2.1(1 - \frac{N_{19}}{20})] = 20$$

The Graph of N_t as a function of t, for the given values of R = 2.1, K = 20, N_0 = 20 is given below,

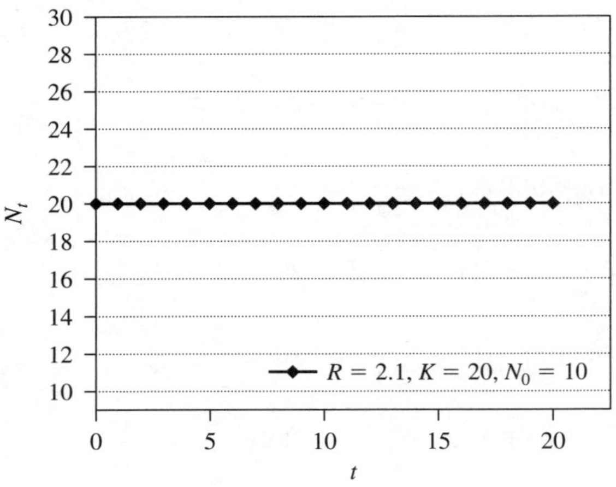

Graph of N_{t+1} as a function of t

(d) We know the behavior of the Ricker's curve is given by:

$$N_{t+1} = N_t exp[R(1 - \frac{N_t}{K})]$$

N_t for $t = 1, 2, 3, 4, \ldots 20$ are:

$$N_1 = N_0 exp[2.1(1 - \frac{N_0}{20})] = 0$$

$$N_2 = N_1 exp[2.1(1 - \frac{N_1}{20})] = 0$$

$$N_3 = N_2 exp[2.1(1 - \frac{N_2}{20})] = 0$$

$$N_4 = N_3 exp[2.1(1 - \frac{N_3}{20})] = 0$$

$$N_5 = N_0 exp[2.1(1 - \frac{N_4}{20})] = 0$$

$$N_6 = N_5 exp[2.1(1 - \frac{N_5}{20})] = 0$$

$$N_7 = N_6 exp[2.1(1 - \frac{N_6}{20})] = 0$$

$$N_8 = N_7 exp[2.1(1 - \frac{N_7}{20})] = 0$$

$$N_9 = N_8 exp[2.1(1 - \frac{N_8}{20})] = 0$$

$$N_{10} = N_9 exp[2.1(1 - \frac{N_9}{20})] = 0$$

$$N_{11} = N_{10} exp[2.1(1 - \frac{N_{10}}{20})] = 0$$

$$N_{12} = N_{11} exp[2.1(1 - \frac{N_{11}}{20})] = 0$$

$$N_{13} = N_{12} exp[2.1(1 - \frac{N_{12}}{20})] = 0$$

$$N_{14} = N_{13} exp[2.1(1 - \frac{N_{13}}{20})] = 0$$

$$N_{15} = N_{14} exp[2.1(1 - \frac{N_{14}}{20})] = 0$$

$$N_{16} = N_{15} exp[2.1(1 - \frac{N_{15}}{20})] = 0$$

$$N_{17} = N_{16} exp[2.1(1 - \frac{N_{16}}{20})] = 0$$

$$N_{18} = N_{17} exp[2.1(1 - \frac{N_{17}}{20})] = 0$$

$$N_{19} = N_{18} exp[2.1(1 - \frac{N_{18}}{20})] = 0$$

$$N_{20} = N_{19} exp[2.1(1 - \frac{N_{19}}{20})] = 0$$

The Graph of N_t as a function of t, for the given values of R = 2.1, K = 20, $N_0 = 0$ is given below,

Graph of N_{t+1} as a function of t

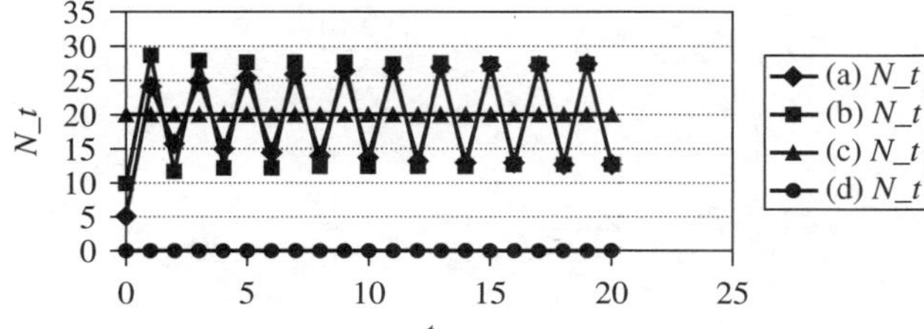

Prob. 55. We have $N_0 = 1$ and $N_1 = 1$,

$$t = 1, N_2 = N_1 + N_0 = 1 + 1 = 2, \frac{N_t}{N_{t-1}} = \frac{N_1}{N_0} = 1$$

$$t = 2, N_3 = N_2 + N_1 = 2 + 1 = 3, \frac{N_t}{N_{t-1}} = \frac{N_2}{N_1} = 2$$

$$t = 3, N_4 = N_3 + N_2 = 3 + 2 = 5, \frac{N_t}{N_{t-1}} = \frac{N_3}{N_2} = \frac{3}{2}$$

$$t = 4, N_5 = N_4 + N_3 = 5 + 3 = 8, \frac{N_t}{N_{t-1}} = \frac{N_5}{N_3} = \frac{5}{3}$$

$$t = 5, N_6 = N_5 + N_4 = 8 + 5 = 13, \frac{N_t}{N_{t-1}} = \frac{N_5}{N_4} = \frac{8}{5}$$

$$t = 6, N_7 = N_6 + N_5 = 13 + 8 = 21, \frac{N_t}{N_{t-1}} = \frac{N_6}{N_5} = \frac{13}{8}$$

$$t = 7, N_8 = N_7 + N_6 = 21 + 13 = 34, \frac{N_t}{N_{t-1}} = \frac{N_7}{N_6} = \frac{21}{13}$$

$$t = 8, N_9 = N_8 + N_7 = 34 + 21 = 55, \frac{N_t}{N_{t-1}} = \frac{N_8}{N_7} = \frac{34}{21}$$

$$t = 9, N_{10} = N_9 + N_8 = 55 + 34 = 89, \frac{N_t}{N_{t-1}} = \frac{N_9}{N_8} = \frac{55}{34}$$

$$t = 10, N_{11} = N_{10} + N_9 = 89 + 55 = 144, \frac{N_t}{N_{t-1}} = \frac{N_{10}}{N_9} = \frac{89}{55}$$

$$t = 11, N_{12} = N_{11} + N_{10} = 144 + 89 = 233, \frac{N_t}{N_{t-1}} = \frac{N_{11}}{N_{10}} = \frac{144}{89}$$

$$t = 12, N_{13} = N_{12} + N_{11} = 233 + 144 = 377, \frac{N_t}{N_{t-1}} = \frac{N_{12}}{N_{11}} = \frac{233}{144}$$

$$t = 13, N_{14} = N_{13} + N_{12} = 377 + 233 = 610, \frac{N_t}{N_{t-1}} = \frac{N_{13}}{N_{12}} = \frac{377}{233}$$

$$t = 14, N_{15} = N_{14} + N_{13} = 610 + 377 = 987, \frac{N_t}{N_{t-1}} = \frac{N_{14}}{N_{13}} = \frac{610}{377}$$

$$t = 15, N_{16} = N_{15} + N_{14} = 987 + 610 = 1597, \frac{N_t}{N_{t-1}} = \frac{N_{15}}{N_{14}} = \frac{987}{610}$$

$$t = 16, N_{17} = N_{16} + N_{15} = 1597 + 987 = 2584, \frac{N_t}{N_{t-1}} = \frac{N_{16}}{N_{15}} = \frac{1597}{987}$$

$$t = 17, N_{18} = N_{17} + N_{16} = 2584 + 1597 = 4181, \frac{N_t}{N_{t-1}} = \frac{N_{17}}{N_{16}} = \frac{2584}{1597}$$

$$t = 18, N_{19} = N_{18} + N_{17} = 4181 + 2584 = 6765, \frac{N_t}{N_{t-1}} = \frac{N_{18}}{N_{17}} = \frac{4181}{2584}$$

$$t = 19, N_{20} = N_{19} + N_{18} = 6765 + 4181 = 10946, \frac{N_t}{N_{t-1}} = \frac{N_{19}}{N_{18}} = \frac{6765}{4181}$$

Prob. 57. This problem works towards finding out the pairs of rabbits produced if each pair reproduces one pair of rabbits at age 1 month and two pairs of rabbits at age 2 months and intially there is one pair of newborn rabbits.

If N_t denotes the number of newborn rabbit pairs at time t (measured in months), then at time 0, there is one pair of rabbits ($N_0 = 1$). At time 1, the pair of rabbits we started with is one month old and produced a pair of newborn rabbits, so $N_1 = 1$. At time 2, there is one pair of two-month-old rabbits and one pair of one-month-old rabbits. Pair of two-month-old rabbits will produce two pairs of rabbits and pair of one-month-old rabbits will produce 1 pair of rabbits, so $N_2 = 3$. At time 3, our original pair of rabbits is now three months old and will stop reproducing; there is then one pair of two-month-old rabbits and three pairs of one-month-old rabbits. Since each pair of one-month-old rabbits produces a pair of newborn rabbits and two-month-old rabbits produces two pairs of newborn rabbits, at time t = 3, there will be 3 + 2 = 5 newborn rabbits.

More generally, to find the number of pairs of newborn rabbits, we need to add up the number of pairs of one-month-old rabbits and 2 times the number of pairs of two-month-old rabbits. The one-month-old rabbits at time t + 1 were newborn rabbits at time t ; the two-month-old rabbits were newborns at time t - 1. So the number of pairs of newborn rabbits at time t + 1 is

$$N_{t+1} = N_t + 2N_{t-1}, t = 1, 2, 3, 4, \ldots, N_0 = 1, N_1 = 1$$

2.5 Review Problems

Prob. 1. The expression 2^{-n} can also be written as $(1/2)^n$. Since $\lim_{n\to\infty}(1/2)^n$ exists, and is equal to 0. We find,

$$\lim_{n\to\infty} 2^{-n} = \lim_{n\to\infty} 1/2^n = 0$$

Prob. 3. The expression $40(1-4^{-n})$ can be written as sum of two terms as, 40 and $40(1/4)^n$. Since $\lim_{n\to\infty}(1/4)^n$ and $\lim_{n\to\infty} 40$ exist, and are equal to 0 and 40. We find,

$$\lim_{n\to\infty} 40(1-4^{-n}) = \lim_{n\to\infty} 40 - \lim_{n\to\infty} 40(1/4)^n = 40 - 0 = 40$$

Prob. 5. Successive terms of a_n,

$$1, a, a_2, a_3, a_4, a_5, a_6, \ldots$$

indicate that the terms continue to grow as $a > 1$. Thus a_n goes to infinity as $n \to \infty$, and we can write $\lim_{n\to\infty} a^n = \infty$. Since infinity ($\infty$) is not a real number, we say that the limit does not exist.

Prob. 7. The expression $\frac{n(n+1)}{n^2-1}$ can be written as, $\frac{n(n+1)}{(n+1)(n-1)}$. Since $\lim_{n\to\infty} \frac{1}{n}$ exists, and is equal to 0. We find,

$$\begin{aligned}\lim_{n\to\infty} \frac{n(n+1)}{n^2-1} &= \lim_{n\to\infty} \frac{n(n+1)}{(n-1)(n+1)} = \lim_{n\to\infty} \frac{n}{n-1} \\ &= \lim_{n\to\infty} \frac{1}{1-\frac{1}{n}} = \frac{\lim_{n\to\infty} 1}{\lim_{n\to\infty} 1 - \lim_{n\to\infty} \frac{1}{n}} = \frac{1}{1-0} = 1\end{aligned}$$

Prob. 9. The expression $\frac{\sqrt{n}}{n+1}$ can be written as, $\frac{\frac{1}{\sqrt{n}}}{1+\frac{1}{n}}$. Since $\lim_{n\to\infty}\frac{1}{n}$ and $\lim_{n\to\infty}\frac{1}{\sqrt{n}}$ exist, and are equal to 0. We find,

$$\lim_{n\to\infty}\frac{\frac{1}{\sqrt{n}}}{1+\frac{1}{n}} = \frac{\lim_{n\to\infty}\frac{1}{\sqrt{n}}}{\lim_{n\to\infty}1+\frac{1}{n}} = \frac{0}{0+1} = 0$$

Prob. 11. Looking at the sequence, we can guess the next terms, namely, $\frac{11}{12}, \frac{13}{14}, \frac{15}{16}, \frac{17}{18}, \frac{19}{20}$ and so on. We thus find

$$a_n = \frac{2n+1}{2n+2}; n = 0, 1, 2, 3, \ldots$$

Prob. 13. Looking at the sequence, we can guess the next terms, namely, $\frac{6}{37}, \frac{7}{50}, \frac{8}{65}, \frac{9}{82}, \frac{10}{101}$ and so on. We thus find

$$a_n = \frac{n+1}{n^2+2n+2}; n = 0, 1, 2, 3, \ldots$$

Prob. 15. The table below lists the value of N_t for $t = 0, 1, 2, \ldots, 10$ for the three cases (a) $R = 2$, (b) $R = 5$, and (c) $R = 10$:

	N_t		
Time	a	b	c
0	20	20	20
1	33.33333	55.55556	71.42857
2	50	86.2069	96.15385
3	66.66667	96.89922	99.60159
4	80	99.36407	99.96002
5	88.88889	99.87216	99.996
6	94.11765	99.97441	99.9996
7	96.9697	99.99488	99.99996
8	98.46154	99.99898	100
9	99.22481	99.9998	100
10	99.61089	99.99996	100

The values are plotted in the graph below:

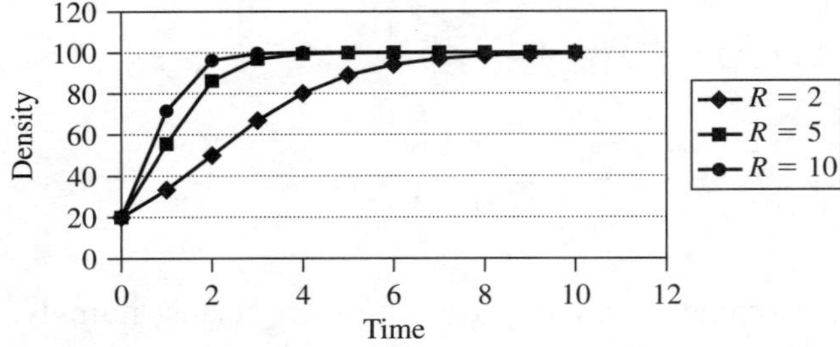

Prob. 17. The following table lists $\hat{R}_t$ for $t = 1, 2, \ldots, 20$

t	$\hat{R}_t$	t	$\hat{R}_t$
1	0.81	11	0.746302
2	0.590762	12	0.829422
3	0.443105	13	0.777593
4	0.708517	14	0.660381
5	0.841117	15	0.718356
6	0.889108	16	0.769141
7	0.856812	17	0.819627
8	0.915157	18	0.876903
9	0.936861	19	0.82846
10	0.733367	20	0.84445

It appears that $\hat{R}_t$ is about 0.8, which is less than 1.

Prob. 19. The following table lists N_t for $t = 0, 1, 2, \ldots, 20$ for the three cases (a) $c = 0.1$, (b) $c = 0.5$, and (c) $c = 0.9$:

	N_t		
Time	a)	b)	c)
0	50	50	50
1	77.99639	52.925	12.92855
2	94.56945	55.20777	3.469201
3	98.77543	56.93542	0.939761
4	99.33615	58.21211	0.255213
5	99.39717	59.13871	0.069357
6	99.40363	59.80235	0.018852
7	99.40431	60.27311	0.005124
8	99.40438	60.60475	0.001393
9	99.40439	60.83726	0.000379
10	99.40439	60.9997	0.000103
11	99.40439	61.11292	2.8E-05
12	99.40439	61.19169	7.61E-06
13	99.40439	61.2465	2.07E-06
14	99.40439	61.28447	5.62E-07
15	99.40439	61.31085	1.53E-07
16	99.40439	61.32916	4.15E-08
17	99.40439	61.34186	1.13E-08
18	99.40439	61.35066	3.07E-09
19	99.40439	61.35677	8.34E-10
20	99.40439	61.361	2.27E-10

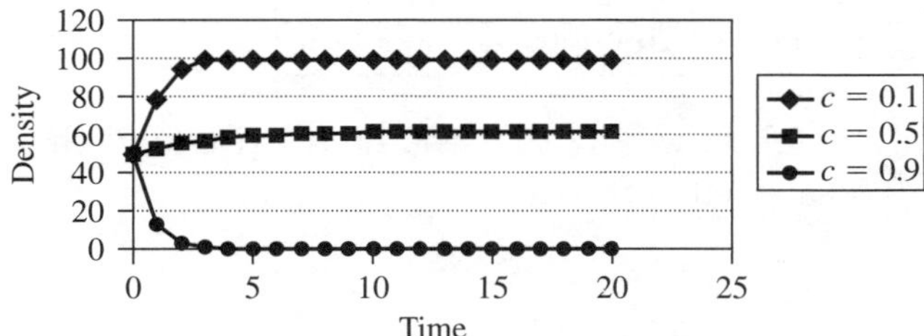

3 Limits and Continuity

3.1 Limits

Prob. 1. The table 1. for $x^2 - 3x + 1$, suspects that the limit of this function is -1 as x approaches 2 (from either side). We compute the values of $x^2 - 3x + 1$ for x close to 2 (but not equal to 2). In the left half of the table, we approach x = 2 from the left ($x \to 2^-$); in the right half of the table, we approach x from the right ($x \to 2^+$).

x	$x^2 - 3x + 1$	x	$x^2 - 3x + 1$
1.9	-1.09	2.1	-0.89
1.99	-1.01	2.01	-0.99
1.999	-1.001	2.001	-0.999
1.9999	-1.0001	2.0001	-0.9999

Since this limit is a finite number, we say that the limit exists and that the function converges to -1 as x tends to 2.

Prob. 3. The table 3. for $\frac{x}{1+x^2}$, suspects that the limit of this function is -0.5 as x approaches -1 (from either side). We compute the values of $\frac{x}{1+x^2}$ for x close to -1 (but not equal to -1). In the left half of the table, we approach x = -1 from the left ($x \to -1^-$); in the right half of the table, we approach x from the right ($x \to -1^+$).

x	$\frac{x}{1+x^2}$	x	$\frac{x}{1+x^2}$
-1.1	-0.498	-0.9	-0.497
-1.01	-0.4999	-0.99	-.4999
-1.001	-0.499999	-0.999	-.499999
-1.0001	-0.49999999	-0.9999	-.49999999

Since this limit is a finite number, we say that the limit exists and that the function converges to -0.5 as x tends to -1.

Prob. 5. The table 5. for $3\cos\frac{x}{4}$, suspects that the limit of this function is 2.12 as x approaches π (from either side). We compute the values of $3\cos\frac{x}{4}$ for x close to π (but not equal to π). In the left half of the table, we approach x = π from the left ($x \to \pi^-$); in the right half of the table, we approach x from the right ($x \to \pi^+$). (The x-values are in degree.)

x	$3\cos\frac{x}{4}$	x	$3\cos\frac{x}{4}$
179.9	2.122	180.1	2.120
179.99	2.121	180.01	2.12
179.999	2.121	180.001	2.12
179.9999	2.121	180.0001	2.12

Since this limit is a finite number, we say that the limit exists and that the function converges to 2.12 as x tends to π.

Prob. 7. The table 7. for $2sec(\frac{x}{3})$, suspects that the limit of this function is 2.3093 as x approaches $\frac{\pi}{2}$ (from either side). We compute the values of $2sec(\frac{x}{3})$ for x close to $\frac{\pi}{2}$ (but not equal to $\frac{\pi}{2}$). In the left half of the table, we approach x = $\frac{\pi}{2}$ from the left ($x \to \frac{\pi}{2}^-$); in the right half of the table, we approach x from the right ($x \to \frac{\pi}{2}^+$). (The x-values are in degree.)

x	$2sec(\frac{x}{3})$	x	$2sec(\frac{x}{3})$
89.9	2.308	90.1	2.3102
89.99	2.3093	90.01	2.30947
89.999	2.3093	90.001	2.3094
89.9999	2.3093	90.0001	2.3094

Since this limit is a finite number, we say that the limit exists and that the function converges to 2.3093 as x tends to $\frac{\pi}{2}$.

Prob. 9. The table 9. for $e^{\frac{-x^2}{2}}$, suspects that the limit of this function is 0.6065 as x approaches -1 (from either side). We compute the values of $e^{\frac{-x^2}{2}}$ for x close to -1 (but not equal to -1). In the left half of the table, we approach x = -1 from the left ($x \to -1^-$); in the right half of the table, we approach x from the right ($x \to -1^+$).

x	$e^{\frac{-x^2}{2}}$	x	$e^{\frac{-x^2}{2}}$
-0.9	0.667	-1.1	0.5461
-0.99	0.6125	-1.01	0.6005
-0.999	0.6071	-1.001	0.6059
-0.9999	0.6066	-1.0001	0.6064

Since this limit is a finite number, we say that the limit exists and that the function converges to 0.6065 as x tends to -1.

Prob. 11. The table 11. for $\ln(x+1)$, suspects that the limit of this function is 0 as x approaches 0 (from either side). We compute the values of $\ln(x+1)$ for x close to 0 (but not equal to 0). In the left half of the table, we approach x = 0 from the left ($x \to 0^-$); in the right half of the table, we approach x from the right ($x \to 0^+$).

x	$\ln(x+1)$	x	$\ln(x+1)$
-0.1	-0.10536	0.1	0.0953
-0.01	-0.01005	0.01	0.00995
-0.001	-0.0010005	0.001	0.000999
-0.0001	-0.000100005	0.0001	0.0000999

Since this limit is a finite number, we say that the limit exists and that the function converges to 0 as x tends to 0.

Prob. 13. The table 13. for $\frac{x^2-16}{x-4}$, suspects that the limit of this function is 7 as x approaches 3 (from either side). We compute the values of $\frac{x^2-16}{x-4}$ for x close to 3 (but not equal to 3). In the left half of the table, we approach x = 3 from the left ($x \to 3^-$); in the right half of the table, we approach x from the right ($x \to 3^+$).

x	$\frac{x^2-16}{x-4}$	x	$\frac{x^2-16}{x-4}$
2.9	6.9	3.1	7.1
2.99	6.99	3.01	7.01
2.999	6.999	3.001	7.001
2.9999	6.9999	3.0001	7.0001

Since this limit is a finite number, we say that the limit exists and that the function converges to 7 as x tends to 3.

Prob. 15. The table 15. for $\frac{2}{x}$, suspects that the limit of this function is 0 as x approaches ∞ (from either side). We compute the values of $\frac{2}{x}$ for x close to ∞ (but not equal to ∞). In the left half of the table, we approach x $= \infty$ from the left $(x \to \infty^-)$; We can not approach infinity from the right side, but we can see that for a very large value of x, function goes closer to 0.

x	$\frac{2}{x}$	x	$\frac{2}{x}$
1000000	0.000002	999999	0.000002
100000	0.00002	99999	0.00002
10000	0.0002	9999	0.0002
1000	0.002	999	0.002

Since this limit is a finite number, we say that the limit exists and that the function converges to 0 as x tends to ∞.

Prob. 17. Simplify the expression,

$$f(x) = \frac{x}{1+x} = \frac{1}{\frac{1}{x}+1}$$

now it is easy to see that the limit when $x \to \infty$ is 1.

Prob. 19. The graph of $y = e^x$ immediately shows that the limit of e^x is 0 as x approaches $-\infty$.

Prob. 21. The graph of $f(x) = \frac{2}{x-4}, x = 4^-$, reveals that

$$\lim_{x\to 4^-} \frac{2}{x-4} = -\infty$$

We arrive at the same conclusion when we compute values of f(x) for x close to 4. When x is slightly smaller than 4, f (x) is negative and decreases without bound as x approaches 4 from the left. We conclude that f(x) diverges as x approaches 4 from the left.

Prob. 23. The graph of $f(x) = \frac{2}{1-x}, x = 1^-$, shows

$$\lim_{x\to 1^-} \frac{2}{1-x} = \infty$$

Prob. 25. The graph of $f(x) = \frac{1}{1-x^2}, x = 1^-$, shows that

$$\lim_{x\to 1^-} \frac{1}{1-x^2} = \infty$$

Prob. 27. The graph of $f(x) = \frac{1}{(1-x)^2}, x = 1$, shows

$$\lim_{x\to 1^+} \frac{1}{(1-x)^2} = \infty; \lim_{x\to 1^-} \frac{1}{(1-x)^2} = \infty$$

Prob. 29. The graph of $f(x) = \frac{\sqrt{x^2+9}-3}{x^2}, x = 0$, indicates that the limit exists and, based on the graph, we conjecture that it is equal to 0.16667. If, instead, we use a calculator to produce a table for values of f(x) close to 0, something strange seems to happen.

x	$f(x)$	x	$f(x)$
0.1	0.1666	0.00001	0.15
0.01	0.1666	0.000001	0.1
0.001	0.1666	0.0000001	0
0.0001	0.1666	0.00000001	0

As we get closer to 0, we first find that f(x) gets closer to 0.16667, but when we get very close to 0, f(x) seems to drop to 0. What is going on? First, before you worry too much, $\lim_{x\to 0} f(x) = 0.16666$. In the next section, we will learn how to compute this limit without resorting to the (somewhat dubious) help of the calculator. The strange behavior of the calculated values happens since when x is very small, the difference in the numerator is so close to 0 that the calculator can no longer accurately determine its value. The calculator can only accurately compute a certain number of digits, which is good enough for most cases. Here, however, we need greater accuracy. The same strange thing happens when you try to graph this function on a graphing calculator. When the x range of the viewing window is too small, the graph is no longer accurate.

Prob. 31. The graph of $f(x) = \frac{1-\sqrt{1-x^2}}{x^2}, x = 0$, indicates that the limit exists and, based on the graph, we conjecture that it is equal to 0.5.

Prob. 33.

(a) The graph and the table, for $\frac{2}{x^2}$, suspects that the limit of this function is 0 as x approaches ∞. We compute the values of $\frac{2}{x^2}$ for x close to ∞ (but not equal to ∞). In the table, we approach x = ∞ from the left ($x \to \infty^-$).

x	$\frac{2}{x^2}$
100	0.02
10000	0.0002
1000000	0.000002
100000000	0.00000002

Since this limit is a finite number, we say that the limit exists and that the function converges to 0 as x tends to ∞.

(b) The graph and the table, for $\frac{2}{x^2}$, suspects that the limit of this function is 0 as x approaches $-\infty$. We compute the values of $\frac{2}{x^2}$ for x close to $-\infty$ (but not equal to $-\infty$). In the table, we approach x = $-\infty$ from the right ($x \to -\infty^+$).

x	$\frac{2}{x^2}$
-100	-0.02
-10000	-0.0002
-1000000	-0.000002
-100000000	-0.00000002

Since this limit is a finite number, we say that the limit exists and that the function converges to 0 as x tends to $-\infty$.

(c) A graph of $f(x) = \frac{2}{x^2}$, $x \neq 0$, reveals that f(x) increases without bound as $x \to 0$. We can also suspect this when we plug in values close to 0.

By choosing values sufficiently close to 0, we can get arbitrarily large values of $\frac{2}{x^2}$.

x	$\frac{2}{x^2}$	x	$\frac{2}{x^2}$
-0.1	100	0.1	100
-0.01	10,000	0.01	10,000
-0.001	10^6	0.001	10^6
-0.0001	10^8	0.0001	10^8

This indicates that $\lim_{x\to 0} \frac{1}{x^2}$ does not exist.

Prob. 35. Simply using a calculator and plugging in values to find limits can yield wrong answers if one does not exercise proper caution. If we produced a table of values of $f(x) = \sin\frac{1}{x-1}$ for x = 1.1, 1.01, 1.001, . . . , we would find $\sin\frac{1}{1.1-1} = -0.5440, \sin\frac{1}{1.01-1} = -0.5064, \sin\frac{1}{1.001-1} = 0.8269$, and so on. (Note that we measure angles in radians.) This might prompt us to conclude that the limit of the function is uncertain. But let's look at its graph, which is shown in Figure below. The graph does not support our calculator-based conclusion. What we find instead is that the values of f (x) oscillate infinitely often between -1 and +1 as $x \to 1$. We can see why as follows: As $x \to 1^+$, the argument in the sine function goes to infinity (likewise, as $x \to 1^-$, the argument goes to negative infinity).

$$\lim_{x\to 1^+} \frac{1}{x-1} = \infty; \lim_{x\to 1^-} \frac{1}{x-1} = -\infty$$

As the argument of the sine function goes to $+\infty$ or $-\infty$, the function values oscillate between -1 and $+1$. Therefore, $\sin\frac{1}{x-1}$ continues to oscillate between -1 and $+1$ as $x \to 1$.

Prob. 37. Using Rules 1 and 2, this becomes

$$\lim_{x\to -1} x^3 + 7 \lim_{x\to -1} x - \lim_{x\to -1} 1$$

provided the individual limits exist. For the first term, we use Rule 3,

$$\lim_{x\to -1} x^3 = (\lim_{x\to -1} x)(\lim_{x\to -1} x)(\lim_{x\to -1} x)$$

provided $\lim_{x\to -1} x$ exists. Using (3.3), it follows that $\lim_{x\to -1} x = -1$ and we find that

$$(\lim_{x\to -1} x)(\lim_{x\to -1} x)(\lim_{x\to -1} x) = (-1)(-1)(-1) = -1$$

To compute the second term, we use (3.3) again, $\lim_{x\to -1} x = -1$. For the last term, we find $\lim_{x\to -1} 1 = 1$. Now that we showed that the individual limits exist, we can use Rules 1 and 2 to evaluate

$$\lim_{x\to -1} [x^3 + 7x - 1] = \lim_{x\to -1} x^3 + 7 \lim_{x\to -1} x - \lim_{x\to -1} 1 = -1 + (7)(-1) - 1 = -9 \tag{3.1}$$

Prob. 39. Using Rules 1 and 2, this becomes

$$\lim_{x\to -5} 4 + 2 \lim_{x\to -5} x^2$$

provided the individual limits exist. For the first term, we use Rule 3,

$$\lim_{x\to-5} x^2 = (\lim_{x\to-5} x)(\lim_{x\to-5} x)$$

provided $\lim_{x\to-5} x$ exists. Using (3.3), it follows that $\lim_{x\to-5} x = -5$ and we find that

$$(\lim_{x\to-5} x)(\lim_{x\to-5} x) = (-5)(-5) = 25$$

To compute the second term, we use (3.3) again, $\lim_{x\to-5} 4 = 4$. Now that we showed that the individual limits exist, we can use Rules 1 and 2 to evaluate

$$\lim_{x\to-5}[4 + 2x^2] = \lim_{x\to-5} 4 + 2 \lim_{x\to-5} x^2 = 4 + 2(-5)(-5) = 54 \qquad (3.2)$$

Prob. 41. Using Rules 1 and 4, this becomes

$$2 \lim_{x\to3} x^2 - \frac{\lim_{x\to3} 1}{\lim_{x\to3} x}$$

provided the individual limits exist. For the first term, we use Rule 3,

$$\lim_{x\to3} x^2 = (\lim_{x\to3} x)(\lim_{x\to1} x)$$

provided $\lim_{x\to3} x$ exists. Using (3.3), it follows that $\lim_{x\to3} x = 3$ and we find that

$$(\lim_{x\to3} x)(\lim_{x\to3} x) = (3)(3) = 9$$

To compute the second term, we use (3.3) again, $\lim_{x\to3} x = 3$. For the last term, we find $\lim_{x\to3} 1 = 1$. Now that we showed that the individual limits exist, we can use Rules 1, 2 and 4 to evaluate

$$\lim_{x\to3}[2x^2 - \frac{1}{x}] = 2 \lim_{x\to3} x^2 - \frac{\lim_{x\to3} 1}{\lim_{x\to3} x} = 2(3)(3) - \frac{1}{3} = \frac{53}{3} \qquad (3.3)$$

Prob. 43. Using Rule 4, we find

$$\lim_{x\to-3} \frac{x^3 - 20}{x + 1} = \frac{\lim_{x\to-3}(x^3 - 20)}{\lim_{x\to-3}(x + 1)}$$

provided the limits in the numerator and denominator exist and the limit in the denominator is not equal to 0. Using Rules 2 and 3 in the numerator, we find

$$\lim_{x\to-3}(x^3 - 20) = \lim_{x\to-3} x^3 - \lim_{x\to-3} 20 = (-3)(-3)(-3) - 20 = -47$$

Breaking up the limit of the sum in the numerator into a sum of limits is only justified once we show that the individual limits exist. Using Rules 1 and 2 in the denominator, we find

$$\lim_{x \to -3} (x + 1) = \lim_{x \to -3} x + \lim_{x \to -3} 1 = -3 + 1 = -2$$

Again using the limit laws is only justified once we demonstrate that the individual limits exist. Since the limits in both the denominator and the numerator exist and the limit in the denominator is not equal to 0, we obtain

$$\lim_{x \to -3} \frac{x^3 - 20}{x + 1} = \frac{-47}{-2} = 23.5$$

Prob. 45. Using Rule 4, we find

$$\lim_{x \to 3} \frac{3x^2 + 1}{2x - 3} = \frac{\lim_{x \to 3}(3x^2 + 1)}{\lim_{x \to 3}(2x - 3)}$$

provided the limits in the numerator and denominator exist and the limit in the denominator is not equal to 0. Using Rules 2 and 3 in the numerator, we find

$$\lim_{x \to 3} (3x^2 + 1) = 3 \lim_{x \to 3} x^2 + \lim_{x \to 3} 1 = 3(3)(3) + 1 = 28$$

Breaking up the limit of the sum in the numerator into a sum of limits is only justified once we show that the individual limits exist. Using Rules 1 and 2 in the denominator, we find

$$\lim_{x \to 3} (2x - 3) = 2 \lim_{x \to 3} x - \lim_{x \to 3} 3 = 2(3) - 3 = 3$$

Again using the limit laws is only justified once we demonstrate that the individual limits exist. Since the limits in both the denominator and the numerator exist and the limit in the denominator is not equal to 0, we obtain

$$\lim_{x \to 3} \frac{3x^2 + 1}{2x - 3} = \frac{28}{3} = 9.33$$

Prob. 47. The function $f(x) = \frac{1-x^2}{1-x}$ is a rational function, but since $\lim_{x \to 1}(1 - x) = 0$, we cannot use Rule 4. Instead, we need to simplify f(x) first.

$$\lim_{x \to 1} \frac{1 - x^2}{1 - x} = \lim_{x \to 1} \frac{(1 - x)(1 + x)}{1 - x}$$

Since $x \neq 1$, we can cancel 1 - x in the numerator and denominator, which yields

$$\lim_{x \to 1}(1+x) = 2$$

where we used that 1 + x is a polynomial when computing the limit.

Prob. 49. The function $f(x) = \frac{x^2-2x-3}{x-3}$ is a rational function, but since $\lim_{x\to 3}(x-3) = 0$, we cannot use Rule 4. Instead, we need to simplify f(x) first.

$$\lim_{x \to 3} \frac{x^2-2x-3}{x-3} = \lim_{x \to 3} \frac{(x-3)(x+1)}{x-3}$$

Since $x \neq 3$, we can cancel x - 3 in the numerator and denominator, which yields

$$\lim_{x \to 3}(x+1) = 4$$

where we used that x + 1 is a polynomial when computing the limit.

Prob. 51. The function $f(x) = \frac{3-x}{x^2-9}$ is a rational function, but since $\lim_{x\to 3}(x^2-9) = 0$, we cannot use Rule 4. Instead, we need to simplify f(x) first.

$$\lim_{x \to 3} \frac{3-x}{x^2-9} = \lim_{x \to 3} \frac{3-x}{(x-3)(x+3)}$$

Since $x \neq 3$, we can cancel x - 3 in the numerator and denominator, which yields

$$\lim_{x \to 3} \frac{-1}{x+3} = \frac{-1}{6}$$

where we used that $\frac{-1}{x+3}$ is a polynomial when computing the limit.

Prob. 53. The function $f(x) = \frac{2x^2+3x-2}{x+2}$ is a rational function, but since $\lim_{x\to -2}(x+2) = 0$, we cannot use Rule 4. Instead, we need to simplify f(x) first.

$$\lim_{x \to -2} \frac{2x^2+3x-2}{x+2} = \lim_{x \to -2} \frac{(2x-1)(x+2)}{x+2}$$

Since $x \neq -2$, we can cancel x + 2 in the numerator and denominator, which yields

$$\lim_{x \to -2}(2x-1) = -5$$

where we used that 2x - 1 is a polynomial when computing the limit.

3.2 Continuity

Prob. 1. We must check all three conditions.

- $f(x)$ is defined at x = 2 since $f(2) = 2^3 - 2.2 + 1 = 5$.
- We use that $\lim_{x \to c} x = c$ to conclude that $\lim_{x \to 2} f(x)$ exists.
- Using the limit laws, we find that $\lim_{x \to 2} f(x) = 5$. This is same as $f(2)$. Since all three conditions are satisfied, $f(x) = x^3 - 2x + 1$ is continuous at x = 2.

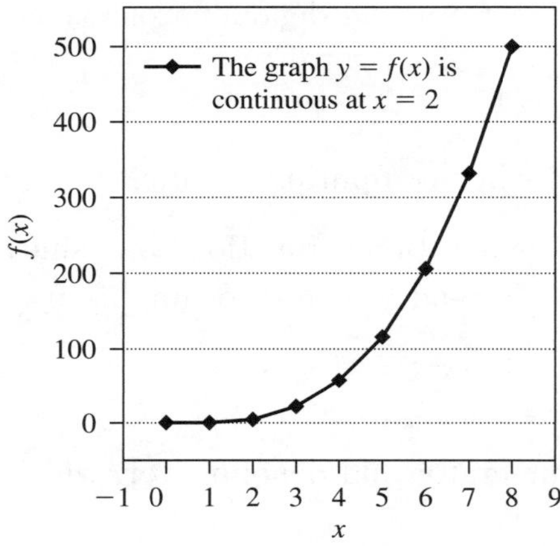

Prob. 3. We must check all three conditions.

- $f(x)$ is defined at $x = \frac{\pi}{4}$ since $f(\frac{\pi}{4}) = \sin(\frac{2\pi}{4}) = 1$.
- We use that $\lim_{x \to c} x = c$ to conclude that $\lim_{x \to \frac{\pi}{4}} f(x)$ exists.
- Using the limit laws, we find that $\lim_{x \to \frac{\pi}{4}} f(x) = 1$. This is same as $f(\frac{\pi}{4})$. Since all three conditions are satisfied, $f(x) = \sin(2x)$ is continuous at $x = \frac{\pi}{4}$.

Prob. 5. The graph of f(x) is shown below. Since equation $\frac{x^2-x-2}{x-2}$ can be written as $\frac{(x+1)(x-2)}{x-2}$, which is equal to $x + 1$. We therefore get:

$$\lim_{x \to 2} f(x) = 2 + 1 = 3$$

which is equal to $f(2)$. So, from the above given explaination and figure, we can conclude that the function is continuous at $x = 2$.

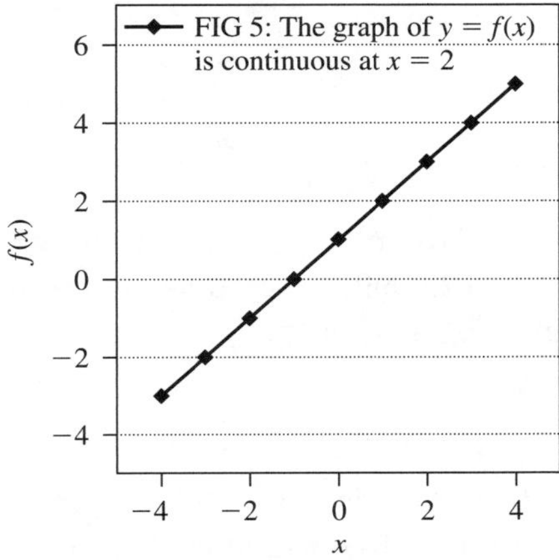

Prob. 7. To compute

$$\lim_{x\to 3} \frac{x^2-9}{x-3}$$

we factor the numerator, $x^2 - 9 = (x+3)(x-3)$. Hence, since $x \neq 3$,

$$\lim_{x\to 3} \frac{x^2-9}{x-3} = \lim_{x\to 3} \frac{(x-3)(x+3)}{x-3} = \lim_{x\to 3}(x+3) = 6 \tag{3.4}$$

To ensure that $f(x)$ is continuous at x = 3, we require that

$$\lim_{x\to 3} f(x) = f(3)$$

We therefore need to choose 6 for a. This is the only choice for a that would make $f(x)$ continuous. Any other value of a would result in f(x) being discontinuous.

Prob. 9. To show that $f(x) = \frac{1}{x-3}$ is discontinuous at x = 3, we show that

$$\lim_{x\to 3^+} \frac{1}{x-3} = \infty \quad \text{and} \quad \lim_{x\to 3^-} \frac{1}{x-3} = -\infty$$

Limit does not exist and it also does not match with the given value of $f(3)$. Because ∞ is not a real number, we will say f(x) is discontinuous at $x = 3$.

Prob. 11. To show that $f(x) = \frac{x^2-3x+2}{x-2}$ is discontinuous at x = 1, we show that

$$\lim_{x\to 1^+} \frac{x^2-3x+2}{x-2} = \lim_{x\to 1^+} \frac{(x-1)(x-2)}{x-2} = \lim_{x\to 1^+} \frac{1}{x-1} = \infty$$

Limit does not exist and it also does not match with the given value of $f(1)$. Because ∞ is not a real number, equal to the given $f(x)$, we will say f(x) is discontinuous at x = 1.

Prob. 13. The floor function:

$$f(x) = \lfloor x \rfloor =$$

the largest integer less than or equal to x. To explain the function, we compute a few values: $f(3.1) = 3, f(3) = 3, f(2.9999) = 2$. The function jumps whenever x is an integer. let k be an integer; then $f(k) = k$ and

$$\lim_{x \to k^+} f(x) = k, \lim_{x \to k^-} = k - 1$$

That is, only when x approaches an integer from the right is the limit equal to the value of function. The function therefore discontinuous at integer values and discontinuity can not be removed. As 2.5 is not an integer, $f(2.5)$ is continuous at x = 2.5, while at x = 3 $f(3)$ is discontinuous.

Prob. 15.

(a) f is continuous from the right as a square root of a positive polynomial function.

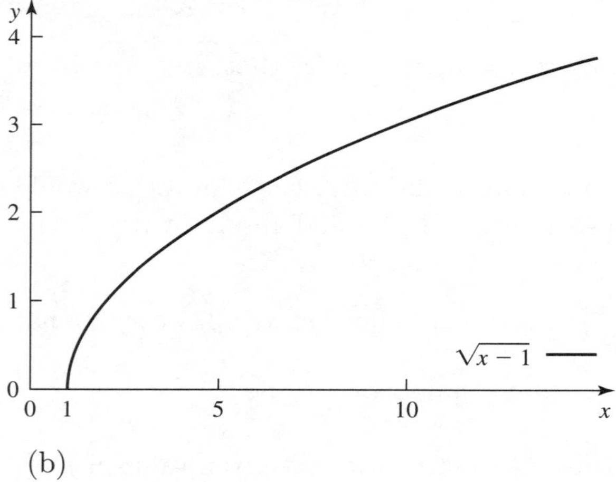

(b)

(c) f is not defined for $x < 1$, so it does not make sense to look at continuity at the left at x = 1.

Prob. 17. $f(x) = 3x^4 - x^2 + 4$ f(x) is a polynomial and is defined for all $x \in \mathbf{R}$.

Prob. 19. $f(x) = \frac{x^2+1}{x-1}$ is a rational function. It is defined for all $x \neq 1$; it is therefore continuous for all $x \neq 1$.

Prob. 21. Set $g(x) = -|x|$ and $f(x) = e^x$; then h(x) = (f o g)(x). Since g(x) is continuous for all $x \in \mathbf{R}$, and the range of $g(x)$ is $(-\infty, 0)$. $f(x)$ is continuous for all values in the range of g(x) [in fact, $f(x)$ is continuous for all $x \in \mathbf{R}$]. It therefore follows that h(x) is continuous for all $x \in \mathbf{R}$.

Prob. 23. f(x) is a logarithmic function. $f(x) = \ln \frac{x}{x+1}$ is defined as long as $\frac{x}{x+1} > 0$ which happens for $x < -1$ or $x > 0$. It is therefore continuous for all $x < -1$ or $x > 0$.

Prob. 25. $f(x)$ is a trigonometric function. The tangent function is defined for all $x \neq \frac{\pi}{2} + k\pi$, where k is an integer. It is therefore continuous for all $x \neq \frac{1}{4} + \frac{k}{2}$, where k is an integer.

Prob. 27.

(a) Graph for the functions are shown in figure below. For $c = 1$,

$$\lim_{x \to 0^+} x + c = c = 1$$

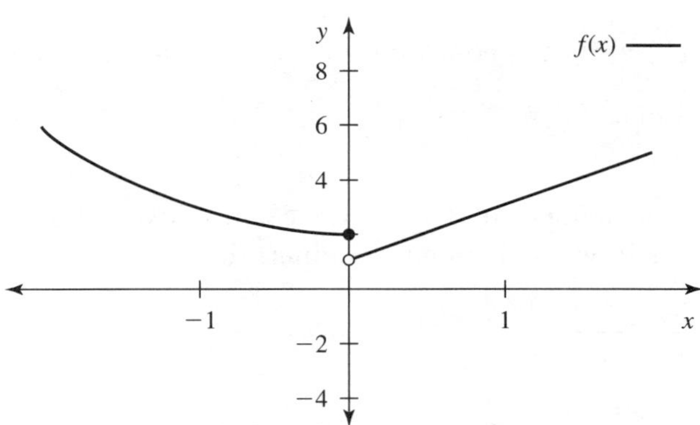

We know that the $f(0) = 2$, which is different from $\lim_{x \to 0^+} = 1$. So we can say that the function is not continuos at x = 0.

(b) If we choose a value of c = 2, then the function will be continuous for all $x \in (-\infty, \infty)$.

Prob. 29. Since $f(x)$ is continuous at $x = \frac{\pi}{3}$,

$$\lim_{x \to \frac{\pi}{3}} \sin(\frac{x}{2}) = \sin(\frac{\pi}{6}) = 0.5$$

Therefore $\lim_{x \to \frac{\pi}{3}} f(x)$ exists and is equal to 0.5.

Prob. 31. We use $\sin^2 x + \cos^2 x = 1$. Hence,

$$\lim_{x \to \frac{\pi}{2}} \frac{\cos^2 x}{1 - \sin^2 x} = \lim_{x \to \frac{\pi}{2}} \frac{\cos^2 x}{\cos^2 x} = 1 \tag{3.5}$$

So the limit for this function exists at $x = \frac{\pi}{2}$.

Prob. 33. The function $f(x) = \sqrt{1 + 8x^4}$ is continuous at x = 1. Hence,

$$\lim_{x \to 1} \sqrt{1 + 8x^4} = \sqrt{1 + 8} = 3$$

Prob. 35. The function $f(x) = \sqrt{x^2 + 2x + 2}$ is continuous at $x = -1$. Hence,

$$\lim_{x \to -1} \sqrt{x^2 + 2x + 2} = \sqrt{1} = 1$$

Prob. 37. The function $f(x) = e^{\frac{-x^2}{2}}$ is continuous at x = 0. Hence,

$$\lim_{x \to 0} e^{\frac{-x^2}{2}} = e^0 = 1$$

Prob. 39. The function $f(x) = e^{x^2 - 4}$ is continuous at x = 2. Hence,

$$\lim_{x \to 2} e^{x^2 - 4} = e^0 = 1$$

Prob. 41. $f(x) = \frac{e^{2x} - 1}{e^x - 1}$ is not defined at $x = 0$. We use a trick that will allow us to find the limit: namely we will split the numerator.

$$\lim_{x \to 0} \frac{e^{2x} - 1}{e^x - 1} == \lim_{x \to 0} \frac{(e^x - 1)(e^x + 1)}{e^x - 1} = \lim_{x \to 0} (e^x + 1) = e^0 + 1 = 2 \tag{3.6}$$

Prob. 43. The function $f(x) = \frac{1}{\sqrt{5x^2 - 4}}$ is continuous at $x = -2$. Hence,

$$\lim_{x \to -2} \frac{1}{\sqrt{5x^2 - 4}} = \frac{1}{4} = 0.25$$

Prob. 45. $f(x) = \frac{\sqrt{x^2 + 9} - 3}{x^2}$ is not defined at $x = 0$. We use a trick that will allow us to find the limit: namely we will rationalize the numerator.

$$\frac{\sqrt{x^2 + 9} - 3}{x^2} = \frac{\sqrt{x^2 + 9} - 3}{x^2} \cdot \frac{\sqrt{x^2 + 9} + 3}{\sqrt{x^2 + 9} + 3} = \frac{x^2}{x^2 \cdot (\sqrt{x^2 + 9} + 3)} = \frac{1}{\sqrt{x^2 + 9} + 3} \tag{3.7}$$

Now we can find the limit for f(x)

$$\lim_{x\to 0} \frac{\sqrt{x^2+9}-3}{x^2} = \lim_{x\to 0} \frac{1}{\sqrt{x^2+9}+3} = \frac{1}{6} \tag{3.8}$$

Prob. 47. The function $f(x) = \ln(1-x)$ is continuous at x = 0. Hence,

$$\lim_{x\to 0} \ln(1-x) = \ln(1) = 0$$

3.3 Limits and Infinity

Prob. 1. The degree of the denominator is greater than the degree of the numerator, so the limit is 0.

Prob. 3. The degree of the numerator is greater than the degree of the denominator, so the limit does not exist. The function tends to ∞ since it behaves as $\frac{x^3}{x} = x^2$.

Prob. 5. The degree of the numerator equals that of the numerator, so the limit is $\frac{3}{1} = 3$.

Prob. 7. The degree of the numerator is greater than that of the numerator, so the limit does not exist. The function tends to ∞ since it behaves as $\frac{x^2}{2x} = \frac{x}{2}$.

Prob. 9. The degree of the numerator is greater than that of the numerator, so the limit does not exist. The function tends to ∞ as $x \to -\infty$ since it behaves as $\frac{x^2}{-x} = -x$.

Prob. 11. The degree of the numerator is greater than that of the numerator, so the limit does not exist. The function tends to ∞ as $x \to -\infty$ since it behaves as $\frac{x^2}{-x} = -x$.

Prob. 13. Using the limit laws,

$$\lim_{x\to\infty} \frac{4}{1+e^{-2x}} = \frac{4}{1+0} = 4$$

since e^{-2x} tends to 0 as $x \to \infty$.

Prob. 15. We have

$$\lim_{x\to\infty} \frac{2e^x}{e^x+3} = \lim_{x\to\infty} \frac{2}{1+3e^{-x}} = 2.$$

Prob. 17. The limit is 0 since $\exp(x)$ is the same function as e^x.

Prob. 19. Since $|\sin x| \leq 1$, we have $|e^{-x}\sin x| \leq e^{-x}$, which tends to 0 as $x \to \infty$. So $|e^{-x}\sin x| \to 0$ as $x \to \infty$. This implies that $e^{-x}\sin x$ tends to 0 as $x \to \infty$ because $|e^{-x}\sin x|$ is the distance from $e^{-x}\sin x$ to 0.

Prob. 21. The numerator has limit 3 and the denominator has limit 2, so the limit is $\frac{3}{2}$.

Prob. 23. We have

$$\lim_{x\to-\infty} \frac{e^x}{1+x} = \lim_{x\to-\infty} e^x \lim_{x\to-\infty} \frac{1}{1+x} = 0 \cdot 0 = 0.$$

Prob. 25. We have

$$\lim_{N\to\infty} r(N) = \lim_{N\to\infty} a\frac{N}{k+N} = a \lim_{N\to\infty} \frac{N}{k+N} = a.$$

Prob. 27. (a)

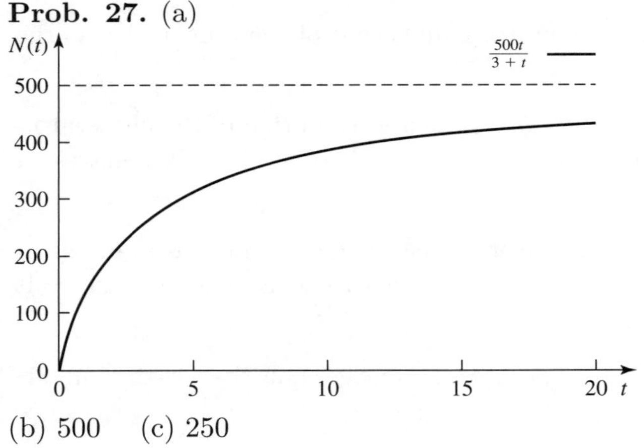

(b) 500 (c) 250

Prob. 29. (a)

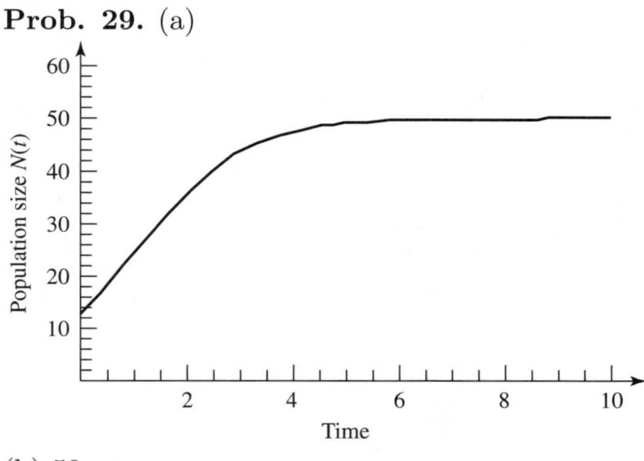

(b) 50

3.4 The Sandwich Theorem and Some Trigonometric Limits

Prob. 1.

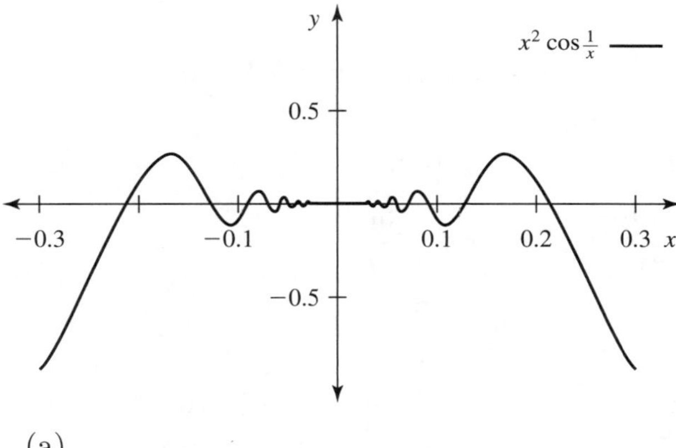

(a)

(b) Since $-1 \leq \cos\frac{1}{x} \leq 1$ for any x such that $\frac{1}{x}$ is defined (that is, for $x \neq 0$), we can multiply both inequalities by the positive number x^2, obtaining

$$-x^2 \leq x^2 \cos\frac{1}{x} \leq x^2.$$

(c) The role of $g(x)$ in the Sandwich Theorem is played by $x^2\cos\frac{1}{x}$, and the sandwiching functions $f(x)$ and $h(x)$ are, respectively, $-x^2$ and x^2. Since $\lim_{x\to 0} -x^2 = \lim_{x\to 0} x^2 = 0$, we conclude that $\lim_{x\to 0} x^2\cos\frac{1}{x} = 0$.

Prob. 3.

(a)

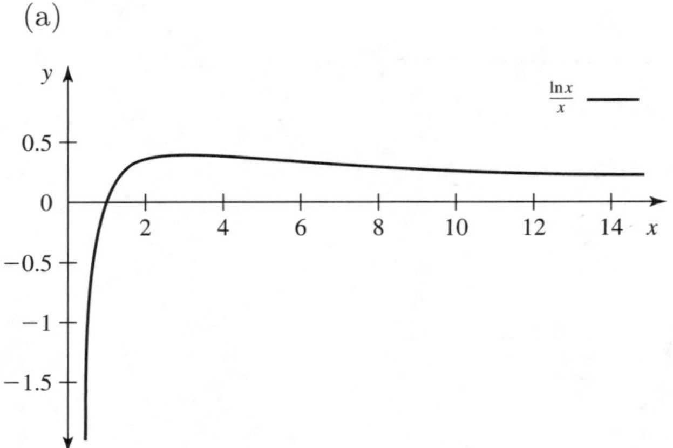

(b) The inequalities hold for $x \geq e$.

(c) We apply the Sandwich Theorem with $f(x) = \frac{1}{x}$, $g(x) = \frac{\ln x}{x}$, $h(x) = \frac{1}{\sqrt{x}}$.

Since $\lim_{x\to\infty} \frac{1}{x} = \lim_{x\to\infty} \frac{1}{\sqrt{x}} = 0$, we deduce that $\lim_{x\to\infty} \frac{\ln x}{x} = 0$.

Prob. 5. Use the substitution $y = 2x$ to obtain

$$\begin{aligned}\lim_{x\to 0} \frac{\sin(2x)}{2x} &= \lim_{(y/2)\to 0} \frac{\sin y}{y} \\ &= \lim_{y\to 0} \frac{\sin y}{y} = 1.\end{aligned}$$

The last equality was proved in Section 3.4. The second is justified because, when y goes to 0, so does $y/2$.

Prob. 7. Set $y = 5x$. Then

$$\begin{aligned}\lim_{x\to 0} \frac{\sin(5x)}{x} &= \lim_{x\to 0} \cdot \frac{\sin(5x)}{\frac{1}{5}(5x)} \\ &= 5 \lim_{y\to 0} \frac{\sin y}{y} = 5.\end{aligned}$$

Prob. 9.

$$\lim_{x\to 0} \frac{\sin \pi x}{x} = \lim_{x\to 0} \frac{\sin(\pi x)}{\frac{1}{\pi}(\pi x)}$$

$$= \pi \lim_{x\to 0} \frac{\sin(\pi x)}{\pi x} = \pi \lim_{y\to 0} \frac{\sin y}{y} = \pi.$$

Prob. 11. We relate $\frac{\sin \pi x}{\sqrt{x}}$ to a function whose limit at 0 we already know, as follows:

$$\frac{\sin \pi x}{\sqrt{x}} = \frac{\sin \pi x}{x} \cdot \sqrt{x} \quad \text{for } x \neq 0.$$

Hence we must find $\lim\limits_{x\to 0} \frac{\sin \pi x}{\pi x} \cdot \pi\sqrt{x}$. But both $\frac{\sin \pi x}{\pi x}$ and $\pi\sqrt{x}$ have a limit as $x \to 0$; they are 1 and 0 respectively. We are therefore allowed to say that the limit of the product is the product of the limits, namely, $1 \cdot 0 = 0$.

Prob. 13. We decompose the function into a product of three factors, each of which has a limit at

$$0 = \lim_{x\to 0} \frac{\sin x \cos x}{x(1-x)} = \lim_{x\to 0} \frac{\sin x}{x} \cdot \cos x \cdot \frac{1}{1-x}$$

$$\lim_{x\to 0} \frac{\sin x}{x} \cdot \lim_{x\to 0} \cos x \cdot \lim_{x\to 0} \frac{1}{1-x}.$$

The last equality is allowed because each of the limits on the right-hand side exists. All three limits equal 1, so our desired limit is 1.

Prob. 15.

$$\lim_{x\to 0} \frac{1-\cos x}{2x} = \frac{1}{2} \lim_{x\to 0} \frac{1-\cos x}{x} = \frac{1}{2} \cdot 0 = 0.$$

(The limit of $\frac{1-\cos x}{x}$ as $x \to 0$ was proved in Section 3.4.)

Prob. 17.

$$\begin{aligned} &\lim_{x\to 0} \tfrac{\sin x(1-\cos x)}{x^2} \\ &= \lim_{x\to 0} \tfrac{\sin x}{x} \cdot \tfrac{(1-\cos x)}{x} \\ &= \lim_{x\to 0} \tfrac{\sin x}{x} \cdot \lim_{x\to 0} \tfrac{1-\cos x}{x} \\ &= 1 \cdot 0 = 0. \end{aligned}$$

(since both limits on the right exist)

3.5 Properties of Continuous Functions

Prob. 1. (a)

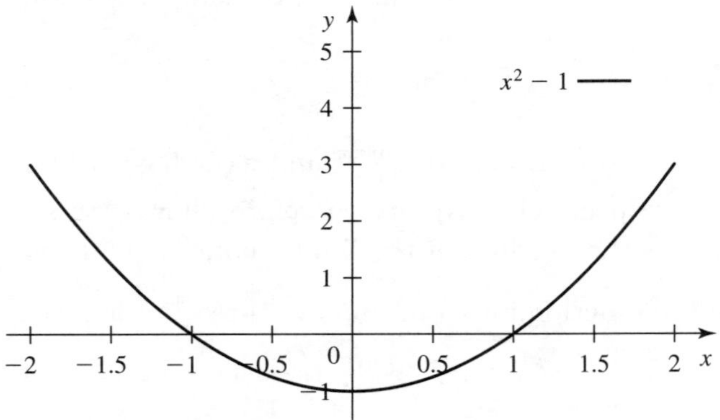

(b) We have $f(0) = -1$ and $f(2) = 3$. The function is continuous since it a polynomial. Since $f(0) < 0 < f(2)$, there is some value c of the variable such that $0 < c < 2$ and $f(c) = 0$, thanks to the Intermediate Value Theorem.

Prob. 3. (a)

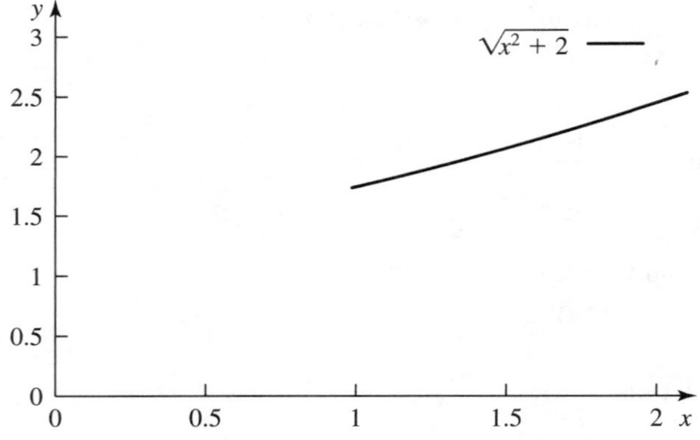

(b) We have $f(1) = \sqrt{1+2} = \sqrt{3} < 2$ and $f(2) = \sqrt{2^2+2} = \sqrt{6} > 2$. Since $f(1) < 2 < f(2)$, the Intermediate Value Theorem, together together with the fact that f is continuous (being a square root of a polynomial that only takes positive values), implies that there is some value c of the variable such that $1 < c < 2$ and $f(c) = 2$.

Prob. 5. A solution to $e^{-x} = x$ is a value of x for which the function $f(x) = e^{-x} - x = 0$. Now, $f(1) = e^{-1} - 1$ is negative, since $e^{-1} = 1/e$ is less than 1, while $f(0) = e^0 - 0 = 1$ is positive. Since the function is continuous as the difference of two continuous functions and $f(1) < 0 < f(0)$ we can apply the Intermediate Value Theorem to learn that there is a value c of the variable such that $0 < c < 1$ and $f(c) = 0$, that is, $e^{-c} = c$.

Prob. 7. We define the function $f(x) = e^{-x} - x$, which is continuous. We are looking for a zero of this function. We have $f(0) > 0$ and $f(1) < 0$ by calculation (or see exercise 5). Therefore there must be a zero in $(0, 1)$, by the intermediate value theorem. We test $x = 0.5$, for which $f(x) = 0.106531$. Since this is positive, the function changes sign between $x = 0.5$ and $x = 1$. We then test $x = 0.75$, which gives $f(x) = -0.277633$. We have narrowed the location of a solution to the interval $(0.5, 0.75)$. We next test a point between 0.5 and 0.75 – we are not forced to choose the middle point, and in fact it may speed things up if we choose a point closer to 0.5 than to 0.75, because the value of f is closer to 0 at 0.5 than it is at 0.75. So let's take $x = 0.6$. This gives $f(x) = -0.051188$. We have narrowed the interval to $(0.5, 0.6)$. We have $f(0.55) = 0.026950$, narrowing it further to $(0.55, 0.6)$. We have $f(0.575) = -0.012295$, narrowing it further to $(0.55, 0.575)$. We have $f(0.565) = 0.003360$, narrowing it further to $(0.565, 0.575)$. Any number within this interval can be written 0.56 to two decimal places. Therefore our solution is 0.56.

Prob. 9.

(a) We must start somewhere, so we test $f(0)$, which gives 2, and $f(1)$, which gives exactly 0. Wow! What good luck! The solution $x = 1$ is accurate to infinitely many decimal places.

(c) The graph has a tangency with the x-axis at $x = 1$, and $f(x)$ is positive near $x = 1$ on both sides. So, if we had not found the solution $x = 1$ already, we would not be able to find it using the bisection method, since we would not suspect (by looking at the sign of $f(x)$ at $x = 0$ and $x = 2$, say) that there is a solution in between. On the other hand, having found the solution $x = 1$ and having plotted the graph, we can easily find the other root using the bisection method, starting (say) with the interval $[-1, 0]$.

Prob. 11.

(a) The number should be reported as 23 individuals.

(b) We must decide to what accuracy formula (3.7) models the population. It is reasonable to think that the coefficients 54 and 13 that appear in the formula have 2 significant digits only (otherwise they would be written 54.0 and the like). Thus the result of formula (3.7) is also limited to two significant digits, and he correct way to express the result at $t = 10$ is 23 million individuals, or 2.3×10^7 individuals.

(c) A small population (such as given in case (a)) is not very well approximated by a continuous function. In particular, formula (3.7) purports to be valid for $t \geq 0$. At the lower end of this interval, $t = 0$, we have $N(t) = 4.15385$, and for some slightly bigger t we have $N(t) = 4.5$. Since the real population is measured by integers, this value of $N(t)$ involves a relative error of at least 0.5 in 4.5, or 11%.

In case (b), on the other hand, the size of the population is large and the issue of the discreteness of the population does not play a significant role in the approximation. Other sources of error, including counting errors and the empirical nature of the fit contained in equation (3.7), would instead the the limiting factors in the precision of the approximation.

Prob. 13. Explanation #1: We know the two roots; they're obviously -2 and 2.

Explanation #2: A second degree equation of positive discriminant has two roots.

Explanation #3: For x large in absolute value (say $x = \pm 100$), $x^2 - 4$ is positive, while for $x = 0$, it is negative. Therefore, by the intermediate value theorem, there is a root in the interval $(0, 100)$ and a root in the interval $(-100, 0)$.

3.6 Formal Definition of Limits

Prob. 1. Two inequalities must be satisfied:

$$2x - 1 < 0.01$$

and

$$2x - 1 > -0.01.$$

The first equation reduces to $x < 0.505$ and the second to $x > 0.495$. Therefore the desired values of x are those in the interval $(0.495, 0.505)$.

Prob. 3. To find the x values that satisfy $\left|3x^2 + 1\right| < 0.1$, we solve,

$$-0.1 < 3x^2 + 1 < 0.1$$

$$-1.1 < 3x^2 < -0.9$$

We see that this does not have a solution. Hence there are no x values that satisfy this inequality.

Prob. 5. (a)

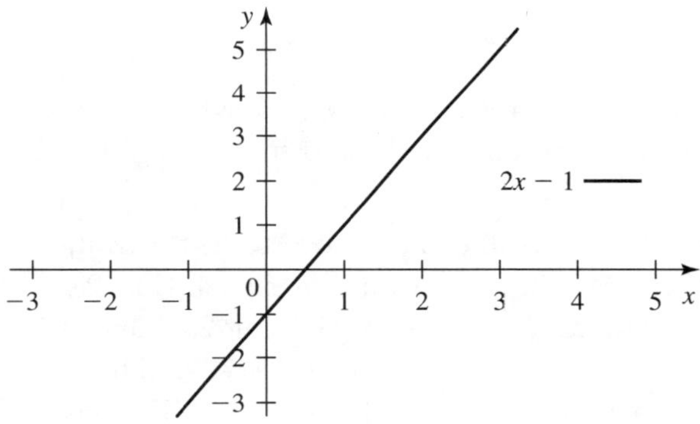

(b) We must find all x such that $f(x) - 3$ is at most 0.1 in absolute value. The inequalities to be satisfied are $(2x-1)-3 < 0.1$ and $(2x-1)-3 > -0.1$. The first gives $x < 2.05$ and the second gives $x > 1.95$. Thus the desired values of x are those in the interval $(1.95, 2.05)$.

Prob. 7.

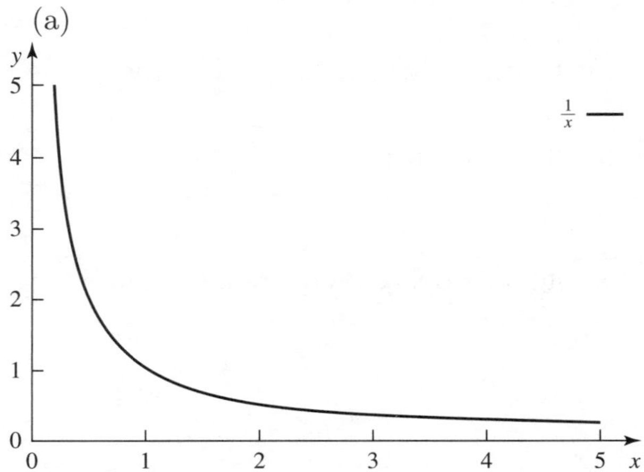

(b) We must find all x such that $f(x) > 4$, that is, $1/x > 4$, which amounts to $x < 1/4$. Since $x > 0$ by assumption, the desired interval is $(0, 1/4)$.

Prob. 9. We must show that for any x sufficiently close to 2, the value of $2x - 1$ is close to 3. Take $\epsilon > 0$ and let's find the values of x such that $2x - 1$ is ϵ-close to 3 (that is, lies in the interval $(3-\epsilon, 3+\epsilon)$. Solving $2x - 1 < 3 + \epsilon$ gives $x < 2 + \epsilon/2$ and solving $2x - 1 > 3 - \epsilon$ gives $x > 2 - \epsilon/2$. Thus any $x \in (2 - \epsilon/2, 2 + \epsilon/2)$ will map in the interval $y \in (3 - \epsilon, 3 + \epsilon)$. This shows that we can choose δ (namely $\epsilon/2$) such that any x that is δ-close to 2 maps ϵ-close to 3.

Prob. 11. We must show that for any x sufficiently close to 9, the value of $\sqrt{x}$ is close to 3. Take $\epsilon > 0$ and let's find the values of x such that $\sqrt{x}$ is ϵ-close to 3. Solving $\sqrt{x} < 3 + \epsilon$ gives $x < (3 + \epsilon)^2 = 9 + 6\epsilon + \epsilon^2$ and solving $\sqrt{x} > 3 - \epsilon$ gives $x > (3 - \epsilon)^2 = 9 - 6\epsilon + \epsilon^2$. To avoid having to deal with ϵ^2, we reason as follows: If ϵ is not too large, say $\epsilon < 1$, the condition $x > 9 - 5\epsilon$ is stronger than $x > 9 - 6\epsilon + \epsilon^2$ (because $-5\epsilon > -5\epsilon - \epsilon(1 - \epsilon)$); moreover the condition $x < 9 + 5\epsilon$ is always stronger than $x < 9 + 6\epsilon + \epsilon^2$. Thus any $x \in (9 - 5\epsilon, 9 + 5\epsilon)$ will map in the interval $y \in (3 - \epsilon, 3 + \epsilon)$, provided that $\epsilon < 1$. Naturally, if ϵ is large (≥ 1), we can simply take $x \in (9 - 5, 9 + 5)$ keeping the interval independent of ϵ. This shows that if we choose $\delta = \min(5\epsilon, 5)$, any x that is δ-close to 9 maps ϵ-close to 3.

Prob. 13. We must show that for any x sufficiently close to 0, the value of $f(x)$ is very large and positive. Take $M > 0$ and let's find the values of

x such that $4/x^2 > M$. The result is $|x| < 2/\sqrt{M}$. This shows that if we choose $\delta = 2/\sqrt{M}$, any x that is δ-close to 0 (and different from 0) will map to a number greater than M.

Prob. 15. We must show that for any x sufficiently close to 0, the value of $f(x)$ is very large and positive. Take $M > 0$ and let's find the values of x such that $1/x^4 > M$. The result of course is $|x| < M^{-1/4}$. This shows that if we choose $\delta = M^{-1/4}$, any x that is δ-close to 0 (and different from 0) will map to a number greater than M.

Prob. 17. We must show that for any x sufficiently large and positive, the value of $f(x)$ is close to 0. Take $\epsilon > 0$ and let's find the values of x such that $|3/x^2 - 0| < \epsilon$. The result is $|x| < \sqrt{3/\epsilon}$. This shows that if we choose $L = sqrt3/\epsilon$, any $x > L$ maps ϵ-close to 0.

Prob. 19. We must show that for any x sufficiently large and positive, the value of $f(x)$ is close to 1. Take $\epsilon > 0$ and let's find the values of x such that $|x/(x+1) - 1| < \epsilon$. This unfolds into two inequalities to be solved: $x/(x+1) < 1+\epsilon$ and $x/(x+1) > 1-\epsilon$. Since we are interested in x large we will assume that $x+1 > 0$, so we can clear denominators in good conscience. Then the first inequality turns into

$$x < (x+1)(1+\epsilon) \iff x > -\frac{1}{\epsilon} - 1,$$

and the second into

$$x < (x+1)(1-\epsilon) \iff x > \frac{1}{\epsilon} - 1.$$

Evidently the second condition is the more stringent one and we conclude that the desired values of x are those satisfying $x > \frac{1}{\epsilon} - 1$. In other words, if we choose $L = \frac{1}{\epsilon} - 1$, any $x > L$ maps ϵ-close to 1.

Prob. 21. Given $\epsilon > 0$, we must find δ such that any x that is δ-close to c maps to a y that is ϵ-close to $mc + b$. Because the function $mx + b$ is linear, this is very easy: all intervals are stretched by a factor of $|m|$. So we just take $\delta = \epsilon/|m|$. Verification: if $|x-c| < \epsilon/|m|$, we have $(mx+b)-(mc+b) = m(x-c) < \epsilon$, as desired.

3.8 Review Problems

Prob. 1. The function is continuous for every value of x. This is because it is a composition of three continuous functions: $f_1(x) = |x|$, $f_2(y) = -y$, and $f_3(z) = e^z$.

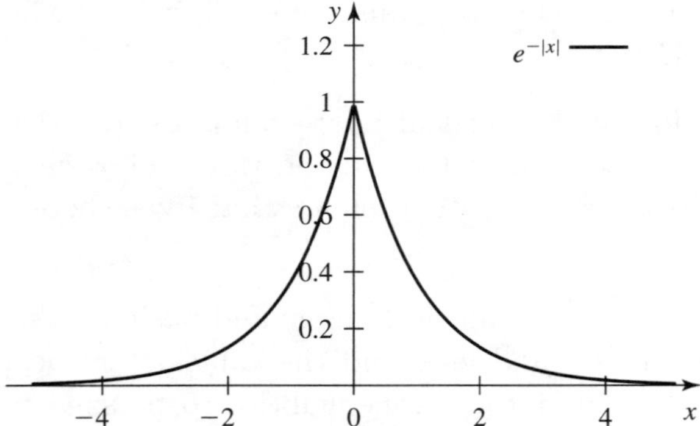

Prob. 3. The function is continuous for every value of x. This is because the function that equals the denominator, $h(x) = e^x + e^{-x}$, is continuous for every x (as the sum of two continuous functions), and furthermore it takes only positive values. Now regard the function $f(x) = g(h(x))$ as the composition of h and g, where $g(y) = 2/y$ is regarded as a function defined on positive numbers, and therefore continuous. The composition of two continuous functions is continuous.

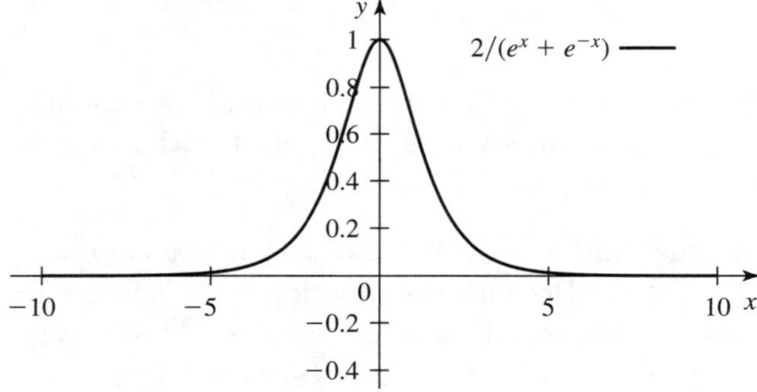

Prob. 5. An example of such a function is

$$f(x) = \begin{cases} 0 & \text{for } x < 0, \\ 1 & \text{for } x \geq 0. \end{cases}$$

and the graph of yet another one is:

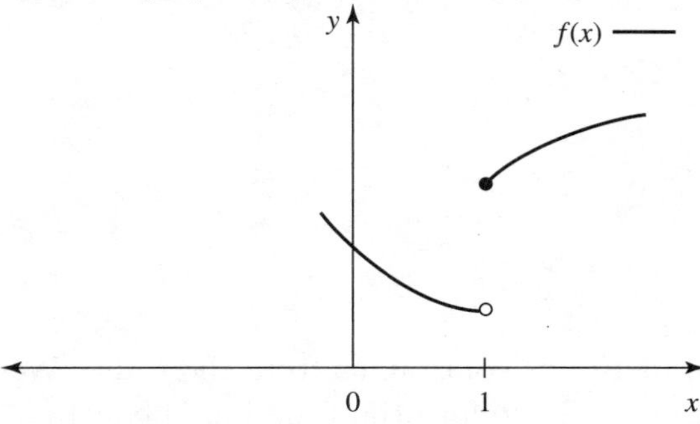

Prob. 7. An example of such a function is $(e^x - e^{-x})/(e^x + e^{-x})$, with graph:

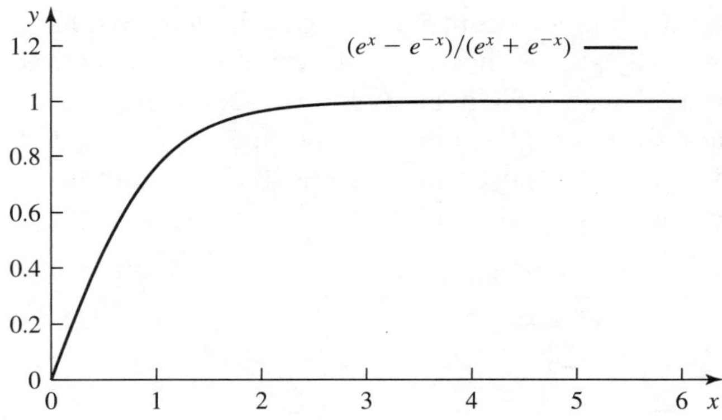

Another one is give by the graph:

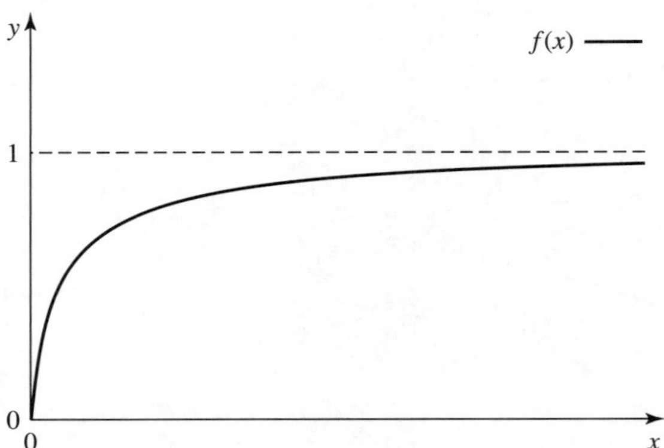

Prob. 9. First we show that the function is continuous from the right. We have $f(x) = -2$ for every $x \in [-2, -1)$. Therefore the function is constant and equal to -2 in an interval whose left endpoint is -2; in particular, $\lim_{x \to -2^+} f(x) = -2$. Since we also have $f(-2) = -2$, the function is continuous from the right at -2.

Next we show that the function is discontinuous from the left. We have $f(x) = -3$ for every $x \in [-3, -2)$. Therefore the function is constant and equal to -3 in an interval whose right endpoint is -2; in particular, $\lim_{x \to -2^-} f(x) = -3$. Since we have $f(-2) = -2$, the limit from the left does not agree with the function at the point -2, and so the function is discontinuous from the left there.

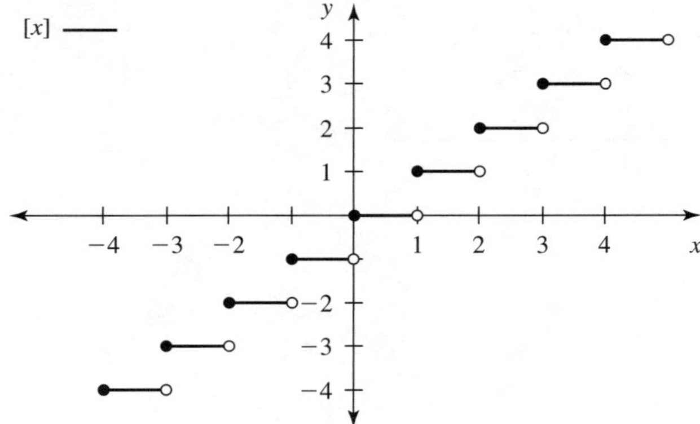

Prob. 11. The limit of $N(t)$ as $t \to \infty$ equals a, because we can rewrite $N(t)$ as

$$N(t) = \frac{a}{\frac{k}{t} + 1},$$

and as $t \to \infty$ the denominator has the limit 1. Therefore $a = 1.24 \times 10^6$. We also know that at $t = 5$ the value of $N(t)$ is half of its limiting value, which is to say, $\frac{1}{2}a$. Thus we have

$$\frac{1}{2}a = N(5) = a\frac{5}{k+5},$$

which reduces to $1/2 = 5/(k+5)$; solving for k gives $k = 5$.

Prob. 13. The function $g(t)$ should take the value 1 when $s(t) \geq 1/2$ and the value 0 when $s(t) < 1/2$. Now, $s(t) = \sin(\pi t)$ is periodic of period 2, and within the interval $[0, 2)$ it is less than $1/2$ whenever t lies between $1/6$ and $5/6$ (corresponding to $\sin \pi/6 = \sin 30° = 1/2$ and $\sin 5\pi/6 = \sin 150° = 1/2$).

Therefore one way to encode the function g is the following. Given the argument t, first reduce it to the interval $[0, 2)$ by subtracting multiples of 2. This can be done by subtracting from t (which is positive) as great a multiple of 2 as will fit: in formulas, $t - 2\lfloor t/2 \rfloor$. Then check if this reduced value lies in the interval $[1/6, 5/6]$. If so, assign g the value 1 (because $s(t) \geq 1/2$); if not, assign g the value 0 (because $s(t) < 1/2$). To summarize: $g(t) = 1$ if $t - 2\lfloor t/2 \rfloor \in [1/6, 5/6]$, and $g(t) = 0$ otherwise.

(a) Another way to describe the function is:

$$g(t) = \begin{cases} 1 \text{ for } \frac{1}{6} + 2k \leq x \leq \frac{5}{6} + 2k, k = 0, 1, 2, \ldots \\ 0 \text{ otherwise} \end{cases}$$

(b) The function $s(t)$ is continuous (since the sine function is continuous). But $g(t)$ is discontinuous (not surprising, since it only takes two values, 0 and 1; notice also that the floor function is discontinuous).

Prob. 15. (a)

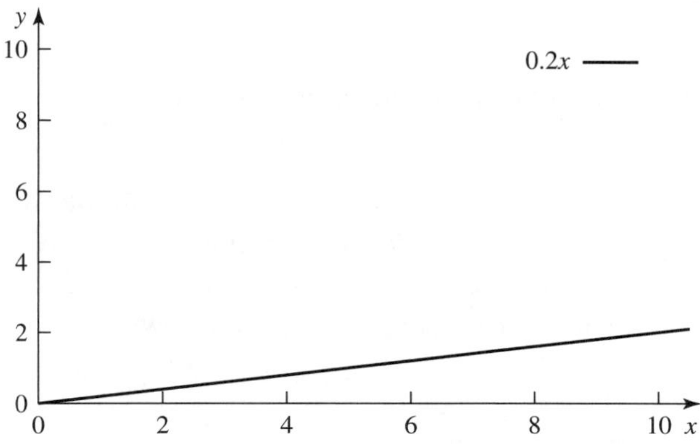

This graph is linear.

(b) In the modified model in which prey handling time is taken into account, the actual searching time is

$$T - T_h \frac{N_e}{P},$$

and this is what is proportional to the number of prey encounters:

$$\frac{N_e}{P} = a\left(T - T_h \frac{N_e}{P}\right)N.$$

This is a first-degree equation in N_e, and its solution is

$$N_e = P \frac{aTN}{1 + aT_h N}.$$

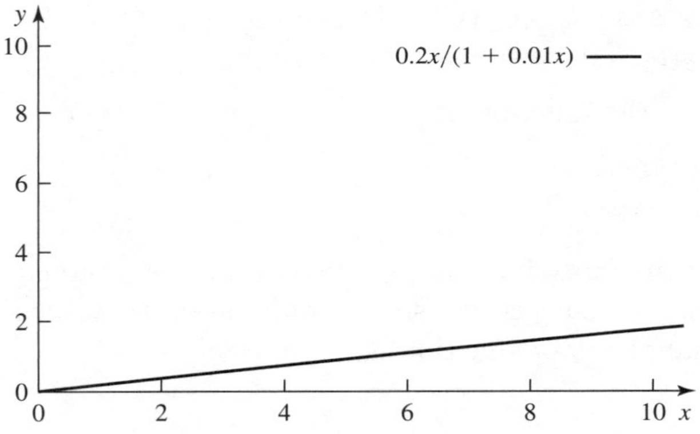

(c) This is clear because the denominator of the right-hand side of (3.11) becomes 1.

(d) When $T_h = 0$, we are in the situation of (3.10) and the limit of the right-hand side as $N \to \infty$ is ∞ (the function grows linearly with N). On the other hand, when $T_h > 0$, the limit of the right-hand side of (3.11) is T/T_h; this can be seen by rewriting (3.11) as follows:

$$\frac{N_e}{P} = \frac{aT}{\frac{1}{N} + aT_h}$$

and observing that, as $N \to \infty$, the denominator approaches aT_h. The interpretation of the difference is that, if we disregard the prey-handling time and the prey becomes ridiculously easy to find, there will be a ridiculously large number of prey encounters. But if we take into account the prey-handling time, things change: even though another prey can be found immediately after one is eaten, the total number of prey that can be eaten is limited by the quotient of the available time T by the per-prey handling time T_h.

Prob. 17.

(a) The function sinh and cosh must be continuous because they are defined by taking sums, differences, and constant multiples of continuous functions. To see that tanh is continuous we must also observe that the denominator $e^x + e^{-x}$ never vanishes; the quotient of a continuous numerator by a continuous, nonvanishing denominator is continuous.

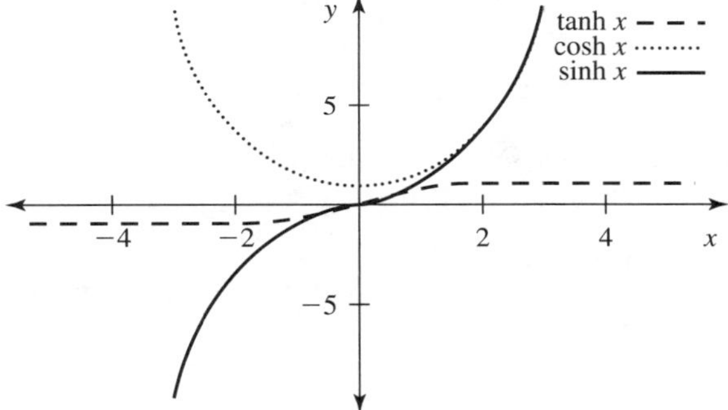

(b) As $x \to \infty$, the term e^{-x} in the definition of the hyperbolic sine becomes very small, and the term e^x very large, so the difference is very large and the limit is ∞. On the other hand, when $x \to -\infty$, the term e^{-x} becomes very large and the term e^x becomes very small; so the difference is negative and very large in absolute value, and the limit is $-\infty$.

With the hyperbolic cosine we have a sum instead of a difference, and either term (e^x or e^{-x}) gets very large, while the other gets very small. Therefore the sum gets very large whether we take $x \to \infty$ or $x \to -\infty$. The limit is ∞ in both cases.

With the hyperbolic tangent we cannot use this simple reasoning, because both numerator and denominator get very large. Instead we rewrite the definition as follows:

$$\tanh x = \frac{1 - e^{-2x}}{1 + e^{-2x}}$$

(by dividing numerator and denominator by e^x). This shows that, as $x \to \infty$, the limit is 1, because the limit of both numerator and denominator in the new expression is 1. To find the limit as $x \to -\infty$ we must do yet another rewriting:

$$\tanh x = \frac{e^{2x} - 1}{e^{2x} + 1}$$

(this time we multiplied top and bottom by e^x). This shows that, as $x \to -\infty$, the limit is -1, because the limit of the numerator is -1 while the limit of the denominator is 1.

(c) $\cosh^2 x - \sinh^2 x = \frac{1}{4}(e^x + e^{-x})^2 - \frac{1}{4}(e^x - e^{-x})^2 = \frac{1}{4}((e^{2x} + 2 + e^{-2x}) - (e^{2x} - 2 + e^{-2x})) = \frac{1}{4}4 = 1.$

The equality $\tanh x = \sinh x / \cosh x$ is obvious from the definitions.

(d) Saying that $\sinh x$ is an odd function is saying that $\sinh(-x) = -\sinh x$. This can be seen as follows:

$$\sinh(-x) = \frac{1}{2}(e^{-x} - e^{-(-x)}) = \frac{1}{2}(e^{-x} - e^{x}) = -\frac{1}{2}(e^{x} - e^{-x}) = -\sinh x.$$

Similarly,

$$\cosh(-x) = \frac{1}{2}(e^{-x} + e^{-(-x)}) = \frac{1}{2}(e^{-x} + e^{x}) = \frac{1}{2}(e^{x} + e^{x}) = \cosh x.$$

For tanh, use the fact that the quotient of an odd function and an even function is odd.

4 Differentiation

4.1 Formal Definition of the Derivative

Prob. 1. 0

Prob. 3. 2

Prob. 5. 0

Prob. 7. $f'(0) = 0$

Prob. 9. $-2h$

Prob. 11. $\sqrt{4+h} - 2$

Prob. 13. (a) $f'(-1) = -10$ (b) $y = -10x - 5$

Prob. 15. (a) $f'(2) = -12$ (b) $y = \frac{1}{12}x - \frac{43}{6}$

Prob. 17. $f'(x) = \frac{1}{2\sqrt{x}}$

Prob. 19. $y = 4x - 2$

Prob. 21. $y = \frac{1}{4}x + 1$

Prob. 23. $y = -\frac{1}{6}x - \frac{19}{6}$

Prob. 25. $y = -\frac{1}{4}x + \frac{5}{4}$

Prob. 27. $f(x) = 2x^2$ and $x = a$

Prob. 29. $f(x) = \frac{1}{x^2+1}$, $a = 2$

Prob. 31. (a)

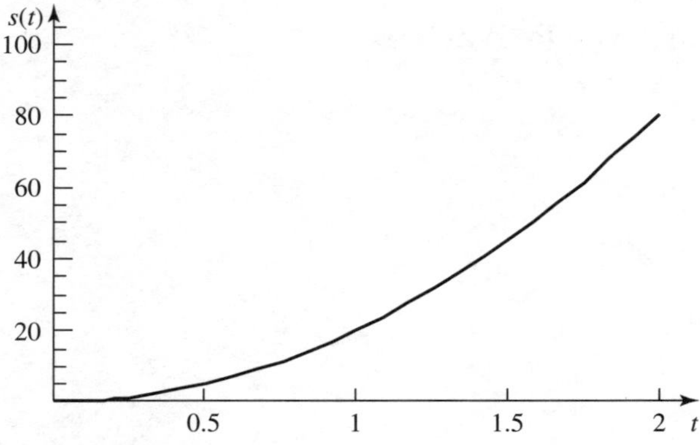

(b) 40 km/hr, (c) 40 km/hr

Prob. 33. (a) $s\left(\frac{3}{4}\right) = 30$, $s(1) = \frac{160}{3}$ (b) $\frac{280}{3}$ (c) $v\left(\frac{3}{4}\right) = 80$, $\left|v\left(\frac{3}{4}\right)\right| = 80$

Prob. 35. $R^{\star} = 1.25$

Prob. 37. $f(N) = 0$ for $N = 0$ and $N = 20$

Prob. 39. $x = 7$ or $x = 4$; reaction ceases when $x = 4$.

Prob. 41. $\frac{dN}{dt} = 0$ for $N = 0$ or $N = K$

Prob. 43. B

Prob. 45.

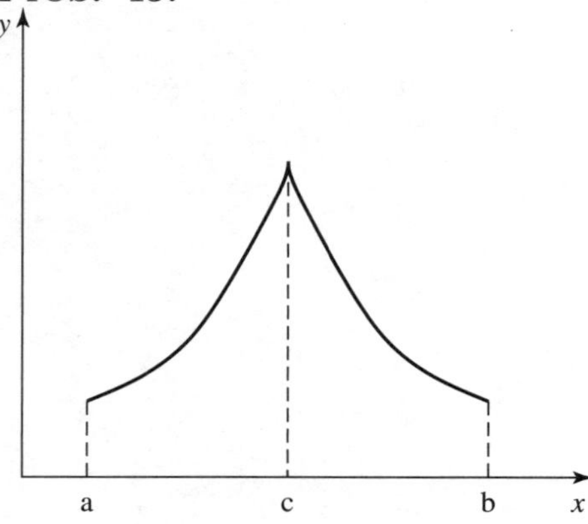

Prob. 47. $x = -5$

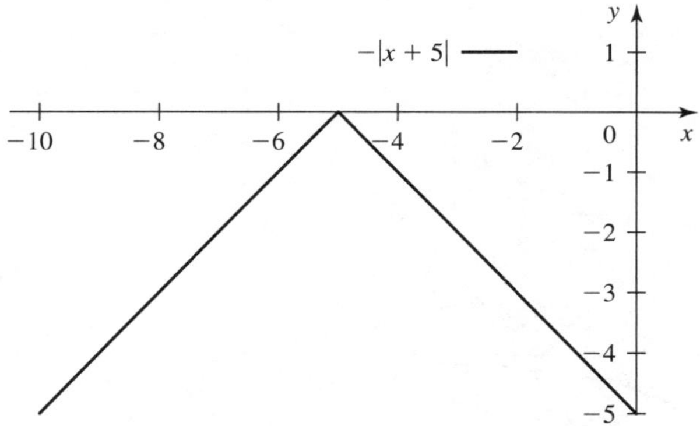

Prob. 49. $x = -2$

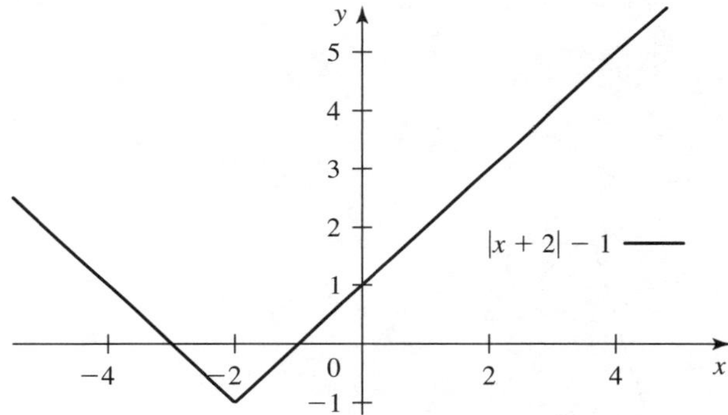

Prob. 51. $x = 3$

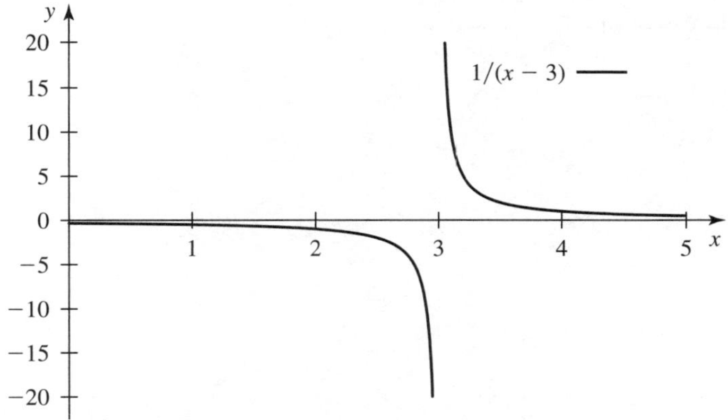

Prob. 53. $x = -1$

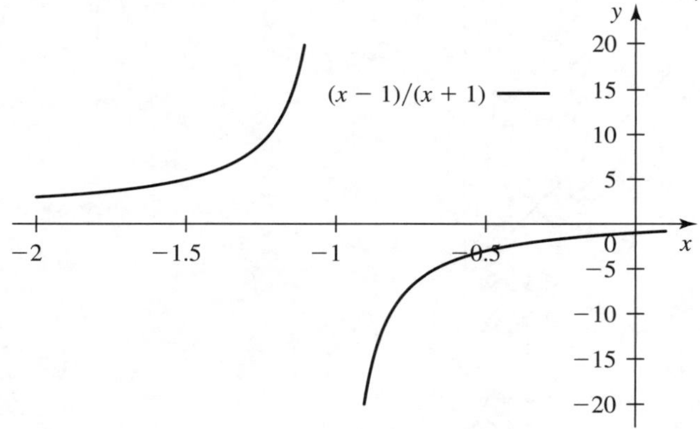

Prob. 55. $x = \pm\sqrt{1/2}$

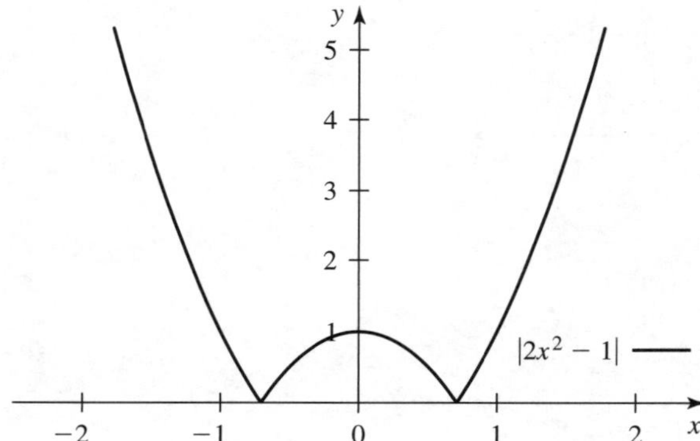

Prob. 57. $x = 1$

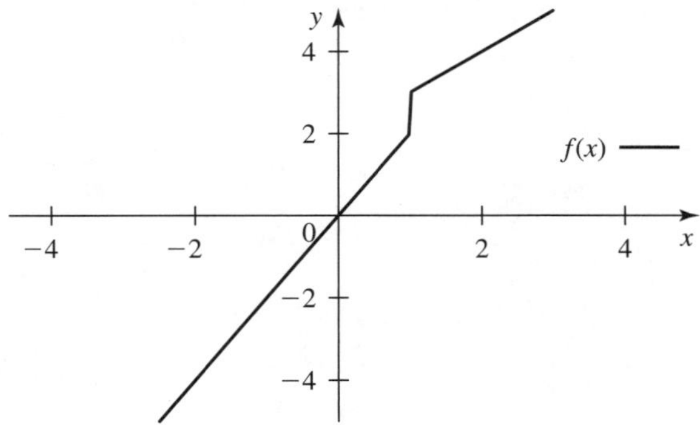

Prob. 59. $x = 0$

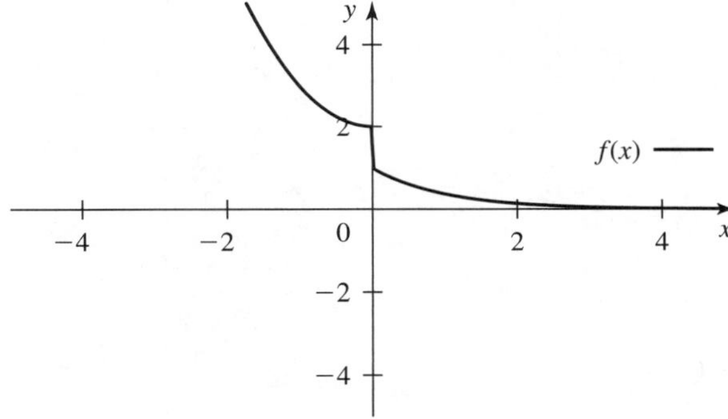

4.2 Basic Rules of Differentiation and Derivatives of Polynomials

Prob. 1. $6x^2 - 3$

Prob. 3. $-10x^4 + 7$

Prob. 5. $-4 - 10x$

Prob. 7. $-4 - 10x$

Prob. 9. $2t^3 + 4$

Prob. 11. $2x \sin \frac{\pi}{3}$

Prob. 13. $-12x^3 \tan \frac{\pi}{6}$

Prob. 15. $3t^2 e^{-2} + 1$

Prob. 17. $3s^2 e^3$

Prob. 19. $60x^2 - 24x^5 + 72x^7$

Prob. 21. $24x^2 + \frac{1}{8}$

Prob. 23. $3ax^2$

Prob. 25. $2ax$

Prob. 27. $2rs$

Prob. 29. $3rs^2x^2 - r$

Prob. 31. $4(b-1)N^3 - \frac{2N}{b}$

Prob. 33. $a^3 - 3at^2$

Prob. 35. $V_0\gamma$

Prob. 37. $1 - \frac{2N}{K}$

Prob. 39. $2rN - 3\frac{r}{K}N^2$

Prob. 41. $4\frac{2\pi^5}{15}\frac{k^4}{c^2h^3}T^3$

Prob. 43. $-191x + y - 377 = 0$

Prob. 45. $8x - \sqrt{2}y - 18 = 0$

Prob. 47. $y - 61x + 96 = 0$

Prob. 49. $3y + x + 4 = 0$

Prob. 51. $-x + 24y - 73\sqrt{3} = 0$

Prob. 53. $\sqrt{3}y - \frac{1}{2}x + \frac{5}{2} = 0$

Prob. 55. $2ax - y - a = 0$

Prob. 57. $(a^2+2)y - 4ax + 4a = 0$

Prob. 59. $\frac{1}{3a}x + y + a + \frac{1}{3a} = 0$

Prob. 61. $2a(a+1)y + \frac{1}{2}(a+1)^2x - 8a^2 - (a+1)^2 = 0$

Prob. 63. $(0, 0)$

Prob. 65. $\left(\frac{3}{2}, \frac{9}{4}\right)$

Prob. 67. $(0, 0)$ and $\left(\frac{2}{9}, -\frac{4}{243}\right)$

Prob. 69. $(0, 0)$, $\left(-\frac{1}{2}, -\frac{17}{96}\right)$, and $\left(4, -\frac{160}{3}\right)$

Prob. 71. $(0, 4)$; only point

Prob. 73. $\left(-\frac{1}{2}, \frac{15}{4}\right)$; only point

Prob. 75. $\left(\frac{1}{3}\sqrt{3}, \frac{7}{9}\sqrt{3}+2\right)$, $\left(-\frac{1}{3}\sqrt{3}, -\frac{7}{9}\sqrt{3}+2\right)$

Prob. 77. Tangent line: $y = 2x - 1$

Prob. 79. $y = 2ax - a^2$ and $y = -2ax - a^2$

Prob. 81. $P'(x)$ is a polynomial of degree 3.

4.3 Product Rule and Quotient Rules

Prob. 1. Write $f(x) = g(x)h(x)$ with $g(x) = 2x - 1$ and $h(x) = 2 - x^2$. Then $f'(x) = g'(x)h(x) + g(x)h'(x) = 2(2-x^2) + (2x-1)(-2x) = 4 + 2x - 6x^2$.

Prob. 3. $\frac{d}{dx}((x^3+17)(3x-14x^2)) = 3x^2(3x-14x^2) + (x^3+17)(3-28x) = 51 - 476x + 12x^3 - 70x^4$.

Prob. 5. $\frac{d}{dx}((\frac{1}{2}x^2-1)(2x+3x^2)) = x(2x+3x^2) + (\frac{1}{2}x^2-1)(2+6x) = -2 - 6x + 3x^2 + 6x^3$.

Prob. 7. $\frac{d}{dx}\frac{(x-1)(x+1)}{5} = \frac{1}{5}(1(x+1)+(x-1)1) = \frac{2x}{5}$.

Prob. 9. $\frac{d}{dx}(3x-1)^2 = \frac{d}{dx}(3x-1)(3x-1) = 3(3x-1)+(3x-1)3 = 6(3x-1)$.

Prob. 11. We write $f(x) = 3uv$ where $u = v = 1 - 2x$. Since the two factors of the same, $uv' = u'v = 2uu'$ and we get $f'(x) = 3(2(1-2x) \times (-2)) = -12(1-2x)$.

Prob. 13. $\frac{d}{ds}(2s^2-5s)^2 = 2(2s^2-5s)\frac{d}{ds}(2s^2-5s) = 2(2s^2-5s)(4s-5)$.

Prob. 15. $\frac{d}{dt}(3(2t^2-5t^4)^2) = 3(2(2t^2-5t^4)(4t-20t^3))$.

Prob. 17. $f'(x) = -2 - 6x + 9x^2$, so $f'(1) = 1$ and $f(1) = 0$. The tangent at 1 is given by $y = f(1) + f'(1)(x-1)$, that is, $y = x - 1$.

Prob. 19. $f'(x) = -8(-6 + 9x^2 - 16x^3 + 12x^5)$, so $f'(-1) = -56$ and $f(-1) = -8$. The tangent at -1 is given by $y = f(-1) + f'(-1)(x - (-1))$, that is, $y = -56x - 64$.

Prob. 21. $f'(x) = 3x^2 - 2x - 2$, so $f'(2) = 6$ and $f(2) = 2$. The normal at 2 is given by $y = f(2) - (1/f'(2))(x - 2)$, that is, $y = (14 - x)/6$.

Prob. 23. $f'(x) = -5(4x + 1)$, so $f'(0) = -5$ and $f(0) = 2$. The normal at 0 is $y = 2 + x/5$.

Prob. 25. $f(x) = uvw$, with $u = 2x - 1$, $v = 3x + 4$, $w = 1 - x$. We have $(uvw)' = u'(vw) + u(vw)' = u'vw + uv'w + uvw'$. Therefore $f'(x) = 2(3x+4)(1-x) + (2x-1)3(1-x) + (2x-1)(3x+4)(-1) = -18x^2 + 2x + 9$.

Prob. 27. $f'(x) = (\frac{d}{dx}(x - 3))(2x^2 + 1)(1 - x^2) + (x - 3)(\frac{d}{dx}(2x^2 + 1)) \times (1 - x^2) + (x - 3)(2x^2 + 1)(\frac{d}{dx}(1 - x^2)) - 10x^4 + 24x^3 + 3x^2 - 6x + 1$.

Prob. 29. Because a is a constant ("positive" is irrelevant), we can treat it as a number, taking the derivative of $(x - 1)(2x - 1)$ and multiplying by a. The result is $f'(x) = a(4x - 3)$.

Prob. 31. Because a is a constant ("positive" is irrelevant), the only product we need to treat using the product rule for derivatives is $x^2 - a$ times itself. We get $f'(x) = 2a(2(x^2 - a)(2x)) = 8ax(x^2 - a)$.

Prob. 33. $g'(t) = 2a(at + 1)$.

Prob. 35. $(fg)' = f'g + fg'$; in particular, at the point 2, $(fg)'(2) = f'(2)g(2) + f(2)g'(2)$. Substituting the known values we get $(fg)'(2) = 3 + 8 = 11$.

Prob. 37. $dy/dx = 2(\frac{dx}{dx}f(x) + x\frac{df(x)}{dx}) = 2(f(x) + xf'(x))$.

Prob. 39. $dy/dx = -5(\frac{dx^3}{dx}f(x) + x^3\frac{df(x)}{dx}) - 2 = -5(3x^2f(x) + x^3f'(x)) - 2 = -5x^2(3f(x) + xf'(x)) - 2$.

Prob. 41. $dy/dx = 3(f'(x)g(x) + f(x)g'(x))$.

Prob. 43. $dy/dx = (f'(x) + 2g'(x))g(x) + (f(x) + 2g(x))g'(x) = f'(x)g(x) + f(x)g'(x) + 4g'(x)g(x)$.

Prob. 45. The absolute growth rate is B times the specific growth rate: $Bg(B)$. Note that B is itself a function of t, but the growth rates (absolute and specific) are being thought of as functions of B alone, and t plays no real role in this problem. We are being asked to relate the slope of the graph of the absolute growth rate $Bg(B)$ versus B (at the point $B = 0$) with the specific growth rate (again at the point $B = 0$).

The slope of the graph of $Bg(B)$ is, of course, the derivative of $Bg(B)$ with respect to B, which, by the product rule, equals $g(B) + Bg'(B)$. When $B = 0$, this reduces to $g(B)$, as was to be shown.

Prob. 47. Since $f(N) = r(aN - N^2)(1 - N/K)$, we apply the product rule (treating r, a and K as constants) and get $f'(N) = r((a - 2N) \times (1 - N/K) + (aN - N^2)(-1/K))$.

Prob. 49. Write $f(x) = g(x)/h(x)$ with $g(x) = 2x + 1$ and $h(x) = x + 1$. Then $f'(x) = (g'(x)h(x) - g(x)h'(x))/h'(x)^2 = (2(x+1) - (2x+1))/(x+1)^2 = 1/(x+1)^2$.

Prob. 51. Here $g(x) = 1 - 2x^2$ and $h(x) = 1 - x$. Then $f'(x) = (g'(x)h(x) - g(x)h'(x))/h'(x)^2 = (-4x(1 - x) - (1 - 2x^2)(-1))/(1 - x)^2 = (2x^2 - 4x + 1)/(1 - x)^2$.

Prob. 53. $f'(x) = \frac{2x^3 - 3x^2 + 3}{(x-1)^2}$.

Prob. 55. $h'(t) = \frac{t^2 + 2t - 4}{(t+1)^2}$.

Prob. 57. $f'(x) = 2(x^2 - 2x + 2))/(1 - x)^2$.

Prob. 59. $\frac{d}{dx}(\sqrt{x}(x - 1)) = (\frac{d}{dx}\sqrt{x})(x - 1) + \sqrt{x}\frac{d}{dx}(x - 1) = \frac{x-1}{2\sqrt{x}} + \sqrt{x} = \frac{3x-1}{2\sqrt{x}}$.

Prob. 61. $\frac{d}{dx}(\sqrt{3x}(x^2 - 1)) = \sqrt{3}\frac{d}{dx}(\sqrt{x}(x^2 - 1)) = \sqrt{3}\Big((\frac{d}{dx}\sqrt{(}x^2 - 1)) + \sqrt{x}\frac{d}{dx}(x^2 - 1)\Big) = \sqrt{3}(\frac{x^2-1}{2\sqrt{x}} + 2x\sqrt{x}) = \sqrt{3}\frac{5x^2-1}{2\sqrt{x}}$.

Prob. 63. Since $1/x^3 = x^{-3}$, we have $f'(x) = 3x^2 - (-3x^{-4}) = 3(x^2 + x^{-4})$.

Prob. 65. We can treat the fraction term either as a quotient or as a product (of $3x - 1$ with x^{-3}). We choose to do the latter. Then $f'(x) = 4x - (3x^{-3} + (3x - 1)(-3)x^{-4}) = (4x^5 + 6x - 3)/x^4$. For yet another approach see exercise 66.

Prob. 67. Using the fact that derivatives of $s^{1/3}$ and $s^{2/3}$ are respectively $\frac{1}{3}s^{-2/3}$ and $\frac{2}{3}s^{-2/3}$, we have

$$g'(s) = \frac{\frac{1}{3}s^{-2/3}(s^{2/3} - 1) - \frac{2}{3}s^{-1/3}(s^{1/3} - 1)}{(s^{2/3} - 1)^2}.$$

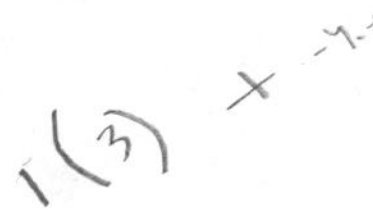

The numerator simplifies to $-\frac{1}{3}(s^{1/3}-1)s^{-2/3}$. An even simpler expression for the derivative can be obtained if we observe that the original function is in fact equal to $1/(s^{1/3}+1)$, and treat this as a quotient of the constant function 1 by $s^{1/3}-1$. The derivative then turns out to be $\frac{1}{3}(s^{1/3}+1)^{-2}s^{-2/3}$. (This is of course equivalent to the result previously found.)

Prob. 69. We write $\sqrt{2x}$ as $\sqrt{2}x^{1/2}$, with derivative $(1/\sqrt{2})x^{-1/2}$, and we write $2/\sqrt{x}$ as $2x^{-1/2}$, with derivative $-x^{-3/2}$. Applying the product rule then gives $f'(x) = -x^{-3/2} + (\frac{\sqrt{2}}{2} - 2)x^{-1/2} - 3\sqrt{2}x^{1/2}$.

Prob. 71. The given function is $f(x) = \frac{x^2+3}{x^3+5}$. We want to find the tangent line at $x = -2$. Now, $f(-2) = \frac{4+3}{-8+5} = -\frac{7}{3}$. Differentiating $f(x)$ using the quotient rule, we find,

$$f'(x)\frac{2x(x^3+5)-3x^2(x^2+3)}{(x^3+5)^2} = \frac{2x^4+10x-3x^4-9x^2}{(x^3+5)^2}$$

Evaluating the first derivative at $x = -2$, we find $f(-2) = -8$. Hence, the equation of the tangent line is

$$\begin{aligned} y \quad + \quad \frac{7}{3} &= (-8)(x+2) \\ y \quad &= \quad -8x - \frac{55}{3} \end{aligned}$$

Prob. 73. $f'(x) = (15-4x)/x^4$, so $f'(2) = \frac{7}{16}$ and $f(2) = -\frac{1}{8}$. The tangent at 2 is given by $y = f(2) + f'(2)(x-2)$, that is, $y = -1 + \frac{7}{16}x$.

Prob. 75. We treat a as a number and apply the quotient rule. The result is $3a/(x+3)^2$.

Prob. 77. An application of the quotient rule gives $f'(x) = 8ax/(4+x^2)^2$.

Prob. 79. We differentiate with respect to R, treating k as a constant. The result is $(2R(k+R) - R^2)/(k+R)^2 = R(2k+R)/(k+R)^2$.

Prob. 81. We write $h(t) = \sqrt{a}\sqrt{t}(t-a) + at$. Treating a as a number and applying the power rule to $\sqrt{t}$ and then the product rule, we get

$$h'(t) = \sqrt{at} + \frac{\sqrt{a}(t-a)}{2\sqrt{t}} + a.$$

Prob. 83. We have $(f/2g)' = \frac{1}{2}(f/g)' = \frac{1}{2}(f'g - fg')/g^2$; specializing to the point 2, we get

$$(f/2g)'(2) = \frac{1}{2}(1 \cdot 3 - (-4) \cdot (-2))/3^2 = -\frac{5}{18}.$$

Prob. 85. We apply the quotient rule with numerator $g(x) = x^2 + 4f(x)$ and denominator $h(x) = f(x)$. The result is $(g'(x)h(x) - g(x)h'(x))/h(x)^2 = ((2x + 4f'(x))f(x) - (x^2 + 4f(x))f'(x))/f(x)^2 = (2xf(x) - x^2 f'(x))/f(x)^2$. (Some time could have been saved by observing in the beginning that $y = x^2/f(x) + 4$.)

Prob. 87. We observe that $y = 1 - \frac{x}{f(x)+x}$, so

$$y' = -\frac{1(f(x) + x) - x(f'(x) + 1)}{(f(x) + x)^2} = -\frac{f(x) - xf'(x)}{(f(x) + x)^2}.$$

Prob. 89. The derivative of the denominator is $2g(x)g'(x)$. Therefore

$$y' = \frac{f'(x)g(x)^2 - 2f(x)g(x)g'(x)}{g(x)^4}.$$

Prob. 91. The derivative of $f(x)g(x)$ is $f'(x)g(x) + f(x)g'(x)$. Applying the product rule again to the factors $\sqrt{x}$ and $f(x)g(x)$ we get $y' = 1/(2\sqrt{x})f(x)g(x) + \sqrt{x}(f'(x)g(x) + f(x)g'(x))$.

Prob. 93. We write $y = c/x$ and take the derivative $y' = -c/x^2$. The tangent at the point (x_1, y_1) is given by the equation $y = y_1 + y'(x_1) \times (x - x_1)$, or

$$y = y_1 - \frac{c(x - x_1)}{x_1^2} = \frac{c(2x_1 - x)}{x_1^2},$$

where for the second equality we used the fact that $y_1 = c/x_1$. The intersection of the equation with the x-axis happens when $y = 0$, and solving for x gives $x = 2x_1$, which does not depend on c. (Note that each pair (x_1, y_1) belongs to only one hyperbola, and the problem does not state which variable is to be regarded as the independent variable as c varies. If y_1 were the independent variable, the x-value of the intersection would be $x = 2c/y_1$, which does depend on c.)

$$\frac{(4+x^2)(2ax) - (ax^2)(2x)}{(4+x^2)^2}$$

4.4 The Chain Rule and Higher Derivatives

Prob. 1. Let $u = x - 2$. By the chain rule,

$$f'(x) = \frac{d}{dx}u^2 = 2u\frac{du}{dx} = 2(x-2)$$

Prob. 3. Let $u = 1 - 3x^2$. By the chain rule,

$$f'(x) = \frac{d}{dx}u^4 = 4u^3\frac{du}{dx} = 4u^3 \cdot (-6x) = -24x(1-3x^2)^3.$$

Prob. 5. Let $u = x^2 + 3$. By the chain rule,

$$f'(x) = \frac{d}{dx}\sqrt{u} = \frac{1}{2\sqrt{u}}\frac{du}{dx} = \frac{1}{2\sqrt{u}} \cdot (2x) = \frac{x}{\sqrt{x^2+3}}.$$

Prob. 7. Let $u = 3 - x^3$. By the chain rule,

$$f'(x) = \frac{d}{dx}\sqrt{u} = \frac{1}{2\sqrt{u}}\frac{du}{dx} = \frac{1}{2\sqrt{u}} \cdot (-3x^2) = -\frac{3x^2}{2\sqrt{3-x^3}}.$$

Prob. 9. Let $u = x^3 - 2$. By the chain rule,

$$f'(x) = \frac{d}{dx}\frac{1}{u^4} = -\frac{4}{u^5}\frac{du}{dx} = -\frac{4}{u^5} \cdot 3x^2 = -\frac{12x^2}{(x^3-2)^5}.$$

Prob. 11. Let $u = 2x^2 - 1$. Using the chain rule and the quotient rule, we have

$$\begin{aligned} f'(x) &= \frac{d}{dx}\frac{3x-1}{\sqrt{u}} \\ &= \frac{\sqrt{u} \cdot 3 - (3x-1) \cdot \frac{1}{2\sqrt{u}}\frac{du}{dx}}{(\sqrt{u})^2} \\ &= \frac{3u - \frac{1}{2}(3x-1)(4x)}{u^{\frac{3}{2}}} \\ &= \frac{3(2x^2-1) - 2x(3x-1)}{u^{\frac{3}{2}}} \\ &= \frac{2x-3}{(2x^2-1)^{\frac{3}{2}}}. \end{aligned}$$

Prob. 13. Let $u = 2x - 1$, $v = x - 1$. Using the chain rule and the quotient rule, we have

$$\begin{aligned} f'(x) &= \frac{d}{dx}\frac{\sqrt{u}}{v^2} \\ &= \frac{v^2 \cdot \frac{1}{2\sqrt{u}}\frac{du}{dx} - \sqrt{u} \cdot 2v\frac{dv}{dx}}{v^4} \\ &= \frac{\frac{v}{2}\frac{du}{dx} - 2u\frac{dv}{dx}}{v^3\sqrt{u}} \\ &= \frac{(x-1) - 2(2x-1)}{v^3\sqrt{u}} \\ &= \frac{-3x+1}{(x-1)^3\sqrt{2x-1}} \end{aligned}$$

Prob. 15. Let $u = s + \sqrt{s}$. By the chain rule,

$$f'(s) = \frac{d}{ds}\sqrt{u} = \frac{1}{2\sqrt{u}}\frac{du}{ds} = \frac{1}{2\sqrt{s+\sqrt{s}}}\left(1 + \frac{1}{2\sqrt{s}}\right).$$

Prob. 17. Let $u = t/(t-3)$. Using the chain rule and the quotient rule, we have

$$g'(t) = \frac{d}{dt}u^3 = 3u^2\frac{du}{dt} = 3u^2 \cdot \frac{(t-3)-t}{(t-3)^2} = -\frac{9t^3}{(t-3)^4}.$$

Prob. 19. Let $u = r^2 - r$, $v = r + 3r^3$. Using the chain rule and the product rule, we have

$$\begin{aligned} f'(r) &= \frac{d}{dr}(u^3v^{-4}) \\ &= 3u^2\frac{du}{dr} \cdot v^{-4} + u^3\left(-4v^{-5}\frac{dv}{dr}\right) \\ &= u^2v^{-5}\left(3v\frac{du}{dr} - 4u\frac{dv}{dr}\right) \\ &= \frac{(r^2-r)^2}{(r+3r^3)^5} \cdot \left[3(r+3r^3)(2r-1) - 4(r^2-r)(1+9r^2)\right] \end{aligned}$$

Prob. 21. Let $u = 3 - x^4$. By the chain rule,

$$\begin{aligned} h'(x) &= \frac{d}{dx}u^{\frac{1}{5}} = \frac{1}{5}u^{-\frac{4}{5}}\frac{du}{dx} \\ &= \frac{1}{5}u^{-\frac{4}{5}}(-4x^3) \\ &= -\frac{4}{5}x^3(3-x^4)^{-\frac{4}{5}}. \end{aligned}$$

Prob. 23. Notice that

$$f(x) = \sqrt[7]{x^2 - 2x + 1} = \sqrt[7]{(x-1)^2}.$$

Let $u = x - 1$. By the chain rule,

$$f'(x) = \frac{d}{dx}u^{\frac{2}{7}} = \frac{2}{7}u^{-\frac{5}{7}}\frac{du}{dx} = \frac{2}{7}(x-1)^{-\frac{5}{7}}.$$

Prob. 25. Let $u = 3s^7 - 7s$. By the chain rule,

$$\begin{aligned} g'(s) &= \frac{d}{ds}u^{\pi} = \pi u^{\pi-1}\frac{du}{ds} \\ &= \pi u^{\pi-1}(21s^6 - 7) \\ &= 7\pi(3s^7 - 7s)^{\pi-1}(3s^6 - 1). \end{aligned}$$

Prob. 27. Let $u = 3t + 3/t$. By the chain rule,

$$\begin{aligned} h'(t) &= \frac{d}{dt}u^{\frac{2}{5}} = \frac{2}{5}u^{-\frac{3}{5}}\frac{du}{dt} \\ &= \frac{2}{5}u^{-\frac{3}{5}}\left(3 - \frac{3}{t^2}\right) \\ &= \frac{6}{5}\left(3t + \frac{3}{t}\right)^{-\frac{3}{5}}\left(1 - \frac{1}{t^2}\right). \end{aligned}$$

Prob. 29. Let $u = ax + 1$. By the chain rule,

$$\begin{aligned} f'(x) &= \frac{d}{dx}u^3 = 3u^2\frac{du}{dx} \\ &= 3u^2 \cdot a = 3a(ax+1)^2. \end{aligned}$$

Prob. 31. Let $u = k + N$. Then, using the chain rule and the product rule, we have

$$\begin{aligned} g'(N) &= \frac{d}{dN}(bNu^{-2}) \\ &= b\left(u^{-2} + N \cdot -2u^{-3}\frac{du}{dN}\right) \\ &= bu^{-3}(u - 2N) \\ &= \frac{b(k-N)}{(k+N)^3}. \end{aligned}$$

Prob. 33. Let $u = T_0 - T$. Then, by the chain rule,

$$g'(T) = \frac{d}{dT}au^3 - b = 3au^2\frac{du}{dT} = -3a(T_0 - T)^2.$$

Prob. 35.

(a) By the chain rule,

$$\begin{aligned} \frac{d}{dx}[f(x^2+3)] &= f'(x^2+3)\frac{d}{dx}(x^2+3) \\ &= \frac{1}{x^2+3} \cdot 2x \\ &= \frac{2x}{x^2+3}. \end{aligned}$$

(b) By the chain rule,

$$\begin{aligned} \frac{d}{dx}[f(\sqrt{x-1})] &= f'(\sqrt{x-1})\frac{d}{dx}\sqrt{x-1} \\ &= \frac{1}{\sqrt{x-1}} \cdot \frac{1}{2\sqrt{x-1}} \\ &= \frac{1}{2(x-1)}. \end{aligned}$$

Prob. 37. Let $u = f(x)/g(x) + 1$. By the chain rule and the quotient rule,

$$\begin{aligned} \frac{d}{dx}\left(\frac{f(x)}{g(x)} + 1\right)^2 &= \frac{d}{dx}u^2 = 2u\frac{du}{dx} \\ &= 2\left(\frac{f}{g} + 1\right)\frac{gf' - fg'}{g^2}. \end{aligned}$$

Prob. 39. Let $v = g(2x) + 2x$.

$$\begin{aligned}
\frac{d}{dx}\frac{[f(x)]^2}{g(2x)+2x} &= \frac{d}{dx}\frac{f^2}{v} \\
&= \frac{v\frac{d}{dx}f^2 - f^2\frac{dv}{dx}}{v^2} \\
&= \frac{1}{v^2}\left\{v \cdot 2f\frac{df}{dx} - f^2\frac{d}{dx}[g(2x)+2x]\right\} \\
&= \frac{1}{v^2}\left\{2vff' - f^2[g'(2x)\cdot 2 + 2]\right\} \\
&= \frac{2}{[g(2x)+2x]^2}\left\{f(x)f'(x)[g(2x)+2x] - f(x)^2[g'(2x)+1]\right\}.
\end{aligned}$$

Prob. 41. Let $u = x^3 - 3x$, $v = \sqrt{u} + 3x = \sqrt{x^3 - 3x} + 3x$. Using the chain rule,

$$\begin{aligned}
\frac{dy}{dx} &= \frac{d}{dx}v^3 \\
&= 3v^2\frac{d}{dx}(\sqrt{u} + 3x) \\
&= 3v^2\left[\frac{1}{2\sqrt{u}}\frac{d}{dx}(x^3 - 3x) + 3\right] \\
&= 3(\sqrt{x^3 - 3x} + 3x)^2 \cdot \left[\frac{3x^2 - 3}{2\sqrt{x^3 - 3x}} + 3\right]
\end{aligned}$$

Prob. 43. Let $u = 3x^2 - 1$, $v = 1 + u^3 = 1 + (3x^2 - 1)^3$. By the chain rule,

$$\begin{aligned}
\frac{dy}{dx} &= \frac{d}{dx}v^2 \\
&= 2v\frac{d}{dx}(1 + u^3) \\
&= 2v \cdot 3u^2\frac{d}{dx}(3x^2 - 1) \\
&= 36x(3x^2 - 1)^2\left(1 + (3x^2 - 1)^3\right)
\end{aligned}$$

Prob. 45. Let

$$\begin{aligned}
u &= 2x + 1, \quad v = x^3 - 1, \\
w &= 3v^3 - 1 = 3(x^3 - 1)^3 - 1.
\end{aligned}$$

Using the chain rule,

$$\begin{aligned}\frac{dy}{dx}=\frac{d}{dx}\left(\frac{u}{w}\right)^3 &= 3\left(\frac{u}{w}\right)^2\frac{d}{dx}\frac{u}{w}\\ &= \frac{3u^2}{w^4}\left[w\frac{d}{dx}(2x+1)-u\frac{d}{dx}(3v^3-1)\right]\\ &= \frac{3u^2}{w^4}\left[2w-9uv^2\frac{d}{dx}(x^3-1)\right]\\ &= \frac{3u^2}{w^4}(2w-27x^2uv^2)\\ &= 3\left(\frac{2x+1}{3(x^3-1)^3-1}\right)^2\frac{6(x^3-1)^3-2-27x^2(2x-1)(x^3-1)^2}{(3(x^3-1)^3-1)^2}\end{aligned}$$

Prob. 47. In problems **47** to **52**, differentiate the given equations with respect to x regarding y as a function of x. Various differentiation rules (chain, product, quotient) are used when needed without further comment.

$$2x+2y\frac{dy}{dx}=0\Rightarrow\frac{dy}{dx}=-\frac{x}{y}$$

Prob. 49.

$$\frac{3}{4}x^{-\frac{1}{4}}+\frac{3}{4}y^{-\frac{1}{4}}\frac{dy}{dx}=0\Rightarrow\frac{dy}{dx}=-\frac{x^{-\frac{1}{4}}}{y^{-\frac{1}{4}}}=-\sqrt[4]{\frac{y}{x}}$$

Prob. 51.

$$\frac{1}{2\sqrt{xy}}\left(y+x\frac{dy}{dx}\right)=2x\Rightarrow\frac{dy}{dx}=4\sqrt{xy}-\frac{y}{x}$$

Prob. 53. Let us first rewrite the given equation as $x^2=y^2$. Differentiation then yields

$$2x=2y\frac{dy}{dx}\Rightarrow\frac{dy}{dx}=+\frac{x}{y}.$$

Prob. 55. We first find dy/dx using implicit differentiation:

$$2x+2y\frac{dy}{dx}=0\Rightarrow\frac{dy}{dx}=-\frac{x}{y}$$

The slope of the tangent is given by

$$\left.\frac{dy}{dx}\right|_{(4,-3)}=-\frac{4}{-3}=\frac{4}{3},$$

(a) The slope of the normal is thus $-3/4$. Hence, the equation of the tangent line at $(4, -3)$ is

$$y + 3 = \frac{4}{3}(x - 4) \Rightarrow y = \frac{4}{3}x - \frac{25}{3},$$

(b) The equation for the normal line is:

$$y + 3 = -\frac{3}{4}(x - 4) \Rightarrow y = -\frac{3}{4}x.$$

Prob. 57. We first find dy/dx using implicit differentiation:

$$\frac{2x}{25} - \frac{2y}{9}\frac{dy}{dx} = 0 \Rightarrow \frac{dy}{dx} = \frac{9x}{25y}$$

The slope of the tangent is given by

$$\left.\frac{dy}{dx}\right|_{(\frac{25}{3},4)} = \frac{9 \cdot \frac{25}{3}}{25 \cdot 4} = \frac{3}{4},$$

and the slope of the normal is thus $-4/3$.

(a) Hence, the equation of the tangent line at $(25/3, 4)$ is

$$y - 4 = \frac{3}{4}(x - \frac{25}{3}) \Rightarrow y = \frac{3}{4}x - \frac{9}{4},$$

(b) and that of the normal line is

$$y - 4 = -\frac{4}{3}(x - \frac{25}{3}) \Rightarrow y = -\frac{4}{3}x + \frac{136}{9}.$$

Prob. 59.

(a) Differentiating the equation $x^{2/3} + y^{2/3} = 4$ with respect to x yields

$$\frac{2}{3}x^{-\frac{1}{3}} + \frac{2}{3}y^{-\frac{1}{3}}\frac{dy}{dx} = 0 \Rightarrow \frac{dy}{dx} = -\sqrt[3]{\frac{y}{x}}.$$

Thus, the value of dy/dx at $(-1, 3\sqrt{3})$ is

$$-\sqrt[3]{\frac{3\sqrt{3}}{-1}} = \sqrt{3}.$$

(b) The curve $x^{2/3} + y^{2/3} = 4$:

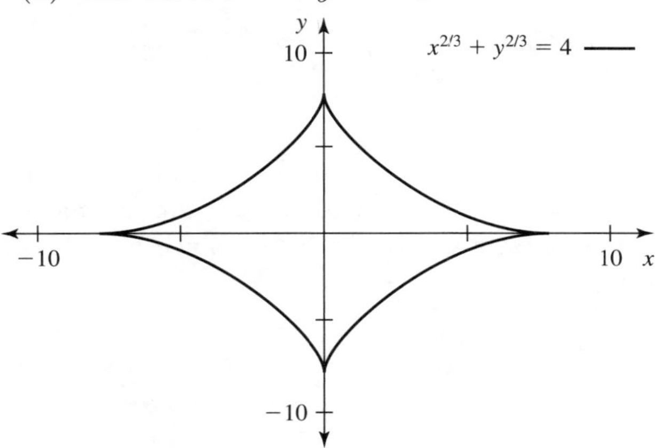

Prob. 61. To simplify notations, in problems **61** to **64**, we use $x', y', \ldots$ to denote derivatives with respect to t. Differentiating $x^2 + y^2 = 1$ with respect to t yields

$$2xx' + 2yy' = 0 \Rightarrow y' = -\frac{xx'}{y}.$$

We already have $x = 1/2$ and $x' = 2$, so we only need to find y. From $x^2 + y^2 = 1$ and the assumption that $y > 0$, we obtain

$$\left(\frac{1}{2}\right)^2 + y^2 = 1 \Rightarrow y = \frac{\sqrt{3}}{2}.$$

Therefore, we have

$$y' = -\frac{\frac{1}{2} \cdot 2}{\frac{\sqrt{3}}{2}} = -\frac{2}{\sqrt{3}}.$$

Prob. 63. Since $x^2y = 1$ and $x = 2$, we have $y = 1/4$. Differentiating $x^2y = 1$ with respect to t yields

$$2xx'y + x^2y' = 0 \Rightarrow y' = -\frac{2x'y}{x} = -\frac{2 \cdot 3 \cdot \frac{1}{4}}{2} = -\frac{3}{4}.$$

Prob. 65. The chain rule gives us

$$\frac{dV}{dt} = \frac{d(x^3)}{dt} = 3x^2\frac{dx}{dt}.$$

Prob. 67. The chain rule gives us

$$\frac{dS}{dt} = \frac{d(4\pi r^2)}{dt} = 8\pi r \frac{dr}{dt}.$$

Prob. 69. Let us express all volumes in m^2. In mathematical terms, the statement that water is drained at a rate of 250 liters per minute means $dV/dt = -0.25$, and the quantity of interest now is dh/dt. Since $V = 25\pi h$, by differentiating with respect to t, we have

$$\frac{dV}{dt} = 25\pi \frac{dh}{dt},$$

and thus,

$$\frac{dh}{dt} = \frac{-0.25}{25\pi} \approx -0.0032.$$

In words, the height is dropping at a rate of about 3.2cm per minute.

Prob. 71. Let x be the distance traveled by the eastbound biker in miles and y be that of the southbound biker. The distance between is $\sqrt{x^2+y^2}$. The rate at which this distance is changing is given by

$$\begin{aligned} \frac{d}{dt}\sqrt{x^2+y^2} &= \frac{1}{2\sqrt{x^2+y^2}}\left(2x\frac{dx}{dt} + 2y\frac{dy}{dt}\right) \\ &= \frac{1}{\sqrt{x^2+y^2}}\left(x\frac{dx}{dt} + y\frac{dy}{dt}\right). \end{aligned}$$

After 20 minutes, $x = 5$ and $y = 6$. Also, we always have $dx/dt = 15$ and $dy/dt = 18$. Thus, the above expression gives us the first answer

$$\frac{1}{\sqrt{5^2+6^2}}(5 \cdot 15 + 6 \cdot 18) \approx 23.4.$$

After 40 minutes, $x = 10$ and $y = 12$, so the second answer is

$$\frac{1}{\sqrt{10^2+12^2}}(10 \cdot 15 + 12 \cdot 18) \approx 23.4$$

In fact, the distance between the two bikers is increasing at a constant rate.

Prob. 73.

$$f'(x) = 3x^2 - 6x, \quad f''(x) = 6x - 6$$

Prob. 75. Note that

$$\frac{x-1}{x+1} = \frac{x+1-2}{x+1} = 1 - \frac{2}{x+1}.$$

Hence we have

$$\begin{aligned} g'(x) &= \frac{2}{(x+1)^2} \\ g''(x) &= -\frac{4}{(x+1)^3}. \end{aligned}$$

Prob. 77.

$$\begin{aligned} g'(t) &= \frac{1}{2\sqrt{3t^3+2t}} \cdot (9t^2+2) \\ &= \frac{1}{2}(3t^3+2t)^{-\frac{1}{2}}(9t^2+2) \\ g''(t) &= -\frac{1}{4}(3t^3+2t)^{-\frac{3}{2}}(9t^2+2)^2 + \frac{1}{2}(3t^3+2t)^{-\frac{1}{2}} \cdot 18t \\ &= \frac{1}{4}(3t^3+2t)^{-\frac{3}{2}}\Big[-(9t^2+2)^2 + 36t(3t^3+2t)\Big] \\ &= \frac{1}{4}(3t^3+2t)^{-\frac{3}{2}} \cdot (27t^4+36t^2-4) \end{aligned}$$

Prob. 79.

$$\begin{aligned} f'(s) &= \frac{1}{2\sqrt{s^{\frac{3}{2}}-1}} \cdot \frac{3}{2}s^{\frac{1}{2}} = \frac{3}{4}\left(\frac{s}{s^{\frac{3}{2}}-1}\right)^{\frac{1}{2}} \\ f''(s) &= \frac{3}{8}\left(\frac{s}{s^{\frac{3}{2}}-1}\right)^{-\frac{1}{2}} \cdot \frac{(s^{\frac{3}{2}}-1) - s\cdot\frac{3}{2}s^{\frac{1}{2}}}{(s^{\frac{3}{2}}-1)^2} = -\frac{3(\frac{1}{2}s^{\frac{3}{2}}+1)}{8\sqrt{s}(s^{\frac{3}{2}}-1)^{\frac{3}{2}}} \end{aligned}$$

Prob. 81.

$$\begin{aligned} g'(t) &= -\frac{5}{2}t^{-7/2} - \frac{1}{2}t^{-1/2} \\ g''(t) &= \frac{35}{4}t^{-9/2} + \frac{1}{4}t^{-3/2} \end{aligned}$$

Prob. 83. By repeated differentiation, the first ten derivatives of x^5 are:

$$f'(x) = 5x^4, \quad f''(x) = 20x^3, \quad f'''(x) = 60x^2, \quad f^{(4)}(x) = 120x,$$
$$f^{(5)}(x) = 120, \quad f^{(6)}(x) = \cdots = f^{(10)}(x) = 0.$$

Prob. 85. Let $p(x) = ax^2 + bx + c$. Its first two derivatives are $p'(x) = 2ax + b$ and $p''(x) = 2a$. Thus we have

$$p(0) = c, \quad p'(0) = b, \quad p''(0) = 2a.$$

The condition $p(0) = 3$, $p'(0) = 2$ and $p''(0) = 6$ means

$$c = 3, \quad b = 2, \quad a = 3,$$

and hence the polynomial is $p(x) = 3x^2 + 2x + 3$.
Prob. 87.

(a) The velocity is

$$v(t) = h'(t) = v_0 - gt,$$

while the acceleration is

$$a(t) = v'(t) = -g.$$

(b) By the result in (a), $v(t) = v_0 - gt = 0$ when

$$t = \frac{v_0}{g}.$$

Before this moment, i.e. when $t < v_0/g$, $v(t)$ is positive, so the object is traveling upward. After this moment, i.e. when $t > v_0/g$, $v(t)$ is negative, so the object is traveling downward.

4.5 Derivatives of Trigonometric Functions

Prob. 1. $2\cos x + \sin x$

Prob. 3. $3\cos x - 5\sin x - 2\sec x \tan x$

Prob. 5. $\sec^2 x + \csc^2 x$

Prob. 7. $3\cos(3x)$

Prob. 9. $6\cos(3x + 1)$

Prob. 11. $4\sec^2 x$

Prob. 13. $4\sec(1+2x)\tan(1+2x)$

Prob. 15. $6x\cos(x^2)$

Prob. 17. $2x\sec(x^2-3)\tan(x^2-3)$

Prob. 19. $6\sin x\cos x$

Prob. 21. $-8x\sin(x^2)$

Prob. 23. $-8\cos x\sin x$

Prob. 25. $-4x\sec^2(1-x^2)$

Prob. 27. $-18\tan^2(3x-1)\sec^2(3x-1)$

Prob. 29. $2x\cos(2x^2-1)/\sqrt{\sin(2x^2-1)}$

Prob. 31. $-\sin s/2\sqrt{\cos s}$

Prob. 33.

$$\frac{2\cos 2t(\cos 6t-1)+6\sin 6t(\sin 2t+1)}{(\cos 6t-1)^2}$$

Prob. 35. Note that

$$f(x)=\frac{\sin(x^2+1)}{\cos(x^2+1)}.$$

Hence we have

$$f'(x)=\frac{2x}{\cos^2(x^2-1)}\Big[\cos(x^2-1)\cos(x^2+1)+\sin(x^2-1)\sin(x^2+1)\Big]$$

Prob. 37. $2\cos(2x-1)\cos(3x+1)-3\sin(2x-1)\sin(3x+1)$

Prob. 39. $6x[\sec^2(3x^2-1)\cot(3x^2+1)-\tan(3x^2-1)\csc^2+1]$

Prob. 41. Note that $f(x)=\tan x$. Hence $f'(x)=\sec^2 x$.

Prob. 43.

$$-\frac{\sec x}{\sec x+\tan x}=\frac{-1}{1+\sin x}$$

Prob. 45.

$$-\frac{6x\cos(3x^2-1)}{\sin^2(3x^2-1)}$$

Prob. 47. Note that $g(x) = \sin(1-5x^2)$. Hence we have

$$g'(x) = -10x\cos(1-5x^2).$$

Prob. 49.

$$-\frac{3(2\sec^2 2x - 1)}{(\tan 2x - x)^2}$$

Prob. 51. Since $h(s) = 1$ is a constant, $h'(s) = 0$.

Prob. 53. $\frac{2(1+x^2)\cos 2x - 2x\sin 2x}{(1+x^2)^2}$

Prob. 55. $-\frac{1}{x^2}\sec^2\frac{1}{x}$.

Prob. 57. Rewrite $f(x)$ as $\cos^2 x/\cos x^2$. Then we have

$$f'(x) = \frac{1}{\cos^2 x^2}\Big[-2\cos x^2 \sin x \cdot \cos x + 2x\sin x^2\cos^2 x\Big].$$

Prob. 59. First, compute

$$\frac{dy}{dx} = \frac{\pi}{3}\cos\left(\frac{\pi}{3}x\right).$$

The tangent is horizontal when $dy/dx = 0$, which is, $\frac{\pi}{3}x = n\pi + \frac{\pi}{2}$, or

$$x = 3n + \frac{3}{2},$$

where n is any integer.

Prob. 61. By definition,

$$\begin{aligned}\frac{d}{dx}\cos x &= \lim_{h\to 0}\frac{\cos(x+h)-\cos x}{h}\\ &= \lim_{h\to 0}\frac{1}{h}\Big[\cos x\cos h - \sin x \sin h - \cos x\Big]\\ &= \cos x \lim_{h\to 0}\frac{\cos h - 1}{h} - \sin x \lim_{h\to 0}\frac{\sin h}{h}.\end{aligned}$$

The limit in the second term is 1, as is discussed earlier in the book. To evaluate the other limit, consider the trick

$$\begin{aligned}\frac{\cos h - 1}{h} &= \frac{\cos h - 1}{h}\cdot\frac{\cos h + 1}{\cos h + 1}\\ &= \frac{\cos^2 h - 1}{h(\cosh + 1)} = -\frac{\sin^2 h}{h(\cos h + 1)}\\ &= -\left(\frac{\sin h}{h}\right)\cdot \sin h \cdot \left(\frac{1}{\cos h + 1}\right).\end{aligned}$$

As $h \to 0$, the first factor approaches 1, as mentioned just above, while the second and the third approach 0 and 1/2 respectively. Hence, the limit of the whole things is 0. Going back to the calculation we start with, this implies we have

$$\frac{d}{dx}\cos x = \cos x \cdot 0 - \sin x \cdot 1 = -\sin x.$$

Prob. 63. By the chain rule, we have

$$\begin{aligned}\frac{d}{dx}\sec x &= \frac{d}{dx}\frac{1}{\cos x} \\ &= -\frac{1}{\cos^2 x}\cdot(-\sin x) \\ &= \frac{1}{\cos x}\cdot\frac{\sin x}{\cos x} \\ &= \sec x \tan x.\end{aligned}$$

Prob. 65.

$$\frac{x}{\sqrt{x^2+1}}\cos\sqrt{x^2+1}$$

Prob. 67.

$$\frac{9x^2+3}{2\sqrt{3x^2+3x}}\cos\sqrt{3x^3+3x}$$

Prob. 69. $4x\sin(x^2-1)\cos(x^2-1)$

Prob. 71. $27x^2\tan^2(3x^3-3)\sec^2(3x^3-3)$

Prob. 73.

(a) $\frac{dc}{dt} = \frac{\pi}{2}\cos(\frac{\pi}{2}t)$

(b) The graph of $c(t)$ and dc/dt:

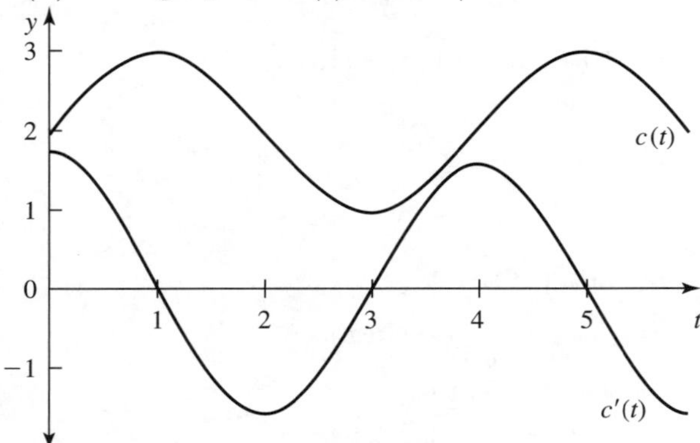

(c) (i) Whenever $c(t)$ reaches a maximum, $dc/dt = 0$.

(ii) When dc/dt is positive, $c(t)$ is increasing.

(iii) In this case, when $dc/dt = 0$, $c(t)$ either reaches a maximum or a minimum. More generally speaking, the value of $c(t)$ is stationary/ the tangent of the graph is horizontal.

4.6 Derivatives of Exponential Functions

Prob. 1. $2e^{2x}$

Prob. 3. $-12e^{1-3x}$

Prob. 5. $(-4x+3)e^{-2x^2+3x-1}$

Prob. 7. $(21x^2 - \frac{1}{2\sqrt{x}})e^{7x^3-\sqrt{x}+3}$

Prob. 9. $(1+x)e^x$

Prob. 11. $(2x-x^2)e^{-x}$

Prob. 13. Using the quotient rule and after simplifications, we have for $f'(x)$:

$$\frac{(1-x)^2e^x - 2x}{(1+x^2)^2}$$

Prob. 15. Using the quotient rule and after simplifications, we have for $f'(x)$:

$$\frac{2(e^x - 1 - e^{-x})}{(2 + e^x)^2}$$

Prob. 17. $\cos x e^{\sin x}$

Prob. 19. $2x \cos(x^2 - 1)e^{\sin(x^2-1)}$

Prob. 21. $e^x \cos(e^x)$

Prob. 23. $(2e^{2x} + 1)\cos(e^{2x} + x)$

Prob. 25. $(1 - \cos x)e^{x - \sin x}$

Prob. 27. $2s \sec s^2 \tan s^2 e^{\sec s^2}$

Prob. 29. $(\sin x + x\cos x)e^{x \sin x}$

Prob. 31. $-3(2x + \sec^2 x)e^{x^2 + \tan x}$

Prob. 33. $(\ln 2)2^x$

Prob. 35. $(\ln 2)2^{x+1}$

Prob. 37. $2(\ln 5)5^{2x-1}$

Prob. 39. $2(\ln 2)x2^{x^2+1}$

Prob. 41. $2(\ln 2)t2^{t^2-1}$

Prob. 43. $\frac{\ln 2}{2\sqrt{x}}2^{\sqrt{x}}$

Prob. 45. $\frac{x \ln 2}{\sqrt{x^2-1}}2^{\sqrt{x^2-1}}$

Prob. 47. $\frac{\ln 5}{2\sqrt{t}}5^{\sqrt{t}}$

Prob. 49. $-2(\ln 2)\sin x 2^{2\cos x}$

Prob. 51. $\frac{\ln 3}{5}r^{-4/5}3^{r^{1/5}}$

Prob. 53.

$$\lim_{h\to 0}\frac{e^{2h} - 1}{h} = \lim_{h\to 0}\frac{e^{2h} - e^0}{h} = \frac{d}{dx}e^{2x}\bigg|_{x=0} = 2e^{2x}|_{x=0} = 2$$

Prob. 55.

$$\lim_{h\to 0+}\frac{e^h-1}{\sqrt{h}}=\lim_{h\to 0+}\frac{e^h-e^0}{h}\cdot\sqrt{h}=\left(\lim_{h\to 0+}\frac{e^h-e^0}{h}\right)\cdot(\lim_{h\to 0+}\sqrt{h})=\frac{d}{dx}e^x\bigg|_{x=0}\cdot 0=0$$

Prob. 57. Let c be the subtangent. We have

$$\frac{2}{c}=\frac{dy}{dx}\bigg|_{x=1}=\frac{d(2^x)}{dx}\bigg|_{x=1}=(\ln 2)2^x|_{x=1}=2\ln 2.$$

Hence, $c=1/\ln 2$.

Prob. 59.

(a) $N(0)=1$

(b)

$$\frac{dN}{dt}=\frac{d(e^{2t})}{dt}=2e^{2t}=2N$$

Prob. 61. The rate of growth is

$$\frac{d}{dt}[N(0)2^t]=N(0)\ln 2\cdot 2^t=(\ln 2)N(t),$$

which is proportional to the population size.

Prob. 63.

(a) Let $a=\frac{K}{N(0)}-1$. We have

$$\frac{dN}{dt}=\frac{d}{dt}\frac{K}{1+ae^{-rt}}=\frac{Kare^{-rt}}{(1+ae^{-rt})^2}$$

(b) Compute

$$rN\left(1-\frac{N}{K}\right)=r\cdot\frac{K}{1+ae^{-rt}}\cdot\left(1-\frac{1}{1+ae^{-rt}}\right)=\frac{rK}{1+ae^{-rt}}\cdot\frac{ae^{-rt}}{1+ae^{-rt}}.$$

From the result in (a), we see that

$$\frac{dN}{dt}=rN\left(1-\frac{N}{K}\right).$$

(c) By the result in (b), the per capita growth rate is

$$\frac{1}{N}\frac{dN}{dt} = r\left(1 - \frac{N}{K}\right).$$

Its graph as a function of N is as below

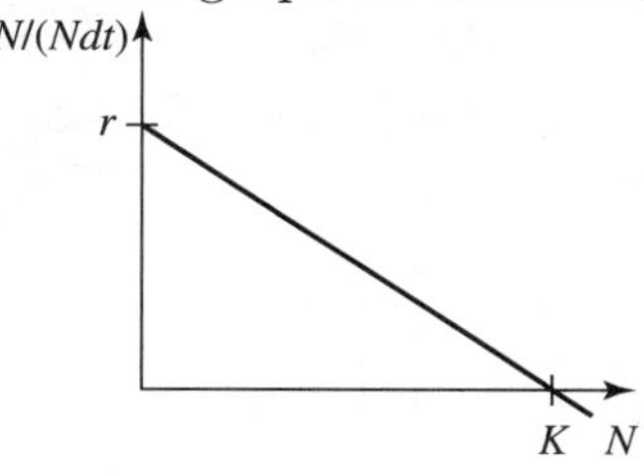

Prob. 65.

(a) The graphs of $L(x) = 10 - 9e^{-x}$ and $L(x) = 10 - 9e^{-0.1x}$:

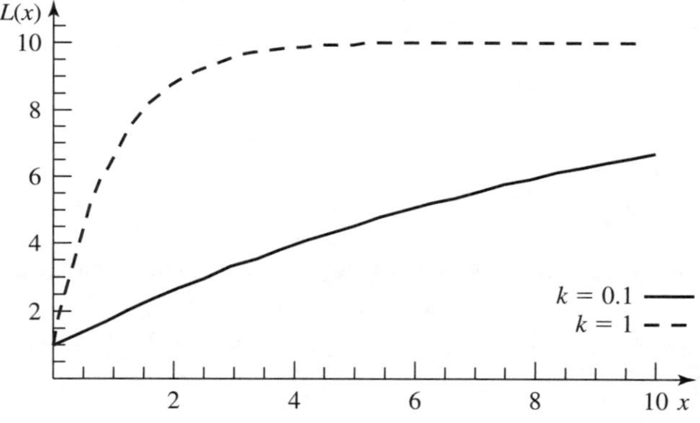

(b) Consider

$$\begin{aligned} L(0) &= L_\infty - (L_\infty - L_0) \cdot 1 \\ &= L_0, \\ L(\infty) &= \lim_{x\to\infty} L(x) \\ &= L_\infty - (L_\infty - L_0) \cdot 0 \\ &= L_\infty. \end{aligned}$$

Hence, L_0 is the length of a fish when it is just born and L_∞ is what its length approaches to when it grows older and older.

(c) The fish reaches $L = 5$ faster for $k = 1.0$.

(d) A straightforward computation shows

$$\begin{aligned}\frac{dL}{dx} &= k(L_\infty - L_0)e^{-kx} \\ &= k(L_\infty - L(x)).\end{aligned}$$

As a fish grows up, its length keeps increasing, thus the quantity on the right hand side is decreasing. Therefore, the rate of growth is also decreasing with age.

Prob. 67. The differential equation is

$$\frac{dW}{dt} = -4W(t)\frac{1}{\text{days}}.$$

Prob. 69. We follow along the lines of the solution to **Prob. 68**. $W(t)$ satisfies the differential equation

$$W'(t) = -\frac{\ln 2}{5}W(t).$$

Prob. 71.

(a) The differential equation implies that $W(t)$ is an exponential function. Its exact form is $W(t) = 6e^{-3t}$. Check that it in deed satisfies the differential equation and the initial condition. Thus, at $t = 4$, the amount of the material left is

$$W(4) = 6e^{-12} \approx 2.5 \times 10^{-3}.$$

(b) The half life is τ if and only if $W(\tau)/W(0) = 1/2$. Since we have

$$\frac{1}{2} = \frac{W(\tau)}{W(0)} = \frac{6e^{-3\tau}}{6} = e^{-3\tau}.$$

Taking logarithm of both sides yields

$$-\ln 2 = -3\tau \quad \Rightarrow \quad \tau = \frac{\ln 2}{3}.$$

Prob. 73.

(a) After every unit time, the value of $W(t)$ drops to 2/5 of its original value. Since $W(0) = 5$, we have for general t

$$W(t) = 5\left(\frac{2}{5}\right)^t.$$

Differentiation gives

$$W'(t) = 5\left(\ln\frac{2}{5}\right)\left(\frac{2}{5}\right)^t.$$

Hence, we have the differential equation $W'(t) = (\ln\frac{2}{5})W(t)$.

(b) By the result in (a), $W(3) = 5(\frac{2}{5})^3 = 0.32$.

(c) The half life τ is determined by

$$\frac{1}{2} = \frac{W(\tau)}{W(0)} = \frac{5(2/5)^\tau}{5} = \left(\frac{2}{5}\right)^\tau.$$

Taking logarithm of both sides gives

$$\ln\frac{1}{2} = \tau\ln\frac{2}{5},$$

i.e. $\tau = \ln\frac{1}{2}/\ln\frac{2}{5} \approx 0.76$.

4.7 Derivatives of Inverse and Logarithmic Functions

Prob. 1. $x = f^{-1}(y) = y^2/2$

(i) $(f^{-1})'(y) = y$

(ii) $f'(x) = 1/\sqrt{2x}$. Hence, $(f^{-1})'(y) = 1/f'(x) = \sqrt{2x} = y$.

Prob. 3. $x = f^{-1}(y) = \sqrt{y/3}$

(i) $(f^{-1})'(y) = \frac{1}{6\sqrt{y/3}}$

(ii) $f'(x) = 6x$. Hence, $(f^{-1})'(y) = 1/f'(x) = \frac{1}{6x} = \frac{1}{6\sqrt{y/3}}$.

Prob. 5. $x = f^{-1}(y) = \sqrt[3]{\frac{3}{2} - \frac{y}{2}}$

(i) $(f^{-1})'(y) = -\frac{1}{6}(\frac{3}{2} - \frac{y}{2})^{-2/3}$

(ii) $f'(x) = -6x^2$. Hence, $(f^{-1})'(y) = 1/f'(x) = -1/6x^2 = -\frac{1}{6}(\frac{3}{2} - \frac{y}{2})^{-2/3}$.

Prob. 7. We have $f'(x) = 4x$, hence $(f^{-1})'(f(x)) = 1/f'(x) = 1/4x$. If we put $x = 1$, we have $(f^{-1})'(0) = (f^{-1})'(f(1)) = 1/4$.

Prob. 9. We have $f'(x) = \frac{1}{2\sqrt{x+1}}$, hence $(f^{-1})'(f(x)) = 1/f'(x) = 2\sqrt{x+1}$. If we put $x = 3$, we have $(f^{-1})'(2) = (f^{-1})'(f(3)) = 4$.

Prob. 11. We have $f'(x) = 1 + e^x$, hence $(f^{-1})'(f(x)) = 1/f'(x) = \frac{1}{1+e^x}$. If we put $x = 0$, we have $(f^{-1})'(1) = (f^{-1})'(f(0)) = 1/2$.

Prob. 13. We have $f'(x) = 1 - \cos x$, hence $(f^{-1})'(f(x)) = 1/f'(x) = \frac{1}{1-\cos x}$. If we put $x = \pi$, we have $(f^{-1})'(\pi) = (f^{-1})'(f(\pi)) = 1/2$.

Prob. 15. We have $f'(x) = 2x + \sec^2 x$, hence $(f^{-1})'(f(x)) = 1/f'(x) = \frac{1}{2x+\sec^2 x}$. If we put $x = 0$, we have $(f^{-1})'(0) = (f^{-1})'(f(0)) = 1$.

Prob. 17. We have $f'(x) = \cos x$, hence $(f^{-1})'(f(x)) = 1/f'(x) = \sec x$. If we put $x = \pi/6$, we have $(f^{-1})'(1/2) = (f^{-1})'(f(\pi/6)) = 2/\sqrt{3}$.

Prob. 19. We have $f'(x) = 5x^4+1$, hence $(f^{-1})'(f(x)) = 1/f'(x) = \frac{1}{5x^4+1}$. If we put $x = 0$, we have $(f^{-1})'(1) = (f^{-1})'(f(0)) = 1$.

Prob. 21. We have $f'(x) = -xe^{-x^2/2} + 2$, hence $(f^{-1})'(f(x)) = 1/f'(x) = \frac{1}{2-xe^{-x^2/2}}$. If we put $x = 0$, we get $(f^{-1})'(1) = (f^{-1})'(f(0)) = 1/2$.

Prob. 23. $1/(x+1)$

Prob. 25. $-2/(1-2x)$

Prob. 27. $2/x$

Prob. 29. $3/x$

Prob. 31. $2\ln x/x$

Prob. 33. $8\ln x/x$

Prob. 35. Note that $f(x) = \frac{1}{2}\ln(1+x^2)$. Its derivative is $x/(x^2+1)$.

Prob. 37. Note that $f(x) = \ln x - \ln(x+1)$. Its derivative is $\frac{1}{x} - \frac{1}{x+1} = \frac{1}{x(x+1)}$.

Prob. 39. Note that $f(x) = \ln(1-x) - \ln(1+2x)$. Its derivative is $\frac{-1}{1-x} - \frac{2}{1+2x} = \frac{-3}{(1-x)(1+2x)}$.

Prob. 41. $(1-\frac{1}{x})e^{x-\ln x} = (1-\frac{1}{x})\frac{e^x}{x}$

Prob. 43. $\cot x$

Prob. 45. $\frac{2x\sec^2 x^2}{\tan x^2} = \frac{2x}{\sin x^2 \cos x^2}$

Prob. 47. $\ln x + 1$

Prob. 49. $(1-\ln x)/x^2$

Prob. 51. $\cos(\ln 3t)/t$

Prob. 53. $2x/(x^2-3)$

Prob. 55. $\frac{-2x}{\ln 10(1-x^2)}$

Prob. 57. $\frac{1}{\ln 10} \cdot \frac{3x^2-3}{x^3-3x}$

Prob. 59. $\frac{1}{\ln 3} \cdot \frac{4u^3}{3+u^4}$

Prob. 61.

(a) Let $f(x) = \ln x$. By definition of the derivative, we have

$$f'(1) = \lim_{h\to 0} \frac{\ln(1+h) - \ln 1}{h} = \lim_{h\to 0} \frac{\ln(1+h)}{h},$$

since $\ln 1 = 0$.

(b) We know that $f'(x) = 1/x$ and hence $f'(1) = 1$. Also, using the property of logarithmic functions, we have $\frac{\ln(1+h)}{h} = \ln(1+h)^{1/h}$. Thus the result in (a) becomes

$$1 = \lim_{h\to 0} \ln(1+h)^{1/h} = \ln\left(\lim_{h\to 0}(1+h)^{1/h}\right),$$

where the second equality comes from the fact ln is a continuous function.

(c) If we let $h = 1/n$, then the limit $h \to 0$ corresponds to $n \to \infty$, and thus the result in (b) becomes

$$\ln\left[\lim_{n\to\infty}\left(1+\frac{1}{n}\right)^n\right] = 1.$$

Applying the function exp to both side will yield the desired result.

Prob. 63. We have $\ln y = x \ln x$, and differentiation yields

$$\frac{y'}{y} = \ln x + 1,$$

or $y' = x^x(\ln x + 1)$.

Prob. 65. We have $\ln y = x \ln(\ln x)$, and differentiation yields

$$\frac{y'}{y} = \ln(\ln x) + \frac{1}{\ln x},$$

or $y' = (\ln x)^{x-1}[\ln x \ln(\ln x) + 1]$.

Prob. 67. We have $\ln y = (\ln x)^2$, and differentiation yields

$$\frac{y'}{y} = \frac{2\ln x}{x},$$

or $y' = 2(\ln x)x^{\ln x - 1}$.

Prob. 69. We have $\ln y = \ln x/x$, and differentiation yields

$$\frac{y'}{y} = \frac{1-\ln x}{x^2},$$

or $y' = x^{\frac{1}{x}-2}(1-\ln x)$.

Prob. 71. We have $\ln y = x^x \ln x$, and thus

$$\ln(\ln y) = x\ln x + \ln(\ln x).$$

Differentiation with respect to x then gives

$$\frac{y'}{y\ln y} = \ln x + 1 + \frac{1}{x\ln x}.$$

Finally, we rewrite y and $\ln y$ in terms of x:

$$y' = x^{x^x + x} \ln x \left(\ln x + 1 + \frac{1}{x \ln x} \right).$$

Prob. 73. We have $\ln y = \cos x \ln x$, and differentiation yields

$$\frac{y'}{y} = -\sin x \ln x + \frac{\cos x}{x}.$$

Thus, $y' = x^{\cos x}(-\sin x \ln x + \frac{\cos x}{x})$.

Prob. 75. First take the logarithm of both sides:

$$\ln y = 2x + 3\ln(9x - 2) - \frac{1}{4}\Big[\ln(x^2 + 1) + \ln(3x^3 - 7) \Big].$$

Then differentiate both sides with respect to x:

$$\frac{y'}{y} = 2 + \frac{27}{9x - 2} - \frac{x}{2(x^2 + 1)} - \frac{9x^2}{4(3x^3 - 7)}.$$

Finally, multiply both sides by y and rewrite y in terms of x. Since the final expression is not particularly enlightening, let us leave it here.

4.8 Approximation and Local Linearity

Prob. 1. $f'(x) = 1/2\sqrt{x}$, hence we have

$$\begin{aligned} \sqrt{65} &\approx \sqrt{64} + \frac{1}{2\sqrt{64}}(65 - 64) \\ &= 8 + 1/16 \approx 8.06, \end{aligned}$$

while $\sqrt{65} = 8.0622\ldots$.

Prob. 3. Let $f(x) = \sqrt[3]{x}$. Then, $f'(x) = \frac{1}{3}x^{-2/3}$, hence we have

$$\sqrt[3]{124} \approx \sqrt[3]{125} + \frac{1}{3}125^{-2/3}(124 - 125) = 5 - 1/75 \approx 4.98,$$

while $\sqrt[3]{124} = 4.9866\ldots$.

Prob. 5. Let $f(x) = x^{25}$. Then, $f'(x) = 25x^{24}$, hence we have

$$0.99^{25} \approx 1^{25} + 25 \cdot 1^{24}(0.99 - 1) = 1 - 0.25 = 0.75$$

while $0.99^{25} = 0.7778\ldots$.

Prob. 7. Let $f(x) = \sin x$. Then, $f'(x) = \cos x$, hence we have

$$\sin\left(\frac{\pi}{2} + 0.02\right) \approx \sin\frac{\pi}{2} + \cos\frac{\pi}{2} \cdot 0.02 = 1 + 0 = 1$$

while $\sin(\frac{\pi}{2} + 0.02) = 0.99998\ldots$.

Prob. 9. Let $f(x) = \ln x$. Then, $f'(x) = 1/x$, hence we have

$$\ln 1.01 \approx \ln 1 + \frac{1}{1}(1.01 - 1) = 0 + 0.01 = 0.01$$

while $\ln 1.01 = 0.009950\ldots$.

Prob. 11. $\frac{1}{1+x} \approx 1 - x$ near $x = 0$

Prob. 13. $\frac{1}{1+x} \approx \frac{1}{2} - \frac{1}{4}(x-1)$ near $x = 1$

Prob. 15. $\frac{1}{(1+x)^2} \approx 1 - 2x$ near $x = 0$

Prob. 17. $\ln(1+x) \approx x$ near $x = 0$

Prob. 19. $\ln x \approx 1 + \frac{1}{e}(x - e)$ near $x = e$

Prob. 21. $e^x \approx 1 + x$ near $x = 0$

Prob. 23. $e^{-x} \approx 1 - x$ near $x = 0$

Prob. 25. $e^{x-1} \approx 1 + (x-1)$ near $x = 1$

Prob. 27. $(1+x)^{-n} \approx 1 - nx$ near $x = 0$

Prob. 29. $\tan x \approx x$ near $x = 0$

Prob. 31. Since $N' = 0.03N$, $N'(4) = 0.03N(4) = 0.03 \times 100 = 3$ and thus

$$\begin{aligned} N(4.1) &\approx N(4) + N'(4)(4.1 - 4) \\ &= 100 + 3 \times 0.1 = 100.3. \end{aligned}$$

Prob. 33. Since $B' = 0.01B$, $B'(1) = 0.01B(1) = 0.01 \times 5 = 0.05$ and thus

$$\begin{aligned} B(1.1) &\approx B(1) + B'(1)(1.1 - 1) \\ &= 5 + 0.05 \times 0.1 = 5.005. \end{aligned}$$

Prob. 35. 2 ± 0.2

Prob. 37. 12 ± 1.2

Prob. 39. $e^2 \pm e^2(0.2)$

Prob. 41. The error in x is $\Delta x = 0.02x = 0.03$, hence the error $f(x)$ is $\Delta f = |f'(1.5)|\Delta x = 0.81$ and its percentage error is 6%.

Prob. 43. The error in x is $\Delta x = 0.02x = 0.4$, hence the error $f(x)$ is $\Delta f = |f'(20)|\Delta x = 0.02$ and its percentage error is 0.67%.

Prob. 45. The assumption means $\Delta r/r = 3\%$. Taking logarithm yields

$$\ln V = \ln(\frac{4}{3}\pi) + 3\ln r.$$

A small change on both sides is

$$\frac{\Delta V}{V} = \Delta(\ln V) = 3\Delta(\ln r) = 3\left(\frac{\Delta r}{r}\right).$$

Hence the accuracy in V is $3 \times 3\% = 9\%$.

Prob. 47. We have $N = kL^{2.11}$ for some constant k. Taking logarithm yields

$$\ln N = \ln k + 2.11 \ln L.$$

A small change on both sides is

$$\frac{\Delta N}{N} = \Delta(\ln N) = 2.11\Delta(\ln L) = 2.11\left(\frac{\Delta L}{L}\right).$$

In order that $\Delta N/N = 5\%$, we must have $\Delta L/L = \frac{1}{2.11} \times 5\% \approx 2.4\%$.

Prob. 49. To make use of the fact that $\Delta(\ln R) = \Delta R/R$, let us first take the logarithm of R:

$$\ln R = \ln k + \ln(a - x) + \ln(b - x).$$

A small change in both sides is

$$\begin{aligned}\frac{\Delta R}{R} &= \Delta(\ln R) \\ &= \Delta\ln(a-x) + \Delta\ln(b-x) \\ &= -\frac{\Delta x}{a-x} - \frac{\Delta x}{b-x}\end{aligned}$$

$$= -x\left(\frac{1}{a-x}+\frac{1}{b-x}\right)\cdot\frac{\Delta x}{x}$$
$$= -\frac{x(a+b-2x)}{(a-x)(b-x)}\cdot\frac{\Delta x}{x}.$$

This gives the relation between the percentage error in R and the percentage error in x.

4.10 Review Problems

Prob. 1. $-12x^3 - x^{-3/2}$

Prob. 3. $-\frac{2}{3}(1-t)^{-2/3}(1+t)^{-4/3}$

Prob. 5. $e^{2x}(2\sin\frac{\pi x}{2} + \frac{\pi}{2}\cos\frac{\pi x}{2})$

Prob. 7. Note that
$$f(x) = \frac{\ln(x+1)}{\ln x}.$$
Hence, an application of the quotient rule gives us:
$$f'(x) = \frac{\frac{1}{x+1}\ln x - \frac{1}{x}\ln(x+1)}{(\ln x)^2}.$$

Prob. 9. $f'(x) = -xe^{-x^2/2}$
$f''(x) = (x^2-1)e^{-x^2/2}$

Prob. 11. $h'(x) = 1/(x+1)^2$
$h''(x) - 2/(x+1)^3$

Prob. 13. In problems **13** to **16**, differentiate the given equation with respect to x and collect terms to express $y' = dy/dx$ in x and y.

$$2xy + x^2y' - 2yy'x - y^2 = \cos x \Rightarrow y' = \frac{\cos x + y^2 - 2xy}{x^2 - 2xy}$$

Prob. 15.

$$\frac{1-y'}{x-y} = 2 \Rightarrow y' = 1 - 2x + 2y$$

Prob. 17. Differentiating both sides with respect to x yields

$$2x + 2yy' = 0 \Rightarrow y' = -\frac{x}{y}.$$

The quotient rule then gives us

$$y'' = \frac{xy' - y}{y^2}$$

Prob. 19. Differentiating both sides with respect to x yields

$$y'e^y = \frac{1}{x} \quad \Rightarrow \quad y' = \frac{e^{-y}}{x}.$$

The quotient rule then gives us

$$y'' = -\frac{(xy'+1)e^{-y}}{x^2}$$

Prob. 21. Let x be the distance the birds have flown from when they were directly overhead, and y be our distance from the birds. We have $dx/dt = 6$ at all times and $y^2 = 100^2 + x^2$. Differentiating this equation with respect to t yields

$$2y\frac{dy}{dt} = 2x\frac{dx}{dt} \Rightarrow \frac{dy}{dt} = \frac{x}{y}\frac{dx}{dt}.$$

When $y = 320$, $x = \sqrt{320^2 - 100^2} \approx 304$, and thus $dy/dt = \frac{304}{320} \cdot 6 = 5.7$.

Prob. 23.

(a) $f'(x)e^{f(x)}$

(b) $f'(x)/f(x)$

(c) $2f(x)f'(x)$

Prob. 25.

(a) The graph of $y = \frac{x^2}{1+x^2}$:

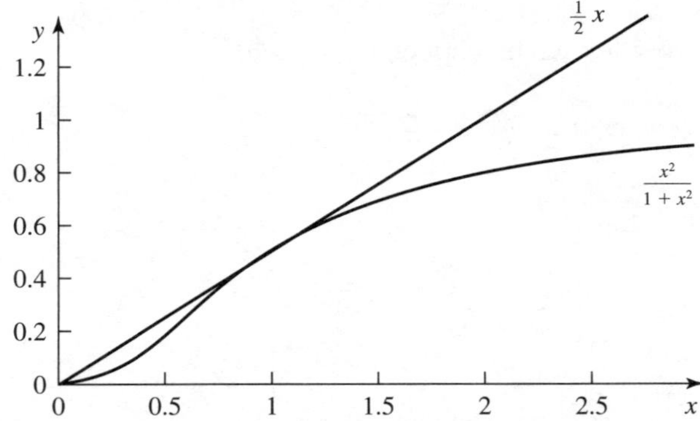

(b) On the one hand, the slope of the line is $f'(c)$. On the other hand, since this line connects $(0,0)$ and $(c, f(c))$, its slope is $f(c)/c$. Thus we have

$$f'(c) = \frac{f(c)}{c},$$

or

$$\frac{2c}{(1+c^2)^2} = \frac{c}{1+c^2}.$$

Solving the equation, we will get $c^2 = 1$, or $c = 1$, since it is assumed that $c > 0$.

This line is also plotted in the above graph.

Prob. 27. $y' = -e^{-x^2}(2x\cos x + \sin x)$. When $x = \pi/3$, it is equal to $-e^{-\pi^2/9}(\frac{\pi}{3} + \frac{\sqrt{3}}{2}) \approx -0.64$. The tangent line is thus

$$y = e^{-\pi^2/9} - 0.64\left(x - \frac{\pi}{3}\right).$$

Prob. 29. Implicit differentiation yields:

$$\ln y + \frac{xy'}{y} = y' \ln x + \frac{y}{x} \Rightarrow y' = \frac{\ln y - y/x}{\ln x - x/y}.$$

When $x = 1$, $y = 1$ and thus the above result tells us $y' = 1$. The tangent line then has the equation

$$y = 1 + (x - 1) = x.$$

Prob. 31. Since $p'(x) = 2ax + b$, $p''(x) = 2a$, the given conditions are

$$\begin{aligned} 6 &= p(-1) = a - b + c \\ 8 &= p'(1) = 2a + b \\ 4 &= p''(0) = 2a, \end{aligned}$$

and the system has the unique solution

$$a = 2, \quad b = 4, \quad c = 8.$$

Prob. 33.

(a) The graph of $s(t)$:

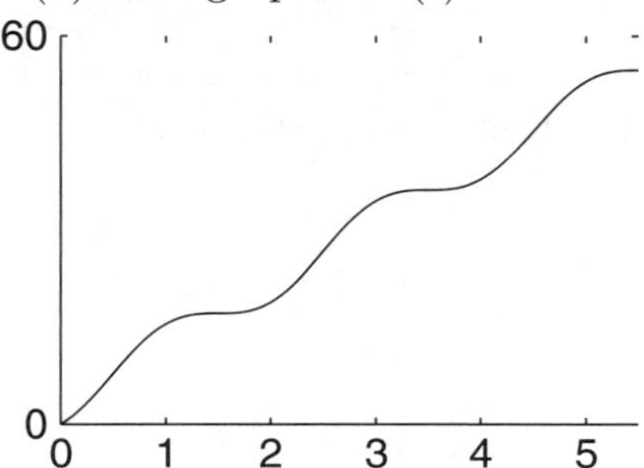

We didn't backtrack because we can see from the graph that $s(t)$ keeps increasing. The distance between the two towns is given by $s(5.5) \approx 17.3$.

(b) $v(t) = s'(t) = 3\pi(1 + \sin \pi t)$ $a(t) = v'(t) = 3\pi^2 \cos \pi t$

(c) The graph of $s(t)$ is shown above. Here are the graphs of $v(t)$

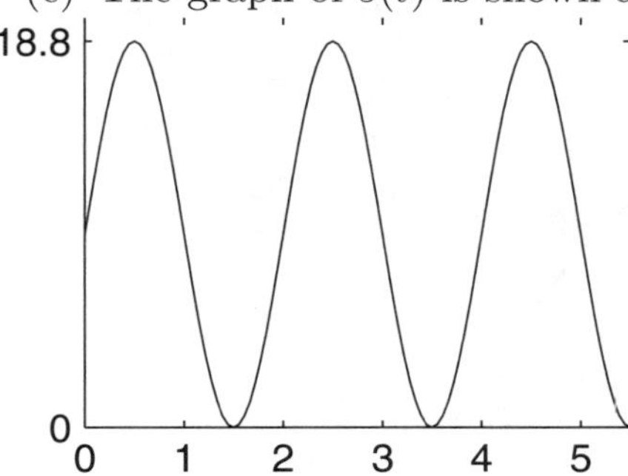

and $a(t)$

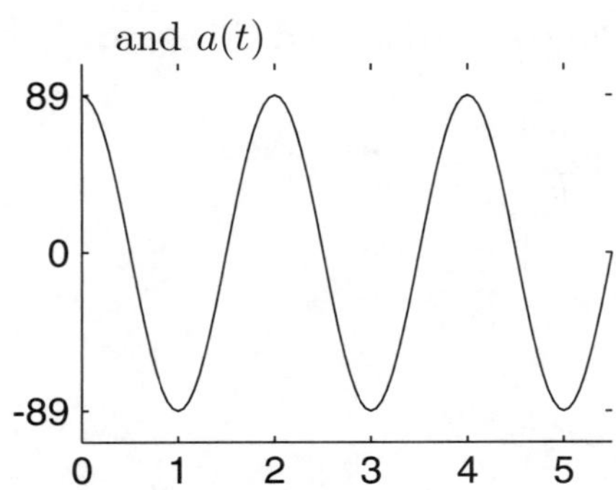

No backtracking means $s(t)$ being an increasing function. This is also equivalent to $v(t) \geq 0$, which is clear from the graph.

(d) At each peak, we switch from slowing down ($a < 0$) to speeding up ($a > 0$). This happens two times, so there are two peaks on the road. Similarly, a changes from positive to negative at each valley, so there are three of them.

Prob. 35.

(a) For the given $N(t)$,

$$\frac{dN}{dt} = -\frac{A\pi}{2T}\sin\frac{\pi t}{2T}.$$

On the other hand, the right hand side of (4.13) equals

$$\frac{\pi}{2T}\left[K-\left(K \quad +A\cos\frac{\pi(t-T)}{2T}\right)\right] = -\frac{\pi}{2T}\cdot A\cos\left(\frac{\pi t}{2T} - \frac{\pi}{2}\right) = -\frac{A\pi}{2T}\sin\frac{\pi t}{2T}.$$

The two computations above show that $N(t)$ satisfies the given differential equation.

(b) The graph of $N(t) = 100 + 50\cos\frac{\pi t}{2}$:

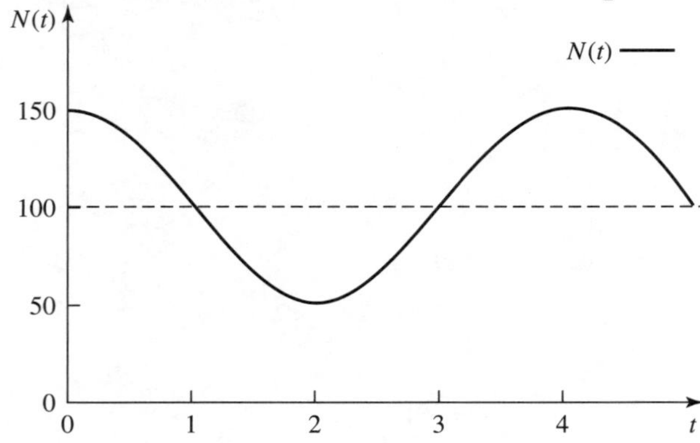

(c) The population grows and shrinks periodically between $K + A$ and $K - A$, with period $4T$.

Prob. 37. The relative errors in S and B are related as follows:

$$\begin{aligned}\frac{\Delta S}{S} &= \Delta \ln S = \Delta \ln[(1.162)B^{0.933}] \\ &= \Delta[\ln 1.162 + (0.933)\ln B] \\ &= 0.933 \cdot \frac{\Delta B}{B}.\end{aligned}$$

Therefore, if we want $\Delta B/B < 10\%$, we must have $\Delta S/S < 0.933 \times 10\% = 0.33\%$.

5 Applications and Differentiation

5.1 Extrema and the Mean Value Theorem

Prob. 1. $y = 2x - 1$, $x \in [0, 1]$:

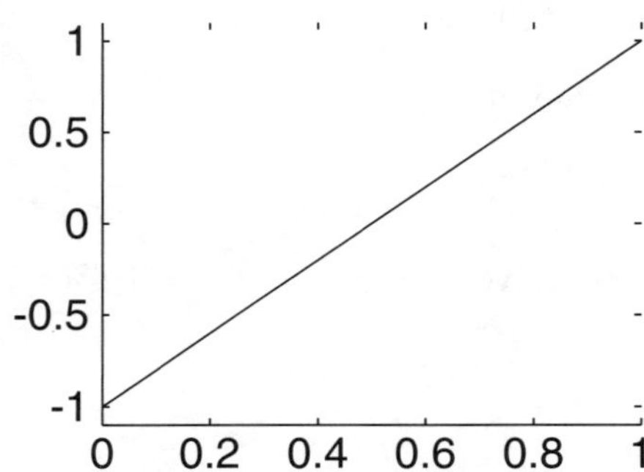

global maximum: $(1, 1)$
global minimum: $(0, -1)$

Prob. 3. $y = \sin x$, $x \in [0, 2\pi]$:

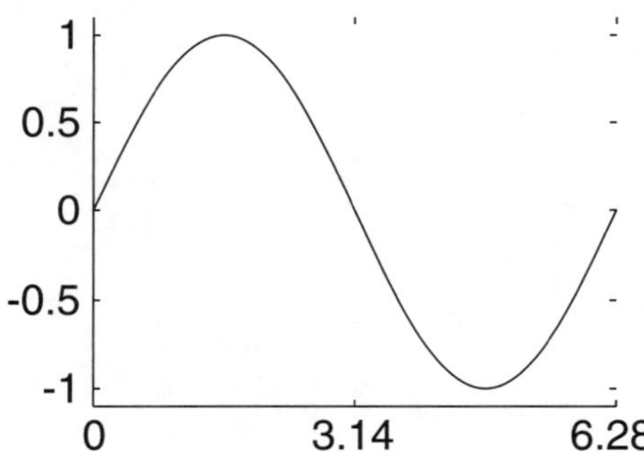

global maximum: $(\pi/2, 1)$
global minimum: $(3\pi/2, -1)$

Prob. 5. $y = |x|$, $x \in [-1, 1]$:

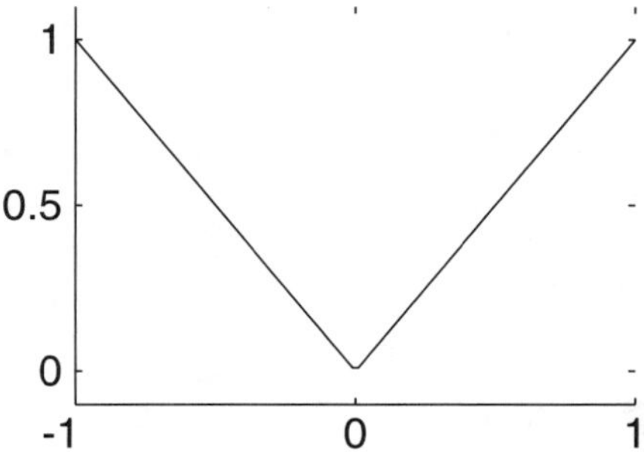

global maxima: $(\pm 1, 1)$
global minimum: $(0, 0)$

Prob. 7. $y = e^{-x}$, $x \in [0, 2]$:

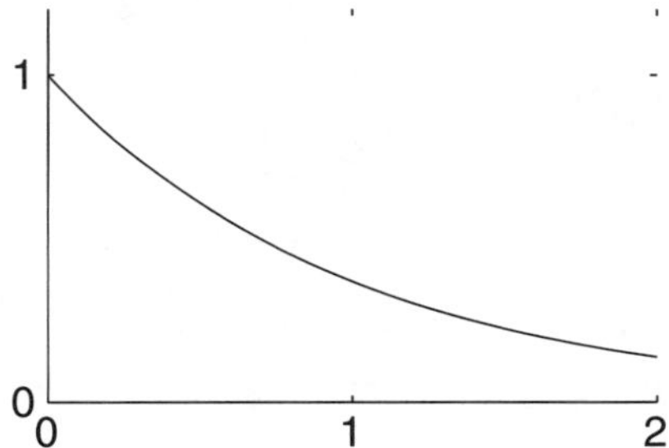

global maximum: $(0, 1)$
global minimum: $(2, e^{-2})$

Prob. 9. An example of a function with the desired properties is $f(x) = \cos(\pi x) - x$:

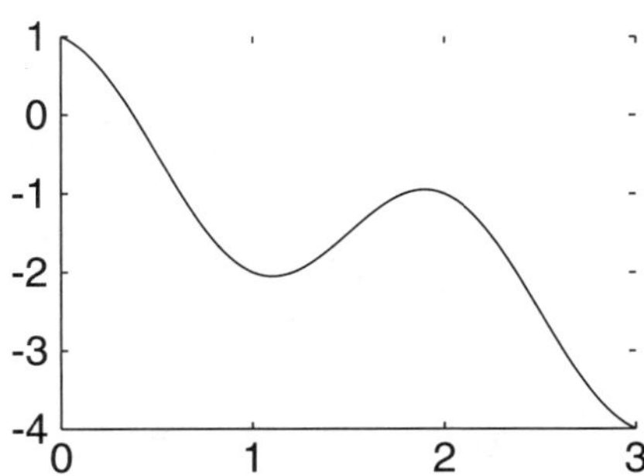

Prob. 11. An example of a function with the desired properties is $f(x) = x^2$:

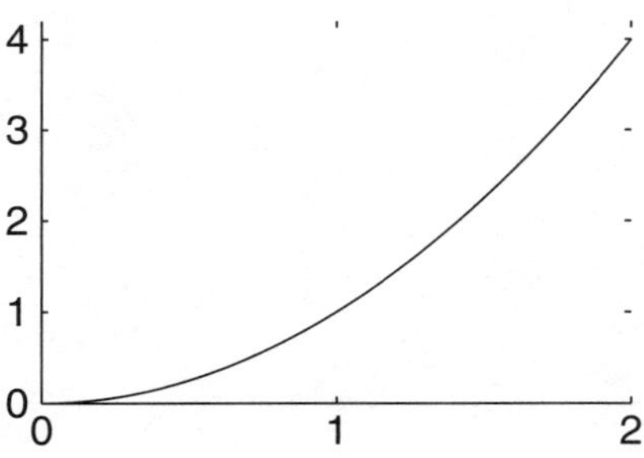

Prob. 13. $y = 2 - x$, $x \in [-1, 3)$:

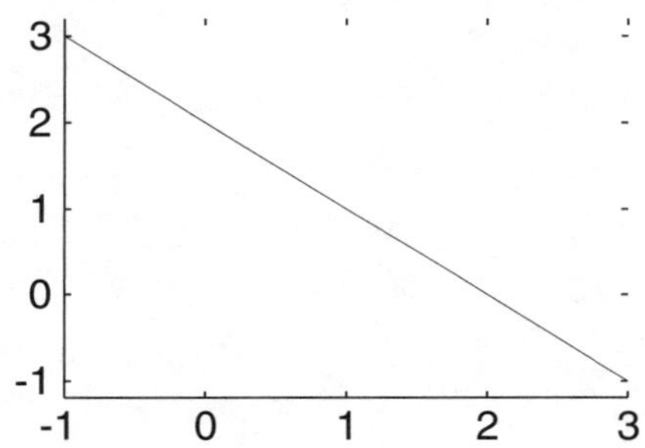

global maximum: $(-1, 3)$
no global minimum

Prob. 15. $y = x^2 - 2$, $x \in [-1, 1]$:

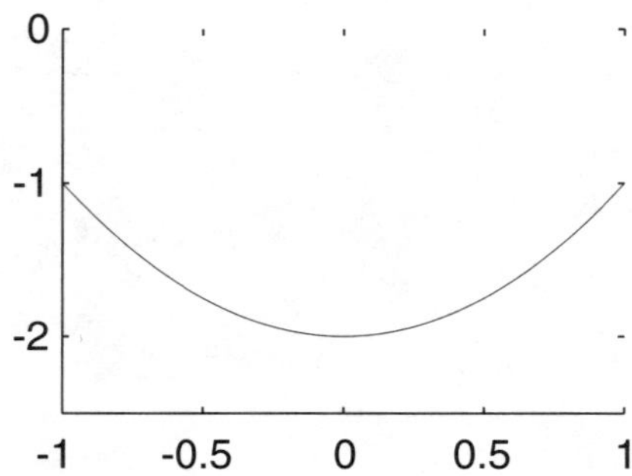

global maxima: $(\pm 1, -1)$
global minimum: $(0, -2)$

Prob. 17. $y = -x^2 + 1$, $x \in [-2, 1]$:

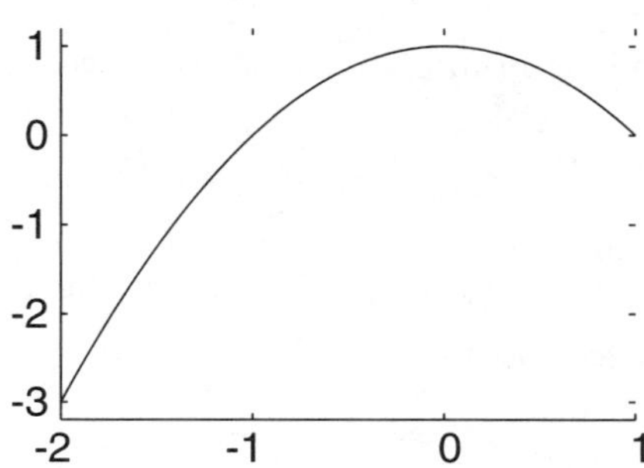

local minimum (1,0)
global maximum: $(0, 1)$
global minimum: $(-2, -3)$

Prob. 19. Since the square of any real number is nonnegative, we have

$$f(x) = x^2 \geq 0 = f(0)$$

for any x. In other words, $x = 0$ is a global minimum, thus in particular a local minimum.

For $f(x) = x^2$, we have $f'(x) = 2x$, and thus $f'(0) = 0$. This is what we expect to happen at the local (global) minimum $x = 0$ according to Fermat's theorem.

Prob. 21. Since the square of any real number is nonnegative, i.e. $x^2 \geq 0$, we have

$$f(x) = -x^2 \leq 0 = f(0)$$

for any x. In other words, $x = 0$ is a global maximum, thus in particular a local maximum.

For $f(x) = -x^2$, we have $f'(x) = -2x$, and thus $f'(0) = 0$. This is what we expect to happen at the local (global) maximum $x = 0$ according to Fermat's theorem.

Prob. 23. For $f(x) = x^3$, we have $f'(x) = 3x^2$, and thus $f'(0) = 0$. However, $x = 0$ is not a local extremum because on its right (i.e. $x > 0$), $f(x) = x^3 > 0 = f(0)$, while its left (i.e. $x < 0$), $f(x) = x^3 < 0 = f(0)$.

Prob. 25. For $f(x) = (x+2)^3$, we have $f'(x) = 3(x+2)^2$, and thus $f'(-2) = 0$. However, $x = -2$ is not a local extremum because on its right (i.e. $x + 2 > 0$), $f(x) > 0 = f(0)$, while its left (i.e. $x + 2 < 0$), $f(x) < 0 = f(0)$.

Prob. 27. The absolute value of any number is nonnegative, so we have

$$f(x) = |x| \geq 0 = f(0),$$

i.e. $x = 0$ is a global, and thus local, minimum.

The graph of $f(x) = |x|$ has an angle at $x = 0$, corresponding to the fact that f is not differentiable there. Here is a formal proof: when $x > 0$, we have $f(x) = |x| = x$, and thus

$$\begin{aligned} & \lim_{x \to 0^+} \frac{f(x) - f(0)}{x - 0} \\ = & \lim_{x \to 0^+} \frac{x}{x} = \lim_{x \to 0^+} 1 = 1. \end{aligned}$$

However, when $x < 0$, we have $f(x) = |x| = -x$, and thus

$$\begin{aligned} & \lim_{x \to 0^-} \frac{f(x) - f(0)}{x - 0} \\ = & \lim_{x \to 0^-} \frac{-x}{x} = \lim_{x \to 0^-} -1 = -1. \end{aligned}$$

Therefore, the derivative

$$f'(0) = \lim_{x \to 0} \frac{f(x) - f(0)}{x - 0}$$

doesn't exist, and f is not differentiable at $x = 0$.

Prob. 29. The absolute value of any number is nonnegative, so we have

$$f(x) = |x^2 - 1| \geq 0 = f(1) = f(-1),$$

i.e. $x = \pm 1$ are global, and thus local, minima.

Let us first consider what happens near $x = 1$. When $x > 1$, $x^2 - 1 > 0$ and thus $f(x) = |x^2 - 1| = x^2 - 1$. Then we have

$$\begin{aligned} \lim_{x \to 1^+} \frac{f(x) - f(1)}{x - 1} &= \lim_{x \to 1^+} \frac{x^2 - 1}{x - 1} \\ &= \lim_{x \to 1^+} (x + 1) \\ &= 2. \end{aligned}$$

When $-1 < x < 1$, $x^2 - 1 = (x+1)(x-1) < 0$ and thus $f(x) = |x^2 - 1| = -x^2 + 1$. So we have

$$\begin{aligned}\lim_{x\to 1^-} \frac{f(x) - f(1)}{x - 1} &= \lim_{x\to 1^-} \frac{-x^2 + 1}{x - 1} \\ &= \lim_{x\to 1^-} -(x+1) \\ &= -2.\end{aligned}$$

Therefore, the derivative

$$f'(1) = \lim_{x\to 1} \frac{f(x) - f(1)}{x - 1}$$

doesn't exist, and f is not differentiable at $x = 1$.

The situation is similar near $x = -1$. When $-1 < x < 1$, $f(x) = -x^2 + 1$ and

$$\lim_{x\to -1^+} \frac{f(x) - f(-1)}{x + 1} = \lim_{x\to -1^+} \frac{-x^2 + 1}{x + 1} = 2.$$

When $x < -1$, $f(x) = x^2 - 1$ and

$$\lim_{x\to -1^-} \frac{f(x) - f(-1)}{x + 1} = \lim_{x\to -1^-} \frac{x^2 - 1}{x + 1} = -2.$$

Therefore, f is not differentiable at $x = -1$.

Prob. 31. The graph of $y = |1 - |x||$ looks like:

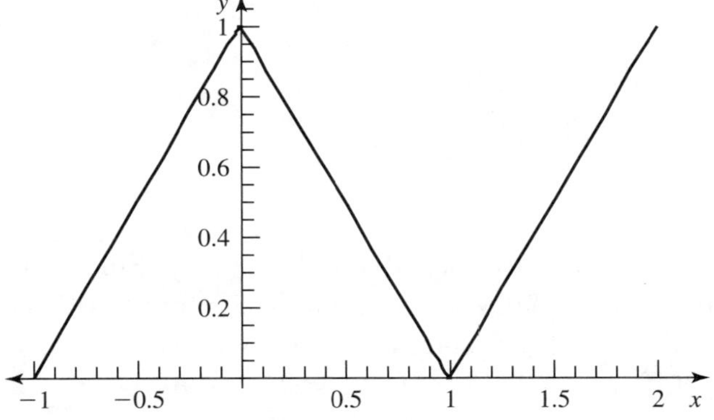

The local maxima are $(0, 1), (2, 1)$ and they are both global. The local minima are $(\pm 1, 0)$ and also, they are both global.

Prob. 33.

(a) The graph of $y = dN/dt = 2N(1 - N/100)$:

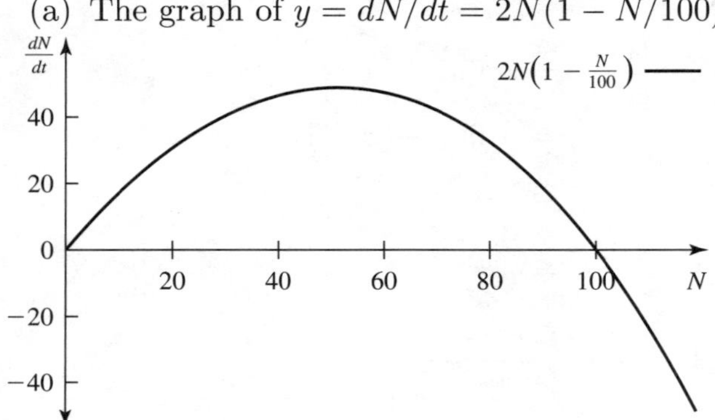

From the graph, we see that the growth rate is maximal when $N = 50$.

(b) Since $f(N)$ is a polynomial in N, it is differentiable throughout any open set in which it is defined. In particular, it is differentiable when $N > 0$. Expanding $f(N) = r(N - \frac{N^2}{K})$, we get

$$f'(N) = r\left(1 - \frac{2}{K}N\right).$$

(c) From the result in (b), we have $f'(N) = 0$ if and only if $N = K/2$. If $K = 100$, this happens at $N = 50$, the same number we obtain in (a).

Prob. 35.

(a) The slope of the line connecting $(0, 0)$ and $(2, 4)$ is

$$\frac{4-0}{2-0} = 2.$$

(b) Notice that the two points in (a) are $(0, f(0))$ and $(2, f(2))$. Since f is continuous on $[0, 2]$ and differentiable throughout $(0, 2)$, by the MVT, there must exist a number $c \in (0, 2)$ such that $f'(c)$ equals the above slope. To find c, we let $f'(c) = 2$. Since $f'(x) = 2x$, we have $2c = 2$, and thus $c = 1$.

Prob. 37. Since $f(x) = x^2$ is differentiable (and continuous) everywhere, and $f(1) = 1 = f(-1)$, by Rolle's theorem, there exists some $c \in (-1, 1)$

such that $f'(c) == 0$,or, the graph of $f(x)$ has a horizontal tangent at $x = c$. Since $f'(x) = 2x$, we have $2c = 0$, or $c = 0$.

Prob. 39. Since $f(x) = x(1 - x)$ is differentiable (and continuous) everywhere, we may apply Rolle's theorem to f, i.e. if $f(a) = f(b)$ for some $a < b$, then there exists $c \in (a, b)$ with $f'(c) = 0$. For example, since $f(0) = 0 = f(1)$, f' vanishes somewhere in $[0, 1]$.

Prob. 41. Since $f(x) = -x^2 + 2$ is differentiable (and continuous) everywhere, by the MVT, there exists some $c \in (-1, 2)$ such that

$$\begin{aligned} f'(c) &= \frac{f(2) - f(-1)}{2 - (-1)} = \frac{-2 - 1}{3} \\ &= -1. \end{aligned}$$

Prob. 43. An example of such a function is $f(x) = x^2$:

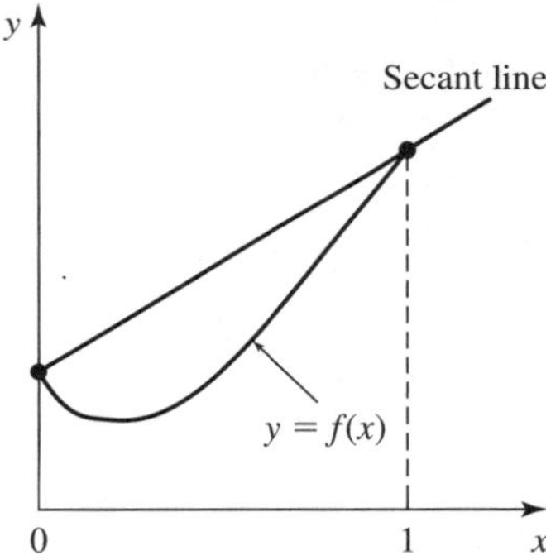

The existence of such a point inside the interval $[0, 1]$ is predicted by the MVT.

Prob. 45.

(a) The slope of the line connecting $(a, f(a))$ and $(b, f(b))$ is

$$\begin{aligned} &\frac{f(b) - f(a)}{b - a} = \frac{b^2 - a^2}{b - a} \\ = &\frac{(a + b)(b - a)}{b - a} = a + b. \end{aligned}$$

(b) Since f is differentiable (and continuous) everywhere, by the mean value theorem, there must exist a number $c \in (a, b)$ such that $f'(c)$ equals the above slope. To find c, let $f'(c) = a + b$. Since $f'(x) = 2x$, we have $2c = a + b$, or $c = (a + b)/2$, i.e. the midpoint in between a and b.

Prob. 47. Since f is not constant on $[a, b]$ and $f(a) = f(b) = 0$, we must have $f(c) \neq 0$ for some $c \in (a, b)$. Assume $f(c) > 0$. (The case $f(c) < 0$ is completely similar.) Note that the MVT also applies to the subintervals $[a, c]$ and $[c, b]$. For the subinterval $[a, c]$, we thus have some $c_1 \in (a, c) \subset (a, b)$ with

$$f'(c_1) = \frac{f(c) - f(a)}{c - a} = \frac{f(c)}{c - a} > 0,$$

since $f(a) = 0$, $f(c) > 0$ and $a < c$. For the subinterval $[c, b]$, we have some $c_2 \in (c, b) \subset (a, b)$ with

$$f'(c_2) = \frac{f(b) - f(c)}{b - c} = \frac{-f(c)}{b - c} < 0,$$

since $f(b) = 0$, $f(c) > 0$ and $b > c$.

Prob. 49.

(a) Between $t = 0$ and $t = 5$, the car moves a distance of $s(5) - s(0) = 1.25$ (meters). Thus, the average velocity is

$$\frac{\text{distance traveled}}{\text{time elapsed}} = \frac{1.25 \text{ meters}}{5 \text{ secs}} = 0.25 \text{ meter per sec.}$$

(b) The instantaneous velocity at time t is given by

$$s'(t) = \frac{3}{100}t^2.$$

(c) From (a) and (b), the instantaneous velocity is equal to the average velocity when $3t^2/100 = 0.25$, or $t = \sqrt{25/3} \approx 2.89$.

Prob. 51. The assumption $|dB/dt| \leq 1$, or

$$-1 \leq B'(t) \leq 1, \qquad 0 \leq t \leq 3,$$

together with Corollary 1 of the MVT, tells us that

$$-1 \cdot (3 - 0) \leq B(3) - B(0) \leq 1 \cdot (3 - 0).$$

Since $B(0) = 3$, this means

$$0 \leq B(3) \leq 6.$$

Prob. 53. Consider any number $x(\neq 2)$. Since f' vanishes identically between 2 and x, Corollary 2 of the MVT implies that f is constant in the interval between 2 and x. Hence, $f(x) = f(2) = 3$.

Prob. 55. The assumption $|f(x) - f(y)| \leq |x - y|^2$ implies

$$\begin{aligned} 0 &\leq \left|\frac{f(x) - f(y)}{x - y}\right| \\ &\leq \frac{|x - y|^2}{|x - y|} = |x - y|. \end{aligned}$$

Regard x as a variable and let it approach y. By the sandwich principle, we have

$$\lim_{x \to y}\left|\frac{f(x) - f(y)}{x - y}\right| = 0,$$

which implies

$$f'(y) = \lim_{x \to y}\frac{f(x) - f(y)}{x - y} = 0.$$

(Fact: $|g(t)| \to 0$ implies $g(t) \to 0$. Proof: use $-|g(t)| \leq g(t) \leq |g(t)|$ and the sandwich principle.) Since y is arbitrary, f' vanishes identically. By Corollary 2 of the MVT, f is a constant function.

5.2 Monotonicity and Concavity

Prob. 1. In problems **1** to **20**, the range of x over which f is increasing (decreasing) is given by solving $f' > 0$ ($f' < 0$), and the range over which f is concave up (down) is given by solving $f'' > 0$ ($f'' < 0$). $f(x) = 3x - x^2$ $f'(x) = 3 - 2x$, $f''(x) = -2$. So f is increasing for $x < 3/2$, decreasing for $x > 3/2$ and concave down for all $x \in \mathbb{R}$

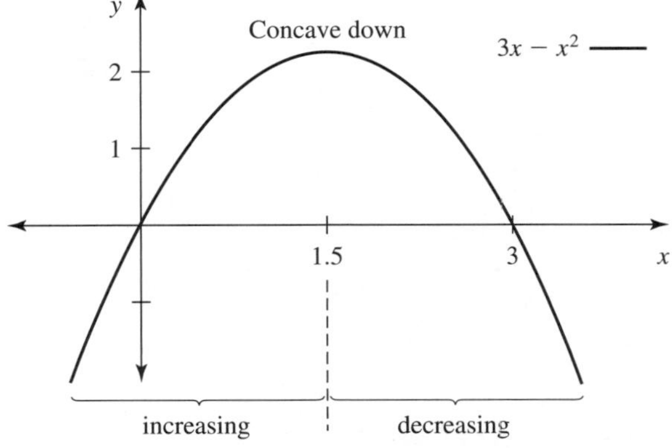

Prob. 3. $f(x) = x^2 + x - 4$ so $f'(x) = 2x + 1$ and $f''(x) = 2$. So f is increasing for $x > -1/2$, decreasing for $x < -1/2$ and concave up for all $x \in \mathbb{R}$

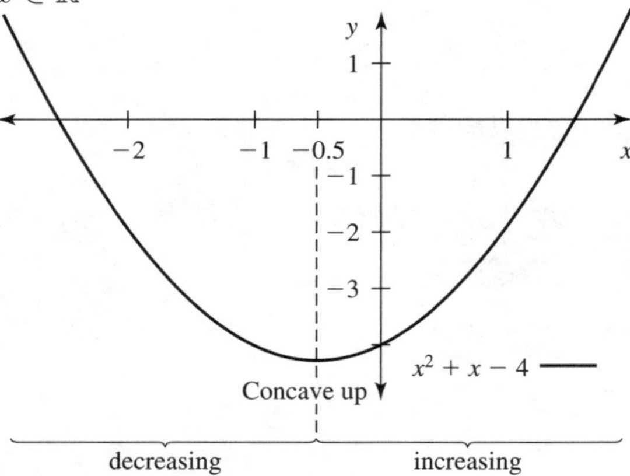

Prob. 5. $f(x) = -\frac{2}{3}x^3 + \frac{7}{2}x^2 - 3x + 4$, so $f'(x) = -2x^2 + 7x - 3 = -(2x-1)(x-3)$ and $f''(x) = -4x + 7$. So f is increasing for $1/2 < x < 3$, decreasing for $x < 1/2$, $x > 3$, concave up: $x < 7/4$ and concave down: $x > 7/4$

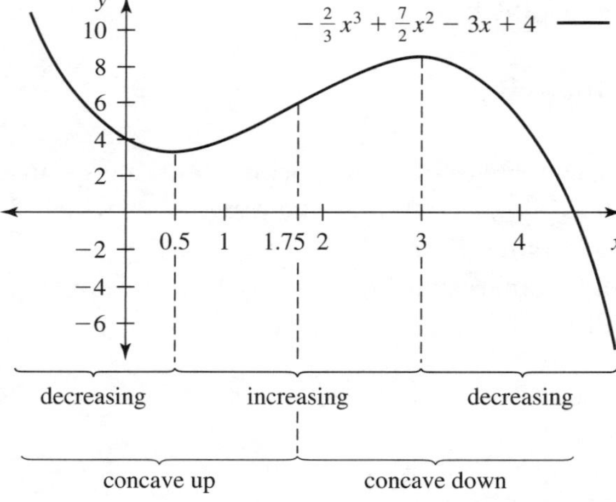

Prob. 7. $f(x) = \sqrt{x+1}$, $x \geq -1$, so $f'(x) = \frac{1}{2}(x+1)^{-\frac{1}{2}}$, $x > 1$ and $f''(x) = -\frac{1}{4}(x+1)^{-\frac{3}{2}}$ So f is increasing for all $x > -1$ and concave down for all $x > -1$

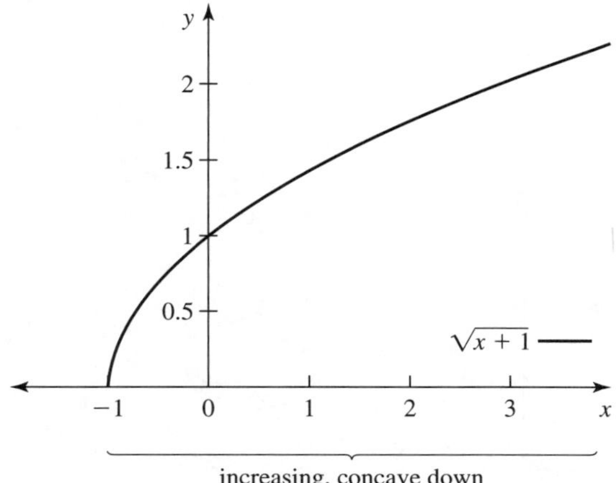

Prob. 9. $f(x) = \frac{1}{x}$, $x \neq 0$, so $f'(x) = -\frac{1}{x^2}$ and $f''(x) = \frac{2}{x^3}$. So f is decreasing: all $x \neq 0$, concave up for $x > 0$ and concave down for $x < 0$.

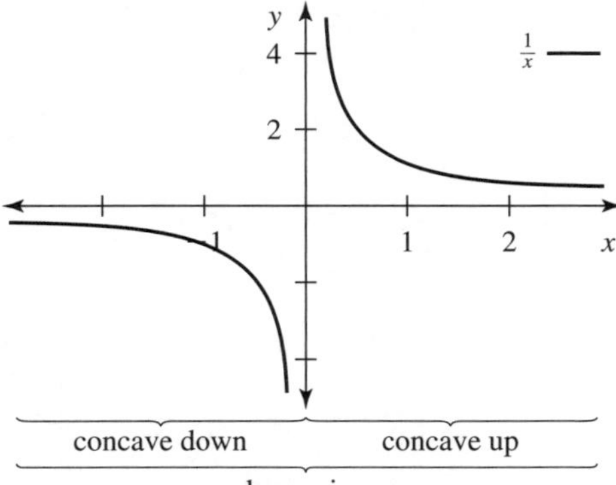

Prob. 11. $f(x) = (x^2+1)^{\frac{1}{3}}$, $x \in \mathbb{R}$ so $f'(x) = \frac{2}{3}x(x^2+1)^{-\frac{2}{3}}$ and $f''(x) = -\frac{2}{9}(x^2-3)(x^2+1)^{-\frac{5}{3}}$. So f is increasing: $x > 0$, decreasing: $x < 0$, concave up: $-\sqrt{3} < x < \sqrt{3}$ and concave down: $x > \sqrt{3}$, $x < -\sqrt{3}$

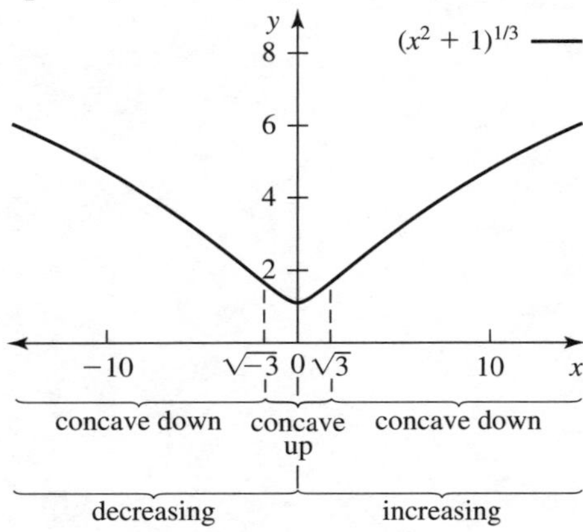

Prob. 13. $f(x) = \frac{1}{(1+x)^2}$, $x \neq -1$ so $f'(x) = -\frac{2}{(1+x)^3}$ and $f''(x) = \frac{6}{(1+x)^4}$. So f is increasing: $x < -1$, decreasing: $x > -1$ and concave up: all $x \neq -1$

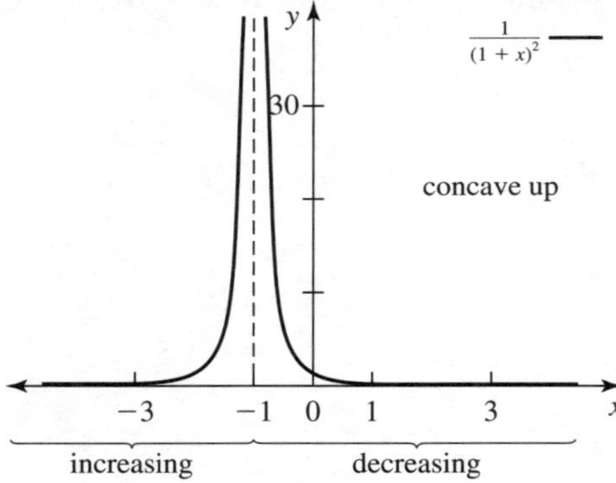

Prob. 15. $f(x) = \sin x$, $0 \le x \le 2\pi$ so $f'(x) = \cos x$ and $f''(x) = -\sin x$. So f is increasing: $\left(0, \frac{\pi}{2}\right) \cup \left(\frac{3\pi}{2}, 2\pi\right)$, decreasing: $\left(\frac{\pi}{2}, \frac{3\pi}{2}\right)$, concave up: $(\pi, 2\pi)$ and concave down: $(0, \pi)$

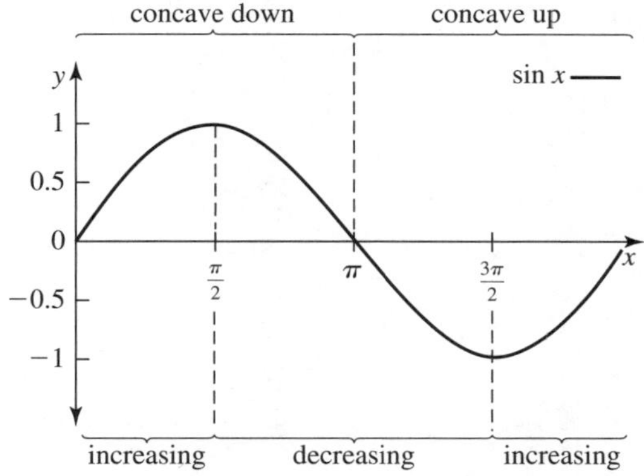

Prob. 17. $f(x) = e^x$ so $f'(x) = f''(x) = e^x$ and f is increasing for all $x \in \mathbb{R}$ and concave up for all $x \in \mathbb{R}$

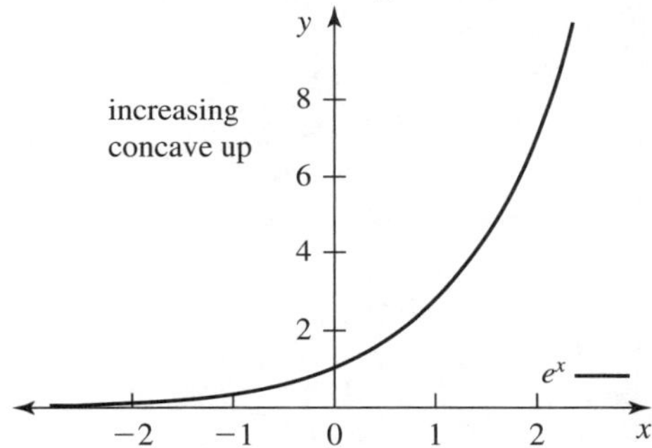

Prob. 19. $f(x) = e^{-x^2/2}$ so $f'(x) = -xe^{-x^2/2}$ and $f''(x) = (x^2 - 1)e^{-x^2/2}$. So f is increasing for $x < 0$, decreasing for $x > 0$, concave up for $x > 1$, $x < -1$ and concave down: $-1 < x < 1$

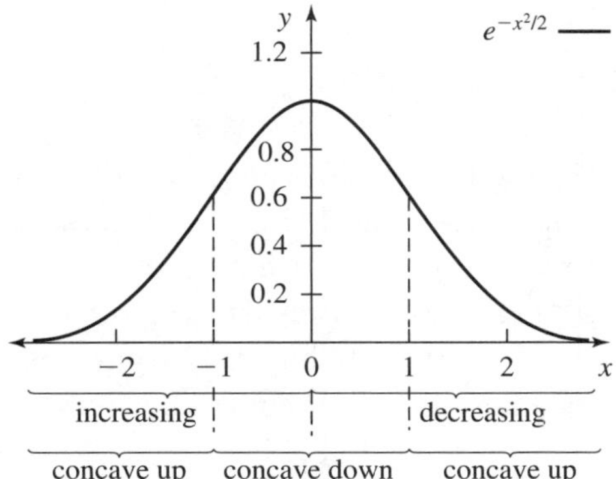

Prob. 21.

(a) The graph of a function increasing at an accelerating rate:

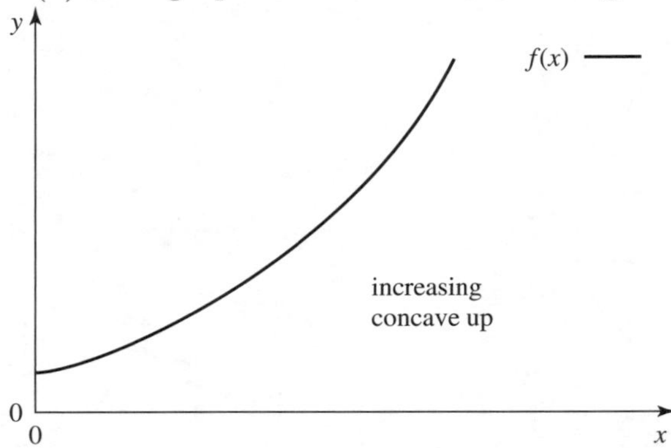

(b) The graph of a function increasing at an decelerating rate:

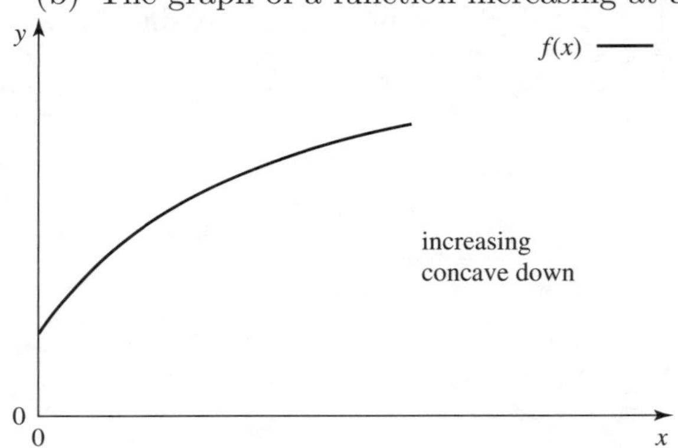

(c) In the situation of (a), the function must have positive first and second derivatives. The graph is concave up. For (b), the function must have a positive first derivative and a negative second derivative. The graph is concave down.

Prob. 23.

(a) If f' is strictly positive (negative) throughout (a, b), the graph of f in between a and b is always climbing up (down), and thus can intersect the x-axis at most once. In other words, $f(x) = 0$ for at most one x in (a, b). Since we already know there exists at least one such x, there is exactly one. (Equivalently, if $f(x) = 0$ has two solutions, Rolle's theorem says f' vanishes somewhere in (a, b), contradicting the assumption.)

(b) Let $f(x) = x^3 - 4x + 1$. Since $f(-1) = 4 > 0$ and $f(1) = -2 < 0$, $f(x) = 0$ has a solution in $(-1, 1)$. But $f'(x) = 3x^2 - 4 < 3(1) - 4 < 0$ in $(-1, 1)$. Part (a) then tells us $f(x) = 0$ has exactly one solution in the interval.

Prob. 25. If $f''(x) < 0$ throughout an interval, the first derivative test tells us f' is decreasing in this interval, which by definition means the function f is concave down.

Prob. 27.

(a) Below is the graph of

$$g(N) = 3\left(1 - \frac{N}{10}\right)$$

versus N:

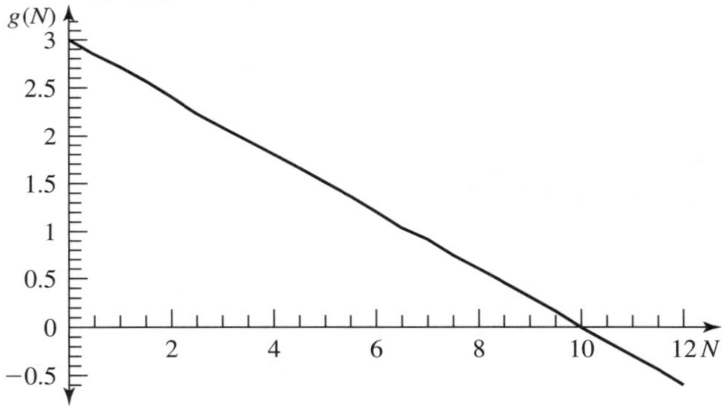

(b) We have $g(N) = r(1 - N/K)$. Thus, $g'(N) = -r/K < 0$, and $g(N)$ is decreasing for all $N > 0$.

Prob. 29. Compute the derivative

$$\begin{aligned} f'(N) &= \left[1 - \left(\frac{N}{K}\right)^{\theta}\right] - N \cdot \frac{\theta N^{\theta-1}}{K^{\theta}} \\ &= 1 - (1+\theta)\left(\frac{N}{K}\right)^{\theta}. \end{aligned}$$

It is easy to deduce that the expression is positive, i.e. the growth rate is increasing, when

$$(0 <)\, N < \frac{K}{(1+\theta)^{\frac{1}{\theta}}},$$

and negative otherwise.

Prob. 31. Since $f'(P) = -ae^{-aP}$ is always negative, f decreases with P.

Prob. 33.

(a) Since

$$y'(x) = \frac{1170}{x^2}e^{-10/x}$$

is always positive, the height y increases with age x. For $x > 0$, we have $-10/x < 0$ and thus $e^{-10/x} < 1$. Therefore, the height $y = 117e^{-10/x}$ never exceeds 117. As the tree ages ($x \to \infty$), $-10/x$ approaches zero and thus $y(x)$ approaches 117.

(b) Compute

$$y''(x) = 1170e^{-10/x}\left(\frac{10 - 2x}{x^4}\right).$$

Hence, the graph is concave up (down), i.e. $y'' > 0$ ($y'' < 0$), when $x < 5$ ($x > 5$).

(c) The graph of $y = 117e^{-10/x}$:

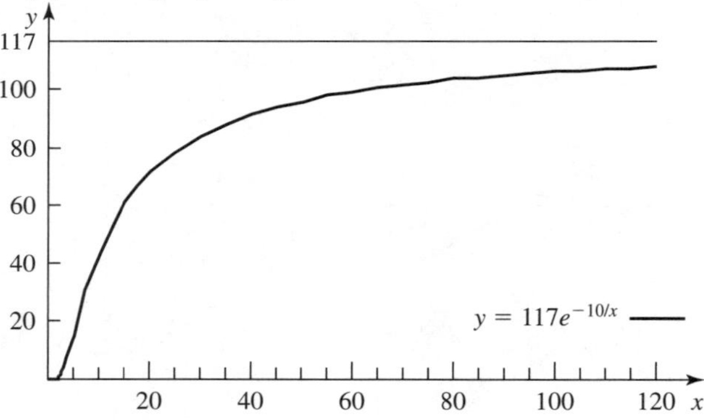

(d) The greatest rate of growth corresponds to the largest slope. Looking at the graph, we notice that the value is largest at $x = 5$ (as expected from the calculation in (b)), and it is exactly where the graph changes from being concave up to being concave down.

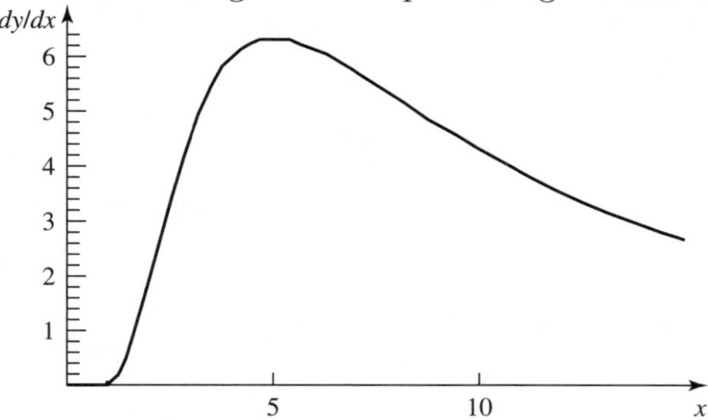

Prob. 35. In terms of the function $X(F)$, we are looking for a condition such that $X'(F) > 0$ and $X''(F) < 0$ for all $F \geq 0$. Since

$$\begin{aligned} X'(F) &= c\gamma F^{\gamma-1}, \\ X''(F) &= c\gamma(\gamma-1)F^{\gamma-2}, \end{aligned}$$

and $c > 0$, the desired condition on γ is $\gamma > 0$ and $\gamma(\gamma - 1) < 0$, or equivalently,

$$0 < \gamma < 1.$$

Prob. 37.

(a) At equilibrium, i.e. $dN/dt = 0$, we have

$$0 = Ne^{-aN} - N^2 = N(e^{-aN} - N).$$

The nontrivial equilibrium ($N^* \neq 0$) thus satisfies $e^{-aN^*} = N^*$.

(b) If we differentiate

$$e^{-aN^*} = N^*$$

with respect to a, regarding N^* as a function of a, we obtain

$$-[a(N^*)' + N^*]e^{-aN^*} = (N^*)',$$

which after rearrangement becomes

$$(N^*)' = -\frac{N^*e^{-aN^*}}{1 + ae^{-aN^*}} < 0.$$

In other words, N^* decreases with a.

Prob. 39.

(a) Compute

$$\frac{d}{dN}\frac{A(N)}{N} = \frac{d}{dN}\frac{S}{1+(aN)^b} = -\frac{Sab(aN)^{b-1}}{[1+(aN)^b]^2}.$$

Since the right hand side is negative, $A(N)/N$ is a decreasing function of N.

(b) Using the fact that $\log(x/y) = \log x - \log y$, we have

$$\begin{aligned} k &= \log NS - \log\left[\frac{NS}{1+(aN)^b}\right] \\ &= \log(NS) - \left[\log(NS) - \log(1+(aN)^b)\right] \\ &= \log[1+(aN)^b]. \end{aligned}$$

(i) Recall that

$$\log N = \log_{10} N = \frac{\ln N}{\ln 10}.$$

Hence we have

$$\frac{d\log N}{dN} = \frac{d}{dN}\left(\frac{\ln N}{\ln 10}\right) = \frac{1}{N\ln 10}.$$

(ii) Using the chain rule and the result of (i), we have

$$\frac{dk}{d\log N} = \frac{dk/dN}{d\log N/dN} = (\ln 10)N\frac{dk}{dN}.$$

Using the first result in (b), we compute

$$\frac{dk}{dN} = \frac{d}{dN}\frac{\ln[1+(aN)^b]}{\ln 10} = \frac{1}{\ln 10}\cdot\frac{ab(aN)^{b-1}}{1+(aN)^b}.$$

Hence, we have

$$\begin{aligned}\frac{dk}{d\log N} &= (\ln 10)N\cdot\frac{1}{\ln 10}\cdot\frac{ab(aN)^{b-1}}{1+(aN)^b}\\ &= \frac{b(aN)^b}{1+(aN)^b}\\ &= \frac{b}{(aN)^{-b}+1}.\end{aligned}$$

(iii) Since $a, b > 0$, as $N \to \infty$, $(aN)^{-b} \to 0$ and thus

$$\lim_{N\to\infty}\frac{dk}{d\log N} = \lim_{N\to\infty}\frac{b}{1+(aN)^{-b}} = b.$$

(iv) Let us consider the relation between $dk/d\log N$ and $A'(N) = dA/dN$. By definition,

$$k = \log(NS) - \log A(N) = \log N - \log S - \log A(N).$$

Using the chain rule, we have

$$\begin{aligned}\frac{dk}{d\log N} &= 1 - \frac{d\log A(N)/dN}{d\log N/dN}\\ &= 1 - \frac{A'(N)}{A(N)\ln 10}\Big/\frac{1}{N\ln 10}\\ &= 1 - \frac{NA'(N)}{A(N)}.\end{aligned}$$

Therefore, if $dk/d\log N > 1\,(< 1)$, then $A'(N)$ is negative (positive), i.e. $A(N)$ is decreasing (increasing). (Note: N and $A(N)$ are positive.)

Since the initial density of seeds is by definition NS and S is a constant, a higher density means increasing N. From the result in (ii), we observe that $dk/d\log N$ increases with N. ($N \uparrow \Rightarrow (aN)^{-b} \downarrow \Rightarrow 1/[1+(aN)^{-b}] \uparrow$) And from what we have just deduced, if $dk/d\log N$ increases beyond 1, the number of surviving plants switches from growing to dropping. Therefore, to summarize, if the initial density of seeds is higher (lower) than some fixed value, there will be less (more) plants in the following year than this year.

(v) From the result in (iv), the case $dk/d\log N = 1$ corresponds to $A'(N) = 0$. Thus, over the range where $dk/d\log N = 1$ holds, the number of plants is in equilibrium.

Prob. 41.

(a) Since

$$\begin{aligned} \frac{dY}{dX} &= \frac{d(bX^a)}{dX} = abX^{a-1}, \\ \frac{d}{dX}\frac{Y}{X} &= \frac{d(bX^{a-1})}{dX} \\ &= (a-1)bX^{a-2}, \end{aligned}$$

the condition we are looking for is $ab > 0$ and $(a-1)b < 0$. Since b is positive, we conclude $0 < a < 1$.

In this case, $Y'' = a(a-1)bX^{a-2}$ is negative, i.e. Y as a function of X is concave down.

(b) Now X is the body length, Y is the skull length and $0 < a < 1$. From (a), we know

$$\frac{d}{dX}\frac{Y}{X} < 0.$$

As a vertebrate grow up (X increases), the size of the skull Y becomes smaller and smaller *compared to the size of the body* X.

Prob. 43. Differentiating both sides of the given equation $y' = ky/x$ with respect to x yields

$$y'' = k\left(\frac{xy' - y}{x^2}\right).$$

If we use the equation $y' = ky/x$, or $xy' = ky$, again, we have

$$y'' = k\left(\frac{ky - y}{x^2}\right) = k(k-1)\frac{y}{x^2}.$$

Since x and y are positive, y as a function of x is concave up, i.e. $y'' > 0$, if and only if $k(k-1) > 0$, or $k > 1$. (k is assumed to be positive.)

5.3 Extrema, Inflection Points and Graphing

Prob. 1. Given,

$$y = 4 - x^2$$

and

$$-2 \leq x \leq 3.$$

To find the maxima or minima, we first find the derivative of the above function,

$$y' = -2x$$

Equating the above derivative to zero, we get

$$y' = 0,$$

$$-2x = 0, x = 0$$

Thus $x = 0$ is probable candidate for extrema. Finding the second derivative of y we get

$$y'' = -2 < 0$$

Hence we have a maxima at $x = 0$. The value of at $x = 0$ is

$$y(0) = 4 - 0 = 4$$

Also checking the boundary values we have

$$y(-2) = 4 - 4$$

$$= 0$$

and for $x = 3$, we have

$$y(3) = 4 - 3^2$$

$$= -5$$

Also, the the function is decreasing in the interval

$$0 \leq x \leq 3$$

and increasing in the interval

$$-2 \leq x \leq 0$$

Hence we have: local max (0,4); abs max (0, 4); local max (-2,0) and (3, -5); abs min (3, -5).

Prob. 3. Given,

$$y = \sin x$$

and

$$0 \leq x \leq 2\pi.$$

To find the maxima or minima, we first find the derivative of the above function,

$$y' = \cos x$$

Equating the above derivative to zero, we get

$$y' = 0,$$

$$\cos x = 0, x = \frac{\pi}{2}, \frac{3\pi}{2}$$

Finding the second derivative, we find that,

$$y'' = -\sin x$$

$$= -1 < 0, x = \frac{\pi}{2}$$

$$= 1 > 0, x = \frac{3\pi}{2}$$

Thus, the second derivative is negative for $x = \frac{\pi}{2}$ and hence is a maxima. It is positive for $x = 3\frac{\pi}{2}$ and hence it has a minima at that point. Also, evaluating the value of function at end points, we get

$$y(0) = 0,$$

$$y(2\pi) = 0$$

Hence, we have from above results, the final result as, at

$$x = \frac{\pi}{2} \Rightarrow Globalmaxima$$

$$x = \frac{3\pi}{2} \Rightarrow Globalminima$$

Prob. 5. To find the maxima or minima, we first find the derivative, $y' = -e^{-x}$ Equating the derivative to zero, we get

$$-e^{-x} = 0$$

Hence its derivative is never equal to zero. Finding the 2nd derivative,

$$y'' = e^{-x}$$

Now,

$$e^{-x} > 0 for all x \geq 0$$

Hence the function is a continuously decreasing function and thus it has a local maxima at the point $x = 0$. Thus we have the following result,

$$x = 0 \Rightarrow Local maxima$$

Prob. 7. Given, $y = (x-1)^3 + 1$ and $x \in \mathbf{R}$. To find the maxima or minima, we first find the derivative of the above function,

$$y' = 3(x-1)^2$$

Equating the above derivative to zero, we get

$$y' = 0,$$

$$3(x-1)^2 = 0$$

$$x = 1$$

Finding the second derivative of the function we have

$$y'' = 6(x-1)$$

$$y'' = \begin{cases} > 0 & \text{for} \quad x > 1; \\ < 0 & \text{for} \quad x < 1; \\ = 0 & \text{for} \quad x = 1. \end{cases}$$

The function continuously increases. Thus the function has no maxima and minima.

Prob. 9. Given,

$$y = \cos x$$

and

$$0 \leq x \leq \pi.$$

To find the maxima or minima, we first find the derivative of the above function,

$$y' = -\sin x$$

Equating the above derivative to zero, we get

$$y' = 0,$$

$$-\sin x = 0, x = 0, \pi$$

Finding the second derivative of the above function,

$$y'' = -\cos x$$

$$y'' = \begin{cases} > 0 \text{ for } \frac{\pi}{2} < x < \pi; \\ < 0 \text{ for } 0 < x < \frac{\pi}{2}. \end{cases}$$

Thus the function has a maxima at $x = 0$ and a minima at $x = \pi$. Hence we have the result,

$$x = 0 \Rightarrow Globalmaxima$$

$$x = \pi \Rightarrow Globalminima$$

Prob. 11. Given, $y = e^{-|x|}$ and $x \in \mathbf{R}$. We redefine the function y as

$$y = \begin{cases} e^{-x} & \text{for} \quad x > 0; \\ e^{x} & \text{for} \quad x < 0. \end{cases}$$

To find the maxima or minima, we first find the derivative of the above function,

$$y' = \begin{cases} -e^{-x} & \text{for} \quad x > 0; \\ e^{x} & \text{for} \quad x < 0. \end{cases}$$

Equating the above derivative to zero, we get

$$y' = 0,$$

So, no value of x satisfies the equation and hence we there is no maxima and minima in the above interval. Evaluating the function t $x = 0$, we find that

$$y(0) = 1$$

which is,in fact a global maxima. Thus, we have one global maxima, which is at $x = 0$.

$$x = 0 \Rightarrow Globalmaxima$$

Prob. 13. Given, $y = \frac{x^3}{3} + \frac{x^2}{2} - 6x + 2$ and $x \in \mathbf{R}$. To find the maxima or minima, we first find the derivative of the above function, $y' = x^2 + x - 6$ Equating the above derivative to zero, we get

$$y' = 0,$$

$$x^2 + x - 6 = 0$$
$$(x+3)(x-2) = 0$$
$$x = -3, 2$$

Finding the second derivative of the above function we have,

$$y'' = 2x + 1$$

$$y'' = \begin{cases} < 0 & \text{for} \quad x < -\frac{1}{2}; \\ > 0 & \text{for} \quad x > \frac{1}{2}. \end{cases}$$

Since -3 lies in the 1st interval and 2 lies in the second interval we have, local maxima and minima at these points.

$$x = -3 \Rightarrow Localmaxima$$

$$x = 2 \Rightarrow Localminima$$

Prob. 15. Given,

$$y = \sin(\pi x^2)$$

and

$$-1 \leq x \leq 1.$$

To find the maxima or minima, we first find the derivative of the above function,

$$y' = 2\pi x \cos(\pi x^2)$$

Equating the above derivative to zero, we get

$$y' = 0,$$

$$2\pi x \cos(\pi x^2) = 0$$

$$x = 0, \frac{1}{\sqrt{2}}, -\frac{1}{\sqrt{2}}$$

Finding the second derivative of the above function,

$$y'' = 2\pi(\cos(\pi x^2) - (2\pi x)^2 \sin(\pi x^2)$$

Also, we check for end points and we find their values. So, we have maxima and minima.

$$x = 0 \Rightarrow Local\ minimum$$

$$x = 1 \Rightarrow Local\ minimum$$

$$x = -1 \Rightarrow \textit{Local minimum}$$
$$x = \frac{1}{\sqrt{2}} \Rightarrow \textit{Local maximum}$$
$$x = -\frac{1}{\sqrt{2}} \Rightarrow \textit{Local maximum}$$

Prob. 17. We have to prove that the $f'(x) = 0$ is not a sufficient condition for the extrema to exist. We start by letting the function to be,

$$f(x) = x^3$$

So finding its derivative,

$$f' = 3x^2$$

Equating it to zero we have,

$$f' = 0$$
$$3x^2 = 0, x = 0$$

So, we should have a extrema here. But, we see that the derivative doesn't change sign at $x = 0$. Thus f' doesn't change its sign and hence $x = 0$ is not a local maxima or minima.

Prob. 19. Given the function as, $f(x) = x^3 - 2$ and $x \in \mathbf{R}$ Since, we have to find the points of inflection, we need to find the first derivative as well as second derivative and check whether the second derivative changes sign at the inflection point. For the point to be inflection, second derivative should change sign .

$$f'(x) = 3x^2$$
$$f''(x) = 6x$$

Thus

$$6x = 0, x = 0$$

Hence we get $x = 0$ as the candidate inflection point. Also, the second derivative changes sign at $x = 0$ as

$$f''(x) = \begin{cases} > 0 & \text{for} \quad x > 0; \\ < 0 & \text{for} \quad x < 0. \end{cases}$$

Thus it is, indeed a point of inflection.

Prob. 21. Given the function as,

$$f(x) = e^{-x^2}$$

and

$$x \geq 0$$

Since, we have to find the points of inflection, we need to find the first derivative as well as second derivative and check whether the second derivative changes sign at the inflection point. For the point to be inflection, second derivative should change sign .

$$f'(x) = -2xe^{-x^2}$$

$$f''(x) = -2e^{-x^2}(1 - 2x^2)$$

Thus

$$-2e^{-x^2}(1 - 2x^2) = 0$$

$$1 - 2x^2 = 0$$

$$x = \frac{1}{\sqrt{2}} \quad \text{since} \quad x \geq 0$$

Hence we get $x = \frac{1}{\sqrt{2}}$ as the candidate inflection point. Also, the second derivative changes sign at $x = 0$ as

$$f''(x) = \begin{cases} > 0 & \text{for} \quad 0x > \frac{1}{\sqrt{2}} x < -\frac{1}{\sqrt{2}}; \\ < 0 & \text{for} \quad -\frac{1}{\sqrt{2}} < x < \frac{1}{\sqrt{2}}. \end{cases}$$

Thus it is, indeed a point of inflection.

Prob. 23. Given the function as,

$$f(x) = \tan x$$

and

$$-\frac{\pi}{2} \leq x \leq \frac{\pi}{2}$$

Since, we have to find the points of inflection, we need to find the first derivative as well as second derivative and check whether the second derivative changes sign at the inflection point. For the point to be inflection, second derivative should change sign .

$$f'(x) = \sec^2 x$$

$$f''(x) = 2\sec^2 x \tan x$$

Thus

$$2\sec^2 x \tan x = 0, x = 0$$

Hence we get $x = 0$ as the candidate inflection point. Also, the second derivative changes sign at $x = 0$ as

$$f''(x) = \begin{cases} > 0 & \text{for} \quad 0 < x < \frac{\pi}{2}; \\ < 0 & \text{for} \quad -\frac{\pi}{2} < x < 0. \end{cases}$$

Thus it is, indeed a point of inflection.

Prob. 25. Given the function as, $f(x) = x^4$ and $x \in \mathbf{R}$ Since, we have to find the points of inflection, we need to find the first derivative as well as second derivative and check whether the second derivative changes sign at the inflection point. For the point to be inflection, second derivative should change sign .

$$f'(x) = 4x^3$$

$$f''(x) = 12x^2$$

Thus

$$12x^2 = 0, x = 0$$

Hence we get $x = 0$ as the candidate inflection point. But it is not a inflection point as the 2nd derivative does not change sign around $x = 0$. Thus it is, not a point of inflection.

Prob. 27. Given the function as

$$y = \frac{2x^3}{3} - 2x^2 - 6x + 2$$

where,

$$-2 \leq x \leq 5$$

To find the local maxima and minima we have to find the first and second derivative. First, we find the first derivative,

$$y' = 2x^2 - 4x - 6$$

Equating it to zero, we have

$$y' = 0$$

$$2x^2 - 4x - 6 = 0$$

$$2(x - 3)(x + 1) = 0$$

$$x = 3, -1$$

Next, we find the second derivative,

$$y'' = 4x - 4$$

From, above we can say that,

$$y'' = \begin{cases} > 0 & \text{for} \quad x \geq 1, \\ < 0 & \text{for} \quad x \leq 1; \end{cases}$$

Thus we can say that the function has local maxima and minima.

$$x = 3 \Rightarrow Localminima$$

$$x = -1 \Rightarrow Localmaxima$$

Also, to find the point of inflection, we equate second derivative to 0,

$$y'' = 0$$

$$4x - 4 = 0$$

$$x = 1$$

From above, we see that the function does change sign at $x = 1$ and hence it is the point of inflection.

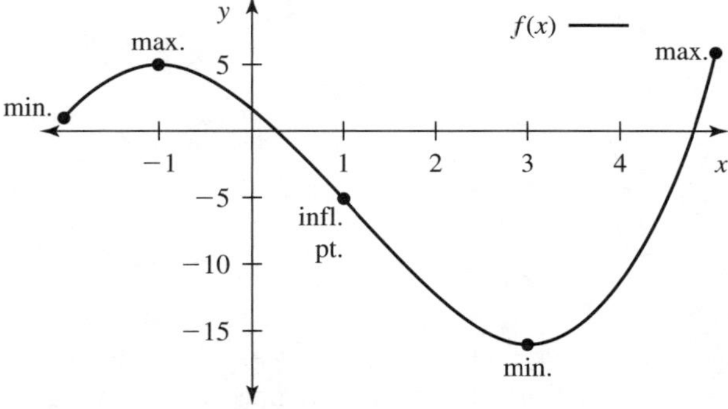

Prob. 29. Given the function as

$$y = |x^2 - 9|$$

where,

$$-4 \leq x \leq 5$$

We can redefine the function as

$$y = \begin{cases} x^2 - 9 & -4 \leq x \leq -3, 3 \leq x \leq 5; \\ 9 - x^2 & -3 < x < 3. \end{cases}$$

To find the local maxima and minima we have to find the first and second derivative. First, we find the first derivative,

$$y' = \{2x - 4 \leq x \leq -3,\ 3 \leq x \leq 5;\ -2x\ \ -3 < x < 3$$

. Thus equating each of them to zero we find that, the second part is only zero for $x = 0$

Next, we find the second derivative,

$$y'' = \begin{cases} 2 - 4 \leq x \leq -3, 3 \leq x \leq 5; \\ \qquad\qquad -2 - 3 < x < 3. \end{cases}$$

Thus the function has a local maxima at $x = 0$.

$$x = 0 \Rightarrow Localmaxima$$

As the second derivative is not equal to zero for any x, we do not have any point of inflection.

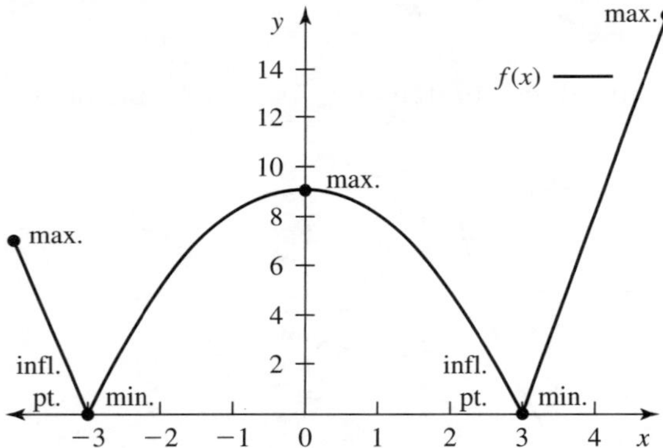

Prob. 31. Given the function as $y = x + \cos x$ where, $x \in \mathbf{R}$ To find the local maxima and minima we have to find the first and second derivative. First, we find the first derivative,

$$y' = 1 - \sin x$$

Equating it to zero, we have

$$y' = 0$$

$$1 - \sin x = 0$$

$$x = \frac{\pi}{2} + 2k\pi,\ k \in \mathbb{Z}$$

Next we find the second derivative of the function,

$$y'' = -\cos x$$

Since $y'' = 0$ for $x = \frac{\pi}{2} + 2k\pi$, $k \in \mathbb{Z}$, and $y' \geq 0$, there are no local extrema, where $n = 1, 2, \ldots$ To find the point of inflection we equate second derivative to 0

$$y'' = 0$$
$$-\cos x = 0$$
$$x = \frac{\pi}{2} + k\pi, \quad k \in \mathbb{Z}$$

Thus the above represents, the point of inflection as we see that the 2nd derivative changes sign at them.

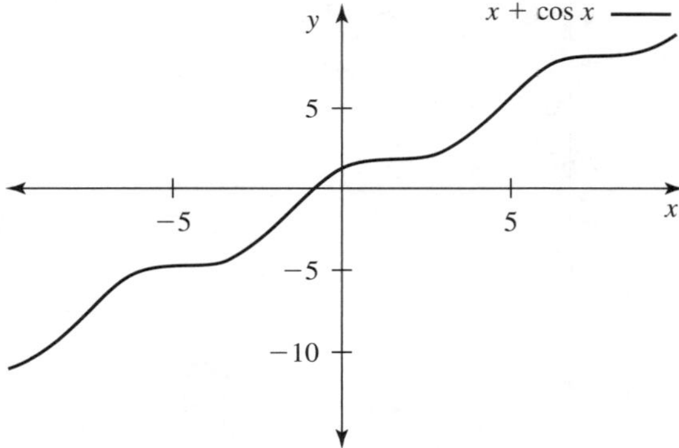

Prob. 33. $y = \frac{x^2-1}{x^2+1}$ where, $x \in \mathbf{R}$ To find the local maxima and minima we have to find the first and second derivative. First, we find the first derivative,

$$y' = \frac{4x}{(x^2+1)^2}$$

Equating it to zero, we have

$$y' = 0$$
$$\frac{4x}{(x^2+1)^2} = 0$$
$$x = 0$$

Next, we find the second derivative, since $y'' > 0$ for $x = 0$, we conclude that the function has a local minimum at $x = 0$ The point of inflection is where $y'' = 0$

$$x = \pm\frac{1}{3}\sqrt{3}$$

since the function changes sign as it passes through these values, there are inflection points at $x = \pm\frac{1}{3}\sqrt{3}$.

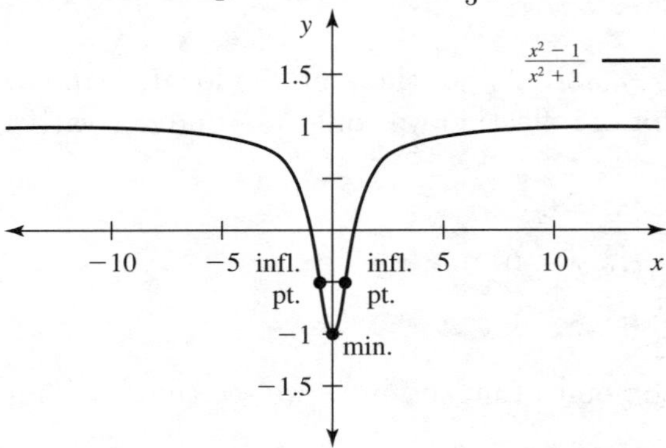

Prob. 35. Given the function,

$$f(x) = \frac{x}{x-1}$$

and

$$x \neq 1$$

(a)

$$\begin{aligned}\lim_{x\to+\infty} f(x) &= \lim_{x\to+\infty} \frac{x}{x-1}\\ &= \lim_{x\to+\infty} \frac{1}{1-\frac{1}{x}}\\ &= \frac{1}{1-0} = 1\end{aligned}$$

And,

$$\begin{aligned}\lim_{x\to-\infty} f(x) &= \lim_{x\to-\infty} \frac{x}{x-1}\\ &= \lim_{x\to-\infty} \frac{1}{1-\frac{1}{x}}\\ &= \frac{1}{1-0} = 1\end{aligned}$$

Thus we have

$$\lim_{x\to+\infty} f(x) = \lim_{x\to-\infty} f(x) = 1$$

(b)

$$\begin{aligned}\lim_{x\to 1^+} f(x) &= \lim_{x\to 1^+} \frac{x}{x-1}\\ &= \lim_{x\to 1^+} \frac{1}{1-\frac{1}{x}}\\ &= \frac{1}{1-(1-\epsilon)} = +\infty\end{aligned}$$

And,

$$\begin{aligned}\lim_{x\to 1^-} f(x) &= \lim_{x\to 1^-} \frac{x}{x-1}\\ &= \lim_{x\to 1^-} \frac{1}{1-\frac{1}{x}}\\ &= \frac{1}{1-(1+\epsilon)} = -\infty\end{aligned}$$

Thus $x = 1$ is vertical asmyptote.

(c) We first compute y' and the y'',

$$y' = -\frac{1}{(x-1)^2}$$

$$y'' = 2\frac{1}{(x-1)^3}$$

The function is decreasing in the interval $x > 1$ and in the interval $x < 1$. The function has no local extrema.

(d) The function is concave up in the interval $x > 1$ and is concave down in the interval $x < 1$. There are no inflection points.

(e)

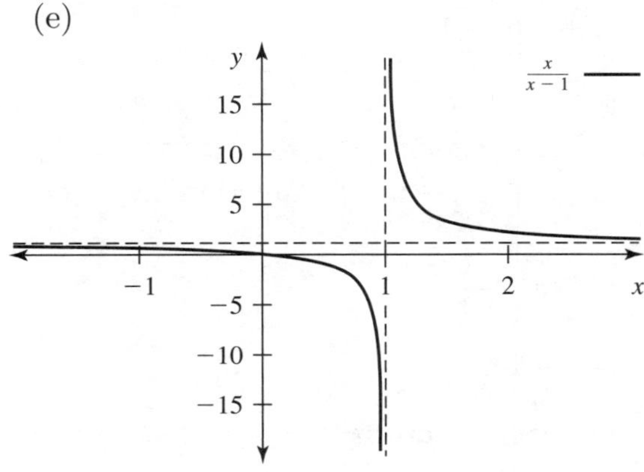

Prob. 37. Given the function,

$$f(x) = \frac{2x^2 - 5}{x + 2}$$

and

$$x \neq -2$$

(a)

$$\lim_{x \to -2^+} f(x) = \lim_{x \to -2^+} \frac{2x^2 - 5}{x + 2}$$
$$= \lim_{x \to -2^+} \frac{2(-2 + \epsilon)^2 - 5}{(-2 + \epsilon) + 2}$$
$$= \infty$$

And,

$$\lim_{x \to -2^-} f(x) = \lim_{x \to -2^-} \frac{2x^2 - 5}{x + 2}$$
$$= \lim_{x \to -2^-} \frac{2(-2 - \epsilon)^2 - 5}{(-2 - \epsilon) + 2}$$
$$= -\infty$$

Thus $x = -2$ is vertical asmyptote.

(b) Next, we find f' and f''.

$$f' = \frac{2x^2 + 8x + 5}{(x + 2)^2}$$

Thus the function has chances of local extrema at

$$x = \frac{-4 + \sqrt{6}}{2}, \frac{-4 - \sqrt{6}}{2}$$

The function is increasing for $x > \frac{-4+\sqrt{6}}{2}$ and $x < \frac{-4-\sqrt{6}}{2}$. It is decreasing in $(\frac{-4-\sqrt{6}}{2}, -2)$ and $(-2, \frac{-4+\sqrt{6}}{2})$. The function has a local minimum and it is at

$$x = \frac{-4 + \sqrt{6}}{2}$$

and a local maximum at $x = \frac{-4-\sqrt{6}}{2}$.

(c) It is concave up for $x > -2$ and concave down for $x < -2$. There are no inflection points.

(d) The oblique asymptote has equation $y = 2x - 4$ since $\frac{2x^2-5}{x+2} = 2x - 4 + \frac{3}{x+2}$.

(e)

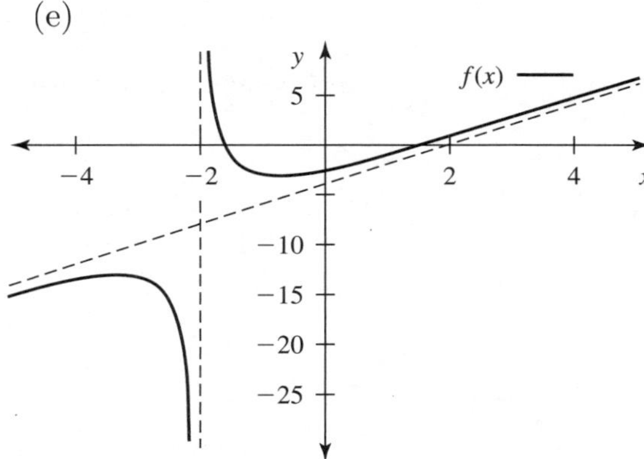

Prob. 39. Given the function, $f(x) = \frac{x^2}{x^2+1}$ and $x \in \mathbf{R}$

(a) First we find the derivative of the above function.

$$f'(x) = \frac{2x}{(1+x^2)^2}$$

Thus the function is increasing for $x > 0$ and decreasing for $x < 0$.

(b) For concavity, we compute $f''(x)$

$$f'' = \frac{2-6x^2}{(1+x^2)^3}$$

To compute

$$f'' = 0$$

we have

$$3x^2 - 1 = 0$$

$$x = \pm\frac{1}{\sqrt{3}}$$

Thus we have

$$y'' = \begin{cases} < 0 & \text{for} \quad x > \frac{1}{\sqrt{3}} \text{ and } x < -\frac{1}{\sqrt{3}}; \\ > 0 & \text{for} \quad -\frac{1}{\sqrt{3}} < x < \frac{1}{\sqrt{3}}. \end{cases}$$

Thus the points of inflection are

$$x = \pm\frac{1}{\sqrt{3}}$$

The graph is concave up for $-\frac{1}{\sqrt{3}} < x < \frac{1}{\sqrt{3}}$ and concave down elsewhere.

(c) The graph has a horizontal asymptote at $y = 1$.

$$\begin{aligned}\lim_{x\to+\infty} f(x) &= \lim_{x\to+\infty} \frac{x^2}{x^2+1} \\ &= \lim_{x\to+\infty} \frac{1}{1+\frac{1}{x^2}} \\ &= 1\end{aligned}$$

And,

$$\begin{aligned}\lim_{x\to-\infty} f(x) &= \lim_{x\to-\infty} \frac{x^2}{x^2+1} \\ &= \lim_{x\to-\infty} \frac{1}{1+\frac{1}{x^2}} \\ &= 1\end{aligned}$$

Thus we have

$$\lim_{x\to+\infty} f(x) = \lim_{x\to-\infty} f(x) = 1$$

Hence $y = 1$ is a horizontal asymptote.

(d)

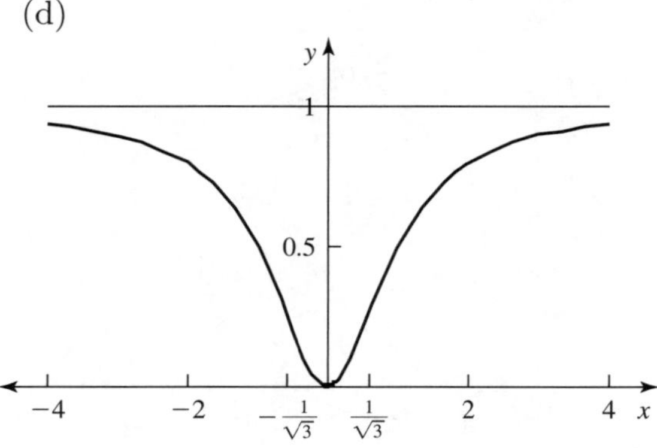

Prob. 41. Given the function, $f(x) = \frac{x}{x+a}$ and $x \geq 0$

(a) First we find the derivative of the above function.

$$f'(x) = \frac{a}{(a+x)^2}$$

Thus the function is increasing for $x \geq 0$. There are no local extrema.

(b) For concavity, we compute $f''(x)$

$$f'' = -\frac{2a}{(a+x)^3}$$

Thus we have $y'' < 0$ for $x > 0$

There are no inflection points. Its concave up for $x > 0$.

(c) The graph has a horizontal asymptote at $y = 1$.

$$\begin{aligned}\lim_{x\to+\infty} f(x) &= \lim_{x\to+\infty} \frac{x}{x+a} \\ &= \lim_{x\to+\infty} \frac{1}{1+\frac{a}{x}} \\ &= 1\end{aligned}$$

(d)

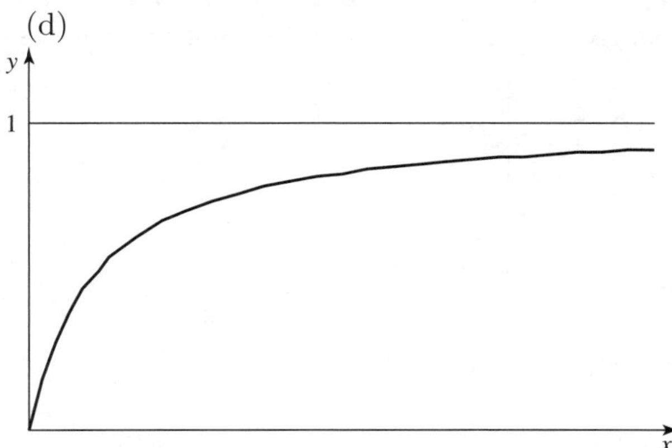

Prob. 43. Given the growth rate of population as

$$f(N) = N(1 - (\frac{N}{K})^{\theta})$$

and

$$N \geq 0$$

So, we have to find N such that $f(N)$ is maximal. Hence we find 1st derivative and evaluate it to zero.

$$f'(N) = 1 - (\frac{N}{K})^{\theta} - \frac{N}{K}\theta(\frac{N}{K})^{\theta-1}$$

$$= 1 - (\frac{N}{K})^{\theta}(1 + \theta)$$

Hence equating it to zero, we have

$$1 - (\frac{N}{K})^{\theta}(1 + \theta) = 0$$

$$(\frac{N}{K})^{\theta} = \frac{1}{1 + \theta}$$

$$N = K(\frac{1}{1 + \theta})^{\frac{1}{\theta}}$$

Thus we have the maximal value at the above population. We, confirm this by taking the second derivative of the function and by noting that the

$$f''(K(\frac{1}{1 + \theta})^{\frac{1}{\theta}}) < 0$$

and thus has a maxima.

5.4 Optimization

Prob. 1. The perimeter of a rectangle is given by

$$p = 2(l + b)$$

where $p =$ perimeter of the rectangle, $l =$ length of the rectangle, $b =$ height of the rectangle. Also given the area of the rectangle is

$$Area = l \times b = 25.$$

Hence from above, we get

$$l = \frac{25}{b}.$$

Substituting in the first equation, we get

$$p = 2(\frac{25}{b} + b), \quad b > 0.$$

We, differentiate the above to get the minimum value of the perimeter. Thus,

$$\frac{dp}{db} = 2(-\frac{25}{b^2} + 1).$$

Equating the above equation to zero, we have

$$2(-\frac{25}{b^2} + 1) = 0 \Rightarrow \frac{25}{b^2} = 1 \Rightarrow b^2 = 25.$$

Thus we get $b = 5$ and to get the length, we use the equation $25 = l \times b$ to obtain $l = 5$, so it is a square of side 5 and the perimeter is 20. Since $\frac{d^2p}{db^2} = 2(\frac{50}{b^3}) > 0$, and $\lim_{b \to 0^+} p = \lim_{b \to \infty} p = \infty$, $b = 5$ yields the smallest perimeter.

Prob. 3. Given the equation of parabola is

$$y = 3 - x^2, \quad 0 \leq x \leq \sqrt{3}$$

Let there be any point on the parabola $(x, y), x, y > 0$, which are the co-ordinates of the endpoint of the rectangle. Hence the area of the rectangle is

$$A = 2xy$$

substituting value of y in terms of x, we get

$$A = 2x(3 - x^2)$$

To get the maximum area, we differentiate the above equation, so we have

$$\frac{dA}{dx} = 2(3 - x^2 + x(-2x))$$

$$= (3 - 3x^2)(2)$$

Equating the above equation to zero, we have

$$3 - 3x^2 = 0$$

$$x^2 = 1$$

$$x = 1$$

Thus the $x = 1$ and from it we have

$$y = 3 - x^2 \Rightarrow y = 3 - 1 \Rightarrow y = 2$$

Thus the height of the rectangle is $y = 2$ and length is $2x = 2$, and the area is 4. Since $\frac{d^2A}{dx^2} = 2(-6x) < 0$ for $x = 1$ and $A(0) = A(\sqrt{3}) = 0$, the largest area is obtained when $x = 1$.

Prob. 5. Let us assume that, where $p =$ perimeter of the rectangle, Given $l =$ length of the rectangle, $b =$ height of the rectangle. Hence

$$p = 2l + b$$

Since it is bounded by river on one side. We know that the area of the rectangle is

$$Area = l \times b$$

So, from 1st equation we get b in terms of l, Hence

$$b = p - 2l$$

Thus substituting in the area equation, we have

$$Area = l(p - 2l), \quad 0 \le l \le \frac{p}{2}$$

Differentiating, the above equation we have,

$$\frac{d(Area)}{dl} = p - 4l$$

Equating the above equation to zero,

$$p - 4l = 0$$

$$4l = p$$

$$l = \frac{p}{4}$$

Given that $p = 320$ So length is $l = 80$ and, the height is

$$b = p - 2l$$
$$= 320 - 160$$
$$= 160$$

Since $\frac{d^2(Area)}{dl^2} = -4 < 0$ and $A(0) = A(\frac{p}{2}) = 0$, the maximum area is when $l = 80$ and $b = 160$.

Prob. 7. We are given a right angled triangle whose side lengths are a and b and we are given that the hypotenuse of the triangle is of length 5. Perimeter of the triangle will be equal to

$$p = a + b + 5$$

where p is the perimeter. Now we know that for a right angled triangle, the length of the hypotenuse is

$$h = \sqrt{a^2 + b^2}$$

Thus we have,

$$a = \sqrt{h^2 - b^2}$$

Substituting we have,

$$p = \sqrt{h^2 - b^2} + b + 5, \quad 0 \le b \le h$$

Differentiating the above function, we have

$$\frac{dp}{db} = -\frac{b}{\sqrt{h^2 - b^2}} + 1 \text{ and } \frac{d^2p}{db^2} = \frac{-h^2}{(h^2 - b^2)^{3/2}} < 0 \text{ for } 0 < b < h$$

Substituting the above equation to zero, we have

$$-\frac{b}{\sqrt{h^2 - b^2}} + 1 = 0$$
$$b = \sqrt{h^2 - b^2}$$
$$2b^2 = h^2$$
$$b = \sqrt{\frac{h^2}{2}}$$

substituting $h = 5$, we have

$$b = \frac{5}{\sqrt{2}}$$

Thus, we have a as

$$a = \sqrt{h^2 - b^2} = \frac{5}{\sqrt{2}}$$

Thus the value for the perimeter is

$$p = \frac{5}{\sqrt{2}} + \frac{5}{\sqrt{2}} + 5$$
$$= 5(\sqrt{2} + 1)$$

Since $\frac{d^2p}{db^2} < 0$ and $p(0) = p(h) = h + 5 = 10 < 5(\sqrt{2} + 1)$, the largest perimeter is for $a = b = 5/\sqrt{2}$.

Prob. 9. We are given a rectangle whose lower left corner is at $(0, 0)$ and whose upper right corner is on the curve

$$y = \frac{1}{x}.$$

We have to find a rectangle, with the minimum possible perimeter. We know that the perimeter of a is,

$$p = 2y + 2x$$

where $p =$ perimeter of the rectangle, $y =$ length of the rectangle, $x =$ height of the rectangle. Thus, from the equation $y = \frac{1}{x}$, we substitute the value of y in the perimeter equation. Thus

$$p = \frac{2}{x} + 2x, \quad x > 0$$

Differentiating the above equation we have,

$$\frac{dp}{dx} = -\frac{2}{x^2} + 2 \text{ and } \frac{d^2p}{dx^2} = \frac{4}{x^3} > 0 \text{ for } x > 0$$

Equating the above equation to zero, we have

$$-\frac{2}{x^2} + 2 = 0$$
$$2x^2 = 2$$
$$x = 1$$

since we are working in 1st quadrant. Thus y is equal to

$$y = \frac{1}{x} = 1$$

Since $\frac{d^2p}{dx^2} > 0$ and $\lim_{x\to 0^+} p(x) = \lim_{x\to\infty} p(x) = \infty$, the minimum perimeter of the rectangle is

$$p = 2x + 2y = 4$$

Prob. 11. Given the point lies on a line, whose equation is,

$$y = 4 - 3x$$

So, the coordinates of any point p on this line is (x, y), where y can be found from above.

(a) So the distance of any point from this line to origin is equal to,

$$distance = \sqrt{x^2 + y^2}$$

Substituting the value of y from above, we get

$$distance(D) = \sqrt{x^2 + (4 - 3x)^2}, \quad x \in \mathbb{R}$$

(b) Now to find a point which is closest to origin, we differentiate the above equation with respect to x,

$$\frac{d(D)}{dx} = \frac{10x - 12}{\sqrt{x^2 + (4 - 3x)^2}} \quad \text{and} \quad \frac{d^2D}{dx^2} = \frac{16}{(x^2 + (4 - 3x)^2)^{3/2}} > 0$$

Equating the above equation to zero, we have

$$\frac{10x - 12}{\sqrt{x^2 + (4 - 3x)^2}} = 0$$

$$10x - 12 = 0$$

$$x = \frac{6}{5}$$

Thus we get y, by substituting x

$$y = 4 - 3x$$

$$y = 4 - 3\frac{6}{5}$$

$$= \frac{2}{5}$$

Since $\frac{d^2D}{dx^2} > 0$ and $\lim_{x\to\pm\infty} D(x) = \infty$, the coordinates of the closest point is,

$$p = (\frac{6}{5}, \frac{2}{5})$$

(c) Now the square of the distance is given by,

$$g(x) = (f(x))^2 = x^2 - (4 - 3x)^2$$

Differentiating the above equation we get,

$$g' = 2x - 24 + 18x \quad \text{and} \quad g''(x) = 20 > 0$$
$$= 20x - 24$$

Equating the above equation to zero, we have

$$24 - 20x = 0$$
$$x = \frac{6}{5}$$

Thus we get the same coordinate, as above.

Prob. 13. Given that $f(x)$ is a positive differential function that has a local minimum at $x = c$. Hence we have

$$\frac{d(f(x))}{dx} = 0$$

at $x = c$. Let there be another function $g(x)$ such that,

$$g(x) = [f(x)]^2$$

Then, we have

$$g'(x) = 2f(x)f'(x) \quad \text{and} \quad g''(x) = 2[f(x)]^2 + 2f(x)f''(x)$$

Thus

$$g'(c) = 2f(c)f'(c) = 0 \quad \text{and} \quad g''(c) = 2f(c)f''(c) > 0$$

as $f'(c) = 0$. Hence we have a local minima for $g(x)$ at $x = c$.

Prob. 15. Let the height of the cylinder by l and the radius be r. The volume is given as $V = \pi r^2 l$ and hence $l = \frac{1000}{\pi r^2}$. Now, the amount of material needed is equal to the surface area, which is given by $P = 2\pi r^2 + \frac{2000}{r}$, $r > 0$. Differentiating this function yields

$$\frac{dP}{dr} = 4\pi r - \frac{2000}{r^2} \quad \text{and} \quad \frac{d^2P}{dr^2} = 4\pi + \frac{4000}{r^3} > 0, \quad r > 0$$

Setting the first derivative equal to 0 to find candidates for local extrema, we get

$$4\pi r = \frac{2000}{r^2}$$
$$r = \left(\frac{500}{\pi}\right)^{1/3}$$

Since the second derivative is always positive, this is a local minimum and we find that the height of the cylinder is

$$l = \frac{1000}{\pi(500/\pi)^{2/3}} = 2\left(\frac{500}{\pi}\right)^{3/2}$$

Since $\lim_{r\to 0+} P(r) = \lim_{r\to\infty} P(r) = \infty$, we find that the surface area is minimized when the diameter is equal to the height, which is equal to $l = 2\left(\frac{500}{\pi}\right)^{1/3}$.

Prob. 17. We are given an sector such that it has an angle θ and the radius is equal to r. We know that area is given by,

$$A = \frac{1}{2}r^2\theta$$

and, perimeter is given by

$$s = r\theta + 2r$$

(a) Thus we have,

$$\theta = \frac{2A}{r^2}$$

Thus, perimeter is equal to

$$s = r\theta + 2r = \frac{2A}{r} + 2r, \quad r > 0$$

We find

$$\frac{ds}{dr} = -\frac{2A}{r^2} + 2$$

and

$$\frac{d^2s}{dr^2} = \frac{4A}{r^3} > 0$$

To find candidates for extrema, we set

$$2 - \frac{2A}{r^2} = 0 \Rightarrow r = \sqrt{A}$$

Since $\frac{d^2s}{dr^2} > 0$ and $\lim_{r\to 0^+} s(r) = \lim_{r\to\infty} s(r) = \infty$, we find the minimum perimeter when substituting $A = 2$: we get $r = \sqrt{2}$ and $\theta = 2$

(b) Thus for $A = 10$ we have, $r = \sqrt{10}$ and $\theta = 2$

Prob. 19. Let the dimensions of the right circular cylinder be l, r, where l = height of the cylinder and, r = radius of the circular base. So, the volume of the cylinder is given by,

$$V = \pi r^2 l$$

Given that $V = 355cm^3$, hence we have,

$$l = \frac{355}{\pi r^2}$$

The total surface area is equal to, $P = 4\pi r^2 + \pi r l$ as cylinder is closed from top and bottom. Thus we have to minimize the total surface area. Substituting l from above we get,

$$p = \pi(4r^2 + \frac{710}{\pi r})$$

Differentiating, we get,

$$\frac{dp}{dr} = \pi(8r - \frac{710}{\pi r^2})$$

Equating it to zero we have,

$$\pi(8r - \frac{710}{\pi r^2}) = 0$$

$$8\pi r^3 = 710$$

$$r = \sqrt[3]{\frac{710}{8\pi}} = \sqrt[3]{\frac{355}{4\pi}}$$

Thus we compute height as,

$$l = \frac{355}{\pi r^2}$$

$$= \frac{355}{\pi \sqrt[\frac{3}{2}]{\frac{355}{4\pi}}}$$

Thus we get the dimensions as

$$l = 4\left(\frac{355}{4\pi}\right)^{1/3} \quad \text{and} \quad r = \sqrt[3]{\frac{355}{4\pi}}$$

Prob. 21.

(a) Differentiating,

$$\frac{dw(t)}{dt} = \frac{f'(t)(c+t) - f(t)}{(c+t)^2}$$

Thus equating to zero, we have

$$\frac{f'(t)(c+t) - f(t)}{(c+t)^2} = 0 \Rightarrow f'(t^*) = \frac{f(t^*)}{c+t^*}$$

Hence if $(-c, 0)$ line is tangential to the curve $f(c)$, then the slope of the line is $f(t^*)/(c+t^*)$, which is equal to $f'(t^*)$. Thus, we get the solution.

(b) Substituting,

$$f(t) = \frac{t}{1+t}$$

and having $c = 2$, we have

$$\frac{f'(t)(c+t) - f(t)}{(c+t)^2} = 0$$

with $f'(t) = \frac{1}{(1+t)^2}$, we get

$$2 - t^2 = 0$$
$$t = \sqrt{2}$$

Thus, for maximum $w(t)$ we should have $t = t = \sqrt{2}$.

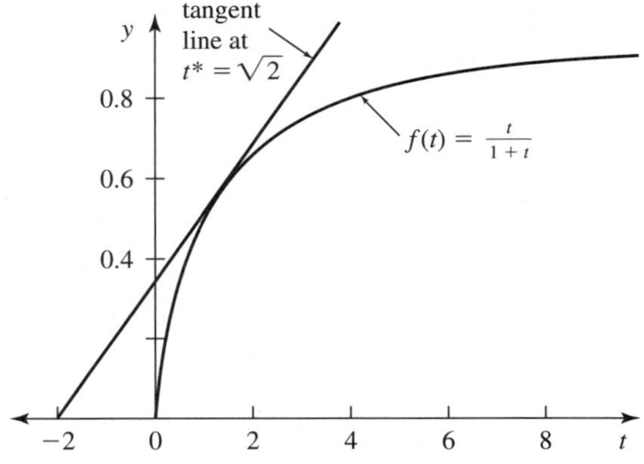

Equation: $y = \frac{1}{(1+\sqrt{2})^2}t + \frac{2}{(1+\sqrt{2})^2}$

Prob. 23. We are given the equation,

$$\frac{e^{-x(r(x)+L)}(1-e^{-kx})^3 c}{1-e^{-(r(x)+L)}} = 1$$

where k,$L$and$c$ are constants.

(a) We have to find the $\frac{dr}{dx}$. To do this, we take log on both the sides. Takinglog we have,

$$-x(r(x)+L) + 3\ln(1-e^{-kx}) = \ln(1-e^{-(r(x)+L)})$$

Differentiating,

$$-L - r(x) - xr'(x) + \frac{3ke^{-kx}}{1-e^{-kx}} = \frac{r'(x)e^{-x(r(x)+L)}}{1-e^{-(r(x)+L)}}$$

$$r'(x)(x + \frac{e^{-x(r(x)+L)}}{1-e^{-(r(x)+L)}}) = -r(x) + \frac{3ke^{-kx}}{1-e^{-kx}} - L$$

$$r'(x) = \frac{-r(x) + \frac{3ke^{-kx}}{1-e^{-kx}} - L}{x + \frac{e^{-x(r(x)+L)}}{1-e^{-(r(x)+L)}}}$$

Thus we get$\frac{dr}{dx}$.

(b) Setting $\frac{dr}{dx=0}$, we have

$$\frac{-r(x) + \frac{3ke^{-kx}}{1-e^{-kx}} - L}{x + \frac{e^{-x(r(x)+L)}}{1-e^{-(r(x)+L)}}} = 0$$

After doing, rearrangement, we have

$$r(x) = \frac{3ke^{-kx}}{1-e^{-kx}} - L$$

Hence proved.

5.5 L'Hospital's Rule

Prob. 1. Applying L'Hospital's rule we have,

$$y = \lim_{x \to 4} \frac{2x}{1} = 2 \times 4 = 8.$$

Prob. 3. Applying L'Hospital's rule we have,

$$y = \lim_{x \to -2} \frac{4x + 1}{1} = -7.$$

Prob. 5. Applying L'Hospital's rule we have, a substituting the value, we have

$$y = \lim_{x \to 0} \frac{1}{\sqrt{2x + 4}} = \frac{1}{2}.$$

Prob. 7. Applying L'Hospital's rule we have, and substituting the value, we have

$$y = \lim_{x \to 0} \frac{\cos x}{\cos x - x \sin x} = 1.$$

Prob. 9. Applying L'Hospital's rule we have,

$$y = \lim_{x \to 0} \frac{\sin x}{\tan x + x \sec^2 x}$$

Differentiating again, and substituting the value, we have

$$y = \lim_{x \to 0} \frac{\cos x}{\sec^2 x + \sec^2 x + 2x \sec^2 x \tan x} = \frac{1}{2}.$$

Prob. 11. Applying L'Hospital's rule we have,

$$y = \lim_{x \to 0^+} \frac{x + 1}{2\sqrt{x}} = \infty$$

Prob. 13. Applying L'Hospital's rule we have, and substituting the value, we have

$$y = \lim_{x \to 0} \frac{(\ln 2) 2^x}{(\ln 3) 3^x} = \frac{\ln 2}{\ln 3}.$$

Prob. 15. Applying L'Hospital's rule we have,

$$y = \lim_{x \to 0} \frac{e^x - 1}{2x}$$

Differentiating again,

$$y = \lim_{x \to 0} \frac{e^x}{2}$$

Substituting the value, we have

$$= \frac{1}{2}.$$

Prob. 17. Applying L'Hospital's rule we have,

$$y = \lim_{x \to \infty} \frac{2 \ln x}{x 2x}$$

Differentiating again,

$$y = \lim_{x \to \infty} \frac{1}{2x^2}$$

Substituting the value, we have

$$= 0.$$

Prob. 19. Applying L'Hospital's rule we have, and substituting the value, we have

$$y = \lim_{x \to (\frac{\pi}{2})^-} \frac{\sec^2 x}{2 \sec^2 x \tan x} = \lim_{x \to (\frac{\pi}{2})^-} \frac{1}{2 \tan x} = 0.$$

Prob. 21. Converting into form so that L'Hospital's rule can be applied,

$$y = \lim_{x \to \infty} \frac{x}{e^x}$$

Applying L'Hospital's rule we have, and substituting the value, we have

$$y = \lim_{x \to \infty} \frac{1}{e^x} = 0.$$

Prob. 23. Converting into form so that L'Hospital's rule can be applied,

$$y = \lim_{x \to 0} \frac{\ln x}{\frac{1}{\sqrt{x}}}$$

Applying L'Hospital's rule we have, and substituting the value, we have

$$y = \lim_{x \to 0} -2x^{\frac{1}{2}} = 0.$$

Prob. 25. Given,

$$y = \lim_{x \to (\frac{\pi}{2})^-} (\frac{\pi}{2} - x) \sec x$$

Converting into form so that L'Hospital's rule can be applied,

$$y = \lim_{x \to (\frac{\pi}{2})^-} \frac{(\frac{\pi}{2} - x)}{\cos x}$$

Applying L'Hospital's rule we have, and substituting the value, we have

$$y = \lim_{x \to (\frac{\pi}{2})^-} \frac{-1}{-\sin x} = 1.$$

Prob. 27. Converting into form so that L'Hospital's rule can be applied, Substituting

$$t = \frac{1}{x}$$

and hence when

$$x \to \infty, t \to 0$$

$$y = \lim_{t \to 0} \frac{\sin t}{\sqrt{t}}$$

Applying L'Hospital's rule we have, and substituting the value, we have

$$y = \lim_{t \to 0} \frac{2 \cos t \sqrt{t}}{1} = 0.$$

Prob. 29. Converting into form so that L'Hospital's rule can be applied,

$$y = \lim_{x \to 0^+} \frac{\cos x - 1}{\sin x}$$

Applying L'Hospital's rule we have, and substituting the value, we have

$$y = \lim_{x \to 0^+} \frac{-\sin x}{\cos x} = 0.$$

Prob. 31. Converting into form so that L'Hospital's rule can be applied,

$$y = \lim_{x \to 0^+} \frac{x - \sin x}{x \sin x}$$

Applying L'Hospital's rule we have,

$$y = \lim_{x\to 0^+} \frac{1-\cos x}{\sin x + x\cos x}$$

Differentiating again and substituting the value, we have

$$y = \lim_{x\to 0^+} \frac{\sin x}{2\cos x - x\sin x} = 0.$$

Prob. 33. Converting into form so that L'Hospital's rule can be applied . Taking ln on both sides,

$$\ln y = \lim_{x\to 0^+} \frac{2\ln x}{\frac{1}{x}}$$

Applying L'Hospital's rule we have, and substituting the value, we have

$$\ln y = \lim_{x\to 0^+} \frac{-2x^2}{x} \ln y = 0. \Rightarrow y = 1$$

Prob. 35. Converting into form so that L'Hospital's rule can be applied . Taking ln on both sides,

$$\ln y = \lim_{x\to\infty} \frac{\ln x}{x}$$

Applying L'Hospital's rule we have, and substituting the value, we have

$$\ln y = \lim_{x\to\infty} \frac{1}{x} = 0 \Rightarrow y = 1$$

Prob. 37. Converting into form so that L'Hospital's rule can be applied . Taking ln on both sides,

$$\ln y = \lim_{x\to\infty} \frac{\ln(1+\frac{3}{x})}{\frac{1}{x}}$$

Substituting

$$t = \frac{1}{x}$$

such that

$$x\to\infty, t\to 0$$

Thus we have,

$$\ln y = \lim_{t\to 0} \frac{\ln(1+3t)}{t}$$

Applying L'Hospital's rule we have, and substituting the value, we have

$$\ln y = \lim_{t\to 0} \frac{3}{1+t} = 3 \Rightarrow y = e^3$$

Prob. 39. Converting into form so that L'Hospital's rule can be applied . Taking ln on both sides,

$$\ln y = \lim_{x\to\infty} \frac{\ln(\frac{x}{1+x})}{\frac{1}{x}}$$

Substituting

$$t = \frac{1}{x}$$

such that

$$x \to \infty, t \to 0$$

Thus we have,

$$\ln y = \lim_{t\to 0} -\frac{\ln(1+t)}{t}$$

Applying L'Hospital's rule we have, and substituting the value, we have

$$\ln y = \lim_{t\to 0} -\frac{1}{1+t} = -1 \Rightarrow y = \frac{1}{e}$$

Prob. 41. Given,

$$y = \lim_{x\to 0^+} xe^x$$

Now the above function is not in L'Hospital rule format, and hence here we can directly get the limit by substituting the value of x in above function . So,

$$y = \lim_{x\to 0^+} xe^x$$

$$y = 0$$

Prob. 43. Given

$$y = \lim_{x\to(\frac{\pi}{2})^+} \tan x + \sec x$$

We rewrite above function as,

$$y = \lim_{x\to(\frac{\pi}{2})^+} \frac{1+\sin x}{\cos x}$$

Here we can not apply L'Hospital Rule. By directly substituting the value of $x = \frac{\pi}{2}$ we get the limit. Thus we have,

$$y = \infty$$

Prob. 45. Given

$$y = \lim_{x\to 1} \frac{x^2+5}{x+1}$$

Here again L'Hospital rule is not applicable, and hence we get the limit by directly substituting the value of $x = 1$ we get the limit. Thus we have,

$$y = \lim_{x\to 1} \frac{1+5}{1+1}$$

$$y = 3$$

Prob. 47. Making the substitution $y = -x$ we can rewrite the limit as

$$\lim_{x\to -\infty} xe^x = \lim_{y\to\infty} -ye^{-y} = -\lim_{y\to\infty} \frac{y}{e^y} = -\lim_{x\to\infty} \frac{1}{e^x} = 0$$

where we applied L'Hospital on the last calculation.

Prob. 49. Converting into form so that L'Hospital's rule can be applied . Taking ln on both sides,

$$\ln y = \lim_{x\to 0^+} \frac{3\ln x}{\frac{1}{x}}$$

Applying L'Hospital's rule we have,

$$\ln y = \lim_{x\to 0^+} \frac{-3x^2}{x}$$

Substituting the value, we have

$$\ln y = 0 \Rightarrow y = 1$$

Prob. 51. Given

$$y = \lim_{x\to 0} \frac{a^x-1}{b^x-1}$$

We can apply L'Hospital rule to above limit. Applying L'Hospital's rule we have,

$$y = \lim_{x\to 0} \frac{a^x \ln a}{b^x \ln b}$$

$$y = \frac{\ln a}{\ln b}$$

Prob. 53. Given,

$$y = \lim_{x \to \infty} (1 + \frac{c}{x^p})^x$$

Converting into form so that L'Hospital's rule can be applied . Taking ln on both sides,

$$\ln y = \lim_{x \to \infty} \frac{\ln(1 + \frac{c}{x^p})}{\frac{1}{x}}$$

Substituting

$$t = \frac{1}{x}$$

such that

$$x \to \infty, t \to 0$$

Thus we have,

$$\ln y = \lim_{t \to 0} \frac{\ln(1 + ct^p)}{t}$$

Now we have different case depending upon $p > 1$, $p < 0$ $p = 1$. when $p > 1$, Applying L'Hospital's rule we have,

$$\ln y = \lim_{t \to 0} \frac{pct^{p-1}}{1 + ct^p}$$

Substituting the value, we have

$$\ln y = 0$$

$$y = 1$$

when $p = 1$, Applying L'Hospital's rule we have,

$$\ln y = \lim_{t \to 0} \frac{c}{1 + ct}$$

Substituting the value, we have

$$\ln y = c$$

$$y = e^c$$

when $p < 0$, Applying L'Hospital's rule we have,

$$\ln y = \lim_{t \to 0} \frac{pct^{p-1}}{1 + ct^p}$$

Substituting the value, we have

$$\ln y = \infty$$

$$y = \infty$$

Thus we can summarize as

$$y = \begin{cases} 1 & \text{for } p > 1 \\ e^c & \text{for } p = 1 \\ \infty & \text{for } p < 0. \end{cases}$$

Prob. 55. Given

$$y = \lim_{x \to \infty} \frac{\ln x}{x^p}$$

Applying L'Hospital Rule,we have

$$y = \lim_{x \to \infty} \frac{1}{px^p}$$

Thus substituting the value, we get,

$$y = 0$$

as denominator tends to ∞.

Prob. 57. Given,

$$y = 121e^{\frac{-17}{x}}$$

where y is the height in feet and x is the age of the tree in years.

(a) Rate of Growth is given by,

$$\frac{dy}{dx} = 2057\frac{e^{-17/x^2}}{x^2}$$

Thus for $x \to 0^+$ we have

$$\lim_{x \to 0^+} \frac{2057}{x^2} e^{\frac{-17}{x}}$$

Let $t = \frac{1}{x}$ Rewriting we have,

$$= \lim_{t \to \infty} 2057(e^{-17t}t^2)$$

Applying L'Hospital's Rule we have and substituting we have

$$y = 0$$

and $\lim_{x \to \infty} \frac{dy}{dx} = 0$.

(b) To find x such that the growth is maximal, we differentiate the below equation, and equate it to zero,

$$y = GrowthRate = 2057\frac{e^{\frac{-17}{x}}}{x^2}$$

Differentiating we have

$$y' = 2057\left(\frac{17e^{\frac{-17}{x}} - 2xe^{\frac{-17}{x}}}{x^4}\right)$$

Hence we have

$$x = \frac{17}{2}$$

(c) For $x < \frac{17}{2}$, the function is increasing at an accelerating rate and above that at a decelerating rate.

(d)

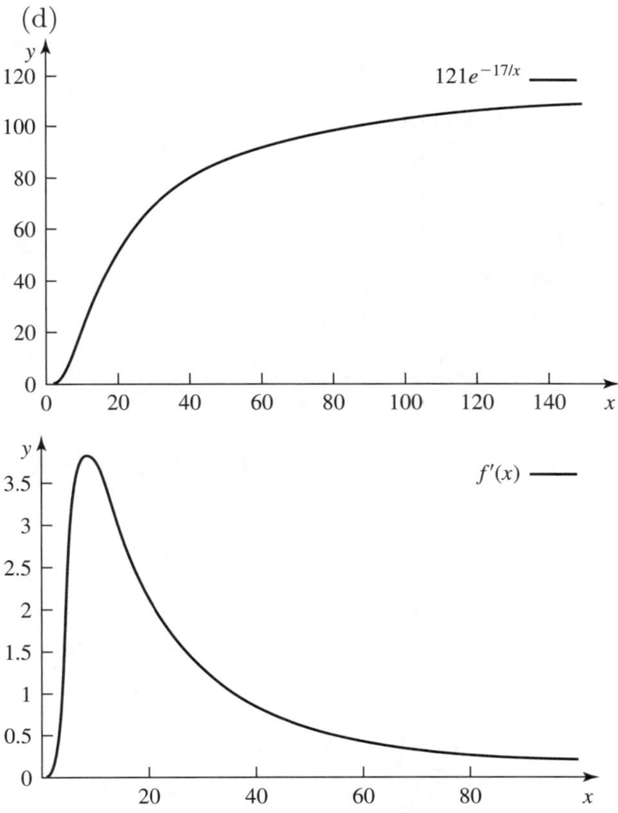

5.6 Difference Equations: Stability

Prob. 1.

(a) Given $N_{t+1} = (1.03)N_t$. This can be written as,

$$N_{t+1} = (1.03)((1.03)N_{t-1})$$
$$\Rightarrow N_{t+1} = (1.03)^{t+1}N_0$$

Given, $N_0 = 10$, Population when the generation, $t = 5$ is

$$\begin{aligned} N_5 &= (1.03)^5 * 10 \\ &= 11.59 \end{aligned}$$

(b) If $N_x = 2N_0$, then $x =$?

$\Rightarrow N_x = 2(10) = 20$. We have $N_{t+1} = (1.03)^{t+1}N_0$.

Place $(t+1)$ as x,

$$\begin{aligned} \Rightarrow \quad N_x &= (1.03)^x N_0 \\ \Rightarrow \quad 20 &= (1.03)^x N_0 \\ \Rightarrow \quad 20 &= (1.03)^x 10 \\ \Rightarrow \quad & \frac{20}{10} = (1.03)^x \end{aligned}$$

applying log on both sides,

$$\begin{aligned} \Rightarrow \quad \log(2) &= log((1.03)^x) \\ \Rightarrow \quad \log(2) &= x \cdot \log(1.03) \\ \Rightarrow \quad x &= \log(2)/\log(1.03) \\ \Rightarrow \quad x &= 23.45 \end{aligned}$$

by round-off, $x = 24$. I.e., It will be in 24th generation, the population size will be twice that of N_0.

Prob. 3.

(a) Given population model is, $N_{t+1} = bN_t$. Also given, population increases by 2% each generation. Therefore, the N term at $t = 1$ is

$$\begin{aligned} & N_1 = bN_0 \\ \Rightarrow \quad & N_0 + (0.02)N_0 = bN_0 \\ \Rightarrow \quad & (1.02)N_0 = bN_0 \\ \text{therefore} \quad & b = 1.02. \end{aligned}$$

(b) If $N_0 = 20$, then $N_t = ?$ when $t = 10$. Given population model can be written as $N_t = b^t N_0$

$$\begin{aligned} \Rightarrow \quad N_{10} &= (1.02)^{10} 20 \\ \Rightarrow \quad N_{10} &= 24.37 \end{aligned}$$

(c) If $N_t = 2N_0$, then $t = ?$

$$\begin{aligned} \Rightarrow \quad 2N_0 &= (1.02) t N_0 \\ \Rightarrow \quad 2 &= (1.02)^t \end{aligned}$$

applying log on both sides,

$$\begin{aligned} \Rightarrow \quad \log(2) &= \log((1.02)^t) \\ \Rightarrow \quad \log(2) &= t \log(1.02) \\ \Rightarrow \quad t &= \log(2)/\log(1.02) \\ \Rightarrow \quad t &= 35 \end{aligned}$$

Prob. 5.

(a) Given population model is, $N_{t+1} = bN_t$. Also given, population increases by $x\%$ each generation. Therefore, the N term at $t = 1$ is

$$\begin{aligned} & N_1 = bN_0 \\ \Rightarrow \quad & N_0 + (x/100)N_0 = bN_0 \\ \Rightarrow \quad & (1 + x/100)N_0 = bN_0 \\ \text{therefore} \quad & b = (100 + x)/100. \end{aligned}$$

(b) If $N_t = 2N_0$, then determine t.

$$\begin{aligned} \Rightarrow \quad & 2N_0 = ((100 + x)/100)^t N_0 \\ \Rightarrow \quad & 2 = ((100 + x)/100)^t \end{aligned}$$

applying log on both sides,

$$\begin{aligned} \Rightarrow \quad & \log(2) = \log(((100 + x)/100)^t) \\ \Rightarrow \quad & \log(2) = t \log((100 + x)/100) \\ \Rightarrow \quad & t = \log(2)/\log((100 + x)/100) \end{aligned}$$

Now, for $x = 0.1$,

$$\begin{aligned} t &= \log(2)/\log((100+0.1)/100) \\ t &= 693.49 \end{aligned}$$

For $x = 0.5$,

$$\begin{aligned} t &= \log(2)/\log((100+0.5)/100) \\ t &= 138.97 \end{aligned}$$

For $x = 1$,

$$\begin{aligned} t &= \log(2)/\log((100+1)/100) \\ t &= 69.66 \end{aligned}$$

For $x = 2$,

$$\begin{aligned} t &= \log(2)/\log((100+2)/100) \\ t &= 35.00 \end{aligned}$$

For $x = 5$,

$$\begin{aligned} t &= \log(2)/\log((100+5)/100) \\ t &= 14.20 \end{aligned}$$

For $x = 10$,

$$\begin{aligned} t &= \log(2)/\log((100+10)/100) \\ t &= 7.27 \end{aligned}$$

Prob. 7.

(a) Given population model is,

$$N_{t+1} = 0.9N_t$$

The equilibrium can be obtained by solving $N = RN$. But the only solution for this is $N^* = 0$. As $0 < R < 1$, N_t will return to equilibrium $N^* = 0$ if $N_0 > 0$. Therefore as $0 < R < 1$, this population model is stable.

(b) The fixed points are found graphically, where the graphs of $N_{t+1} = 1.3N_t$ and $N_{t+1} = N_t$ intersects.

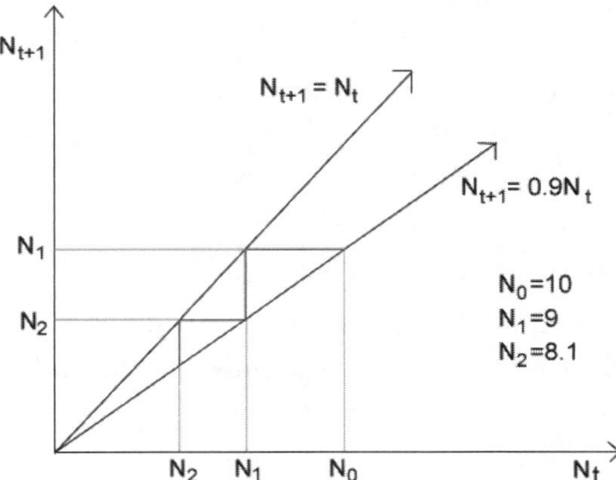

For any N_0, the population growth is converging towards equilibrium $N^* = 0$. Therefore, given population model is stable.

Prob. 9. Given model is, $x_{t+1} = \frac{2}{3} - \frac{2}{3}x_t^2$. To find the equilibrium, we need to solve $x = f(x)$ i.e., to solve $x = \frac{2}{3} - \frac{2}{3}x^2$

$$\begin{aligned} \Rightarrow \quad \frac{2}{3}x^2 + x - \frac{2}{3} &= 0 \\ \Rightarrow \quad 2x^2 + 3x - 2 &= 0 \end{aligned}$$

The left hand side can be factored into $(2x - 1)(x + 2)$, and we find that $(2x - 1)(x + 2) = 0$,

$$\text{therefore} \quad x = \frac{1}{2} \quad \text{or} \quad x = -2$$

To determine stability, we need to evaluate the derivative of $f(x) = \frac{2}{3} - \frac{2}{3}x^2$ at the equilibrium. Now, $f'(x) = -\frac{4}{3}x$ if $x = \frac{1}{2}$, then

$$\left| f'\left(\frac{1}{2}\right) \right| = \left| -\frac{2}{3} \right| = \frac{2}{3} < 1$$

and if $x = -2$, then

$$|f'(-2)| = \left| \frac{8}{3} \right| = \frac{8}{3} > 1$$

Thus $x = \frac{1}{2}$ is locally stable and $x = -2$ is unstable. As $f'\left(\frac{1}{2}\right) = -\frac{2}{3} > 0$, the equilibrium is approached with oscillations.

Prob. 11. Given model is $x_{t+1} = \frac{x_t}{0.5+x_t}$. To find the equilibrium, we need to solve $x = f(x)$ i.e., to solve $x = \frac{x}{0.5+x}$

$$\begin{aligned} \Rightarrow \quad x(0.5+x) &= x \\ \Rightarrow \quad x(x-0.5) &= 0 \end{aligned}$$

Therefore $x = 0$ or $x = 0.5$.

Now,

$$f'(x) = \frac{0.5+x-x}{(0.5+x)^2} = \frac{0.5}{(0.5+x)^2}.$$

Since $f'(0) = \frac{1}{0.5} = 2 > 1$, we conclude that $x^* = 0$ is unstable. Since $f'(0.5) = 0.5 \in (0,1)$, we conclude that $x^* = 0.5$ is locally stable and is approached without oscillations.

Prob. 13.

(a) Given model is $x_{t+1} = \frac{5x_t^2}{4+x_t^2}$. To find the equilibrium, we need to solve $x = f(x)$ i.e., to solve $x = \frac{5x^2}{4+x^2}$

$$\Rightarrow \quad x(x^2 - 5x + 4) = 0.$$

Therefore $x = 0$ or $x = 4$ or $x = 1$.

Now,

$$f'(x) = \frac{40x}{(4+x^2)^2}$$

Since $f'(0) = 0$, we conclude that $x^* = 0$ is locally stable and is approached without oscillations. Since $f'(4) = \frac{2}{5} \in (0,1)$, we conclude that $x^* = 4$ is locally stable and is approached without oscillations. Since $f'(1) = \frac{8}{5} > 1$, we conclude that $x^* = 1$ is unstable.

(b) (i) Considering $x_0 = 0.5$

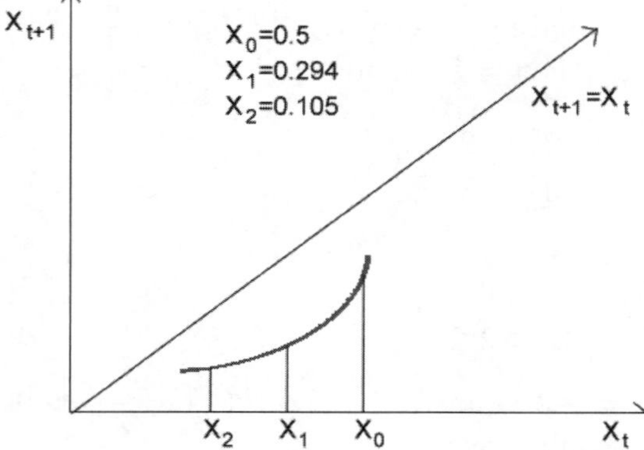

The subsequent terms at $t = 0$, $t = 1$, $t = 2$ shows that the value is converging towards closest fixed point $x^* = 0$

(b) (ii) Considering $x_0 = 2$. The subsequent values at $t = 0$, $t = 1$ and $t = 2$ shows that value x value is reaching towards fixed point $x^* = 4$.

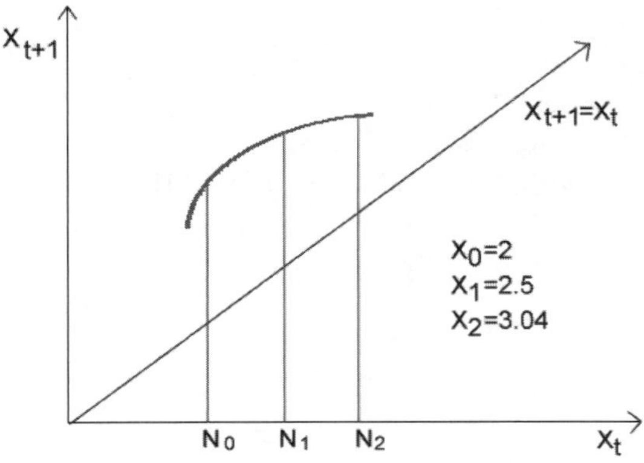

Prob. 15. (b) 0, (c) $1/\beta$, (d) $P = 2/\beta$ is an inflection point,
(e)

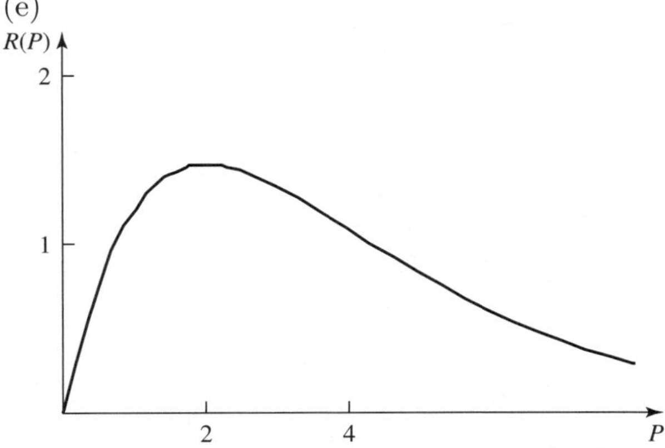

Prob. 17. (b) To find equilibria, solve $N = 10Ne^{-0.01N}$; $N = 0$ is another equilibrium.
(c) Oscillations seem to appear and the system does not seem to converge to the nontrivial equilibrium.

Prob. 19. Equilibria: $N = 0$ and $N = 50$.
(b) Starting from $N = 10$, it appears that the limiting population size is 50. $N = 50$ is locally stable.

5.7 Numerical Methods

Prob. 1. $\sqrt{7} = 2.645751$

Prob. 3. 0.6529186

Prob. 5. 1.895494

Prob. 7. (a) $|x_n| = 2^n x_0$, (b) ∞

Prob. 9. (a) $x_0 = 3, x_1 = 4.166667, x_2 = 4.003333, x_3 = 4.000001$, (b) $x_0 = x_1 = x_2 = \ldots = 4$

5.8 Antiderivatives

Prob. 1. Given the function as,

$$f(x) = 2x^2$$

Integrating, we get

$$\int f(x) = F(x) = \frac{2}{3}x^3 + C.$$

where C is a constant.

Prob. 3. Given the function as,

$$f(x) = x^2 + 2x - 1$$

Integrating, we get

$$\int f(x) = F(x) = \frac{x^3}{3} + x^2 - x + C.$$

where C is a constant.

Prob. 5. Given the function as,

$$f(x) = x^4 - 3x^2 + 1$$

Integrating, we get

$$\int f(x) = F(x) = \frac{x^5}{5} - x^3 + x + C.$$

where C is a constant.

Prob. 7. Given the function as,

$$f(x) = 4x^3 - 2x + 3$$

Integrating, we get

$$\int f(x) = F(x) = x^4 - x^2 + 3x + C.$$

where C is a constant.

Prob. 9. Given the function as,

$$f(x) = x + \frac{1}{x}$$

Integrating, we get

$$\int f(x) = F(x) = \frac{x^2}{2} + \ln|x| + C.$$

where C is a constant.

Prob. 11. Given the function as,

$$f(x) = 1 - \frac{1}{x^2}$$

Integrating, we get

$$\int f(x) = F(x) = x + \frac{1}{x} + C.$$

where C is a constant.

Prob. 13. Given the function as,

$$f(x) = \frac{1}{1+x}$$

Integrating, we get

$$\int f(x) = F(x) = \ln|1+x| + C.$$

where C is a constant.

Prob. 15. Given the function as,

$$f(x) = 5x^4 + \frac{5}{x^4}$$

Integrating, we get

$$\int f(x) = F(x) = x^5 - \frac{5}{3x^3} + C.$$

where C is a constant.

Prob. 17. Given the function as,

$$f(x) = \frac{1}{2+x}$$

Integrating, we get

$$\int f(x) = F(x) = \ln|2+x| + C.$$

where C is a constant.

Prob. 19. Given the function as,

$$f(x) = e^{-3x}$$

Integrating, we get

$$\int f(x) = F(x) = -\frac{e^{-3x}}{3} + C.$$

where C is a constant.

Prob. 21. Given the function as,

$$f(x) = 2e^{2x}$$

Integrating, we get

$$\int f(x) = F(x) = e^{2x} + C.$$

where C is a constant.

Prob. 23. Given the function as,

$$f(x) = \frac{1}{e^{2x}}$$

Integrating, we get

$$\int f(x) = F(x) = -\frac{1/2}{e^{2x}} + C.$$

where C is a constant.

Prob. 25. Given the function as,

$$f(x) = \sin(2x)$$

Integrating, we get

$$\int f(x) = F(x) = -\frac{\cos(2x)}{2} + C.$$

where C is a constant.

Prob. 27. Given the function as,

$$f(x) = \sin(\frac{x}{3})$$

Integrating, we get

$$\int f(x) = F(x) = -3\cos(\frac{x}{3}) + C.$$

where C is a constant.

Prob. 29. Given the function as,

$$f(x) = 2\sin(\frac{\pi x}{2}) - 3\cos(\frac{\pi x}{2})$$

Integrating, we get

$$\int f(x) = F(x) = -\frac{4}{\pi}\cos(\frac{\pi x}{2}) - \frac{6}{\pi}\sin(\frac{\pi x}{2}) + C.$$

where C is a constant.

Prob. 31. Given the function as,

$$f(x) = \sec^2(2x)$$

Integrating, we get

$$\int f(x) = F(x) = \frac{\tan(2x)}{2} + C.$$

where C is a constant.

Prob. 33. Given the function as,

$$f(x) = \sec^2(\frac{x}{3})$$

Integrating, we get

$$\int f(x) = F(x) = 3\tan(\frac{x}{3}) + C.$$

where C is a constant.

Prob. 35. Given the function as,

$$f(x) = \frac{\sec x + \cos x}{\cos x}$$

We can rewrite it as,

$$f(x) = \sec^2 x + 1$$

Integrating, we get

$$\int f(x) = F(x) = \tan x + x + C$$

where C is a constant.

Prob. 37. Given the equation as,

$$\frac{dy}{dx} = \frac{2}{x} - x$$

Thus, we have the general solution by finding the antiderivative of the above function. Hence,

$$y = 2\ln|x| - \frac{x^2}{2} + C$$

where C is a constant.

Prob. 39. Given the equation as,

$$\frac{dy}{dx} = x(1+x)$$

We can rewrite the above function as,

$$\frac{dy}{dx} = x^2 + x$$

Thus, we have the general solution by finding the antiderivative of the above function. Hence,

$$y = \frac{x^3}{3} + \frac{x^2}{2} + C$$

where C is a constant.

Prob. 41. Given the equation as,

$$\frac{dy}{dt} = t(1-t)$$

We can rewrite the above function as,

$$\frac{dy}{dt} = t - t^2$$

Thus, we have the general solution by finding the antiderivative of the above function. Hence,

$$y = \frac{t^2}{2} - \frac{t^3}{3} + C$$

where C is a constant.

Prob. 43. Given the equation as,

$$\frac{dy}{dt} = e^{-\frac{t}{2}}$$

Thus we have the general solution by finding the antiderivative of the above function. Hence,

$$y = -2e^{-\frac{t}{2}} + C$$

where C is a constant.

Prob. 45. Given the equation as,

$$\frac{dy}{dx} = \sin(\pi x)$$

Thus, we have the general solution by finding the antiderivative of the above function. Hence,

$$y = \frac{-1}{\pi}\cos(\pi x) + C$$

where C is a constant.

Prob. 47. Given the equation as,

$$\frac{dy}{dx} = \sec^2 \frac{x}{2}$$

Thus, we have the general solution by finding the antiderivative of the above function. Hence,

$$y = 2\tan\frac{x}{2} + C$$

where C is a constant.

Prob. 49. Given the equation as,

$$\frac{dy}{dx} = 3x^2$$

and given that when,

$$y = 1, x = 0$$

Thus, we have the general solution by finding the antiderivative of the above function. Hence,

$$y = x^3 + C$$

where C is a constant. Substituting the values of y and x we get, $1 = C$ and hence, we have $C = 1$. Thus, we have final equation as

$$y = x^3 + 1$$

Prob. 51. Given the equation as,

$$\frac{dy}{dx} = 2\sqrt{x}$$

and given that when,

$$y = 2, x = 1$$

Thus, we have the general solution by finding the antiderivative of the above function. Hence,

$$y = \frac{4}{3}x^{\frac{3}{2}} + C$$

where C is a constant. Substituting the values of y and x we get,

$$2 = \frac{4}{3} + C$$

Hence, we get

$$C = \frac{2}{3}$$

Thus, we have final equation as

$$y = \frac{4}{3}x^{\frac{3}{2}} + \frac{2}{3}$$

Prob. 53. Given the equation as,

$$\frac{dN}{dt} = \frac{1}{t}$$

and given that when,

$$N(1) = 10$$

Thus, we have the general solution by finding the antiderivative of the above function. Hence,

$$N = \ln t + C$$

where C is a constant. Substituting the values of $N(1)$ we get,

$$10 = \ln 1 + C$$

Hence we get

$$C = 10$$

Thus, we have final equation as

$$N = \ln t + 10$$

Prob. 55. Given the equation as,

$$\frac{dW}{dt} = e^t$$

and given that,

$$W(0) = 1$$

Thus, we have the general solution by finding the antiderivative of the above function. Hence,

$$W = e^t + C$$

where C is a constant. Substituting the values of $N(0)$ we get,

$$1 = 1 + C$$

Hence, we have

$$C = 0$$

Thus, we have final equation as

$$W = e^t$$

Prob. 57. Given the equation as, $\frac{dW}{dt} = e^{-3t}$, $W(0) = \frac{2}{3}$ we have the general solution by finding the antiderivative of the above function. Hence,

$$W = \frac{-1}{3}e^{-3t} + C$$

where C is a constant. Substituting the values of $N(0)$ we get,

$$\frac{2}{3} = \frac{-1}{3} + C$$

Hence, we have

$$C = 1$$

Thus, we have equation as

$$W = \frac{-1}{3}e^{-3t} + 1$$

Prob. 59. Given the equation as,

$$\frac{dT}{dt} = \sin(\pi t)$$

and given that,

$$T(0) = 3$$

Thus, we have the general solution by finding the antiderivative of the above function. Hence,

$$T = -\frac{1}{\pi}\cos(\pi t) + C$$

where C is a constant. Substituting the values of $T(0)$ we get,

$$3 = -\frac{1}{\pi} + C$$

Hence, we have

$$C = 3 + \frac{1}{\pi}$$

Thus, we have equation as

$$T = -\frac{1}{\pi}\cos(\pi t) + 3 + \frac{1}{\pi}$$

Prob. 61. Given the equation as,

$$\frac{dy}{dx} = \frac{e^x + e^{-x}}{2}$$

and given that,

$$y = 0, x = 0$$

We can rewrite the above equation as

$$\frac{dy}{dx} = \frac{e^x}{2} + \frac{e^{-x}}{2}$$

Thus, we have the general solution by finding the antiderivative of the above function. Hence,

$$y = \frac{e^x}{2} - \frac{e^{-x}}{2} + C$$

where C is a constant. Substituting the values of $y = 0$ and $x = 0$ we get,

$$0 = \frac{1}{2} - \frac{1}{2} + C$$

Hence, we have

$$C = 0$$

Thus, we have final equation as

$$y = \frac{e^x}{2} - \frac{e^{-x}}{2}$$

Prob. 63. Given the equation as,

$$\frac{dL}{dx} = e^{-.1x}$$

and $x \geq 0$. Also given that

$$L_\infty = \lim_{x\to\infty} L(x) = 25$$

First, we find, $L(x)$.

$$L(x) = -10e^{-.1x} + C$$

where C is a constant. Substituting this, in above equation, we have

$$L_\infty = \lim_{x\to\infty} (-10e^{-.1x} + C) = 25$$

$$C = 25$$

Thus we have,

$$L(x) = -10e^{-.1x} + 25$$

Hence, we have $L(0)$ as,

$$\begin{aligned} L(0) &= -10 + 25 \\ &= 15 \end{aligned}$$

Prob. 65. $h = 100$ ft, $a = 32\text{ft/s}^2$. With $s(t) = \frac{1}{2}at^2$, we find $t = \sqrt{\frac{2s}{a}} = \sqrt{\frac{(2)(100)}{32}} = 2.5s$. Furthermore, $v(t) = at = (32)(2.5)ft/s = 80ft/s$.

Prob. 67. We are given the volume change at time t is proportional to $t(24 - t)$, and that we water the plant at a constant rate of 4 per hour.

(a) Thus the net rate of loss of water is

$$\frac{dV}{dt} = -at(24 - t) + 4$$

This is because the first part of equation denotes the loss of water due to evaporation and the second part denote the water added per hour. Here a is a constant which has been introduced to convert the proportionality to some constant value dependent on t only.

(b)

$$V(t) = -\frac{a}{2}t^2 \cdot 24 + \frac{a}{3}t^3 + 4t + C$$

With $V(0) = 0$ we thus have $V(t) = -\frac{a}{2}t^2 \cdot 24 + \frac{a}{3}t^3 + 4t$. Since $V(24) = 0$, we can find a, namely,

$$\begin{aligned} 0 &= -\frac{a}{2}24^2 \cdot 24 + \frac{a}{3}24^3 + (4)(24) \\ (24)^2\left(\frac{a}{2} - \frac{a}{3}\right) &= 4 \\ \frac{a}{6} &= \frac{4}{(24)(24)} \\ a &= \frac{1}{24} \end{aligned}$$

5.10 Review Problems

Prob. 1. Given the function as,

$$f(x) = xe^{-x}, x \geq 0$$

(a) For evaluation $f(0)$, we put $x = 0$ in above equation. Substituting, the value we have,

$$f(0) = 0$$

For calculating the limit, we have

$$F = \lim_{x\to\infty} xe^{-x}$$

$$= \lim_{x\to\infty} \frac{x}{e^x}$$

Applying L'Hospital rule, we have,

$$F = \lim_{x\to\infty} \frac{1}{e^x}$$

Substituting the values we have

$$F = 0$$

(b) For local maxima, we differentiate the below function,

$$f(x) = xe^{-x}$$

So, differentiating,

$$f'(x) = e^{-x} - xe^{-x}$$

Substituting to zero,we have

$$f'(x) = 0$$

$$e^{-x} - xe^{-x} = 0$$

$$e^{-x}(1 - x) = 0$$

So we have

$$x = 1$$

Since $f'(x)$ changes from positive to negative at $x = 1$, there is a local max at $(1, e^{-1})$. Thus the function has an absolute maximum at $x = 1$ and an absolute minimum at $x = 0$.

(c) To calculate the inflection point, we differentiate it twice and equate to zero. Given,

$$f(x) = xe^{-x}$$

So,

$$f'(x) = e^{-x} - xe^{-x}$$

Differentiating again,

$$\begin{aligned} f''(x) &= -e^{-x} - e^{-x} + xe^{-x} \\ &= e^{-x}(x-2) \end{aligned}$$

Equating to zero, we have

$$e^{-x}(x-2) = 0$$

Hence,

$$x = 2$$

Since $f''(x)$ changes sign at $x = 2$, this is the point of inflection.

(d)

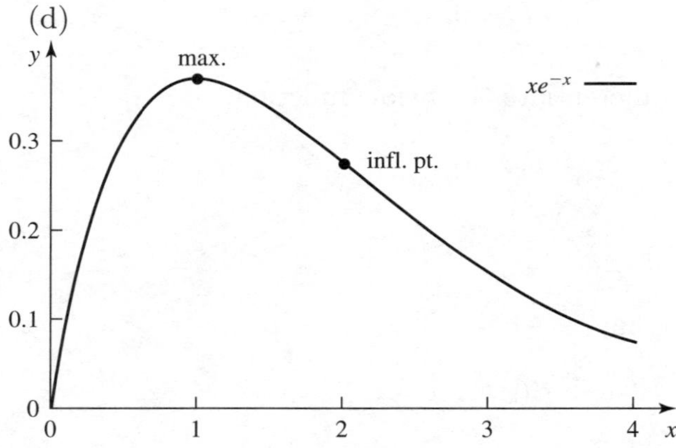

Prob. 3. Given the following definitions,

$$\sinh x = \frac{e^x - e^{-x}}{2}$$

$$\cosh x = \frac{e^x + e^{-x}}{2}$$

$$\tanh x = \frac{e^x - e^{-x}}{e^x + e^{-x}}$$

(a) From above we have,

$$\tanh x = \frac{e^x - e^{-x}}{e^x + e^{-x}}$$

So differentiating it, we have

$$\begin{aligned} f' &= \frac{(e^x + e^{-x})(e^x + e^{-x}) - (e^x - e^{-x})(e^x - e^{-x})}{(e^x + e^{-x})^2} \\ &= \frac{4}{(e^x + e^{-x})^2} \end{aligned}$$

Thus we see that f' is always positive and hence is a strictly increasing function. Also,

$$\begin{aligned} \lim_{x \to \infty} \tanh x &= \lim_{x \to \infty} \frac{e^x - e^{-x}}{e^x + e^{-x}} \\ &= 1 \end{aligned}$$

Again,

$$\begin{aligned} \lim_{x \to -\infty} \tanh x &= \lim_{x \to -\infty} \frac{e^x - e^{-x}}{e^x + e^{-x}} \\ &= 1 \end{aligned}$$

(b) The function $f(x) = \tanh x$ is invertible because the function has the same values at both the limiting points and is strictly increasing with a minimum value at $x = 1$ Its inverse function can be easily calculated. We have,

$$\begin{aligned} y &= \frac{e^x - e^{-x}}{e^x + e^{-x}} \\ e^{-x}(y+1) &= e^x(1-y) \\ e^{-2x} &= \frac{1-y}{1+y} \\ -2x &= \ln(\frac{1-y}{1+y}) \\ x &= \frac{1}{2}\ln(\frac{1+y}{1-y}) \end{aligned}$$

Thus we have,

$$f^{-1}(x) = y = \frac{1}{2}\ln(\frac{1+x}{1-x})$$

(c) To find the differential, we have

$$f^{-1}(x) = y = \frac{1}{2}\ln(\frac{1+x}{1-x})$$

Differentiating,

$$\frac{d}{dx}f^{-1}(x) = \frac{1}{2}(\frac{1-x}{1+x}\frac{+2}{(1-x)^2})$$

Thus we have,

$$\frac{d}{dx}f^{-1}(x) = \frac{1}{1-x^2}$$

Hence, we get the result.

(d) Given that,

$$\tanh x = \frac{\sinh x}{\cosh x}$$

Also given,

$$\cosh^2 x - \sinh^2 x = 1$$

Thus we have to find, $\frac{d(\tanh x)}{dx}$.

$$\begin{aligned}\frac{dy}{dx} &= \frac{1}{\frac{dx}{db}} = 1 - y^2\\ &= 1 - \frac{\sin h^2 x}{\cos h^2 x}\\ &= \frac{\cos h^2 x - \sin h^2 x}{\cos h^2 x}\\ &= \frac{1}{\cos h^2 x}\end{aligned}$$

Prob. 5. We are given a function, such that,

$$R(P) = \alpha P e^{-\beta P}, P \geq 0$$

(a) To, find the first derivative, we differentiate the above function,

$$\begin{aligned}R'(P) &= \alpha e^{-\beta P} - \alpha\beta P e^{-\beta P}\\ &= \alpha e^{-\beta P}(1-\beta P)\end{aligned}$$

To find the second differential, we differentiate the above function,

$$\begin{aligned}R''(P) &= -\beta\alpha e^{-\beta P} - \beta\alpha e^{-\beta P}(1-\beta P)\\ &= -\beta\alpha e^{-\beta P}(2-\beta P)\end{aligned}$$

Hence, we get both the differentials.

(b) We have the differential of $R(P)$ as

$$R'(P) = \alpha e^{-\beta P}(1 - \beta P)$$

Equating it to zero, we have

$$\alpha e^{-\beta P}(1 - \beta P) = 0$$

$$1 - \beta P = 0$$

$$P = \frac{1}{\beta}$$

Thus $R'(P) = 0$ only when $P = \frac{1}{\beta}$ Also $R''(\frac{1}{\beta})$ equals,

$$R''(P) = -\beta\alpha e^{-\beta P}(2 - \beta P)$$

$$R''(\frac{1}{\beta}) = -\beta\alpha e^{-\beta\frac{1}{\beta}}(2 - \beta\frac{1}{\beta})$$

$$= -\beta\alpha e^{-1}$$

$$< 0$$

Thus at $P = \frac{1}{\beta}$, we have a local maxima.

Also, we see from above,

$$R(P) = \alpha P e^{-\beta P}, P \geq 0$$

Substituting $P = \frac{1}{\beta}$, we have

$$R(\frac{1}{\beta}) = \alpha\frac{1}{\beta}e^{-\beta\frac{1}{\beta}}$$

$$= \frac{\alpha}{\beta}e^{-1}$$

(c) To find the global maxima, we have to find $R(0)$ and

$$\lim_{x\to\infty} R(P)$$

So, finding the limit,

$$\lim_{x\to\infty} R(P) = \lim_{x\to\infty} \alpha P e^{-\beta P}$$

Rearranging the function, we get,

$$\lim_{x\to\infty} R(P) = \lim_{x\to\infty} \frac{\alpha P}{e^{\beta P}}$$

Applying L'Hospital Rule,

$$= \lim_{x\to\infty} \frac{\alpha}{\beta e^{\beta P}}$$

Substituting the value we get,

$$\lim_{x\to\infty} R(P) = 0$$

Since $R(0) = 0$, we have a global maxima at $P = \frac{1}{\beta}$.

(d) To find the inflection point we substitute the second derivative to zero. We have

$$R''(P) = -\beta\alpha e^{-\beta P}(2 - \beta P)$$

Equating it to zero, we have

$$-\beta\alpha e^{-\beta P}(2 - \beta P) = 0$$

$$2 - \beta P = 0$$

$$P = \frac{2}{\beta}$$

Also, we see that the second derivative of the function changes sign at this value. Hence $P = \frac{2}{\beta}$ is a point of inflection.

(e)

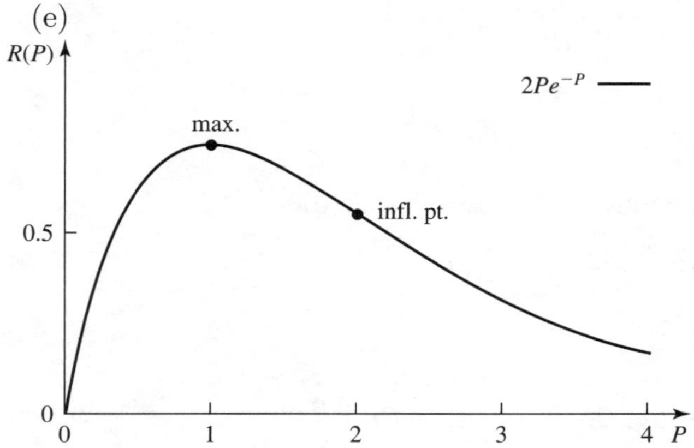

Prob. 7. The Monod growth curve is given by

$$f(x) = \frac{cx}{k+x}$$

where c and k are constants.

(a) To find the horizontal asymptote, we find the limit tending to infinity,

$$\lim_{x\to\infty} f(x) = \lim_{x\to\infty} \frac{cx}{k+x}$$

Applying L'Hospital rule we have,

$$\begin{aligned}\lim_{x\to\infty} f(x) &= \lim_{x\to\infty} c \\ &= c\end{aligned}$$

This is called the saturation value.

(b) For $x \geq 0$ the function is strictly increasing . This is because when we find its derivative we have,

$$\begin{aligned}f'(x) &= \frac{c(k+x) - cx}{(k+x)^2} \\ &= \frac{ck}{(k+x)^2} \\ &> 0\end{aligned}$$

Hence the function is strictly increasing.

(c) For $x = k$ we have

$$\begin{aligned}f(k) &= \frac{ck}{k+k} \\ &= \frac{c}{2}\end{aligned}$$

Thus this value is half of the saturation value.

(d)

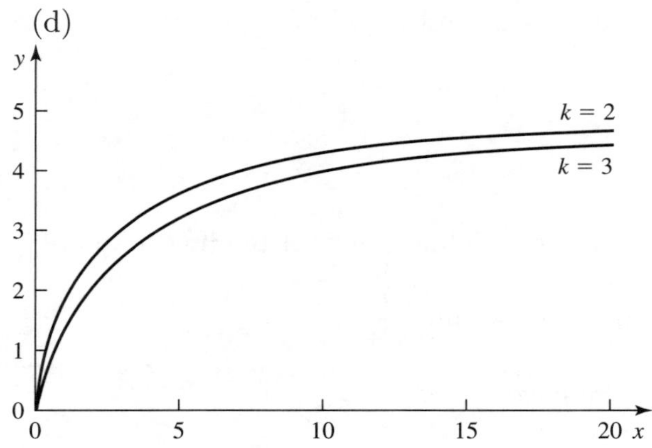

(e) For the three function given below,

$$g(x) = \frac{10x}{3+x}$$
$$h(x) = \frac{10x}{5+x}$$
$$J(x) = \frac{8x}{5+x}$$

We can find their saturation value and half saturation value directly. The higher the saturation value, the larger the function, and when the saturation value is same, we check the half saturation value, the greater it is the smaller the function. Thus, based on this analysis we can write,

$$\frac{10x}{3+x} > \frac{10x}{5+x} > \frac{8x}{5+x}$$

Prob. 9.

(a) $L(\theta) = (\theta^2)^8(2\theta(1-\theta))^6((1-\theta)^2)^3$

(b) $L(\theta)$ is maximal for $\theta = \hat{\theta}$: $\frac{dL}{d\theta} = 0$ at $\theta = \hat{\theta}$ and $\frac{d^2L}{d\theta^2} < 0$ at $\theta = \hat{\theta}$. We define $f(\theta) = \ln L(\theta)$. Then $f'(\theta) = \frac{L'(\theta)}{L(\theta)} = 0$ for $\theta = \hat{\theta}$ and

$$f''(\theta) = \frac{L''(\theta)L(\theta) - (L'(\theta))^2}{L^2(\theta)}$$

Therefore, when $\theta = \hat{\theta}$, $f''(\hat{\theta}) = \frac{L''(\hat{\theta})}{L(\hat{\theta})} < 0$ since $L(\hat{\theta}) > 0$, $L'(\hat{\theta}) = 0$, and $L''(\hat{\theta}) < 0$. Hence, if $L(\theta)$ is maximal for $\theta = \hat{\theta}$, then $f(\theta) = \ln L(\theta)$ also has a maximum at $\theta = \hat{\theta}$.

(c) $\ln L(t) = 22\ln\theta + 12\ln(1-\theta) + 6\ln 2$. Now,

$$\frac{d}{d\theta}\ln L(\theta) = 22\frac{1}{\theta} - 12\frac{1}{1-\theta} = 0$$

for $\frac{22}{\theta} = \frac{12}{1-\theta}$ and hence $\theta = \frac{22}{34} = \frac{11}{17}$.

Prob. 11. Given the equation as,

$$\frac{dc}{dt} = -0.1e^{-0.3t}$$

(a) To find c, we find its autiderivative,

$$c = -\frac{1}{3}e^{-.3t} + K$$

where K is a constant. Given at $t = \infty$, $c = 0$ so $k = 0$. Hence,

$$c(t) = -\frac{1}{3}e^{-.3t}$$

(b) At $t = 0$, we have $c = \frac{1}{3}$. To become the value to half, we have,

$$\frac{1}{2}\frac{1}{3} = +\frac{1}{3}e^{-.3t}$$
$$-\ln 2 = -.3t$$
$$t = \frac{\ln 2}{.3}$$

Thus at $t = \frac{\ln 2}{.3}$ the concentration becomes equal to half its initial amount.

Prob. 13. Given the equation of the height as,

$$h(t) = v_0 t - \frac{1}{2}gt^2$$

(a) To find the time at which, the object reaches the maximum height, we differentiate the below equation.

$$h(t) = v_0 t - \frac{1}{2}gt^2$$
$$h'(t) = v_0 - gt$$

Equating to zero, we have

$$v_0 - gt = 0$$

$$t = \frac{v_0}{g}$$

(b) The height reached at $t = \frac{v_0}{g}$ is,

$$h = v_0\frac{v_0}{g} - \frac{1}{2}g(\frac{v_0}{g})^2$$

$$= \frac{(v_0)^2}{2g}$$

This is the maximum height reached . So maximum height is,

$$\text{Max height} = \frac{(v_0)^2}{2g}$$

(c) Velocity of the particle as it reaches the maximum height is calculated as follows. we know that

$$V = v_0 - gt$$

Substituting the value of $t = \frac{v_0}{g}$, we have $V = 0$. Thus the velocity at the maximum point is zero.

(d) To reach the same height again, the particle will take the amount of time equal to two times the time taken to reach the maximum height. Thus we have from previous calculation,

$$t = \frac{v_0}{g}$$

Thus the time taken to reach the same height again is,

$$T = 2\frac{v_0}{g}$$

6 Integration

6.1 The Definite Integral

Prob. 1. The left endpoints of the four intervals are 0, $\frac{1}{4}$, $\frac{1}{2}$, $\frac{3}{4}$.
The approximation is then $\frac{1}{4} \cdot 0 + \frac{1}{4}\frac{1}{16} + \frac{1}{4}\frac{1}{4} + \frac{1}{4}\frac{9}{16} = \frac{7}{32}$.

Prob. 3. The right endpoints of the four intervals are $\frac{1}{4}$, $\frac{1}{2}$, $\frac{3}{4}$, 1.
The approximation is then $\frac{1}{4}\frac{1}{16} + \frac{1}{4}\frac{1}{4} + \frac{1}{4}\frac{9}{16} + \frac{1}{4} \cdot 1 = \frac{15}{32}$.

Prob. 5.

$$\sum_{k=1}^{4} \sqrt{k} = \sqrt{1} + \sqrt{2} + \sqrt{3} + \sqrt{4}\,.$$

Prob. 7.

$$\sum_{k=0}^{5} 2^k = 2^0 + 2^1 + 2^2 + 2^3 + 2^4 + 2^5\,.$$

Prob. 9.

$$\sum_{k=0}^{4} x^k = x^0 + x^1 + x^2 + x^3 + x^4\,.$$

Prob. 11.

$$\sum_{k=0}^{3} (-1)^k = (-1)^0 + (-1)^1 + (-1)^2 + (-1)^3\,.$$

Prob. 13.

$$\sum_{k=1}^{n} \left(\frac{k}{n}\right)^2 \frac{1}{n} = \left(\frac{1}{n}\right)^2 \frac{1}{n} + \left(\frac{2}{n}\right)^2 \frac{1}{n} + \cdots + \left(\frac{n}{n}\right)^2 \frac{1}{n}\,.$$

Prob. 15. $\sum_{k=1}^{6} k\,.$

Prob. 17. $\sum_{k=1}^{5} \ln k\,.$

Prob. 19. $\sum_{k=1}^{n} 2k\,.$

Prob. 21. $\sum_{k=0}^{n-1} q^k\,.$

Prob. 23.

$$\sum_{k=1}^{20}(3k+2) = 3\sum_{k=1}^{20} k + 2\sum_{k=1}^{20} 1 = 3\frac{20\cdot 21}{2} + 2\cdot 20 = 670\,.$$

Prob. 25.

$$\sum_{k=0}^{6} k(k+1) = \sum_{k=0}^{6}(k^2+k) = \sum_{k=1}^{6} k + \sum_{k=1}^{6} k^2 = \frac{6\cdot 7}{2} + \frac{6\cdot 7\cdot 13}{6} = 112\,.$$

Prob. 27. $\sum_{k=1}^{n} 2k^2 = 2\sum_{k=1}^{n} k^2 = 2\frac{n(n+1)(2n+1)}{6} = \frac{n(n+1)(2n+1)}{3}\,.$

Prob. 29. $\sum_{k=1}^{10}(-1)^k = -1+1-1+1-1+1-1+1-1+1 = 0\,.$

Prob. 31.

(a) All the terms collapse except the first and the last, so

$$\sum_{k=1}^{n}[(k+1)^3 - k^3] = -1^3 + (n+1)^3\,.$$

(b) We have

$$\begin{aligned}\sum_{k=1}^{n}[(k+1)^3 - k^3] &= \sum_{k=1}^{n}(3k^2+3k+1)\\ &= 3\sum_{k=1}^{n} k^2 + 3\sum_{k=1}^{n} k + \sum_{k=1}^{n} 1\\ &= 3\sum_{k=1}^{n} k^2 + 3\frac{n(n+1)}{2} + n\end{aligned}$$

(c) We have $3\sum_{k=1}^{n} k^2 = (n+1)^3 - 1 - 3\frac{n(n+1)}{2} - n$,

therefore, $\sum_{k=1}^{n} k^2 = \frac{n(n+1)(2n+1)}{6}$.

Prob. 33. $0.4[(1-(-0.8)^2)+(1-(-0.4)^2)+(1-0^2)+(1-(0.4)^2)+(1-(0.8)^2)] = 1.36$.

Prob. 35. $(2+(-2)^2)+(2+(-1)^2)+(2+0^2)+(2+1^2) = 14$.

Prob. 37. $\frac{\pi}{2}[\sin\frac{\pi}{2} + \sin\pi + \sin\frac{3\pi}{2}] = 0$.

Prob. 39. $\frac{b^2}{2} - \frac{a^2}{2}$ equals the area of $\triangle(0,0)(b,0)(b,b)$ minus the area of $\triangle(0,0)(a,0)(a,a)$ (see previous exercise.)

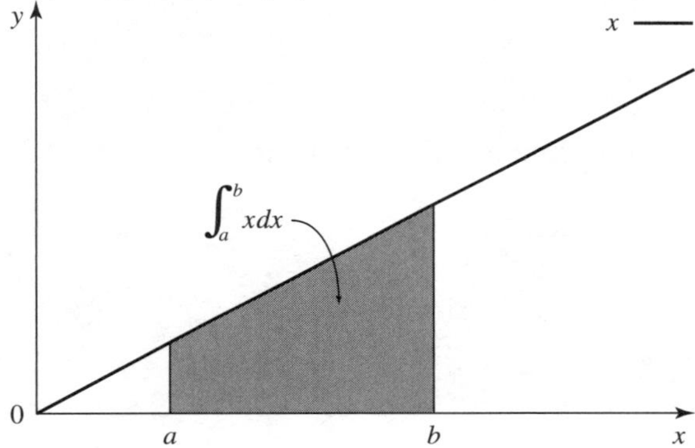

Prob. 41. $\int_0^1 x^2\,dx$.

Prob. 43. $\int_{-3}^{2}(2x-1)\,dx$.

Prob. 45. $\int_{-3}^{3}(x-1)(x+2)\,dx$.

Prob. 47. $\int_{-5}^{2} e^x\,dx$.

Prob. 49. $\lim_{\|P\|\to 0}\sum_{k=1}^{n}\sqrt{c_k+1}\,\Delta x_k$, where P is a partition of $[2,6]$.

Prob. 51. $\lim_{\|P\|\to 0}\sum_{k=1}^{n}\ln c_k\,\Delta x_k$, where P is a partition of $[1,e]$.

Prob. 53. $\lim_{\|P\| \to 0} \sum_{k=1}^{n} g(c_k)\, \Delta x_k$, where P is a partition of $[0, 5]$.

Prob. 55. The integral is the area between x-axis and the graph of $y = x^2 - 1$ from $x = 1$ to $x = 2$ – area between x-axis and the graph of $y = x^2 - 1$ from $x = -1$ to $x = 1$.

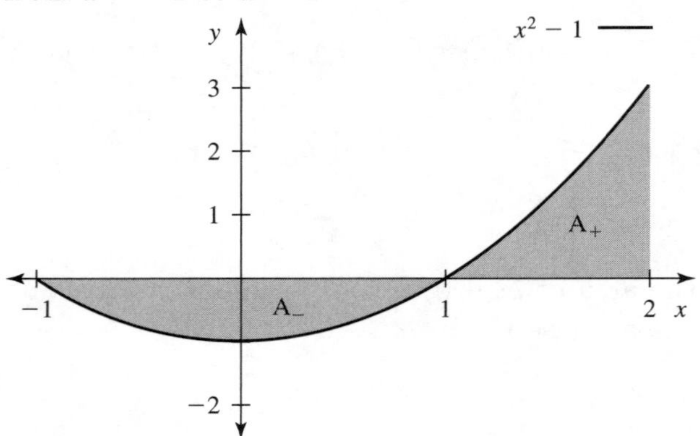

Prob. 57. The integral is the area between x-axis and the graph of $y = e^{-x}$ from $x = 0$ to $x = 5$.

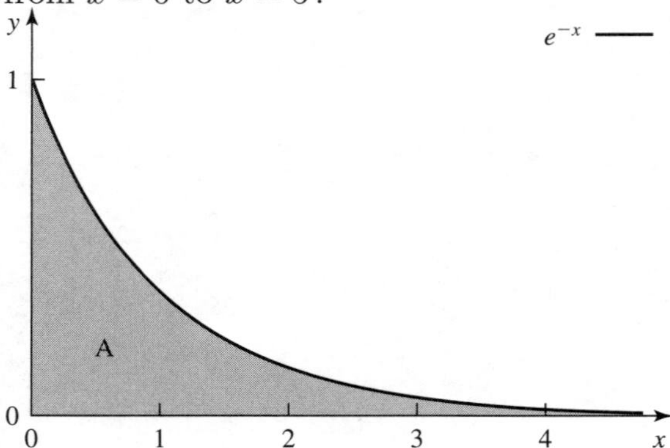

Prob. 59. The integral is the area between x-axis and the graph of $y = \ln x$ from $x = 1$ to $x = 4$ – area between x-axis and the graph of $y = \ln x$ from $x = 1/2$ to $x = 1$.

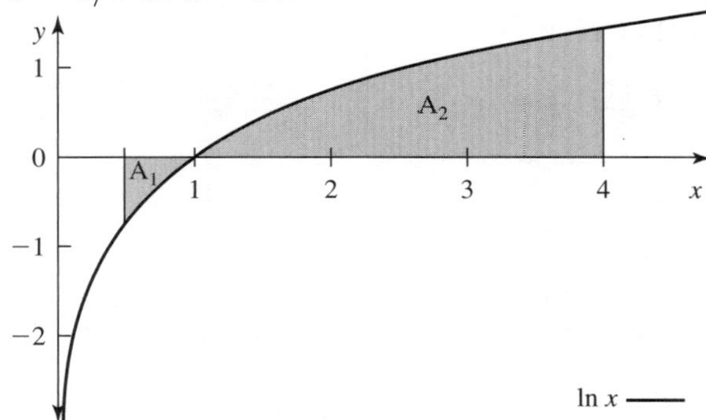

Prob. 61. The integral equals the sum of the areas of the triangles with vertices $(-2, 0), (0, 0), (-2, 2)$ and $(0, 0), (3, 0), (3, 3)$, that is $2 + 4.5 = 6.5$.

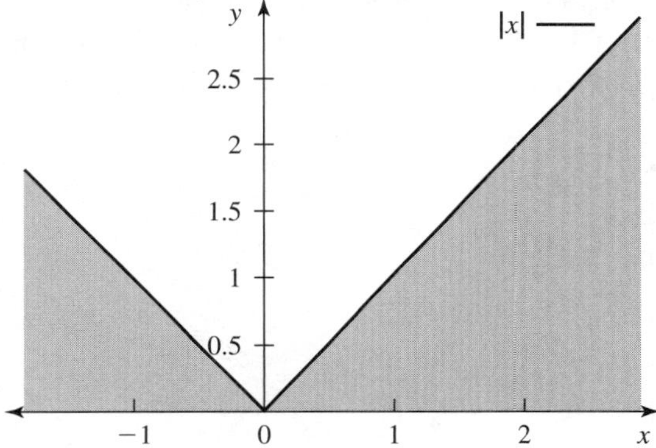

Prob. 63. The integral is negative. Its absolute value equals the area of the trapezoid with basis $3/2$ and 3 and height 3, that is $27/4$. The integral is therefore $-27/4$.

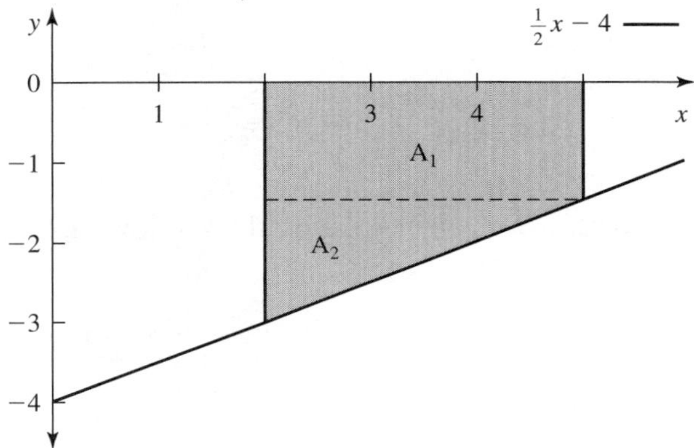

Prob. 65. The integral is negative. Its absolute value equals the area between the x-axis and the graph of the function $\sqrt{4-x^2}-2$. That area can be computed as the area of the rectangle of sides 2 and 4 minus half of the are of the circle of radius 2, that is, $2\pi - 8$.

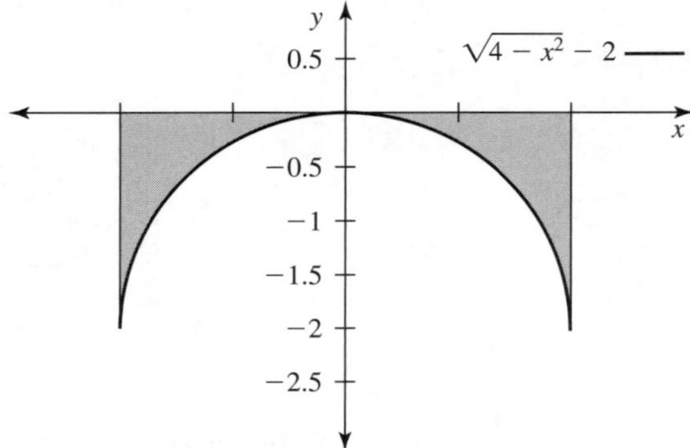

Prob. 67. The integral can be seen as the difference between two areas. One rectangle with vertices $(-3,0),(-3,4),(0,4),(0,0)$ and one fourth of the area of the disc centered at $(0,4)$ with radius 3. Therefore the integral's value is $3\times 4-\frac{9}{4}\pi$.

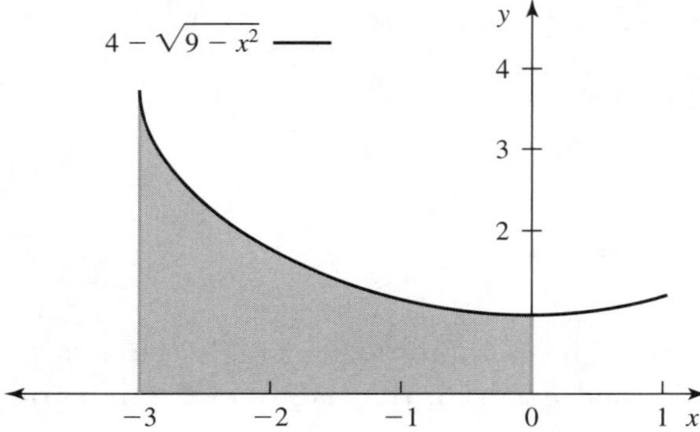

Prob. 69. 0.

Prob. 71. 4.

Prob. 73. 0.

Prob. 75. For $x \in [0,1]$ we have $x^2 \leq x$, so the inequality follows from property (7).

Prob. 77. For $x \in [0,4]$ we have $0 \leq \sqrt{x} \leq 2$, so the inequality follows from property (8).

Prob. 79. The part of the disc centered at the origin with radius 1 contains the square with vertices $(0,0), (1/\sqrt{2},0), (1/\sqrt{2},1/\sqrt{2}), (0,1/\sqrt{2})$ (which has area $1/2$), but is contained in the square with vertices $(0,0), (1,0), (1,1), (0,1)$ (which has area 1).

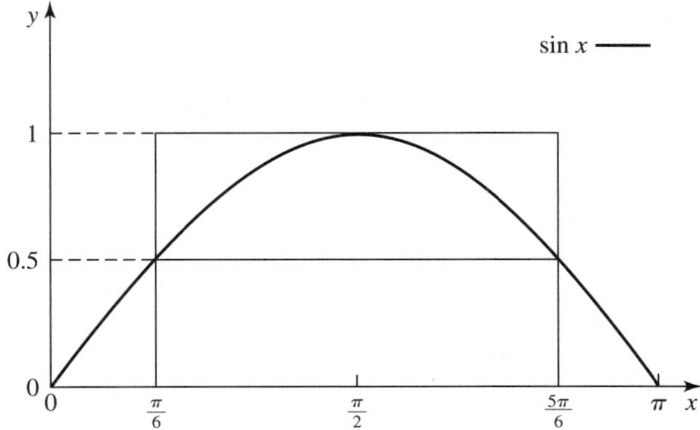

Prob. 81. $\cos x$ becomes negative for $x > \pi/2$, therefore the answer is $a = \pi/2$.

Prob. 83. For $f(x) = (x-2)^3$ we have $f(2+h) = -f(2-h)$, which gives the graph simmetry. We get then $a = 3$.

Prob. 85. We have

$$\begin{aligned}\int_t^{t+1} r(s)\,dt &= \int_0^{t+1} r(s)\,dt - \int_0^t r(s)\,dt = -\ln\frac{N(t+1)}{N(0)} + \ln\frac{N(t)}{N(0)} \\ &= -\ln N(t+1)) + \ln N(0) + \ln N(t) - \ln N(t) - \ln N(0) \\ &= -\ln\frac{N(t+1)}{N(t)}.\end{aligned}$$

6.2 The Fundamental Theorem of Calculus

Prob. 1.

$$y = \int_0^x u^2\,du.$$

Hence

$$\frac{dy}{dx} = \frac{d}{dx}\int_0^x u^2\,du.$$

So, from Fundamental Theorem of Calculus as the function u^2 is continuous between 0 and x

$$\frac{dy}{dx} = x^2.$$

Prob. 3.

$$y = \int_0^x (4u - 3)\, du.$$

Hence

$$\frac{dy}{dx} = \frac{d}{dx} \int_0^x (4u - 3)\, du.$$

So, from Fundamental Theorem of Calculus as the function $4u - 3$ is continuous between 0 and x

$$\frac{dy}{dx} = 4x - 3.$$

Prob. 5.

$$y = \int_0^x \sqrt{2 + u}\, du, x > 0.$$

Hence

$$\frac{dy}{dx} = \frac{d}{dx} \int_0^x \sqrt{2 + u}\, du.$$

So, from Fundamental Theorem of Calculus as the function $\sqrt{2 + u}$ is continuous between 0 and x

$$\frac{dy}{dx} = \sqrt{2 + x}.$$

Prob. 7.

$$y = \int_0^x \sqrt{1 + \sin^2 u}\, du, x > 0.$$

Hence

$$\frac{dy}{dx} = \frac{d}{dx} \int_0^x \sqrt{1 + \sin^2 u}\, du.$$

So, from Fundamental Theorem of Calculus as the function $\sqrt{1 + \sin^2 u}$ is continuous between 0 and x

$$\frac{dy}{dx} = \sqrt{1 + \sin^2 x}.$$

Prob. 9.

$$y = \int_3^x ue^{4u}\, du.$$

Hence

$$\frac{dy}{dx} = \frac{d}{dx}\int_3^x ue^{4u}\,du.$$

Since $f(x) = xe^{4x}$ is continuous for $x > 3$. Applying Fundamental Theorem of Calculus, we get

$$\frac{dy}{dx} = xe^{4x}.$$

Prob. 11.

$$y = \int_{-2}^{x} \frac{1}{u+3}\,du, x > -2.$$

Hence

$$\frac{dy}{dx} = \frac{d}{dx}\int_{-2}^{x} \frac{1}{u+3}\,du.$$

Since $f(x) = 1 \div x + 3$ is continuous for $x > -2$. Applying Fundamental Theorem of Calculus, we get

$$\frac{dy}{dx} = \frac{1}{x+3}.$$

Prob. 13.

$$y = \int_{\frac{\Pi}{2}}^{x} \sin(u^2+1)\,du.$$

Hence

$$\frac{dy}{dx} = \frac{d}{dx}\int_{\frac{\Pi}{2}}^{x} \sin(u^2+1)\,du.$$

Since $f(x) = \sin(x^2+1)$ is continuous for all x. Applying Fundamental Theorem of Calculus, we get

$$\frac{dy}{dx} = \sin(x^2+1).$$

Prob. 15.

$$y = \int_0^{3x} (1+t)\,dt.$$

Here the function $f(x) = 1 + x$ is continuous for all values of x, especially for $x > 0$. Hence we can apply *Leibniz's rule.*

$$h(x) = 3x, g(x) = 0$$

$$h'(x) = 3, g'(x) = 0$$

Hence applying equation (1)

$$\frac{dy}{dx} = 3(1 + x);$$

Prob. 17.

$$y = \int_0^{1-4x} (2t^2 + 1)\, dt.$$

Here the function $f(x) = 2x^2 + 1$ is continuous for all values of x, especially for $x > 0$. Hence we can apply *Leibniz's rule.*

$$h(x) = 1 - 4x, g(x) = 0$$

$$h'(x) = -4, g'(x) = 0$$

Hence applying equation (1)

$$\frac{dy}{dx} = (2(1 - 4x)^2 + 1)(-4);$$

Prob. 19.

$$y = \int_4^{x^2} (\sqrt{t})\, dt, x > 0.$$

Here the function $f(x) = \sqrt{x^2}$ is continuous for all values of $x > 0$. Hence we can apply *Leibniz's rule.*

$$h(x) = x^2, g(x) = 4$$

$$h'(x) = 2x, g'(x) = 0$$

Hence applying equation (1)

$$\frac{dy}{dx} = (x)(2x) = 2x^2;$$

Prob. 21.

$$y = \int_0^{3x} (1 + e^t)\, dt.$$

Here the function $f(x) = 1 + e^{3x}$ is continuous for all values of x, especially for $x > 0$. Hence we can apply *Leibniz's rule.*

$$h(x) = 3x, g(x) = 0$$

$$h'(x) = 3, g'(x) = 0$$

Hence applying equation (1)

$$\frac{dy}{dx} = (1 + e^{3x})3.;$$

Prob. 23.

$$y = \int_1^{3x^2+x} (1 + te^t)\, dt.$$

Here the function $f(x) = 1 + (3x^2 + x)e^{3x^2+x}$ is continuous for all values of x, especially for $x > 0$. Hence we can apply *Leibniz's rule.*

$$h(x) = 3x^2 + x, g(x) = 1$$

$$h'(x) = 6x + 1, g'(x) = 0$$

Hence applying equation (1)

$$\frac{dy}{dx} = (1 + (3x^2 + x)e^{3x^2+x})(6x + 1);$$

Prob. 25.

$$y = \int_x^3 (1 + t)\, dt.$$

Here the function $f(x) = 1 + x$ is continuous for all values x. Hence we can apply *Leibniz's rule.*

$$h(x) = 3, g(x) = x$$

$$h'(x) = 0, g'(x) = 1$$

Hence applying equation (1)

$$\frac{dy}{dx} = -(1 + x);$$

Prob. 27.

$$y = \int_{2x}^3 (1 + \sin t)\, dt.$$

Here the function $f(x) = 1 + \sin x$ is continuous for all values x. Hence we can apply *Leibniz's rule.*

$$h(x) = 3, g(x) = 2x$$

$$h'(x) = 0, g'(x) = 2$$

Hence applying equation (1)

$$\frac{dy}{dx} = -(1 + \sin 2x)(2);$$

Prob. 29.

$$y = \int_x^5 (\frac{1}{u^2})\, du, x > 0.$$

Here the function $f(x) = \frac{1}{x^2}$ is continuous for all values $x > 0$. Hence we can apply *Leibniz's rule.*

$$h(x) = 5, g(x) = x$$

$$h'(x) = 0, g'(x) = 1$$

Hence applying equation (1)

$$\frac{dy}{dx} = -(\frac{1}{x^2});$$

Prob. 31.

$$y = \int_{x^2}^1 (\sec t)\, dt, -1 < x < 1.$$

Here the function $f(x) = \sec x^2$is continuous for all values $-1 < x < 1$. Hence we can apply *Leibniz's rule.*

$$h(x) = 1, g(x) = x^2$$

$$h'(x) = 0, g'(x) = 2x$$

Hence applying equation (1)

$$\frac{dy}{dx} = -(\sec x^2)(2x);$$

Prob. 33.

$$y = \int_x^{2x} (1 + t^2)\, dt.$$

Here the function $f(x) = 1 + x^2$is continuous for all values x. Hence we can apply *Leibniz's rule.*

$$h(x) = 2x, g(x) = x$$

$$h'(x) = 1, g'(x) = 1$$

Hence applying equation (1)

$$\frac{dy}{dx} = (1 + 4x^2)(2) - (1 + x^2) = 1 + 7x^2;$$

Prob. 35.

$$y = \int_{x^2}^{x^3} (\ln t)\, dt, x > 0.$$

Here the function $f(x) = \ln x$ is continuous for all values $x > 0$. Hence we can apply *Leibniz's rule.*

$$h(x) = x^3, g(x) = x^2$$

$$h'(x) = 3x^2, g'(x) = 2x$$

Hence,

$$\frac{dy}{dx} = (\ln x^3)(3x^2) - (\ln x^2)(2x) = (9x^2 \ln x) - (4x \ln x) = (9x - 4)x \ln x$$

Prob. 37.

$$y = \int_{2-x^2}^{x+x^3} (\sin t)\, dt.$$

Here the function $f(x) = \sin x$ is continuous for all values x. Hence we can apply *Leibniz's rule.*

$$h(x) = x + x^3, g(x) = 2 - x^2$$

$$h'(x) = 1 + 3x^2, g'(x) = -2x$$

Hence applying equation (1)

$$\frac{dy}{dx} = (\sin(x + x^3))(1 + 3x^2) + (\sin(2 - x^2))(2x);$$

Prob. 39.

$$y = \int (1 - x^2)\, dx$$

We can write the above integral as

$$y = \int 1\, dx - \int (x^2)\, dx$$

Thus computing the above integral, we get

$$y = x - \frac{x^3}{3} + C$$

where C is a constant.

Prob. 41.

$$y = \int (3x^2 - 2x)\, dx$$

We can write the above integral as

$$y = \int (3x^2)\, dx - \int (2x)\, dx$$

Thus computing the above integral, we get

$$y = x^3 - x^2 + C$$

where C is a constant.

Prob. 43.

$$y = \int (\frac{x^2}{2} + 3x - \frac{1}{3})\, dx$$

We can write the above integral as

$$y = \int (\frac{x^2}{2})\, dx + \int 3x\, dx - \int \frac{1}{3}\, dx$$

Thus computing the above integral, we get

$$y = \frac{x^3}{6} + \frac{3x^2}{2} - \frac{x}{3} + C$$

where C is a constant.

Prob. 45.

$$y = \int (\frac{2x^2 - x}{\sqrt{x}})\, dx$$

We simplify the integrand, $\int (2x^{3/2} - \sqrt{x}) dx$ Therefore,

$$y = \frac{4x^{\frac{5}{2}}}{5} - \frac{2x^{\frac{3}{2}}}{3} + C$$

where C is a constant.

Prob. 47.

$$y = \int (x^2 \sqrt{x})\, dx$$

Hence we can rewrite the above integrand as $x^{\frac{5}{2}}$ Thus computing the integral we have

$$y = \frac{2}{7} x^{\frac{7}{2}} + C$$

where C is a constant.

Prob. 49.

$$y = \int (x^{\frac{7}{2}} + x^{\frac{2}{7}})\, dx$$

We can split the integral as

$$y = \int (x^{\frac{7}{2}})\, dx + \int (x^{\frac{2}{7}})\, dx$$

Now applying standard integration methods to each of the integrals, we get

$$y = \frac{2}{9} x^{\frac{9}{2}} + \frac{7}{9} x^{\frac{9}{7}} + C$$

where C is a constant.

Prob. 51.

$$y = \int (\sqrt{x} + \frac{1}{\sqrt{x}})\, dx$$

Thus we can split the above integral and rewrite as

$$y = \int x^{\frac{1}{2}}\, dx + \int x^{\frac{-1}{2}}\, dx$$

Now applying standard integration methods to each of the integrals, we get

$$y = \frac{2}{3} x^{\frac{3}{2}} + 2x^{\frac{1}{2}} + C$$

where C is a constant.

Prob. 53.

$$y = \int ((x-1)(x+1))\, dx$$

Thus we can expand and split the above integral using algebraic formulae and rewrite as

$$y = \int x^2\, dx - \int 1\, dx$$

Now applying standard integration methods to each of the integrals, we get

$$y = \frac{x^3}{3} - x + C$$

where C is a constant.

Prob. 55.

$$y = \int ((x-2)(3-x))\,dx$$

Thus we can expand and split the above integral using algebraic formulae and rewrite as

$$y = -\int x^2\,dx + \int 5x\,dx - \int 6\,dx$$

Now applying standard integration methods to each of the integrals, we get

$$y = -\frac{x^3}{3} + \frac{5}{2}x^2 - 6x + C$$

where C is a constant.

Prob. 57.

$$y = \int e^{2x}\,dx$$

Computing above integral yields

$$y = \frac{e^{2x}}{2} + C$$

where C is a constant.

Prob. 59.

$$y = \int 3e^{-x}\,dx$$

Computing above integral yields

$$y = -3e^{-x} + C$$

where C is a constant.

Prob. 61.

$$y = \int xe^{\frac{-x^2}{2}}\,dx$$

Computing above integral yields

$$y = -e^{\frac{-x^2}{2}} + C$$

where C is a constant.

Prob. 63.

$$y = \int \sin 2x \, dx$$

We get

$$y = -\frac{\cos 2x}{2} + C$$

where C is a constant.

Prob. 65.

$$y = \int \cos 3x \, dx$$

Computing the above integral from standard Integration formulae, we get

$$y = \frac{\sin(3x)}{3} + C$$

where C is a constant.

Prob. 67.

$$y = \int \sec^2(3x) \, dx$$

Computing the above integral, we get

$$y = \frac{\tan(3x)}{3} + C$$

where C is a constant.

Prob. 69. $\int \frac{\sin x}{1-\sin^2 x} dx = \int \frac{\sin x}{\cos^2 x} dx$ since $\sin^2 x + \cos^2 x = 1$. Now, $\frac{\sin x}{\cos^2 x} = \frac{\sin x}{\cos x}\frac{1}{\cos x} = \tan x \sec x$. Since $\frac{d}{dx} \sec x = \tan x \sec x$, we find $\int \frac{\sin x}{1-\sin^2 x} dx = \sec x + C$

Prob. 71.

$$y = \int \tan(2x) \, dx$$

Write

$$y = \int \frac{\sin(2x)}{\cos(2x)} \, dx$$

Computing the above integral we have

$$y = -\frac{\ln|\cos(2x)|}{2} + C$$

where C is a constant.

Prob. 73.

$$y = \int (\sec^2 x + \tan x)\, dx$$

We can split the above integral as

$$y = \int \sec^2 x\, dx + \int \tan x\, dx$$

Now the first integral is a standard one, so we get

$$y_1 = \tan x + C_1$$

where C_1 is a constant . The second integral can be computed as below, Write the integral as

$$y = \int \frac{\sin(x)}{\cos(x)}\, dx$$

Computing the above integral we have

$$y = -\ln(|\cos(x)| + C_2$$

where C_2 is a constant.

Joining the above two, we get the result as

$$y = \tan x - \ln(|\cos(x)| + C$$

where C is a constant.

Prob. 75.

$$y = \int \frac{4}{1 + x^2}\, dx$$

Here, we take the constant out of the integration sign and what we are left is with standard integration of $\int \frac{1}{1+x^2}$ Thus we can rewrite the above integral as

$$y = 4 \int \frac{1}{1 + x^2}\, dx$$

Computing the integral we have

$$y = 4 \arctan x + C$$

where C is a constant.

Prob. 75.

$$y = \int \frac{1}{\sqrt{1-x^2}}\,dx$$

The above integral is a standard integral of $\arcsin x$ and hence the result can be directly used. Thus using the standard result we get

$$y = \arcsin x + C$$

where C is constant.

Prob. 79.

$$y = \int \frac{1}{x+1}\,dx$$

Hence after integration we have,

$$y = \ln|1+x| + C$$

where C is a constant .

Prob. 81.

$$y = \int \frac{x-1}{x}\,dx$$

Applying algebraic transformation to the above integral, we get

$$y = \int 1\,dx - \int \frac{1}{x}\,dx$$

Now, both parts of the above integral are standard integrals. Thus we can calculate both of them individually and combine them together to get the result. So, let

$$y_1 = \int 1\,dx$$

and

$$y_2 = \int \frac{1}{x}\,dx$$

Hence we have

$$y = y_1 - y_2$$

So, computing both of them we get

$$y_1 = x + C_1$$

$$y_2 = \ln|x| + C_2$$

Hence the result is

$$y = x - \ln|x| + C$$

where C is a constant.

Prob. 83.

$$y = \int \frac{x+3}{x^2-9}\,dx$$

Now we know we can write

$$x^2 - 9 = (x+3)(x-3)$$

Thus writing the denominator of integral as above and carrying out proper algebraic transformation, we can rewrite the integral as

$$y = \int \frac{1}{x-3}\,dx$$

We get

$$y = \ln|x-3| + C$$

where C is a constant .

Prob. 85.

$$y = \int \frac{3-x}{x^2-9}\,dx$$

Now we know we can write

$$x^2 - 9 = -(3-x)(3+x)$$

Thus writing the denominator of integral as above and carrying out proper algebraic transformation, we can rewrite the integral as

$$y = -\int \frac{1}{x+3}\,dx$$

Hence after integration we have,

$$y = -\ln|x+3| + C$$

where C is a constant .

Prob. 87.

$$y = \int \frac{5x^2}{x^2+1}\,dx$$

We can write the numerator of above integral as

$$5x^2 = 5(x^2 + 1) - 5$$

So the above integral takes the form as

$$y = 5\int 1\,dx - 5\int \frac{1}{1+x^2}\,dx$$

Now both of them are standard integrals and hence we can compute them easily. Computing both the integrals individually and combining them we get

$$y = 5x - 5\arctan x + C$$

where C is a constant.

Prob. 89.

$$y = \int 3^x\,dx$$

The above integral is a standard one,and hence applying standard integration formulae to it we get the result as

$$y = \frac{3^x}{\ln 3} + C$$

where C is a constant.

Prob. 91.

$$y = \int 2^{-x}\,dx$$

Applying standard integration formulae, we get the result as

$$y = -\frac{2^{-x}}{\ln 2} + C$$

where C is a constant.

Prob. 93.

$$y = \int (x^2 + 2^x)\,dx$$

We can split the above integral and rewrite it as

$$y = \int x^2\,dx + \int 2^x\,dx$$

Both of the above integrands are standard integrals and hence using standard integration formulae we have the result as

$$y = \frac{x^3}{3} + \frac{2^x}{\ln 2} + C$$

where C is a constant.

Prob. 95.

$$y = \int (\sqrt{x} + \sqrt{e^x})$$

We can re write the above integral as

$$y = \int x^{\frac{1}{2}}\, dx + \int e^{\frac{x}{2}}\, dx$$

Now the first integral can be calculated from standard integration formulae, So

$$y_1 = \frac{2}{3} x^{\frac{3}{2}} + C_1$$

Now to compute the second integral, we get,

$$y_2 = 2e^{\frac{x}{2}} + C_2$$

Combining the above two, we get the result as

$$y = \frac{2}{3} x^{\frac{3}{2}} + 2e^{\frac{x}{2}} + C$$

where C is a constant.

Prob. 97.

$$y = \int_2^4 (3 - 2x)\, dx$$

The above integral can be rewritten as

$$y = \int_2^4 3\, dx - \int_2^4 2x\, dx$$

Hence computing the integral we get

$$y = [3x]_2^4 - [x^2]_2^4$$

Substituting the values we get

$$y = 6 - 12$$

$$y = -6$$

Prob. 99.

$$y = \int_0^1 (x^2 - \sqrt{x})\, dx$$

Computing the two parts of the integral individually we get,

$$y = [\frac{x^3}{3}]_0^1 - [\frac{2}{3}x^{\frac{3}{2}}]_0^1$$

Substituting the values we get, the result as

$$y = \frac{1}{3} - \frac{2}{3}$$

$$y = -\frac{1}{3}.$$

Prob. 101.

$$y = \int_1^8 x^{\frac{-2}{3}}\, dx$$

Computing the above integral using standard integration formulae we have,

$$y = [3x^{\frac{1}{3}}]_1^8$$

Substituting the limits in the calculated integral, we get the result as

$$y = 3$$

Prob. 103.

$$y = \int_0^2 (2t - 1)(t + 3)\, dt$$

Exapnding the above integral using algebraic formulae, we have

$$y = \int_0^2 2t^2\, dt + \int_0^2 5t\, dt - \int_0^2 3\, dt$$

Computing the integral using standard integral formulae, we have

$$y = [\frac{2}{3}t^3 + \frac{5}{2}t^2 - 3t]_0^2$$

Substituting the values we get result as

$$y = \frac{16}{3} + 10 - 6$$

$$y = \frac{28}{3}$$

Prob. 105.

$$y = \int_0^{\frac{\pi}{4}} \sin(2x)\, dx$$

From the standard integration formulae for $\sin x$ we get after doing integration,

$$y = [-\frac{\cos(2x)}{2}]_0^{\frac{\pi}{4}}$$

Substituting the values we get result as

$$y = 0 + \frac{1}{2}$$

$$y = \frac{1}{2}$$

Prob. 107.

$$y = \int_0^{\frac{\pi}{8}} \sec^2(2x)\, dx$$

Using the standard integration formulae for $\sec^2 x$ we get value as

$$y = [\frac{\tan(2x)}{2}]_0^{\frac{\pi}{8}}$$

Substituting the values, we get result as

$$y = \frac{1}{2}[1 - 0] = \frac{1}{2}$$

Prob. 109.

$$y = \int_0^1 \frac{1}{1 + x^2}\, dx$$

The above integral is a standard integral, and hence using standard integration formulae, we have following

$$y = [\arctan x]_0^1$$

Substituting the values we have,

$$y = \frac{\pi}{4}$$

Prob. 111.

$$y = \int_0^{\frac{1}{2}} \frac{1}{\sqrt{1-x^2}}\, dx$$

The above integral is a standard integral, and hence using standard integration formulae, we have following

$$y = [\arcsin x]_0^{\frac{1}{2}}$$

Substituting the values we have,

$$y = \frac{\pi}{6}$$

Prob. 113.

$$y = \int_0^{\frac{\pi}{6}} \tan(2x)\, dx$$

Writing the integral as

$$y = \int_0^{\frac{\pi}{6}} \frac{\sin(2x)}{\cos(2x)}\, dx$$

Computing the above integral we have

$$y = -\frac{\ln|\cos(2x)|}{2}\Bigg|_0^{\pi/6},$$

$$y = -[-\frac{\ln(2)}{2} - 0]$$

$$y = \frac{\ln(2)}{2}$$

Prob. 115.

$$y = \int_{-1}^{0} e^{3x}\, dx$$

Now using standard integration results, we get

$$y = \frac{1}{3}[e^{3x}]_{-1}^{0}$$

Substituting the limits, we get the result as

$$y = \frac{1 - e^{-3}}{3}$$

Prob. 117.

$$y = \int_{-1}^{1} |x|\, dx$$

Since the above integrand is even, we can transform the above integral as,

$$y = 2\int_{0}^{1} x\, dx$$

Now using standard integration formulae, ewe have

$$y = 2[\frac{x^2}{2}]_0^1$$

Substituting the limits, we get

$$y = 1$$

Prob. 119.

$$y = \int_{1}^{e} \frac{1}{x}\, dx$$

Above integral is a standard one and hence using standard integration formulae, we have

$$y = [\ln |x|]_1^e$$

Substituting the limits we have the result as,

$$y = 1 - 0$$

$$y = 1$$

Prob. 121.

$$y = \int_{-2}^{-1} \frac{1}{1-u}\, du$$

Using standard integration formulae, we have

$$y = -\ln|1-u|_{-2}^{-1}$$

Substituting the limits we have the result as,

$$y = -\ln 2 + \ln 3$$

$$y = \ln \frac{3}{2}$$

Prob. 123.

$$\lim_{x \to 0} \frac{1}{x^2} \int_0^x \sin t \, dt$$

Using L'Hospital rule, we first differentiate the numerator. We have the value after differentiation as $\sin x$ Similarly doing the differentiation of numerator we have value as $2x$ Thus our limit becomes as

$$\lim_{x \to 0} \frac{\sin x}{2x}$$

Differentiating it once again, we get the value as

$$\lim_{x \to 0} \frac{\cos x}{2}$$

Thus, we can now substitute the limit, and hence the result is

$$\lim_{x \to 0} \frac{\cos x}{2} = \frac{1}{2}$$

Prob. 125. Given

$$\int_0^x f(t) \, dt = 2x^2$$

Thus differentiating the two sides with respect to x we get

$$f(x) = 4x$$

Thus, we have the result

$$f(x) = 4x$$

6.3 Applications of Integration

Prob. 1. Given

$$y = e^x, y = -x, x = 0, x = 2$$

We first plot the graph for the above curves. We see that both e^x and $-x$ are continuous between $[0, 2]$ and $e^x > -x$. Hence, the area under the curve is

$$\begin{aligned} Area &= \int_0^2 (e^x) - (-x) \, dx \\ &= \int_0^2 e^x + x \, dx \end{aligned}$$

$$= [e^x + \frac{x^2}{2}]_0^2$$

$$= [(e^2 - 1) + (2 - 0)]$$

$$= e^2 + 1$$

Prob. 3. Given

$$y = x^2 - 1, y = x + 1$$

First, we find the point of intersection by equating the two equation .

$$x^2 - 1 = x + 1$$

$$x^2 - x - 2 = 0$$

$$(x - 2)(x + 1) = 0$$

Hence we get,

$$x = 2 and x = -1$$

Thus we compute the area between the two as,

$$Area = \int_{-1}^{2} (x + 1) - (x^2 - 1)\, dx$$

$$= \int_{-1}^{2} x - x^2 + 2\, dx$$

$$= [\frac{x^2}{2} - \frac{x^3}{3} + 2x]_{-1}^2$$

$$= [\frac{3}{2} - 3 + 6]$$

$$= \frac{9}{2}.$$

Prob. 5. Given

$$y = x^2 + 1, y = 4x - 2$$

First, we find the point of intersection by equating the two equation .

$$x^2 + 1 = 4x - 2$$

$$x^2 - 4x + 3 = 0$$

$$(x - 1)(x - 3) = 0$$

Hence we get values as,

$$4x = 1, x = 3$$

Thus the area under the curve in the 1st quadrant is,

$$Area = \int_1^3 (4x - 2) - (x^2 + 1)\, dx$$

$$= \int_1^3 -x^2 + 4x - 3\, dx$$

$$= [-\frac{x^3}{3} + 2x^2 - 3x]_1^3$$

$$= [-\frac{26}{3} + 16 - 6]$$

$$= 10 - \frac{26}{3}.$$

Prob. 7. Given

$$y = x^2, y = \frac{1}{x}, y = 4$$

Here we see that the two functions $y = x^2$ and $y = \frac{1}{x}$ are continuous in 1st quadrant except at $x = 0$. We compute the point of intersection of the 1st two equations,

$$x^2 = \frac{1}{x}$$

$$x^3 = 1$$

$$x = 1$$

Also the point of intersection of the curve $y = \frac{1}{x}$ and $y = 4$ is $x = \frac{1}{4}$ and that of $y = x^2$ and $y = 4$ is $x = 2$. Hence we can calculate the area enclosed in the 1st quadrant by the figure as,

$$Area = \int_{1/4}^1 (4 - \frac{1}{x})dx + \int_1^2 (4 - x^2)dx$$

$$= [4x - \ln|x|]_{1/4}^1 + [4x - \frac{1}{3}x^3]_1^2$$

$$= \frac{14}{3} - \ln 4.$$

Prob. 9. Given $y = \frac{1}{x}, y = 1$ and $\frac{1}{2} < x < \frac{3}{2}$. We can compute the area of enclosed figure by splitting the integral into two parts and adding them,

$$Area = \int_{\frac{1}{2}}^{1} (\frac{1}{x} - 1\, dx + \int_{1}^{\frac{3}{2}} (1 - \frac{1}{x})\, dx$$

$$= [\ln|x| - x]_{\frac{1}{2}}^{1} + [x - \ln|x|]_{1}^{\frac{3}{2}}$$

$$= -1 + \ln 2 + \frac{1}{2} + \frac{3}{2} - \ln\frac{3}{2} - 1$$

$$= 2\ln 2 - \ln 3$$

Prob. 11. Given

$$y = x^2, y = x^3$$

and

$$0 < x < 2.$$

The above two equations intersect at

$$x = 0 \textit{ and } x = 1.$$

Also for $0 < x < 1$ we have

$$x^2 > x^3$$

and for $1 < x < 2$ we have

$$x^3 > x^2.$$

Thus the area can be computed by splitting it into two parts and calculating area for each of them.

$$Area = \int_{0}^{1} x^2 - x^3\, dx + \int_{1}^{2} x^3 - x^2\, dx$$

$$= [\frac{x^3}{3} - \frac{x^4}{4}]_0^1 + [\frac{x^4}{4} - \frac{x^3}{3}]_1^2$$

$$= \frac{1}{3} - \frac{1}{4} + \frac{15}{4} - \frac{7}{3}$$

$$= \frac{7}{2} - 2$$

Prob. 13. Given

$$y = x^2, y = (x-2)^2, y = 0$$

where,

$$0 < x < 2$$

Since, we have to integrate in terms of y, we first calculate y in terms of x.

$$x = \pm\sqrt{y}, x = 2 + \pm\sqrt{y}, y = 0$$

We calculate the point of intersection of the curves

$$x = \pm\sqrt{y}, x = 2 + \pm\sqrt{y}$$

$$\sqrt{y} = 2 - \sqrt{y}$$

$$2\sqrt{y} = 2$$

$$y = 1$$

Hence area under the curve can be calculated as,

$$Area = \int_0^1 (2 - \sqrt{y}) - \sqrt{y}\, dy$$

$$= 2\int_0^1 1 - \sqrt{y}\, dy$$

$$= 2[y - \frac{2}{3}y^{\frac{3}{2}}]_0^1$$

$$= 2[1 - \frac{2}{3}]$$

$$= \frac{2}{3}$$

Prob. 15. Given

$$x = (y-1)^2 + 3, x = 1 - (y-1)^2$$

where

$$0 < y < 2$$

We have to find the area enclosed by the figure, only for 1st quadrant. First we find the point of intersection for the two curves

$$x = (y-1)^2 + 3, x = 1 - (y-1)^2$$

Equating the two values, we get

$$(y-1)^2 + 3 = 1 - (y-1)^2$$

$$2(y-1)^2 + 2 = 0$$
$$(y-1)^2 + 1 = 0$$

Thus the two curves never meet. Also,

$$(y-1)^2 + 3 > 1 - (y-1)^2$$

for

$$0 < y < 2$$

Hence the area can be directly computed as

$$\begin{aligned} Area &= \int_0^2 (y-1)^2 + 3 - (1-(y-1)^2)\,dy \\ &= 2\int_0^2 (y-1)^2 + 1\,dy \\ &= 2[\frac{(y-1)^3}{3} + y]_0^2 \\ &= 2[2\frac{1}{3} + 2] \\ &= \frac{16}{3} \end{aligned}$$

Prob. 17. Given

$$\frac{dN}{dt} = e^{-t}$$

(a) To find $N(t)$ we have to integrate the above equation with respect to t. Integrating

$$N(t) = \int e^{-t}\,dt$$
$$N(t) = -e^{-t} + C$$

where C is a constant. To evaluate this constant, we have to apply boundary condition. Given that at

$$t = 0, N(0) = 100$$

Hence,

$$N(0) = C - 1 = 100$$
$$C = 101$$

Hence we get the result as

$$N(t) = 101 - e^{-t}$$

(b) To find the cumulative value of $N(t)$ between $0 < t < 5$, we compute

$$N(5) - N(0)$$
$$= 101 - e^{-5} - 101 + 1$$
$$= 1 - e^{-5}$$

which is the cumulative change in the population.

(c) For any time t, the cumulative change in population between that period can be written as

$$CumulativePopulation = \int_0^t e^{-x}\, dx$$

where x is a variable . The above integral means that at any time t, the total population is sum total of all the population increase in that period. Geometrically its the area bounded by the curve e^{-x} and the x axis.

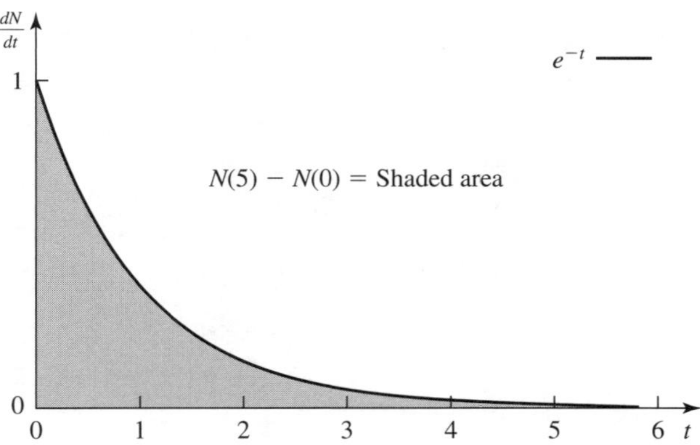

Prob. 19. Given

$$v(t) = -(t-2)^2 + 1$$

for

$$0 \leq t \leq 5$$

(c) To find the position of the particle $s(t)$ at any time t, we know that,

$$\frac{v(t) = dx}{dt}$$

Hence

$$x = \int dx$$

Thus we get,

$$s(t) = \int dx$$

$$s(t) = \int_0^t v(x)\,dx$$

$$s(t) = \int_0^t -(x-2)^2 + 1\,dx$$

$$s(t) = [-\frac{(x-2)^3}{3} + x]_0^t$$

$$s(t) = -\frac{(t-2)^3}{3} + t - \frac{8}{3}$$

Thus $s(t)$ gives the distance traveled by the particle at time t and $s(5) = -\frac{20}{3}$. Rightmost position when $t = 0$ and 3: $s(0) = s(3) = 0$.

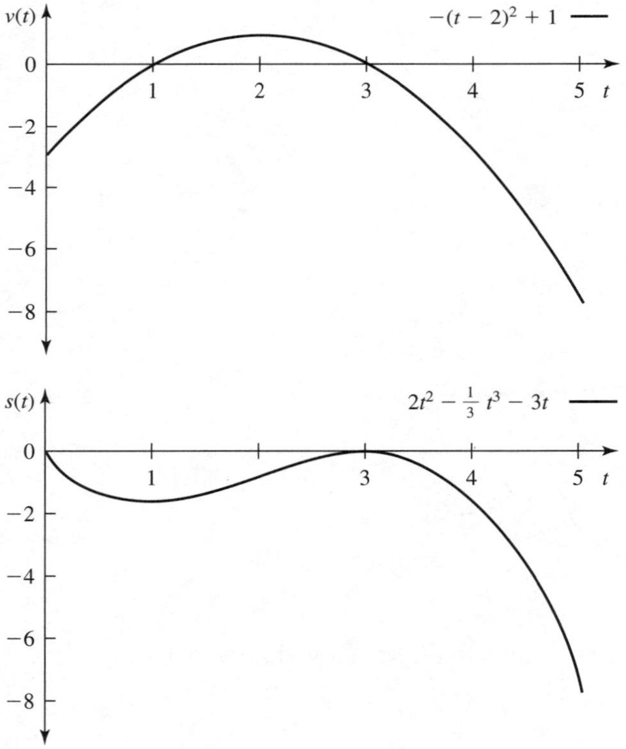

Prob. 21. Given that

$$\frac{dl}{dt}$$

represents the growth of an organism. Hence,

$$\frac{\int_2^7 dl}{dt\, dt}$$

represents the cumulative growth of the organism between the months [2, 7]. Or in other words, it represents the cumulative growth in 5 months .

Prob. 23. Given that

$$\frac{db}{dt}$$

represents the rate of change of biomass at time t. Then

$$\int_1^6 \frac{dB}{dt}\, dt$$

represents the cumulative biomass between time 1 to 6 . That is, total biomass accumulated between time units [1, 6].

Prob. 25. Given that

$$f(x) = x^2 - 2$$

and

$$0 \leq x \leq 2.$$

To compute the average value of $f(x)$ we proceed as,

$$Avg(f(x)) = \frac{1}{2-0} \int_0^2 f(x)\, dx$$

$$= \frac{1}{2} \int_0^2 x^2 - 2\, dx$$

$$= \frac{1}{2}[\frac{x^3}{3} - 2x]_0^2$$

$$= \frac{1}{2}[\frac{8}{3} - 4]$$

$$= -\frac{2}{3}$$

Thus the average value of $f(x)$ is $-\frac{2}{3}$

Prob. 27. We are given that the temperature t in Fahrenheit varies as,

$$T(t) = 68 + \sin\frac{\pi t}{12}$$

and that

$$0 \leq t \leq 24.$$

To compute the average value of the temperature, we proceed as follows,

$$\begin{aligned} Avg(T) &= \frac{1}{24-0}\int_0^{24} T(t)\,dt \\ &= \frac{1}{24}\int_0^{24} 68 + \sin\frac{\pi t}{12}\,dt \\ &= \frac{1}{24}[68t - \frac{12}{\pi}\cos\frac{\pi t}{12}]_0^{24} \\ &= \frac{1}{24}[1632 - \frac{12}{\pi}(1-1)] \\ &= 68 \end{aligned}$$

Hence the average temperature is $6d$.

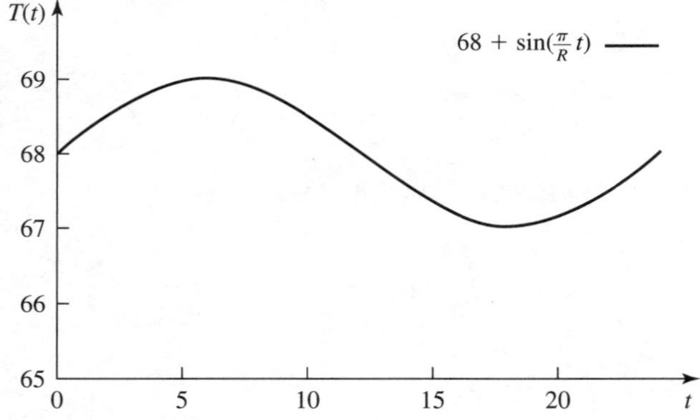

Prob. 29. The function $\tan x$ is odd, so for every positive value it assumes from $[0, 1]$, there is an equivalent negative value being assumed in the interval $[-1, 0]$, so the integral

$$\frac{1}{2}\int_{-1}^{1} \tan x dx = 0.$$

Prob. 31. Given

$$f(x) = 2x$$

and

$$0 \leq x \leq 2$$

Hence average value of $f(x)$ can be computed as,

$$Avg(f) = \frac{1}{2-0}\int_0^2 f(x)\,dx$$

$$= \frac{1}{2}\int_0^2 2x\,dx$$

$$= \frac{1}{2}[x^2]_0^2$$

$$= 2$$

Hence the average value of $f(x)$ is 4. Now to find x such that $f(x)$ is equal to avg(x). We, do the following,

$$f(x) = 2$$

$$2x = 2$$

$$x = 1$$

Hence the value of $x = 2$ gives the result same as average value of $f(x)$.

Prob. 33. We have to find the volume of a right circular cone, whose base radius is r and whose height is h. Let at any point in between the tip and the base, $height = x, radius = y$. From the triangle property we can write,

$$\frac{h-x}{h} = \frac{y}{r}$$

$$x = h - \frac{hy}{r}$$

Hence,

$$dx = \frac{h}{r}dy$$

Thus the volume of that small circular section can be written as

$$dV = \pi y^2 dx$$

Writing dx in terms of dy, we have

$$V = \int_0^r \pi y^2 \frac{h}{r}\, dy$$

$$= \frac{h}{r}[\pi \frac{y^3}{3}]_0^r$$

$$= \frac{1}{3}\pi r^2 h$$

Thus, we get the volume of cone in terms of $randx$ as

$$V = \frac{1}{3}\pi r^2 h.$$

Prob. 35. Given

$$y = 4 - x^2, y = 0, x = 0$$

When we plot the graph, we find that that the graph intersects the x-axis at $x = 2$. Thus when we rotate the figure we find that the area at any point x can be written as

$$area = \pi y^2$$

Thus the volume can be calculated as,

$$V = \int_0^2 \pi(4 - x^2)^2\, dx$$

$$= \pi \int_0^2 16 + x^4 - 8x^2\, dx$$

$$= \pi[16x + \frac{x^5}{5} - \frac{8x^3}{3}]_0^2$$

$$= \pi[32 + \frac{32}{5} - \frac{64}{3}]$$

$$= \frac{256}{15}\pi$$

Above gives the volume bounded by the curve, which is

$$V = \frac{256}{15}\pi.$$

Prob. 37. Given $y = \sqrt{\sin x}, y = 0$ and $0 \le x \le \pi$. When we draw the curve and rotate it about the x-axis we get sphere whose diameter is π. Area at any point x can be written as

$$area = \pi y^2$$

We can then easily compute the volume, as follows,

$$V = \int y^2\,dx$$

$$= \int_0^{\pi} \sin x\,dx$$

$$= \pi[-\cos x]_0^{\pi}$$

$$= 2\pi$$

Thus the volume of the figure, when rotated is $V = 2$.

Prob. 39. Given

$$y = \sec x, y = 0$$

and

$$\frac{-\pi}{3} \le x \le \frac{\pi}{3}$$

When we draw the above curve, we find that the area is symmetric about the y-axis. Thus we need to compute the volume of only one part and double it. Area at any point x can be written as

$$area = \pi y^2$$

We can then easily compute the volume, as follows,

$$V = \pi \int y^2\,dx$$

$$= \pi \int_0^{\frac{\pi}{3}} \sec^2 x\,dx$$

$$= \pi[\tan x]_0^{\frac{\pi}{3}}$$

$$= \pi[\sqrt{3} - 0]$$

$$= \pi\sqrt{3}$$

Hence the volume of the enclosed figure is twice this volume. Thus the volume is

$$V = \pi 2\sqrt{3}.$$

Prob. 41. Given $y = x^2, y = x$ and $0 \leq x \leq 1$ When we draw the curve we see that for $0 \leq x \leq 1$, $x > x^2$. Also to compute the volume we can compute the volume of each of them and then subtract them to get the enclosed volume . Volume of a small disc can be written as

$$dV = \pi x^2 dx - \pi x^4 dx$$

$$\begin{aligned} V &= \pi \int_0^1 x^2 - x^4 \, dx \\ &= \pi[\frac{x^3}{3} - \frac{x^5}{5}]_0^1 \\ &= \pi[\frac{1}{3} - \frac{1}{5}] \\ &= \frac{2}{15}\pi \end{aligned}$$

Thus the volume of the enclosed figure, when rotated around the x-axis is $V = \frac{2}{15}\pi$.

Prob. 43. $y = e^x, y = e^{-x}$ and $0 \leq x \leq 2$ When we draw the curve we see that for $0 \leq x \leq 2$, $e^x > e^{-x}$. Also to compute the volume we can compute the volume of each of them and then subtract them to get the enclosed volume . Volume of a figure can be written as

$$\begin{aligned} V &= \pi \int_0^2 e^{2x} - e^{-2x} \, dx \\ &= \pi[\frac{1}{2}e^{2x} + \frac{1}{2}e^{-2x}]_0^2 \\ &= \pi\frac{1}{2}[e^4 - 1 + e^{-4} - 1] \\ &= \frac{\pi}{2}(e^4 + e^{-4} - 2) \end{aligned}$$

Thus the volume of the enclosed figure, when rotated around the x-axis is

$$V = \frac{\pi}{2}(e^4 + e^{-4} - 2).$$

Prob. 45. Given

$$y = \sqrt{\cos x}, y = 1$$

and

$$0 \leq x \leq \frac{\pi}{2}$$

When we draw the curve we see that for $0 \leq x \leq \frac{\pi}{2}$,

$$1 > \sqrt{\cos x}.$$

Also to compute the volume we can compute the volume of each of them and then subtract them to get the enclosed volume . Volume of a small disc can be written as

$$\begin{aligned} V &= \pi \int_0^{\frac{\pi}{2}} 1 - (\sqrt{\cos x})^2 \, dx \\ &= \pi \int_0^{\frac{\pi}{2}} 1 - \cos x \, dx \\ &= \pi [x - \sin x]_0^{\frac{\pi}{2}} \\ &= \pi [\frac{\pi}{2} - 1] \\ &= (\frac{\pi}{2} - 1)\pi \end{aligned}$$

Thus the volume of the enclosed figure, when rotated around the x-axis is

$$V = (\frac{\pi}{2} - 1)\pi.$$

Prob. 47. Given

$$y = \sqrt{x}, y = 2, x = 0$$

We have to find the volume by rotating around the y-axis. Hence we find y in terms of x.

$$x = y^2$$

Hence the volume can be calculated as

$$\begin{aligned} V &= \pi \int_0^2 (y^2)^2 \, dy \\ &= \pi \int_0^2 y^4 \, dy \end{aligned}$$

$$= \pi[\frac{y^5}{5}]_0^2$$
$$= \pi[\frac{32}{5} - 0]$$
$$= \frac{32\pi}{5}$$

Hence the volume of the enclosed figure, when rotated around the y-axis is

$$V = \frac{32\pi}{5}.$$

Prob. 49. Given

$$y = \ln(x+1), y = \ln 3, x = 0$$

Since, the enclosed figure has to be rotated around y-axis, we get x as a function of y. Hence,

$$x = e^y - 1$$

and

$$0 \leq y \leq \ln 3$$

Hence the volume can be calculated as

$$V = \pi \int_0^{\ln(3)} (e^y - 1)^2 \, dy$$
$$= \pi \int_0^{\ln(3)} e^{2y} - 2e^y + 1 \, dy$$
$$= \pi[\frac{e^{2y}}{2} - 2e^y + y]_0^{\ln 3}$$
$$= \pi[4 - 4 + \ln 3]$$
$$= \pi \ln 3$$

Hence the volume of the enclosed figure, when rotated around the y-axis is

$$V = \pi \ln 3.$$

Prob. 51. Given

$$y = x^2, y = \sqrt{x}$$

and

$$0 \leq x \leq 1$$

Since, the enclosed figure has to be rotated around y-axis, we get x as a function of y. Hence,

$$x = \sqrt{y}, x = y^2$$

and

$$0 \leq y \leq 1$$

Also

$$\sqrt{y} > y^2$$

in the given interval Hence the volume can be calculated as,

$$V = \pi \int_0^1 (\sqrt{y})^2 - (y^2)^2 \, dy$$

$$= \pi \int_0^1 y - y^4 \, dy$$

$$= \pi [\frac{y^2}{2} - \frac{y^5}{5}]_0^1$$

$$= \pi [\frac{1}{2} - \frac{1}{5}]$$

$$= \frac{3\pi}{10}$$

Hence the volume of the enclosed figure, when rotated around the y-axis is

$$V = \frac{3\pi}{10}.$$

Prob. 53. Given, the equation of line as

$$y = 2x$$

and

$$0 \leq x \leq 2$$

(a) Using planar geometry first we calculate both the end points.

$$x = 0, y = 0$$

$$x = 2, y = 4$$

Hence, the length is

$$\sqrt{(2-0)^2 + (4-0)^2} = \sqrt{20}$$

(b) Using integral formulae, we have

$$y' = 2$$

Hence

$$\begin{aligned} length &= \int_0^2 \sqrt{1+2^2}\,dx \\ &= \int_0^2 \sqrt{5}\,dx \\ &= [\sqrt{5}x]_0^2 \\ &= \sqrt{20} \end{aligned}$$

Hence from both the ways, we get the same length.

Prob. 55. Given, the equation of line as

$$y^2 = x^3$$

and

$$1 \leq x \leq 4$$

Hence we calculate y',

$$y' = \frac{3}{2}\sqrt{x}$$

Applying the standard formulae for determining the length,

$$L = \int_a^b \sqrt{1+[f'(x)]^2}\,dx$$

Substituting in above equation,

$$\begin{aligned} L &= \int_1^4 \sqrt{1+[\frac{3}{2}\sqrt{x}]^2}\,dx \\ &= \int_1^4 \sqrt{1+\frac{9x}{4}}\,dx \\ &= [\frac{4}{9}\frac{2}{3}[1+\frac{9x}{4}]^{\frac{3}{2}}]_1^4 \\ &= \frac{8}{27}[10^{\frac{3}{2}} - (\frac{13}{4})^{\frac{3}{2}}] \end{aligned}$$

Above result gives us the length of the line.

Prob. 57. Given, the equation of line as

$$y = \frac{x^3}{6} + \frac{1}{2x}$$

and

$$1 \leq x \leq 3$$

Hence we calculate y',

$$y' = \frac{x^2}{2} - \frac{1}{2x^2}$$

Applying the standard formulae for determining the length,

$$L = \int_a^b \sqrt{1 + [f'(x)]^2}\, dx$$

Substituting in above equation,

$$L = \int_1^3 \sqrt{1 + [\frac{x^2}{2} - \frac{1}{2x^2}]^2}\, dx$$

$$= \int_1^3 \frac{1}{2}\sqrt{(x^4 + 2 + \frac{1}{x^4})}\, dx$$

$$= \frac{1}{2}\int_1^3 x^2 + \frac{1}{x^2}$$

$$= \frac{1}{2}[\frac{x^3}{3} - \frac{1}{x}]_1^3$$

$$= \frac{14}{3}$$

Above result gives us the length of the line.

Prob. 59. Given, the equation of line as

$$y = x^2$$

and

$$-1 \leq x \leq 1$$

Hence we calculate y',

$$y' = 2x$$

Applying the standard formulae for determining the length,

$$L = \int_a^b \sqrt{1 + [f'(x)]^2}\, dx$$

Substituting in above equation,

$$L = \int_{-1}^{1} \sqrt{1 + [2x]^2}\, dx$$

Thus, above equation can be used for computing the length of the equation
.

Prob. 61. Given, the equation of line as

$$y = e^{-x}$$

and

$$0 \leq x \leq 1$$

Hence we calculate y',

$$y' = -e^{-x}$$

Applying the standard formulae for determining the length,

$$L = \int_a^b \sqrt{1 + [f'(x)]^2}\, dx$$

Substituting in above equation,

$$L = \int_0^1 \sqrt{1 + [-e^{-x}]^2}\, dx$$

Thus, above equation can be used for computing the length of the equation
.

Prob. 63. Given the equation of a quarter circle

$$y = \sqrt{1 - x^2}$$

and

$$0 \leq x \leq 1$$

We have to find the length.

(a) First we will find the length by using geometrical formulae. We can easily see that the radius of the circle is 1. Hence, quarter perimeter of the circle is $\pi/2$.

(b) Now, we calculate using integral formulae. For that first we find f'.

$$y' = -\frac{x}{\sqrt{1-x^2}}$$

Thus the length is equal to

$$\begin{aligned} Length &= \int_0^1 \sqrt{1 + [-\frac{x}{\sqrt{1-x^2}}]^2}\, dx \\ &= \int_0^1 \frac{1}{\sqrt{1-x^2}}\, dx \\ &= [\arcsin x]_0^1 \\ &= \pi/2 \end{aligned}$$

Thus from, both ways we get the length as $\pi/2$.

Prob. 65. Given equation is

$$y = \frac{e^x + e^{-x}}{2}$$

and

$$0 \le x \le a.$$

We first calculate y',

$$y' = \frac{e^x - e^{-x}}{2}$$

Thus the length can be computed as

$$\begin{aligned} Length &= \int_0^a \sqrt{1 + [\frac{e^x - e^{-x}}{2}]^2}\, dx \\ &= \int_0^a \sqrt{(\frac{e^x + e^{-x}}{2})^2}\, dx \\ &= \int_0^a \frac{e^x + e^{-x}}{2} \end{aligned}$$

$$= [\frac{e^x - e^{-x}}{2}]_0^a$$

$$= \frac{e^a - e^{-a}}{2}$$

Thus above gives us the length. Also if we substitute a in the derivative of y, then

$$f'(a) = \frac{e^a - e^{-a}}{2}$$

Thus we see that the results are same. Hence, proved.

6.5 Review Problems

Prob. 1. Given the discharge Q is,

$$Q = \int_0^B \overline{V}(b)h(b)\,db$$

where,$h(b)$ is the average velocity at distance b and $h(b)$ is the depth at distance b. Thus to get the total discharge, either we can integrate it or approximate it by taking summation over some interval. Thus,

$$Q = \sum_{b=0}^{b=16} \overline{V}(b)h(b)$$

We are given the table as,

0	0	0
1	0.28	0.172
3	0.76	0.213
5	1.34	0.230
7	1.57	0.256
9	1.42	0.241
11	1.21	0.206
13	0.83	0.187
15	0.42	0.116
16	0	0

Hence we get Q as

$$Q = 3.38$$

as the approximated result.

Prob. 3. The theoretical velocity is given by

$$v(d) = (\frac{D-d}{a})^{\frac{1}{c}})$$

where v(d) is the velocity at depth d below the water surface, c is a constant varying from 5 for coarse beds to 7 for smooth beds, D is the total depth of the channel, and a is a constant that is equal to the distance above the bottom of the channel at which velocity has unit value. Now depending upon the two methods, the second method is more efficient since it takes two different values and computes the average of them. We see that, more the number of samples take, better will be the result, as the error will get reduced. Hence, method 2 is a better method.

(a)

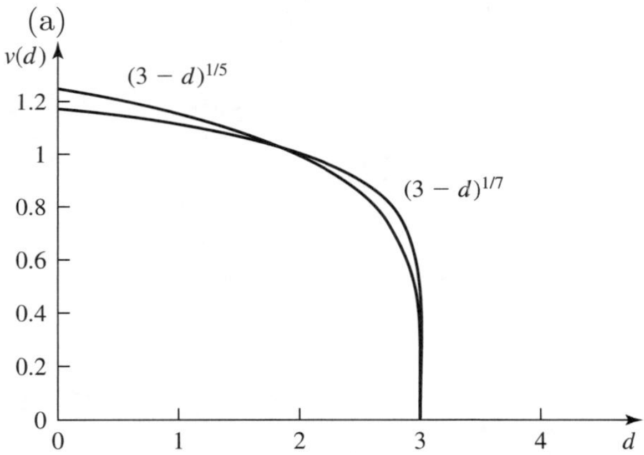

(b) substituting the values in the above equation we see that at $d = D$,

$$V(D) = (\frac{D-D}{a})^{\frac{1}{c}})$$
$$= 0$$

and at $d = 0$, we have

$$v(0) = (\frac{D-0}{a})^{\frac{1}{c}}) = (\frac{D}{a})^{\frac{1}{c}}$$

The derivative of the function is negative, and hence has the maximal value is at $d = 0$.

Prob. 5. Substituting d_1 in the average velocity equation at any depth, we find that

$$v(d_1) = (\frac{D - d_1}{a})^{\frac{1}{c}})$$

Since at this depth $v(d_1) = \overline{v}$, we get

$$\frac{c}{c+1}(\frac{D}{a})^{\frac{1}{c}} = (\frac{D - d_1}{a})^{\frac{1}{c}})$$

Raising each side by the power of c, we have

$$(\frac{c}{c+1})^c \frac{D}{a} = \frac{D - d_1}{a}$$

Thus we have the result,

$$\frac{d_1}{D} = 1 - (\frac{c}{c+1})^c$$

7 Integration Techniques and Computational Methods

7.1 The Substitution Rule

Prob. 1. $\frac{2}{3}(x^2+3)^{3/2}+C$.

Prob. 3. $-\frac{1}{3}(1-x^2)^{3/2}+C$.

Prob. 5. $\frac{5}{3}\sin(3x)+C$.

Prob. 7. $-\frac{7}{12}\cos(4x^3)+C$.

Prob. 9. $\frac{1}{2}e^{2x+3}+C$.

Prob. 11. $-e^{-x^2/2}+C$.

Prob. 13. $\frac{1}{2}\ln|x^2+4x|+C$.

Prob. 15. $3\ln|x+4|+C$.

Prob. 17. $\frac{2}{3}(x+3)^{3/2}+C$.

Prob. 19. $\frac{2}{3}(2x^2-3x+2)^{3/2}+C$.

Prob. 21. $-\ln|1+x-2x^2|+C$.

Prob. 23. $\frac{1}{4}\ln|1+2x^2|+C$.

Prob. 25. $e^{x^2}+C$.

Prob. 27. $e^{1+\ln x}+C$.

Prob. 29. $-\frac{2}{3\pi}\cos\left(\frac{3\pi}{2}x+\frac{\pi}{4}\right)+C$.

Prob. 31. $\frac{1}{2}\tan^2 x+C$.

Prob. 33. $\frac{(\ln x)^3}{3}+C$.

Prob. 35. $\frac{2}{5}(5+x)^{5/2}-\frac{10}{3}(5+x)^{3/2}+C$.

Prob. 37. $\ln|ax^2+bx+c|+C$.

Prob. 39. $\frac{1}{n+1}g^{n+1}(x)+C$.

Prob. 41. $-e^{-g(x)}+C$.

Prob. 43. $\frac{1}{3}(10^{3/2} - 1)$

Prob. 45. $\frac{7}{2025}$

Prob. 47. $1 - e^{-9/2}$.

Prob. 49. $\frac{3}{8}$.

Prob. 51. $\frac{1}{2}$.

Prob. 53. $4 + \ln 27$.

Prob. 55. $\frac{1}{2}$.

Prob. 57. $2e^{-1} - 2e^{-3}$.

Prob. 59. $\int \cot x dx = \int \cos x \sin^{-1} x dx = \ln|\sin x| + C$.

7.2 Integration by Parts

Prob. 1. $x \sin x + \cos x$, $(u = x,\ v' = \cos x)$.

Prob. 3. $\frac{2}{3}x \sin(3x) + \frac{2}{9}\cos(3x)$, $(u = 2x,\ v' = \cos(3x))$.

Prob. 5. $-2x \cos x + 2 \sin x$, $(u = 2x,\ v' = \sin x)$.

Prob. 7. $(x - 1)e^x$, $(u = x,\ v' = e^x)$.

Prob. 9. $(x^2 - 2x + 2)e^x$, $(u = x^2,\ v' = e^x)$.

Prob. 11. $\frac{1}{2}x^2 \ln|x| - \frac{x^2}{4}$, $(u = \ln x,\ v' = x)$.

Prob. 13. $\frac{1}{2}x^2 \ln(3x) - \frac{x^2}{4}$, $(u = \ln(3x),\ v' = x)$.

Prob. 15. $\ln|\cos x| + x \tan x$, $(u = x,\ v' = \sec^2 x)$.

Prob. 17. $\frac{\sqrt{3}}{2} - \frac{\pi}{6}$, $(u = x,\ v' = \sin x)$.

Prob. 19. $2\ln 2 - 1$, $(u = \ln x,\ v' = 1)$.

Prob. 21. $\frac{1}{2}(4\ln 4 - 3)$, $(u = \ln x,\ v' = \frac{1}{2})$.

Prob. 23. $1 - \frac{2}{e}$, $(u = x,\ v' = e^{-x})$.

Prob. 25. $\frac{1}{4}(2 + (\sqrt{3} - 1)e^{\pi/3})$, $(u = \sin x,\ v' = e^x)$,$u = -\cos x,\ v' = e^x$.

Prob. 27. $\frac{2e^{-3x}\left(-6\cos\left(\frac{\pi x}{2}\right)+\pi\sin\left(\frac{\pi x}{2}\right)\right)}{36+\pi^2}$, $(u = \cos\left(\frac{\pi}{2}x\right), v' = e^{-3x})$, $(u = \frac{\pi}{2}\sin\left(\frac{\pi}{2}x\right), v' = \frac{-e^{-3x}}{3})$.

Prob. 29. $\frac{1}{2}x(\sin(\ln x) - \cos(\ln x))$, $(u = \sin(\ln x), v' = 1)$, $(u = \cos(\ln x), v' = 1)$.

Prob. 31. We have

$$\begin{aligned}\int \cos^2 x dx &= \sin x \cos x + \int \sin^2 x dx \\ &= \sin x \cos x + \int (1 - \cos^2 x) dx \\ &= \sin x \cos x + \int 1 dx - \int \cos^2 x dx\end{aligned}$$

therefore $2\int \cos^2 x dx = \sin x \cos x + x$ and we get $\int \cos^2 x dx = \frac{1}{2}(\sin x \cos x + x)$.

Prob. 33. We have

$$\begin{aligned}\int \arcsin x dx &= x \arcsin x - \int \frac{x}{\sqrt{1-x^2}} dx \\ &= x \arcsin x - \int \frac{-1}{2\sqrt{u}} du \quad (1 - x^2 = u) \\ &= x \arcsin x + \sqrt{u} \\ &= x \arcsin x + \sqrt{1-x^2}\,.\end{aligned}$$

Prob. 35. Taking $u = \ln x$, $v' = \frac{1}{x}$, we get

$$\int \frac{1}{x} \ln x dx = (\ln x)^2 - \int \frac{1}{x} \ln x dx\,.$$

Collecting similar terms we conclude that $\int \frac{1}{x} \ln x dx = \frac{1}{2}(\ln x)^2$.

Prob. 37.

(a) Taking $u = x^n$, $v' = e^{ax}$, we get

$$\int x^n e^{ax} dx = \frac{1}{a} x^n e^{ax} - \frac{n}{a} \int x^{n-1} e^{ax} dx\,.$$

(b) We use the previous part with $n=2$, $a=-3$. We have

$$\begin{aligned}\int x^2e^{-3x}dx &= -\frac{1}{3}x^2e^{-3x}+\frac{2}{3}\int xe^{-3x}dx\\ &= -\frac{1}{3}x^2e^{-3x}+\frac{2}{3}(-\frac{1}{3}xe^{-3x}+\frac{1}{3}\int e^x dx)\\ &= -\frac{1}{3}x^2e^{-3x}+\frac{2}{3}(-\frac{1}{3}xe^{-3x}+\frac{1}{3}e^x)\\ &= (-\frac{x^2}{3}-\frac{-2x}{9}-\frac{2}{27})e^{-3x}.\end{aligned}$$

Prob. 39. $2\cos\sqrt{x}+2\sqrt{x}\sin\sqrt{x}$, $(\sqrt{x}=y)$, $(u=\cos y, v'=2u)$.

Prob. 41. $-e^{-x^2/2}(2+x^2)$, $(\frac{x^2}{2}=y)$ and see problem 8.

Prob. 43. $e^{\sin x}(\sin x-1)$, $(\sin x=y)$, $(u=y, v'=e^y)$.

Prob. 45. $2e^{\sqrt{x}}(\sqrt{x}-1)$, $(\sqrt{x}=y)$, $(u=y, v'=e^y)$.

Prob. 47. $\sqrt{x}-\frac{x}{2}+(x-1)\ln(1+\sqrt{x})$, $(\sqrt{x}+1=y)$ and see problem 11.

7.3 Practicing Integration and Partial Fractions

Prob. 1. $(-\frac{1}{4}-\frac{x}{2})e^{-2x}$, $(u=x, v'=e^{-2x})$.

Prob. 3. $\ln|\sin x|$, $(\sin x=y)$.

Prob. 5. $-\cos(x^2)$, $(x^2=y)$.

Prob. 7. $\frac{1}{4}\arctan\frac{x}{4}$, $(\frac{x}{4}=y)$.

Prob. 9. $x-3\ln|x+3|$.

Prob. 11. $\frac{1}{2}\ln|x^2+3|$, $x^2+3=y)$.

Prob. 13.

(a) Taking $u=\ln x, v'=1$, we get $\int\ln x dx = x\ln x-\int 1dx = x\ln x - x$.

(b) Using $\ln x=y$, we have $x=e^y$, $dx=e^ydy$, therefore, $\int \ln x dx = \int ye^ydy$. Using integration by parts, $(u=y, v'=e^y)$, we see that this integral is ye^y-e^y. As $y=\ln x$, we get $x\ln x-x$.

Prob. 15. $(\frac{4x^2}{9}-\frac{8x}{45}-\frac{64}{45})(x-2)^{1/4}$, $((x-2)^{1/4}=y)$.

Prob. 17. $2e^2$.

Prob. 19. $\frac{\pi}{2}$.

Prob. 21. $\frac{1}{2}$.

Prob. 23.

(a) $a = -\frac{1}{2}, b = \frac{1}{2}$.

(b) $\frac{1}{2}\ln|\frac{x-2}{x}|$.

Prob. 25. $\frac{1}{3}\arctan(\frac{1}{3}(x-2))$.

Prob. 27. $\frac{1}{5}\ln\left|\frac{x-3}{x+2}\right| + C$

Prob. 29. $\frac{1}{6}\ln\left|\frac{x-3}{x+3}\right| + C$

Prob. 31. $\frac{1}{3}\ln\left|\frac{x-2}{x+1}\right| + C$

Prob. 33. $2\ln|x+1| - 5\ln|x+2| + x$.

Prob. 35. $x + 2\ln|\frac{x-2}{x+2}|$.

Prob. 37. $2 + \ln\frac{3}{5}$.

Prob. 39. $\frac{\ln 2}{2}$.

Prob. 41. $-\ln 2$.

Prob. 43. $\frac{\pi}{4} - \frac{1}{2}\ln 2$.

Prob. 45. $\frac{1}{1+x} + \ln|\frac{x}{x+1}|$.

Prob. 47. $-\frac{2}{x+1} + \ln|\frac{x+1}{x-1}|$.

Prob. 49. $-\frac{x}{18(x^2-9)} + \frac{1}{108}\ln|\frac{x+3}{x-3}|$.

Prob. 51. $-\frac{1}{x} - \arctan x$.

7.4 Improper Integrals

Prob. 1. The integral is improper because it has infinite upper limit. Its value is 1.

Prob. 3. The integral is improper because it has infinite upper limit. Its value is π.

Prob. 5. The integral is improper because it has infinite upper limit. Its value is 2.

Prob. 7. The integral is improper because it has infinite limits. Its value is 2.

Prob. 9. The integral is improper because it has infinite limits. Its value is 0.

Prob. 11. The integral is improper because the integrand has infinite limit at upper endpoint. Its value is 6.

Prob. 13. The integral is improper because the integrand has infinite limit at lower endpoint. Its value is 2.

Prob. 15. The integral is improper because the integrand is discontinuous in the interval. Its value is -2.

Prob. 17. The integral is convergent. Its value is $\frac{1}{2}$.

Prob. 19. The integral is divergent.

Prob. 21. The integral is convergent. Its value is 3.

Prob. 23. The integral is divergent.

Prob. 25. The integral is divergent.

Prob. 27. The integral is convergent. Its value is 0.

Prob. 29. The integral is divergent.

Prob. 31. $c = 3$.

Prob. 33.

(a) We have, for $p > 1$, $\int \frac{1}{x^p} dx = \frac{x^{1-p}}{1-p}$. Also $\int \frac{1}{x} dx = \ln x$, and the expressions for $A(z)$ follow.

(b) We have, for $0 < p \leq 1$, $\lim_{z\to\infty} A(z) = \infty$ because $\lim_{z\to\infty} z^{1-p} = \infty$.

(c) We have, for $p > 1$, $\lim_{z\to\infty} A(z) = \frac{1}{p-1}$ because $\lim_{z\to\infty} z^{1-p} = 0$.

Prob. 35.

(a) The exponential function is increasing in all its domain, and, for $x \geq 1$, we have $-x^2 \leq -x$.

(b) We know that $\int_1^\infty e^{-x} dx$ converges, therefore, by part a), so does $\int_1^\infty e^{-x^2} dx$.

Prob. 37.

(a) For $x \geq 1$ we have $4x^2 \geq 1 + x^2$, then, exctracting square roots, we get $2x \geq \sqrt{1+x^2}$, therefore $\frac{1}{\sqrt{1+x^2}} \geq \frac{1}{2x}$.

(b) We know that $\int_1^\infty \frac{1}{2x} dx$ diverges, therefore, by part a), so does $\int_1^\infty \frac{1}{\sqrt{1+x^2}}$.

Prob. 39. Convergent, compare with $\int_{-\infty}^\infty e^{-|x|} dx$.

Prob. 41. Divergent, compare with $\int_1^\infty \frac{1}{\sqrt{2x}} dx$.

Prob. 43.

(a) We have $\lim_{x\to\infty} \frac{\ln x}{\sqrt{x}} = \lim_{x\to\infty} \frac{\sqrt{x}}{2x} = 0$.

(b) For $x \geq 75$ we have $2\ln x \leq \sqrt{x}$.

(c) It is enough to study the integral over the interval $[75, \infty]$, since $\int_0^{75} e^{-\sqrt{x}} dx$ offers no problems. We have, from part b), $\ln x^2 \leq \sqrt{x}$, so $\ln \frac{1}{x^2} \geq -\sqrt{x}$, and $\frac{1}{x^2} \geq e^{-\sqrt{x}}$. As $\int_0^\infty \frac{1}{x^2} dx$ converges, so does $\int_0^\infty e^{-\sqrt{x}}$.

The graph of $e^{-\sqrt{x}}$ stays, from some point on, below the graph of $\frac{1}{x^2}$, therefore it defines a finite area with the x-axis.

7.5 Numerical Integration

Prob. 1. 2.32813.

Prob. 3. 0.629204.

Prob. 5. 0.69122. The exact value is 0.6931...

Prob. 7. 5.38382. The exact value is 5.3333...

Prob. 9. 2.34375.

Prob. 11. 0.637963.

Prob. 13. 20.32. The exact value is 20.

Prob. 15. 1.81948. The exact value is 1.88562...

Prob. 17. 82.

Prob. 19. $|f''(x)| = |e^{-x^2/2}(x^2-1)| \le (1)(4-1) = 3$ for $0 \le x \le 2$; $n \ge 100$.

Prob. 21. $|f''(x)| = |e^{-x}| \le 1$ for $0 \le x \le 1$; $n \ge 92$.

Prob. 23. $|f''(x)| = \left|\frac{e^x[(x-1)^2+1]}{x^3}\right| \le \frac{2e^2}{1}$; $n \ge 111$.

Prob. 25.

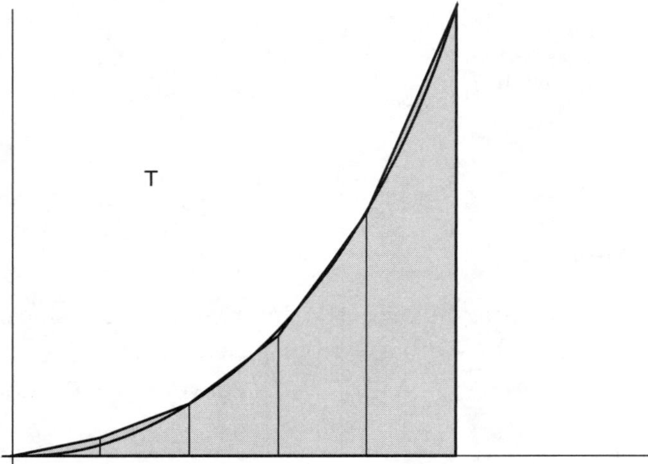

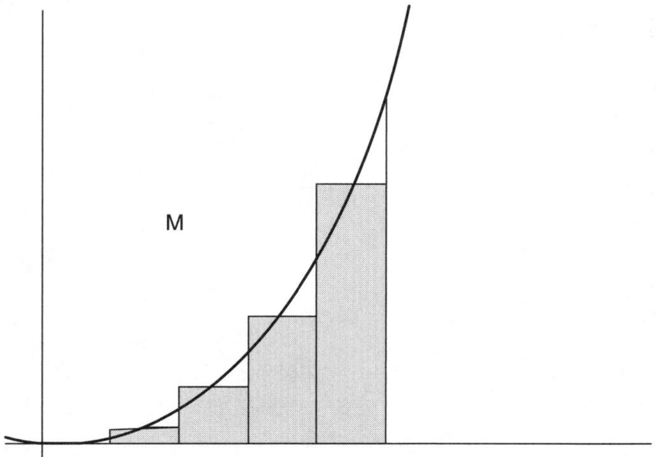

(a) We have $M_5 = 0.245$, $T_5 = 0.268$, the exact value being 0.25.

(b) The same reason as in a) applies.

(c) Reversing the concavity reverses the inequalities above. We find $0.6433 \leq \int_0^1 \sqrt{x}dx \leq 0.6730$.

7.6 Tables of Integrals

Prob. 1. $\frac{x}{2} + \frac{3}{4}\ln|2x-3| + C$.

Prob. 3. $\frac{1}{2}x\sqrt{x^2-16} - 8\ln|x+\sqrt{x^2-16}| + C$.

Prob. 5. $6 - \frac{16}{e}$.

Prob. 7. $\frac{1}{9}(1+2e^3)$.

Prob. 9. $\frac{e^{\pi/6}}{2} + \frac{1}{4} - \frac{\sqrt{3}}{4}$.

Prob. 11. $-e^{-x/2}(2x^2+8x+14) + C$.

Prob. 13. $\frac{1}{10}(5x-3) + \frac{1}{20}\sin(10x-6) + C$.

Prob. 15. $\frac{x}{2}\sqrt{9+4x^2} + \frac{9}{4}\ln|x+\frac{1}{2}\sqrt{9+4x^2}| + C$.

Prob. 17. $\frac{2e^{2x+1}}{\pi^2+16}(4\sin\frac{\pi x}{2} - \pi\cos\frac{\pi x}{2}) + C$.

Prob. 19. 2 ln 2

Prob. 21. $\frac{x}{2}(\cos(\ln|3x|) + \sin(\ln|3x|) + C$.

7.7 The Taylor Approximation

Prob. 1. $1 + 2x$.

Prob. 3. $1 + x$.

Prob. 5. $\ln 2$.

Prob. 7. $1 - \frac{x^2}{2} + \frac{x^4}{24}$.

Prob. 9. x^5.

Prob. 11. $P_3(x) = \sqrt{2} + \frac{x}{2\sqrt{2}} - \frac{x^2}{16\sqrt{2}} + \frac{x^3}{64\sqrt{2}}$. $\sqrt{2+0.1} = 1.4491376...$, $P_3(0.1) = 1.44913800...$

Prob. 13. $P_5(x) = x - \frac{x^3}{6} + \frac{x^5}{120}$. $\sin 1 = 0.841471$, $P_5(1) = 0.841667....$

Prob. 15. $P_2(x) = x$. $\tan 0.1 = 0.100335...$, $P_2(0.1) = 0.1$.

Prob. 17.

(a) $P_3(x) = x - \frac{x^3}{6}$.

(b) $\lim_{x\to 0} \frac{\sin x}{x}$ and $\lim_{x\to 0} \frac{x - \frac{x^3}{6}}{x} = 1$ "should" be the same.

Prob. 19. $P_3(x) = 1 + \frac{x-1}{2} - \frac{1}{8}(x-1)^2 + \frac{1}{16}(x-1)^3$. $\sqrt{3} = 1.4142135623...$, $P_3(2) = 1.4375$.

Prob. 21. $P_3(x) = \frac{\sqrt{3}}{2} - \frac{1}{2}(x - \pi/6) - \frac{1}{4}\sqrt{3}(x - \pi/6)^2 + \frac{1}{12}(x - \pi/6)^3$. $\cos\frac{\pi}{7} = 0.900968867...$, $P_3(2) = 0.9009677287...$

Prob. 23. $P_3(x) = e^2 + e^2(x-2) + \frac{e^2}{2}(x-2)^2 + \frac{e^2}{6}(x-2)^3$. $e^{2.1} = 8.1661...$, $P_3(2.1) = 8.1661...$

Prob. 25. Let g be defined by $g(N) = rN(1 - \frac{N}{K})$ where r and K are constants. The linearization near 0 of g is rN.

Prob. 27. 10.

Prob. 29. 2.

Prob. 31. $P_2(x) = 0$. Error: $|e^{-1/a}|$.

Prob. 33.

(a) Let $g(x) = \tan^{-1} x$. We have $g(0) = 0$, $g'(0) = 0$, $g'''(0) = -2$, and so on ($g^{(ek)(0)} = 0$, $g^{(2k+1)}(0) = (-1)^k(2k)!$. The expression is just the Taylor polynomial for g.

(b) Noting that $\tan^{-1} 1 = \frac{\pi}{4}$ and plugging in $x = 1$ in the expression in part a), we get the given equality.

7.9 Review Problems

Prob. 1. $-\frac{1}{9}(1 - x^3)^3 + C$.

Prob. 3. $-2e^{-x^2} + C$.

Prob. 5. $\frac{6}{7}(1 + \sqrt{x})^{7/3} - \frac{3}{2}(1 + \sqrt{x})^{4/3} + C$.

Prob. 7. $\frac{1}{6}\tan(3x^2) + C$.

Prob. 9. $\frac{x^2}{4}(2\ln x - 1) + C$.

Prob. 11. $\tan x(\ln|\tan x| - 1) + C$.

Prob. 13. $\frac{1}{2}\arctan\frac{x}{2} + C$.

Prob. 15. $-\ln|\cos x| + C$.

Prob. 17. $\frac{e^{2x}}{5}(2\sin x - \cos x) + C$.

Prob. 19. $2\sqrt{e^x} + C$.

Prob. 21. $\frac{1}{2}(x - \cos x \sin x) + C$.

Prob. 23. $\ln\left|\frac{x-1}{x}\right| + C$.

Prob. 25. $x - 5\ln|x + 5| + C$.

Prob. 27. $\ln|x + 5| + C$.

Prob. 29. $\frac{1}{2}x(x + 6) + 4\ln|x - 1|$.

Prob. 31. $4 + \ln 3$.

Prob. 33. $1 - \frac{1}{\sqrt{e}}$.

Prob. 35. $\frac{\pi}{8}$.

Prob. 37. 4.

Prob. 39. $\frac{\pi}{6}$.

Prob. 41. Diverges.

Prob. 43. Diverges.

Prob. 45. 2.

Prob. 47. $-\frac{1}{4}$.

Prob. 49. $e - e^{1/\sqrt{2}}$.

Prob. 51. $M_4 = 0.625$, $T_4 = 0.75$.

Prob. 53. $M_5 = 0.631068$, $T_5 = 0.634226$.

Prob. 55. $2x - \frac{4x^3}{3}$.

Prob. 57. $(x-1) - \frac{1}{2}(x-1)^2 + \frac{1}{3}(x-1)^3$.

Prob. 59. The area to be shaded is the one below the horizontal line $y = K$, above the graph of $y = f_{avg}(x)$.

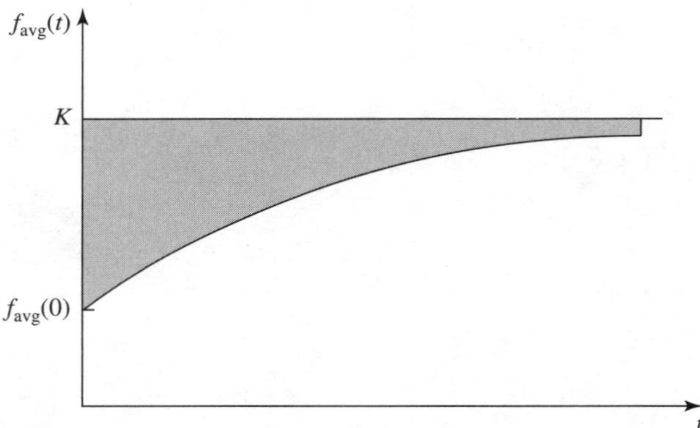

8 Differential Equations

8.1 Solving Differential Equations

Prob. 1.

$$\begin{aligned} y(x) &= y(0) + \int_0^x (u + \sin u) du \\ &= \frac{U^2}{2} - \cos U]_0^x \\ &= \frac{x^2}{2} - \cos x + \cos 0 \\ &= 1 + \frac{x^2}{2} - \cos x \end{aligned}$$

Prob. 3.

$$y(x) = y(1) + \int_1^x \frac{1}{u} du = \ln u \Big|_1^x = \ln x - \ln 1 = \ln x$$

Prob. 5.

$$x(t) = x(0) + \int_0^t \frac{1}{1-u} du = 2 - \ln(1-u) \mid_0^t = 2 - \ln(1-t)$$

Prob. 7.

$$\begin{aligned} s(t) = s(0) + \int_0^t \sqrt{3u+1} du &= 1 + [\frac{2}{3}(3u+1)^{3/2} \cdot \frac{1}{3}]\Big|_0^t \\ &= 1 + \frac{2}{9}(3t+1)^{3/2} - \frac{2}{9} \\ &= \frac{7}{9} + \frac{2}{9}(3t+1)^{3/2} \end{aligned}$$

Prob. 9.

$$v(t) = v(0) + \int_0^t \cos u du = 5 + \sin u \Big|_0^t = 5 + \sin t$$

Prob. 11.

$$\frac{dy}{3y} = dx$$

so $\int \frac{dy}{3y} = \int dx$, then $\frac{1}{3}\ln|3y| = x + c_1$ so $|3y| = e^{3(x+c_1)}$ and taking out the absolute value signs

$$3y = \pm e^{3(x+c_1)}$$

Now using the condition that the point $(0,2)$ belongs to the function

$$3y = ce^{3x}$$

we find $c = 6$, so the solution is

$$y = 2e^{3x}$$

Prob. 13. $\int \frac{dx}{-2x} = \int dt$, so $-\frac{1}{2}\int \frac{dx}{x} = \int dt$ and integrating both sides we have

$$\begin{aligned} -\frac{1}{2}\ln|x| &= t + c_1 \\ \ln|x| &= -2t + c_2 \\ |x| &= e^{-2t+c_2} \\ x &= \pm e^{-2t+c_2} \\ x &= ce^{-2t}. \end{aligned}$$

Now using the condition that $x(1) = 5$, we get $5 = ce^{-2}$, that is, $c = 5e^2$, the solution is

$$x(t) = 5e^2 e^{-2t}$$

or

$$x(t) = 5e^{2-2t}$$

Prob. 15. $\int \frac{dh}{2h+1} = \int ds$, so integrating both sides we have

$$\begin{aligned} \frac{\ln|2h+1|}{2} &= s + c_1 \\ \ln|2h+1| &= 2s + 2c_1 \\ |2h+1| &= e^{2s+c_2} \\ 2h+1 &= \pm e^{2s+c_2} \\ 2h+1 &= c_3 e^{2s} \\ 2h &= -1 + c_3 e^{2s} \\ h &= -\frac{1}{2} + ce^{2s} \end{aligned}$$

with the initial condition $h(0) = 4$, we have:

$$4 = -\frac{1}{2} + c \quad \text{or} \quad c = \frac{9}{2}$$

and the final solution is:

$$h = -\frac{1}{2} + \frac{9}{2}ce^{2s}$$

Prob. 17. $\int \frac{dN}{0.3N} = \int dt$, so integrating both sides

$$\begin{aligned}
\frac{\ln|0.3N|}{0.3} &= t + c_1 \\
|0.3N| &= e^{0.3t+c_2} \\
0.3N &= \pm e^{0.3t+c_2} \\
N(t) &= ce^{0.3t}
\end{aligned}$$

with $N(0) = 20$, the final solution is $c = 20$ and

$$N(t) = 20e^{0.3t}$$

so the population at time $= 5$ is

$$\begin{aligned}
N(5) &= 20 \cdot e^{0.3 \cdot 5} \\
&= 20 \cdot e^{1.5} \\
&\simeq 89 \text{ individuals.}
\end{aligned}$$

Prob. 19.

(a) $\int \frac{dN}{N} = rdt$, so integrating both sides we have

$$\begin{aligned}
\ln|N| &= rt + c_1 \\
|N| &= e^{rt+c_1} \\
N &= \pm e^{rt+c_2} \\
N(t) &= ce^{rt}
\end{aligned}$$

(b) Transforming the coordinates into a semilog, the equation changes to

$$\log N(t) = \log c + (rt) \log e$$

so the slope of the line in the $\log N(t) - t$ graph will be $r \log e$.

(c) Find the best possible line that fits the data on the semilog plot described above, and $r \log e$ will be the slope of the line.

Prob. 21. $N^{-2}dN = \frac{1}{100}dt$ so integrating both sides

$$-\frac{1}{N} = \frac{t}{100} + c_1$$

so

$$N(t) = \frac{1}{C - \frac{t}{100}}$$

with $N(0) = 10$ we get $10 = \frac{1}{c}$, or $c = \frac{1}{10}$, and the final equation is

$$N(t) = \frac{1}{\frac{1}{10} - \frac{t}{100}}$$

or

$$N(t) = \frac{100}{10 - t}$$

(b)

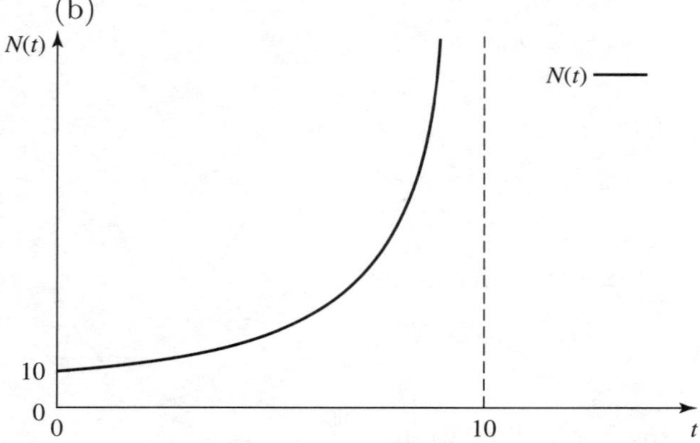

When $t \to 10$, $N(t)$ will approach $+\infty$, meanig that the population will grow without bounds.

Prob. 23.

(a) In this case $L_\infty = 123$, and the equation becamos

$$\frac{dh}{dt} = k(123 - L(t)) \quad \text{with} \quad L(0) = 1$$

and the solution according to Section 8.1.2 is

$$L(t) = 123\left[1 - \left(1 - \frac{1}{123}\right)e^{-kt}\right]$$

now substituting $L(27) = \frac{123}{2} = 61.5$ we get

$$\begin{aligned} \frac{123}{2} &= 123\left[1-\left(1-\frac{1}{123}\right)e^{-27k}\right] \\ \frac{1}{2} &= \left(1-\frac{1}{123}\right)e^{-27k} \\ \frac{123}{2\times 122} &= e^{-27k} \end{aligned}$$

and taking ln on both sides we have

$$\begin{aligned} \ln\frac{2\times 122}{123} = 27k& \\ k &\simeq 0.025 \end{aligned}$$

(b)

$$\begin{aligned} L(10) &= 123\left[1\left(1-\frac{1}{123}\right)e^{-0.025\times 10}\right] \\ &= 123\left[1-\frac{122}{123}\cdot e^{-0.25}\right] \\ &\simeq 28\text{inches} \end{aligned}$$

(c) 90% of the asymptotic length will be

$$\begin{aligned} \frac{9}{10}123 &= 123\left[1\left(1-\frac{1}{123}\right)e^{-0.025t}\right] \\ \frac{1}{10} &= \left(1-\frac{1}{123}\right)e^{-.025t} \\ \frac{123}{1220} &= e^{-0.25t} \\ \ln\frac{1220}{123} &= 0.025t \\ t &\simeq 92\text{months.} \end{aligned}$$

Prob. 25. We will use below the fact that

$$\frac{1}{(y-a)(y-b)} = \frac{1}{a-b}\left[\frac{1}{y-a}-\frac{1}{y-b}\right], \quad a\neq b$$

so the equation becomes

$$\int \frac{dy}{y(1+y)} = \int dx$$

on the equation $yy + 1 = ce^x$ we can easily see that $c = \frac{2}{3}$, so the final solution is

$$y = \frac{(2.3)e^x}{1-(2/3)e^x} = \frac{2e^x}{3-2e^x}$$

Prob. 27. $\frac{1}{y(y-5)} = \frac{-1}{5}\left[\frac{1}{y} - \frac{1}{y-5}\right]$ so

$$\frac{-1}{5}\int \left(\frac{1}{y} - \frac{1}{y-5}\right) dy = \int dx$$

$$-\ln|y| + \ln|y-5| = 5x + c_1$$

$$\ln\left|\frac{y}{y-5}\right| = -5x + c_1$$

$$\frac{y}{y-5} = ce^{-5x}$$

with the initial condition we can see that $c = -\frac{1}{4}$ and the solution can be written as

$$\frac{y}{y-5} = -\frac{1}{4}e^{-5x}$$

cross multiplying and solving for y in terms of x, we have

$$y = (y-5)(-\frac{1}{4})e^{-5x}$$

$$4y = (5-y)e^{-5x}$$

$$4y + ye^{-5x} = 5e^{-5x}$$

$$y(4+e^{-5x}) = 5e^{-5x}$$

and finally the solution is

$$y = \frac{5e^{-5x}}{4+e^{-5x}} = \frac{5}{1+4e^{5x}}$$

Prob. 29. We use the partial fraction method to solve $\frac{dy}{dx} = 2y(3-y)$ with $y_0 = 5$ for $x_0 = 1$. Separating variables yields $\frac{dy}{y(3-y)} = 2dx$. The partial fraction decomposition is

$$\frac{1}{y(3-y)} = \frac{A}{y} + \frac{B}{3-y} = \frac{3A - Ay + By}{y(3-y)}$$

which yields $A = 1/3$ and $B = 1/3$. Hence

$$\int \frac{dy}{y(3-y)} = 2\int dx$$

$$\frac{1}{3}(\ln|y| - \ln|3-y|) = 2x + C_1$$

$$\ln\left|\frac{y}{3-y}\right| = 6x + 3C_1$$

$$\left|\frac{y}{3-y}\right| = e^{6x+3C_1}$$

$$\frac{y}{3-y} = Ce^{6x}.$$

We use the initial condition to determine the value of C: $\frac{5}{-2} = Ce^6$ and hence $C = -\frac{5}{2}e^{-6}$.

We thus have

$$\frac{y}{3-y} = -\frac{5}{2}e^{-6}e^{6x}$$

$$y\left(1 - \frac{5}{2}e^{6(x-1)}\right) = -\frac{15}{2}e^{6(x-1)}$$

$$y = \frac{3}{1 - \frac{2}{5}e^{-6(x-1)}}$$

Prob. 31. $\frac{dy}{y(1+y)} = dx$, so using the fact that

$$\frac{1}{y(y+1)} = \frac{1}{y} - \frac{1}{y+1}$$

we get

$$\int \left(\frac{1}{y} - \frac{1}{y+1}\right) = \int dx,$$

so integrating both sides

$$\begin{aligned} \ln|y| - \ln|y+1| &= x + c_1 \\ \ln\left|\frac{y}{y+1}\right| &= x + c_1 \\ \frac{y}{y+1} &= ce^x \end{aligned}$$

solving for y in terms of x, we get

$$y = \frac{ce^x}{1 - ce^x}$$

Prob. 33. $\frac{dy}{(1+y)^3} = dx$, is integrating both sides we get

$$\begin{aligned} \int \frac{dy}{(1+y)^3} &= \int dx \\ \frac{1}{-2(1+y)^2} &= x + c_1 \\ (1+y)^2 &= -\frac{1}{x+c_1} \\ 1 + y &= \pm\sqrt{1\frac{1}{x+c}} \quad \text{so} \\ y &= -1 \pm \sqrt{-\frac{1}{x+c}} \end{aligned}$$

Prob. 35.

(a)

$$\begin{aligned} \int \frac{du}{u^2 - a^2} &= \int \frac{du}{(u+a)(u-a)} \\ &= \frac{1}{2a} \int \left(\frac{1}{u-a} - \frac{1}{u+a} \right) du \\ &= \frac{1}{2a} (\ln|u-a| - \ln|u+a|) \\ &= \frac{1}{2a} \ln\left|\frac{u-a}{u+a}\right| + c \end{aligned}$$

(b) The differential equation can be changed to $\frac{dy}{y^2-4} = dx$ and integrating both sides we get $\int \frac{dy}{y^2-4} = \int dx$ and using the integral we have

$$\begin{aligned} \frac{1}{4}\ln\left|\frac{y-2}{y+2}\right| &= x + c_1 \\ \ln\left|\frac{y-2}{y+2}\right| &= 4x + c_2 \\ \left|\frac{y-2}{y+2}\right| &= ce^{4x}. \end{aligned}$$

So for the first point $(0,0)$, we can calculate $c = -1$; for $(0,2)$ we get $c = 0$ and for $(0,4)$ we get $c = \frac{1}{3}$.

Now cross multiplying and solving for y in terms of x we obtain

$$\begin{aligned} y - 2 &= ce^{4x}(y+2) \\ y &= \frac{2(ce^{4x}+1)}{1 - ce^{4x}} \end{aligned}$$

so the different solutions are

(i) $(0,0) \to c = -1 \to y = \frac{2(1-e^{4x})}{1+e^{4x}}$

(ii) $(0,2) \to c = 0 \to y = 2$

(iii) $(0,4) \to c = \frac{1}{3} \to y = 2\frac{3+e^{4x}}{3-e^{4x}}$

Prob. 37. This is the Logistic Equation (8.27) with $r = 0.34$ and $K = 200$, so the solution is

$$N(t) = \frac{200}{1 + \left(\frac{200}{50} - 1\right)e^{-0.34t}} = \frac{200}{1 + 3e^{-0.34t}}$$

and

$$\lim_{t\to\infty} N(t) = k = 200.$$

Prob. 39.

(a) The solution is given by

$$N(t) = \frac{k}{1 + \left(\frac{k}{N_0} - 1\right)e^{-rt}}$$

so for $k = 50$, $r = 1.5$ as given we have

$$N(t) = \frac{50}{1 + \left(\frac{50}{10} - 1\right) e^{-1.5t}} = \frac{50}{1 + 4e^{-1.5t}}$$

(b)

$$N(t) = \frac{50}{1 + \left(\frac{50}{90} - 1\right) e^{-1.5t}} = \frac{50}{1 - \frac{4}{9}e^{-1.5t}}$$

(c)

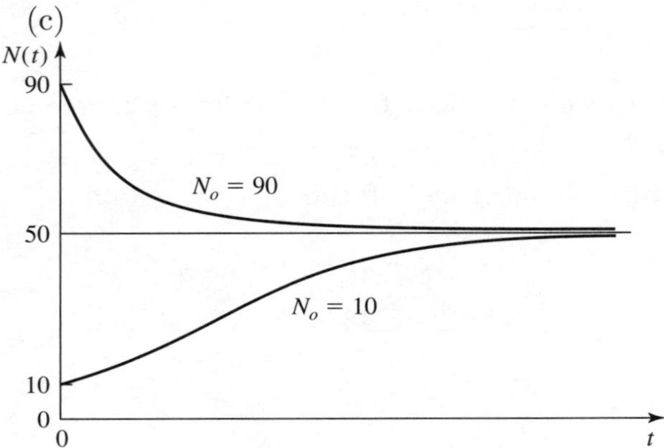

Prob. 41.

(a) If $r = 5$ and $k = 30$, the equation will be given by

$$\frac{dN}{dt} = 5N\left(1 - \frac{N}{30}\right)$$

(b)

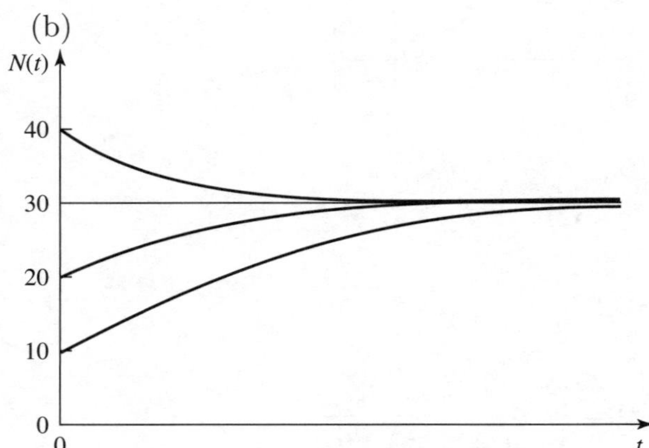

Prob. 43.

(a) From the partial fraction expansion we can see that

$$\frac{1}{p(1-p)} = \frac{1}{p} - \frac{1}{p-1}$$

and using it to separate the variables we have

$$\frac{dp}{p(1-p)} = \frac{s}{2}dt$$

integrating both sides

$$\int \left(\frac{1}{p} - \frac{1}{p-1}\right) dp = \int \frac{s}{2}\, dt$$

$$\ln|p| - \ln|p-1| = \frac{s}{2}t + c_1$$

$$\ln\left|\frac{p}{p-1}\right| = \frac{s}{2}t + c_1$$

$$\frac{p}{p-1} = ce^{\frac{s}{2}t}$$

and substituting the initial condition we can find the value of $c = \frac{p_0}{p_0-1}$, so the solution, can now be computed as

$$p = (p-1)ce^{\frac{s}{2}t}$$

$$p(1 - ce^{-\frac{s}{2}t}) = -ce^{\frac{s}{2}t}$$

so

$$p = \frac{ce^{\frac{st}{2}}}{ce^{\frac{st}{2}} - 1} = \frac{p_0 e^{\frac{st}{2}}}{p_o e^{\frac{st}{2}} + 1 - p_0}$$

(b) For these values the solution becomes

$$p = \frac{0.1e^{0.005t}}{0.1e^{0.005t} + 0.9} = \frac{e^{0.005t}}{e^{0.005t} + 9}$$

and the time can the be solved by

$$0.5 = \frac{e^{0.005t}}{e^{0.005t} + 9}$$

$$0.5e^{0.005t} = 4.5$$
$$e^{0.005t} = 9$$

and applying ln to both sides we have

$$0.005t \simeq 2.1972$$

so $t = 439$ units of time

(c)

$$\lim_{t\to\infty} p(t) = \lim_{t\to\infty} \frac{p_0 e^{\frac{st}{2}}}{p_0 e^{\frac{st}{2}} + 1 - p_0} = 1$$

This means that the frequency of the A_1 will become closer to 1 as times goes by, that is, eventually the population will consist of only $A_1 A_1$ types.

Prob. 45. Separating the variable we have $ydy = (x+1)dx$, so integrating both sides:

$$\frac{y^2}{2} = \frac{(x+1)^2}{2} + c$$

or

$$y^2 = (x+1)^2 + c$$

for our initial conditions ($y_0 = 2$ ir $x_0 = 0$) we find $c = 4 - 1 = 3$, so the final solution is

$$y = \pm\sqrt{x^2 + 2x + 4},$$

and then the initial condition again we see that the branch in case is positive one, that is,

$$y = \sqrt{x^2 + 2x + 4}$$

Prob. 47. Separating the variables we have $\frac{dy}{y+1} = e^{-x}dx$, and integrating both sides

$$\ln|y+1| = -c^{-x} + c_1$$

now substituting the initial condition we find

$$c_1 = 1 + \ln 3$$

and substituting back into the equation

$$\ln|y+1| = 1 + \ln 3 - e^{-x}$$

so

$$|y+1| = \exp[1 + \ln 3 - e^{-x}]$$

giving us two branches $\pm$, but because of the initial condition we are interested in the positive only, that is

$$\begin{aligned} y &= -1 + \exp[1 + \ln 3 - e^{-x}] \\ &= -1 + \exp(1) \cdot \exp(\ln 3) \cdot \exp(-e^{-x}) \\ &= -1 + 3e \cdot \exp(-e^{-x}) \\ &= -1 + 3\exp[1 - e^{-x}] \end{aligned}$$

Prob. 49. Separating the variables we have $\frac{dy}{y+1} = \frac{dx}{x-1}$, and integrating both sides

$$\ln|y+1| = \ln|x-1| + c_1$$

$$\ln\left|\frac{y+1}{x-1}\right| = c_1$$

$$\frac{y+1}{x-1} = c$$

using the initial condition we see that $c = 6$ and the final solution is $y = 6x - 7$.

Prob. 51. Separating the variables we have $\frac{dr}{r} = e^{-t}dt$ and integrating both sides

$$\ln|r| = -c^{-t} + c_1$$

$$|r| = \exp(-e^{-t} + c_1)$$

so

$$r = c\exp(-e^{-t})$$

from the initial conditions we see that $c = e$, and

$$r = \exp[1 - e^{-t}]$$

Prob. 53. The line that relate the two quantities is given by

$$\ln \mathcal{O}_2 = 0.8 \ln m + c$$

and derivating both sides

$$\frac{d\mathcal{O}_2}{\mathcal{O}_2} = 0.8\frac{dm}{m}$$

or the final equation

$$\frac{dO_2}{dm} = 0.8\frac{O_2}{m}$$

Prob. 55. The relation among the logarithm of the two quantities is

$$\ln P = \frac{1}{7.7}\ln p + k$$

where P is the amount of phosphorous in the Daphusia, and p in the algal food

$$\frac{dP/dp}{P} = \frac{1}{7.7}\cdot\frac{1}{p},$$

so

$$\frac{dP}{P} = \frac{1}{7.7}\frac{dp}{p}$$

8.2 Equilibria and Their Stability

Prob. 1. (a) $y = 0, 2$

(b)

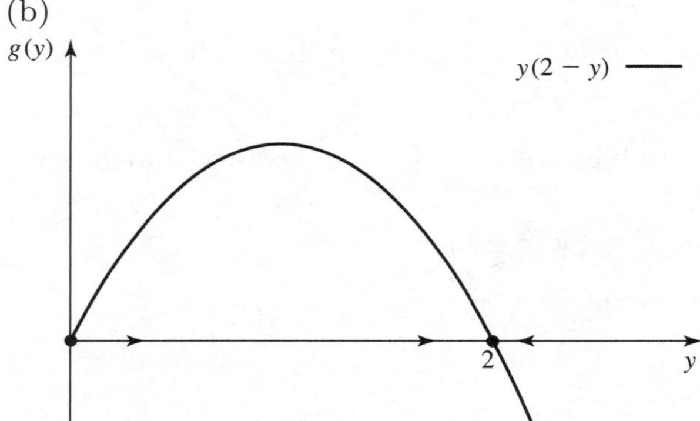

$y = 0$ is unstable; $y = 2$ is locally stable.

(c) Eigenvalue associated with $y = 0$ is $2 > 0$, hence $y = 0$ is unstable; eigenvalue associated with $y = 2$ is $-2 < 0$, hence $y = 2$ is locally stable.

Prob. 3. (a) $y = 0, 1, 2$
(b)

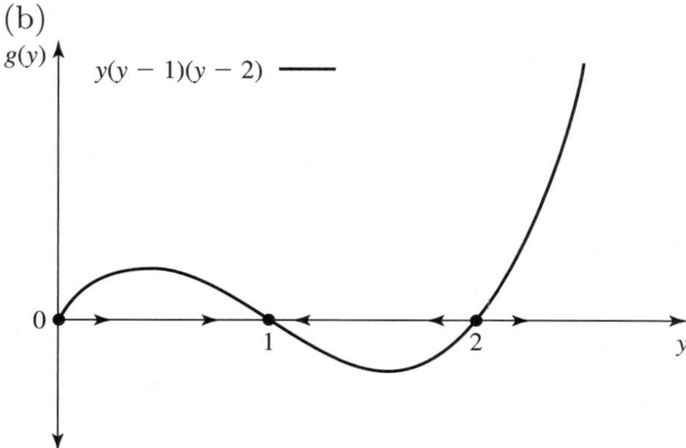

$y = 0$ and $y = 2$ are unstable; $y = 1$ is locally stable.
(c) Eigenvalue associated with $y = 0$ is $2 > 0$, hence $y = 0$ is unstable; eigenvalue associated with $y = 1$ is $-1 < 0$, hence $y = 1$ is locally stable; eigenvalue associated with $y = 2$ is $2 > 0$, hence $y = 2$ is unstable.

Prob. 5. (a) $\frac{dN}{dt} = 1.5N\left(1 - \frac{N}{100}\right)$
(b)

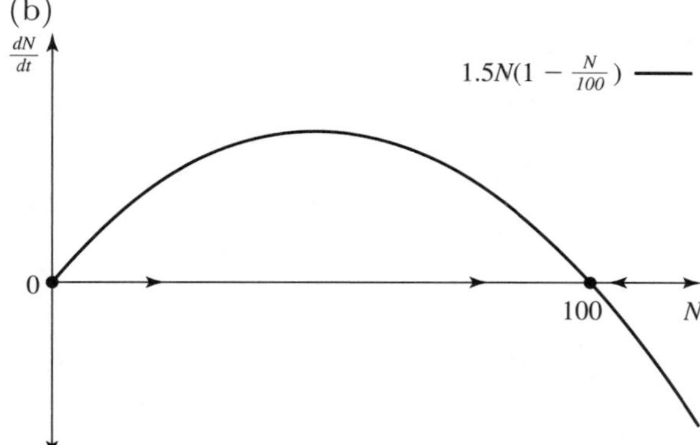

$N = 0$ is unstable; $N = 100$ is locally stable
(c) Eigenvalue associated with $N = 0$ is $1.5 > 0$, hence $N = 0$ is unstable; eigenvalue associated with $N = 100$ is $-1.5 < 0$, hence $N = 100$ is locally stable. Same results as in (b).

Prob. 7. (a) $K = 2000$ (b) $t = \frac{1}{2}\ln 199 \approx 2.65$ (c) 2000

Prob. 9. (a) $N \approx 52.79$ is unstable; $N \approx 947.21$ is locally stable
(b) The maximal harvesting rate is $rK/4$.

Prob. 11.

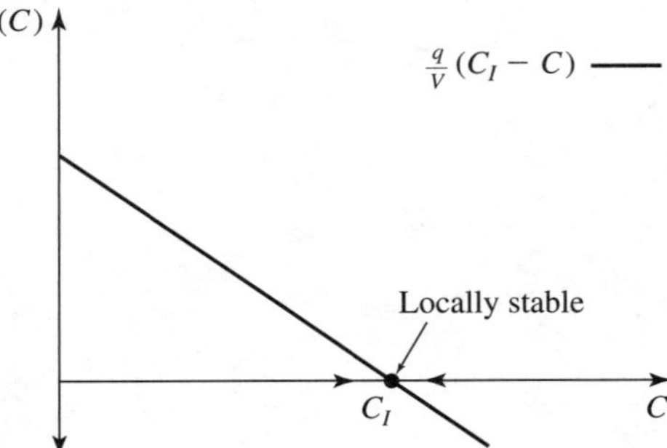

The equilibrium C_I is locally stable.

Prob. 13. (a) $\frac{dC}{dt} = \frac{0.2}{400}(3 - C)$
(b) $C(t) = 3 - 3e^{-t/2000}$, $t \geq 0$; $\lim_{t\to\infty} C(t) = 3$
(c) $C = 3$ is locally stable.

Prob. 15. (a) Equilibrium concentration: $C_I = 254$
(b) $T_R = \frac{1}{0.37} \approx 2.703$
(c) $T_R = \frac{1}{0.37} \approx 2.703$
(d) They are the same.

Prob. 17. We know from the solution of the genaral equantion that

$$C(t) = C_I \left[1 - \left(1 - \frac{C_0}{C_I}\right) e^{-(q/V)t}\right]$$

so multiplying it out, we have:

$$C(t) - C_I = -(C_I - C_0)\, e^{-(q/V)t}$$

which reduces to

$$\frac{C(t) - C_I}{C_0 - C_I} = e^{-(q/V)t}$$

now integrating both sides from 0 to ∞,

$$\int_0^\infty \frac{C(t) - C_I}{C_0 - C_I} dt = \int_0^\infty e^{-(q/V)t} dt = -\frac{V}{q} e^{-(q/V)t}\Big|_0^\infty = -\frac{V}{q}(0{-}1) = \frac{V}{q} = T_R$$

Prob. 19. $T_R = \frac{12.3\times10^9}{220}$seconds ≈ 647.1 days; $C(T_R) \approx 0.806\frac{\text{mg}}{\text{l}}$

Prob. 21. (a)

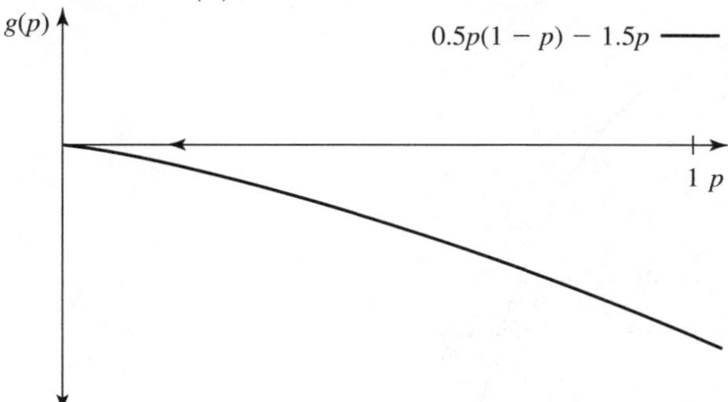

(b) $p = 0$ is locally stable
(c) $g'(0) = -1 < 0$, which implies that 0 is locally stable.

Prob. 23. (a) $\frac{dp}{dt}$ describes the rate of change of $p(t)$; $cp(1-p-D)$ describes colonization of vacant undestroyed patches; $-mp$ describes extinction.
(b)

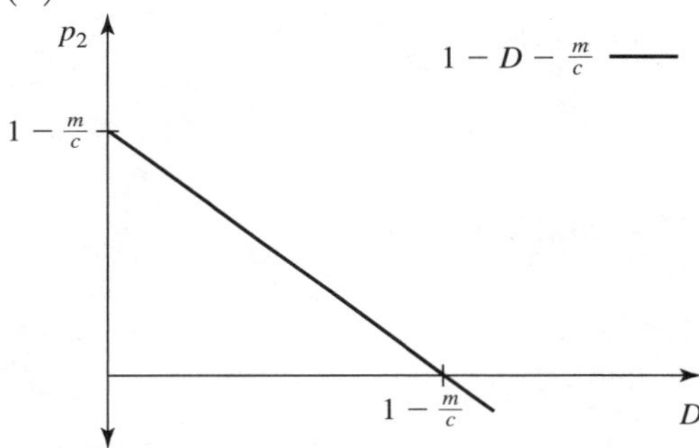

(c) $D < 1 - \frac{m}{c}$; $p_1 = 0$ is unstable; $p_2 = 1 - D - \frac{m}{c}$ is locally stable

Prob. 25. (a) $N = 0$, $N = 17$, and $N = 200$
(b) $N = 0$ is locally stable; $N = 17$ is unstable; $N = 200$ is locally stable

(c)

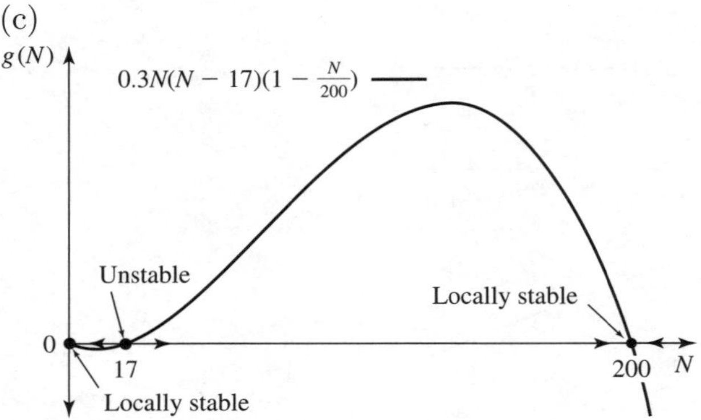

8.3 Systems of Autonomous Equations

Prob. 1. (a) $R_0 = 1.5 > 1$, the disease will spread
(b) $R_0 = \frac{1}{2} < 1$, the disease will not spread

Prob. 3. $R_0 = 0.9999 < 1$, the disease will not spread

Prob. 5. (a)

$$\frac{dN}{dt} = N_I - 5N - 0.02NX + X$$
$$\frac{dX}{dt} = 0.02NX - 2X$$

(b) equilibrium: $(\hat{N}, \hat{X}) = (100, N_I - 500)$, this is a nontrivial equilibrium provided $N_I > 500$

Prob. 7. (a)

$$\frac{dN}{dt} = 200 - N - 0.01NX + 2X$$
$$\frac{dX}{dt} = 0.01NX - 3X$$

(b)

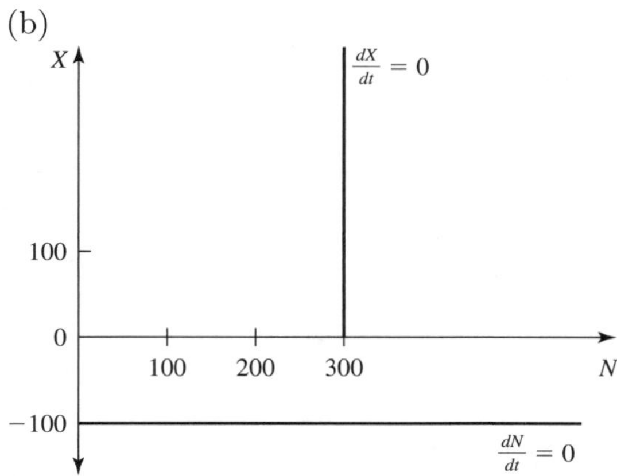

(c) No nontrivial equilibria

Prob. 9. (a) Equilibria: $(0,0)$, $(0,2/3)$, $(1/2,0)$
(b) Since $\frac{dp_2}{dt} < 0$ when $p_1 = 1/2$ and p_2 is small, species 2 cannot invade.

Prob. 11. (a)

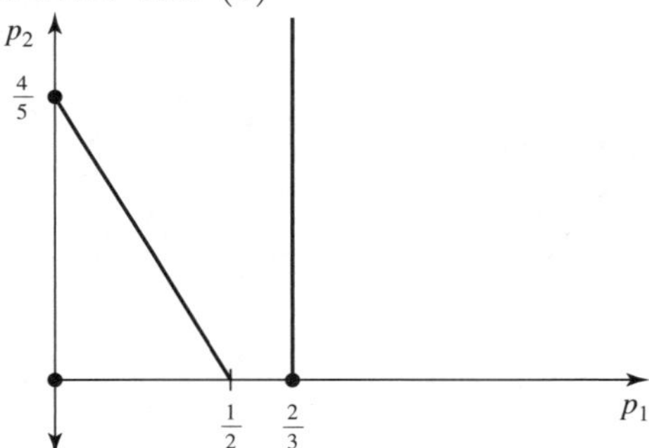

(b) equilibria: $(0,0)$, $(2/3,0)$, $(0,4/5)$

8.5 Review Problems

Prob. 1. (a) $\frac{dT}{dt}$ is proportional to the difference between the temperature of the object and the temperature of the surrounding medium.
(b) $t = \frac{20}{\ln \frac{9}{4}} \ln \frac{9}{4}$ minutes

Prob. 3. (a) $N(t) = N(0)e^{r_e t}$
(b) $N(t) = \frac{K}{1+\left(\frac{K}{N_0}-1\right)e^{-r_l t}}$
(c) $r_e \approx 0.691$; $K = 1001$: $r_l \approx 1.382$; $K = 10{,}000$: $r_l \approx 0.701$

Prob. 5. (a)

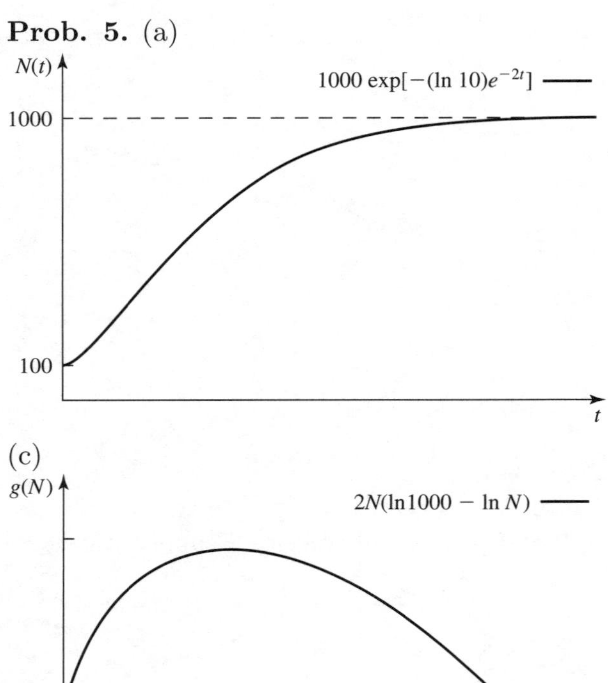

(c)

$N = 0$ is unstable; $N = 1000$ is locally stable
K is the carrying capacity

Prob. 7. (b)

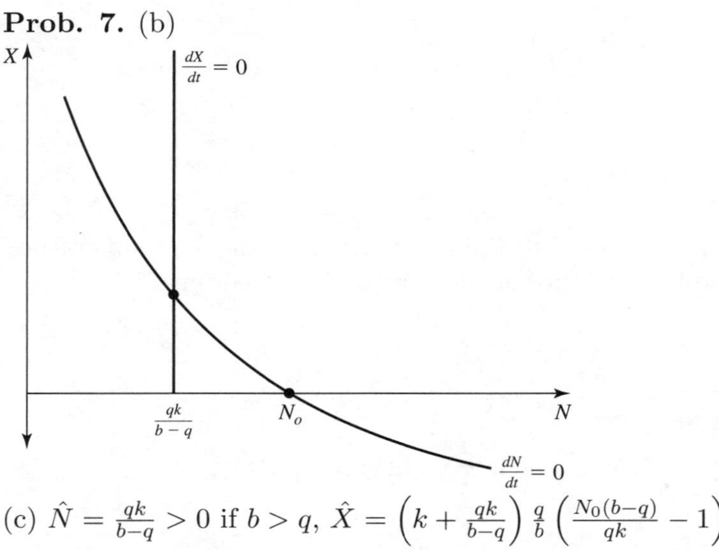

(c) $\hat{N} = \frac{qk}{b-q} > 0$ if $b > q$, $\hat{X} = \left(k + \frac{qk}{b-q}\right) \frac{q}{b} \left(\frac{N_0(b-q)}{qk} - 1\right)$

9 Linear Algebra and Analytic Geometry

9.1 Linear Systems

Prob. 1.

$$x - y = 1 \tag{9.1}$$
$$x - 2y = -2 \tag{9.2}$$

Subtracting (2) from (1) i.e., (1) - (2)

$$y = 3 \tag{9.3}$$

so,

$$x = 4 \tag{9.4}$$

for the eqn (1)	for the eqn (2)
at y = 1, x = 2	at y = 1, x = 0
at y = 2, x = 3	at y = 2, x = 2
at y = 3, x = 4	at y = 3, x = 4

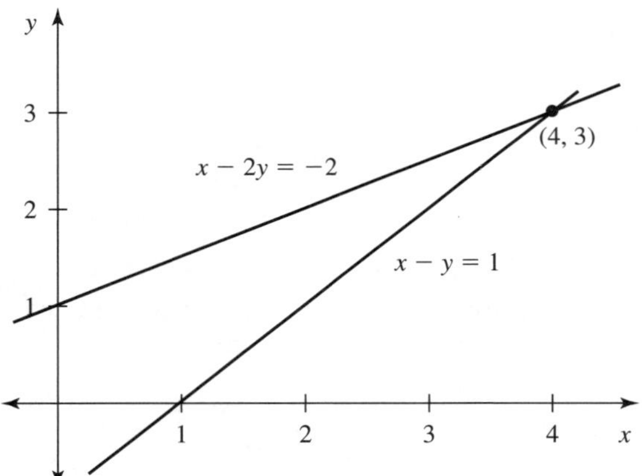

If we draw the graph for the above points, we can get the intersection point of these two lines.

Prob. 3.

$$x - 3y = 6 \tag{9.5}$$
$$y = 3 + \frac{1}{3} \tag{9.6}$$

Multiplying the eqn (9.6) by 3 & rearranging

$$\begin{aligned} -x + 3y &= 9 \\ \textit{That is, } x - 3y &= -9 \end{aligned} \tag{9.7}$$

So eqn (9.5) and eqn (9.7) are contradictory to themselves. So there will not be any solution for this system.
If we give values for both the equations

for the eqn (9.5)	for the eqn (9.7)
at x = 3, y = -1	at x = 3, y = 4
at x = 6, y = 0	at x = 6, y = 5
at x = 9, y = 1	at y = 9, x = 6

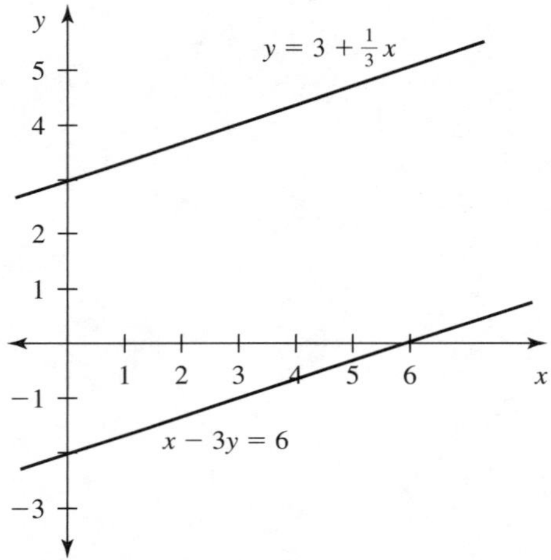

If we draw the graph for the above points, we can see that the lines will not intersect at any point.

Prob. 5.

$$2x - 3y = 5 \tag{9.8}$$

$$4x - 6y = c \tag{9.9}$$

(a) There will be an infinite number of solutions when the lines are identical so, divide the eqn (9.9) by 2

$$2x - 6y = \frac{c}{2} \Rightarrow \frac{c}{2} = 5 \Rightarrow c = 10$$

(b) No solution is only when both lines are parallel and disjoint i.e., $c \neq 10$

(c) There is no possibility for that as the slopes are equal.

Prob. 7. Let us take
x = No. of fish
y = No. of plants

$$x + y = 11 \tag{9.10}$$
$$2.3x + 1.7y = 21.70 \tag{9.11}$$

Solving equations (9.10) & (9.11) 1.7 X (9.10) - (2) gives

$$1.7x + 1.7y = 18.7 \tag{9.12}$$
$$2.3x + 1.7y = 21.70 \tag{9.13}$$

$$\begin{aligned} ---------- &\; - \; ----- \\ (2.3 - 1.7)x + (1.7 - 1.7)y &= 21.7 - 18.7 \\ 6x &= 30 \end{aligned}$$
$$x = 5 \tag{9.14}$$

Substituting eqn (9.14) in eqn (9.10)

$$\begin{aligned} y &= 11 - x \\ y &= 11 - 5 \end{aligned}$$
$$y = 6 \tag{9.15}$$

So he bought 5 fish and 6 plants

Prob. 9. Either a_{12} or a_{22} have to be nonzero, by the condition given, let's assume here $a_{12} \neq 0$ (if a_{22} is the only nonzero of the two terms, the proof will follow similarly). We can then write the expression for

$$x_2 = \frac{-a_{11}x_1}{a_{12}}$$

and substituting that on the equation of the second line we get:

$$a_{21}x_1 - \frac{a_{22}a_{11}}{a_{12}}x_1 = 0$$

$$(a_{21} - \frac{a_{22}a_{11}}{a_{12}})x_1 = 0$$

multiplying both sides by $a_{12} \neq 0$

$$(a_{21}a_{12} - a_{22}a_{11})x_1 = 0$$

since the term in between parentheses is nonzero, x_1 has to be zero. Now substituting back on the first equation we get $x_2 = 0$.

Prob. 11. Solving by augmented matrix method

$$\begin{aligned}
Aug.\ Matrix &= \left[\begin{array}{ccc|c} 5 & -1 & 2 & 6 \\ 1 & 2 & -1 & -1 \\ 3 & 2 & -2 & 1 \end{array}\right] \\
R_3 = R_3 - 3R_2 &= \left[\begin{array}{ccc|c} 5 & -1 & 2 & 6 \\ 1 & 2 & -1 & -1 \\ 0 & -4 & 1 & 4 \end{array}\right] \\
R_2 = 5R_2 - R_1 &= \left[\begin{array}{ccc|c} 5 & -1 & 2 & 6 \\ 0 & 11 & -7 & -11 \\ 0 & -4 & 1 & 4 \end{array}\right] \\
R_3 = 4R_2 + 11R_3 &= \left[\begin{array}{ccc|c} 5 & -1 & 2 & 6 \\ 0 & 11 & -7 & -11 \\ 0 & 0 & -17 & 0 \end{array}\right] \\
-17z &= 0 \\
So, & \\
z &= 0 \qquad (9.16) \\
11y - 7z &= -11
\end{aligned}$$

Substituting (9.16),

$$\begin{aligned}
y &= -1 \qquad (9.17) \\
5x - y + 2z &= 6 \qquad (9.18)
\end{aligned}$$

Substituting (9.16) & (9.17)

$$\begin{aligned}
x &= 1 \\
So & \\
x &= 1 \\
y &= -1 \\
z &= 0
\end{aligned}$$

Prob. 13. By Solution Method:

$$-2x + 4y - z = -1 \quad (9.19)$$
$$x + 7y + 2z = -4 \quad (9.20)$$
$$3x - 2y + 3z = -3 \quad (9.21)$$
$$(9.22)$$

Solving (9.19) and (9.20)
i.e., eqn (9.19) + 2 eqn (9.20)

$$\begin{aligned} -2x + 4y - z &= -1 \\ 2x + 14y + 4z &= -8 \\ \hline 16y + 3z &= -9 \quad (9.23) \\ \hline \end{aligned}$$

$$(9.24)$$

Solving (9.20) and (9.21)
i.e., 3 x eqn (9.20) - eqn (9.21)

$$\begin{aligned} 3x + 21y + 6z &= -12 \\ 3x - 2y + 3z &= -3 \\ \hline 23y + 3z &= -9 \quad (9.25) \\ \hline \end{aligned}$$

$$(9.26)$$

Solving (9.23) and (9.25)
i.e., eqn (9.25) - eqn (9.23)

$$\begin{aligned} 23y + 3z &= -9 \\ 16y + 3z &= -9 \\ \hline -6y &= 0 \\ \hline y &= 0 \quad (9.27) \end{aligned}$$

Substituting eqn (9.27) in eqn (9.25)

$$\begin{aligned} 3z &= -9 \\ z &= -3 \end{aligned} \tag{9.28}$$

Substituting eqn (9.27) & eqn (9.25) in eqn (9.19)

$$\begin{aligned} -2x + 0 + 3 &= -1 \\ -2x &= -4 \\ x &= 2 \end{aligned} \tag{9.29}$$

Prob. 15. Solving by augmented matrix method

$$\begin{aligned}
Aug.\ Matrix &= \begin{bmatrix} -1 & -2 & 3 & | & -9 \\ 2 & 1 & -1 & | & 5 \\ 4 & -3 & 5 & | & -9 \end{bmatrix} \\
R_3 = R_3 - 2R_2 &= \begin{bmatrix} -1 & -2 & 3 & | & -9 \\ 2 & 1 & -1 & | & 5 \\ 0 & -5 & 7 & | & -19 \end{bmatrix} \\
R_2 = R_2 + 2R_1 &= \begin{bmatrix} -1 & -2 & 3 & | & -9 \\ 0 & -3 & 5 & | & -13 \\ 0 & -5 & 7 & | & -19 \end{bmatrix} \\
R_3 = 5R_2 - 3R_3 &= \begin{bmatrix} 1 & -2 & 3 & | & -9 \\ 0 & -3 & 5 & | & -13 \\ 0 & 0 & 4 & | & -8 \end{bmatrix} \\
-4z &= -8 \\
So, & \\
z &= -2 \qquad (9.30) \\
-3y + 5z &= -13
\end{aligned}$$

Substituting (9.30),

$$\begin{aligned} y &= 1 \qquad (9.31) \\ -x - 2y + 3z &= -9 \end{aligned}$$

Substituting (9.30) & (9.31)

$$\begin{aligned} x &= 1 \\ So, & \\ x &= 1 \\ y &= 1 \\ z &= -2 \end{aligned}$$

Prob. 17. Solving by augmented matrix method

$$Aug.\ Matrix = \begin{bmatrix} 1 & 1 & 0 & | & 3 \\ 0 & -1 & 1 & | & -1 \\ 1 & 0 & 1 & | & 2 \end{bmatrix}$$

$$R_3 = R_3 - R_1 = \begin{bmatrix} 1 & 1 & 0 & | & 3 \\ 0 & -1 & 1 & | & -1 \\ 1 & 1 & 0 & | & 3 \end{bmatrix}$$

The first and last row have the same elements. So it has infinite number of solutions.

Prob. 19. This system is underdetermined because it has 3 unknowns and 2 equations.
Solving by augmented matrix method

$$Aug.\ Matrix = \begin{bmatrix} 1 & -2 & 1 & | & 3 \\ 2 & -3 & 1 & | & 8 \end{bmatrix}$$

$$R_2 = R_2 - 2R_1 = \begin{bmatrix} 1 & -2 & 1 & | & 3 \\ 0 & 1 & -1 & | & 2 \end{bmatrix}$$

$$y - z = 2 \tag{9.32}$$

$$x - 2y + z = 3 \tag{9.33}$$

$$So,$$

$$y = 2 + z \tag{9.34}$$

Substituting (9.34) in the eqn. (9.35),

$$x - 2(2 + z) + z = 3$$

$$x - 4 - z = 3$$

$$x = 7 + z \tag{9.35}$$

$$If\ z = t$$

$$\begin{aligned} x &= 7+t \\ y &= 2+t \\ z &= t \end{aligned}$$

Prob. 21. Solving by augmented matrix method

$$\begin{aligned} Aug.\ Matrix &= \begin{bmatrix} 2 & -1 & | & 3 \\ 1 & -1 & | & 4 \\ 3 & -1 & | & 1 \end{bmatrix} \\ R_3 = R_3 + R_2 &= \begin{bmatrix} 2 & -1 & | & 3 \\ 1 & -1 & | & 4 \\ 2 & 0 & | & -3 \end{bmatrix} \\ R_2 = R_1 - R_2 &= \begin{bmatrix} 2 & -1 & | & 3 \\ 1 & 0 & | & -1 \\ 2 & 0 & | & -3 \end{bmatrix} \end{aligned}$$

From the above matrix we are getting the equations like

$$2x = -3 \quad (9.36)$$

$$x = -1 \quad (9.37)$$

Equations (9.36) & (9.37) are giving contradictory solutions
So this is an inconsistent system.

Prob. 23. The composition of three different types of fertilizers are written in the equation form

$$SL1 : 24N + 4P + 8K$$

$$SL2 : 21N + 7P + 12K$$

$$SL1 : 17N + 0P + 0K$$

N for Nitrogen, P for Phosphate & K for Potassium
As given,

$$\begin{gathered} x(24N + 4P + 8K) + y(21N + 7P + 12K) \\ +z(17N + 0P + 0K) = 500N + 100P + 180K \end{gathered}$$

So separating N, P & K

$$\begin{aligned} N(24x+21y+17z) &= 500N \\ P(4x+7y+0z) &= 100P \\ K(8x+12y+0z) &= 180K \end{aligned}$$

If we write the above equations in the matrix form

$$\begin{pmatrix} 24 & 21 & 17 \\ 4 & 7 & 0 \\ 8 & 12 & 0 \end{pmatrix} \begin{pmatrix} x \\ y \\ z \end{pmatrix} = \begin{pmatrix} 500 \\ 100 \\ 180 \end{pmatrix}$$

Solving by augmented matrix method

$$\begin{aligned} Aug.\ Matrix &= \begin{pmatrix} 24 & 21 & 17 & | & 500 \\ 4 & 7 & 0 & | & 100 \\ 8 & 12 & 0 & | & 180 \end{pmatrix} \\ R_3 = 2R_2 - R_3 &= \begin{pmatrix} 24 & 21 & 17 & | & 500 \\ 4 & 7 & 0 & | & 100 \\ 0 & 2 & 0 & | & 20 \end{pmatrix} \end{aligned}$$

$$So,$$

$$2y = 20$$

$$y = 10 \tag{9.38}$$

$$4x + 7y = 100 \tag{9.39}$$

Substituting (9.156) in the above eqn,

$$\begin{aligned} 4x + 70 &= 100 \\ 4x &= 30 \end{aligned}$$

$$x = \frac{15}{2} \tag{9.40}$$

Substituting (9.156) & (9.156) in the eqn,

$$\begin{aligned} 24x + 21y + 17Z &= 500 \\ 180 + 210 + 17z &= 500 \end{aligned}$$

$$So, \tag{9.41}$$

$$z = \frac{110}{17} \tag{9.42}$$

So, $\frac{15}{2}$ amount of SL1, 10 amount of SL2 and $\frac{110}{17}$ amount of SL3 should be applied.

9.2 Matrices

Prob. 1.

$$\begin{aligned} 2C &= \begin{bmatrix} 2 & -4 \\ 2 & -2 \end{bmatrix} \\ A - B + 2C &= \begin{bmatrix} -1-0+2 & 2-1-4 \\ 0-2+2 & -3-4-2 \end{bmatrix} \\ &= \begin{bmatrix} 1 & -3 \\ 0 & -9 \end{bmatrix} \end{aligned}$$

Prob. 3.

$$\begin{aligned} A + B &= 2A - B + D \\ D &= A + B - 2A + B \\ &= -A + 2B \end{aligned}$$

To find D

$$\begin{aligned} 2B &= \begin{bmatrix} 0 & 2 \\ 4 & 8 \end{bmatrix} \\ -A + 2B &= \begin{bmatrix} 1+0 & -2+2 \\ 0+4 & 3+8 \end{bmatrix} \\ &= \begin{bmatrix} 1 & 0 \\ 4 & 11 \end{bmatrix} \end{aligned}$$

Prob. 5. To Prove (A + B)+ C = A + (B + C)

$$\begin{aligned} A + B &= \begin{bmatrix} -1 & 3 \\ 2 & 1 \end{bmatrix} \\ (A + B) + C &= \begin{bmatrix} -1+1 & 3-2 \\ 2+1 & 1-1 \end{bmatrix} \\ &= \begin{bmatrix} 0 & 1 \\ 3 & 0 \end{bmatrix} \\ B + C &= \begin{bmatrix} 0+1 & 1-2 \\ 2+1 & 4-1 \end{bmatrix} \\ &= \begin{bmatrix} 1 & -1 \\ 3 & 3 \end{bmatrix} \end{aligned}$$

$$\begin{aligned} A+(B+C) &= \begin{bmatrix} -1+1 & 2-1 \\ 0+3 & -3+3 \end{bmatrix} \\ &= \begin{bmatrix} -1+1 & 3-2 \\ 2+1 & 1-1 \end{bmatrix} \\ So,\ (A+B)+C &= A+(B+C) \end{aligned}$$

Prob. 7.

$$2A{+}3B{-}C = \begin{bmatrix} 2+15+2 & 6-3-0 & -2+12-4 \\ 4+6-1 & 8+0+3 & 2+3-1 \\ -3-0 & -4-9-0 & 4-9-2 \end{bmatrix} = \begin{bmatrix} 19 & 3 & 6 \\ 9 & 11 & 4 \\ -3 & -13 & -7 \end{bmatrix}$$

Prob. 9.

$$\begin{aligned} D &= -A-B-C \\ D &= -[A+B+C] \\ A+B+C &= \end{aligned}$$

$$\begin{bmatrix} 1+5-2 & 3-1+0 & -1+4+4 \\ 2+2+1 & 4+0-3 & 1+1+1 \\ 0+1+0 & -2-3+0 & 2-3+2 \end{bmatrix}$$

$$\begin{aligned} A+B+C &= \begin{bmatrix} 4 & 2 & 7 \\ 5 & 1 & 3 \\ 1 & -5 & 1 \end{bmatrix} \\ -(A+B+C) &= \begin{bmatrix} -4 & -2 & -7 \\ -5 & -1 & -3 \\ -1 & 5 & -1 \end{bmatrix} \\ D &= \begin{bmatrix} -4 & -2 & -7 \\ -5 & -1 & -3 \\ -1 & 5 & -1 \end{bmatrix} \end{aligned}$$

Prob. 11.

$$A+B = \begin{bmatrix} 1+5 & 3-1 & -1+4 \\ 2+2 & 4+0 & 1+1 \\ 0+1 & -2-3 & 2-3 \end{bmatrix} = \begin{bmatrix} 6 & 2 & 3 \\ 4 & 4 & 2 \\ 1 & -5 & -1 \end{bmatrix}$$

$$B+A = \begin{bmatrix} 5+1 & -1+3 & 4-1 \\ 2+2 & 0+4 & 1+1 \\ 1+0 & -3-2 & -3+2 \end{bmatrix} = \begin{bmatrix} 6 & 2 & 3 \\ 4 & 4 & 2 \\ 1 & -5 & -1 \end{bmatrix}$$

So $A + B = B + A$

Prob. 13. Transpose of A is $[\mathbf{A}]^{\mathbf{t}} = \begin{bmatrix} -1 & 2 \\ 0 & 1 \\ 3 & -4 \end{bmatrix}$

Prob. 15.

$$\begin{aligned} AB &= \begin{bmatrix} -2+0 & -3+0 \\ 2-2 & 3+2 \end{bmatrix} \\ &= \begin{bmatrix} -2 & -3 \\ 0 & 5 \end{bmatrix} \\ BA &= \begin{bmatrix} -2+3 & 0+6 \\ 1+1 & 0+2 \end{bmatrix} \\ &= \begin{bmatrix} 1 & 6 \\ 2 & 2 \end{bmatrix} \end{aligned}$$

So, $AB \neq BA$

Prob. 17.

$$\begin{aligned} AC &= \begin{bmatrix} -1 & 0 \\ 1 & 2 \end{bmatrix} \begin{bmatrix} 1 & 2 \\ 0 & -1 \end{bmatrix} \\ &= \begin{bmatrix} -1+0 & -2+0 \\ 1+0 & 2-2 \end{bmatrix} \\ &= \begin{bmatrix} -1 & -2 \\ 1 & 0 \end{bmatrix} \\ CA &= \begin{bmatrix} 1 & 2 \\ 0 & -1 \end{bmatrix} \begin{bmatrix} -1 & 0 \\ 1 & 2 \end{bmatrix} \\ &= \begin{bmatrix} -1+2 & 0+4 \\ 0-1 & 0-2 \end{bmatrix} \\ &= \begin{bmatrix} 1 & 4 \\ -1 & -2 \end{bmatrix} \end{aligned}$$

So, $AC \neq CA$

Prob. 19. To prove (A+B)C = AC + BC

From the previous problems

$$\begin{aligned} AC &= \begin{bmatrix} -1 & -2 \\ 1 & 0 \end{bmatrix} \\ BC &= \begin{bmatrix} 2 & 1 \\ -1 & -3 \end{bmatrix} \end{aligned}$$

So AC + BC is,

$$AC + BC = \begin{bmatrix} 1 & -1 \\ 0 & -3 \end{bmatrix}$$

To find (A+B)C

$$\begin{aligned} A+B &= \begin{bmatrix} -1 & 0 \\ 1 & 2 \end{bmatrix} + \begin{bmatrix} 2 & 3 \\ -1 & 1 \end{bmatrix} \\ &= \begin{bmatrix} 1 & 3 \\ 0 & 3 \end{bmatrix} \\ [A+B]+C &= \begin{bmatrix} 1 & 3 \\ 0 & 3 \end{bmatrix} \begin{bmatrix} 1 & 2 \\ 0 & -1 \end{bmatrix} \\ &= \begin{bmatrix} 1+0 & 2-3 \\ 0+0 & 0-3 \end{bmatrix} \\ &= \begin{bmatrix} 1 & -1 \\ 0 & -3 \end{bmatrix} \end{aligned}$$

So, $(A+B)C = AC + BC$

Prob. 21. A = 3 x 4 and B = 4 x 2 so AB = 3 x $\underline{4\ 4}$ x 2 = 3 x 2

Prob. 23.

(a) D'= 3 x 4 and BD': = 1 x $\underline{3\ 3}$ x 4, so BD' = 1 x 4

(b) D'A : = 3 x $\underline{4\ 4}$ X 3 and so, D'A = 3 x 3

(c) ACB: = 4 x $\underline{3\ 3}$ x 1 1 x 3 = 4 x $\underline{1\ 1}$ x 3 = 4 x 3

Prob. 25.

$$\begin{aligned} AB &= \begin{bmatrix} 1+6 & 2+3 & 0+9 & -1+0 \\ 0-4 & 0-2 & 0-6 & 0+0 \end{bmatrix} \\ &= \begin{bmatrix} 7 & 5 & 9 & -1 \\ -4 & -2 & -6 & 0 \end{bmatrix} \\ B' &= \begin{bmatrix} 1 & 2 \\ 2 & 1 \\ 0 & 3 \\ -1 & 0 \end{bmatrix} \\ B'A &= \begin{bmatrix} 1+0 & 3-4 \\ 2+0 & 6-2 \\ 0+0 & 0-6 \\ -1+0 & -3+0 \end{bmatrix} \end{aligned}$$

$$= \begin{bmatrix} 1 & -1 \\ 2 & 4 \\ 0 & -6 \\ -1 & -3 \end{bmatrix}$$

Prob. 27.

$$A^2 = \begin{bmatrix} 2 & 1 \\ -1 & -3 \end{bmatrix} \begin{bmatrix} 2 & 1 \\ -1 & -3 \end{bmatrix} = \begin{bmatrix} 4-1 & 2-3 \\ -2+3 & -1+9 \end{bmatrix} = \begin{bmatrix} 3 & -1 \\ 1 & 8 \end{bmatrix}$$

$$A^3 = AA^2 = \begin{bmatrix} 2 & 1 \\ -1 & -3 \end{bmatrix} \begin{bmatrix} 2 & 1 \\ -1 & -3 \end{bmatrix} \begin{bmatrix} 2 & 1 \\ -1 & -3 \end{bmatrix} = \begin{bmatrix} 6+1 & -2+8 \\ -3+3 & 1-24 \end{bmatrix} = \begin{bmatrix} 7 & 6 \\ -6 & -23 \end{bmatrix}$$

$$A^4 = AA^3 = \begin{bmatrix} 2 & 1 \\ -1 & -3 \end{bmatrix} \begin{bmatrix} 7 & 6 \\ -6 & -23 \end{bmatrix} = \begin{bmatrix} 14-6 & 12-23 \\ -7+18 & -6+69 \end{bmatrix} = \begin{bmatrix} 8 & -11 \\ 11 & 63 \end{bmatrix}$$

Prob. 29.

$$\begin{aligned} A &= \begin{bmatrix} 1 & 3 \\ 0 & -2 \end{bmatrix} \\ I_2 &= \begin{bmatrix} 1 & 0 \\ 0 & 1 \end{bmatrix} \\ AI_2 &= \begin{bmatrix} 1+0 & 0+3 \\ 0+0 & 0-2 \end{bmatrix} \\ &= \begin{bmatrix} 1 & 3 \\ 0 & -2 \end{bmatrix} \\ I_2A &= \begin{bmatrix} 1+0 & 3+0 \\ 0+0 & 0-2 \end{bmatrix} \end{aligned}$$

So $AI_2 = I_2A$

Prob. 31. Writing the equations in the matrix form

$$\begin{bmatrix} 2 & 3 & -1 \\ 0 & 2 & 1 \\ 1 & 0 & -2 \end{bmatrix} \begin{bmatrix} x_1 \\ x_2 \\ x_3 \end{bmatrix} = \begin{bmatrix} 0 \\ 1 \\ 2 \end{bmatrix}$$

Prob. 33. Writing the equations in the matrix form

$$\begin{bmatrix} 2 & -3 \\ -1 & 1 \\ 3 & 0 \end{bmatrix} \begin{bmatrix} x_1 \\ x_2 \end{bmatrix} = \begin{bmatrix} 4 \\ 3 \\ 4 \end{bmatrix}$$

Prob. 35. Inverse of matrix A is B :

$$
\begin{aligned}
A &= \begin{bmatrix} 2 & 1 \\ 1 & 1 \end{bmatrix} \\
B &= \begin{bmatrix} 1 & -1 \\ -1 & 2 \end{bmatrix} \\
i.e., & \\
AA^{-1} &= I_2
\end{aligned}
$$

Let us take $A^{-1} = B$

$$
\begin{aligned}
B &= \begin{bmatrix} b_{11} & b_{12} \\ b_{21} & b_{22} \end{bmatrix} \\
So, & \\
\begin{bmatrix} 2 & 1 \\ 1 & 1 \end{bmatrix} \begin{bmatrix} b_{11} & b_{12} \\ b_{21} & b_{22} \end{bmatrix} &= \begin{bmatrix} 1 & 0 \\ 0 & 1 \end{bmatrix}
\end{aligned}
$$

If we write the above matrix in the linear equations form

$$2b_{11} + b_{21} = 1 \tag{9.43}$$
$$1b_{11} + b_{21} = 0 \tag{9.44}$$
$$2b_{12} + b_{22} = 0 \tag{9.45}$$
$$1b_{12} + b_{22} = 1 \tag{9.46}$$

Solving eqn. (9.43) and eqn. (9.44) eqn.9.43 - 2 x eqn. (9.44)

$$-b_{21} = 1$$
$$b_{21} = -1 \tag{9.47}$$
$$\tag{9.48}$$

Substituting the eqn (9.47) in eqn (9.43)

$$2b_{11} - 1 = 1$$
$$2b_{11} = 2 \tag{9.49}$$
$$b_{11} = 1$$

Solving eqn. (9.45) and eqn. (9.46) eqn. (9.45) - 2 x eqn. (9.46)

$$-b_{22} = -2$$
$$b_{22} = 2 \tag{9.50}$$

Substituting the eqn (9.50) in eqn (9.45)

$$\begin{aligned} 2b_{12} &= -2 \\ b_{12} &= -1 \qquad (9.51) \\ \begin{bmatrix} b_{11} & b_{12} \\ b_{21} & b_{22} \end{bmatrix} &= \begin{bmatrix} 2 & -1 \\ -1 & 2 \end{bmatrix} \end{aligned}$$

So, $B = (A)^{-1}$

$$= \begin{bmatrix} 1 & -1 \\ -1 & 2 \end{bmatrix}$$

Prob. 37.

$$\begin{aligned} A &= \begin{bmatrix} -1 & 1 \\ 2 & 3 \end{bmatrix} \\ AA^{-1} &= I_2 \end{aligned}$$

Let us take $A^{-1} = B$

$$B = \begin{bmatrix} b_{11} & b_{12} \\ b_{21} & b_{22} \end{bmatrix}$$

So,

$$\begin{bmatrix} -1 & 1 \\ 2 & 3 \end{bmatrix} \begin{bmatrix} b_{11} & b_{12} \\ b_{21} & b_{22} \end{bmatrix} = \begin{bmatrix} 1 & 0 \\ 0 & 1 \end{bmatrix}$$

If we write the above matrix in the linear equations form

$$-b_{11} + b_{21} = 1 \qquad (9.52)$$

$$2b_{11} + 3b_{21} = 0 \qquad (9.53)$$

$$-b_{12} + b_{22} = 0 \qquad (9.54)$$

$$2b_{12} + 3b_{22} = 1 \qquad (9.55)$$

Solving eqn. (9.52) and eqn. (9.53)
2 x eqn.9.52 - eqn. (9.53)

$$\begin{aligned} 5b_{21} &= 2 \\ b_{21} &= \frac{2}{5} \qquad (9.56) \end{aligned}$$

$$(9.57)$$

From the equation (9.52)

$$\begin{aligned} b_{11} &= b_{21} - 1 \\ &= -\frac{3}{5} \end{aligned} \tag{9.58}$$

Solving eqn. (9.54) and eqn. (9.55) 2 x eqn. (9.54) - eqn. (9.55)

$$\begin{aligned} 5b_{22} &= 1 \\ b_{22} &= \frac{1}{5} \end{aligned} \tag{9.59}$$

From the equation (9.54)

$$\begin{aligned} b_{12} &= b_{22} \\ b_{22} = b_{12} &= \frac{1}{5}\begin{bmatrix} b_{11} & b_{12} \\ b_{21} & b_{22} \end{bmatrix} \\ &= \begin{bmatrix} -\frac{3}{5} & \frac{1}{5} \\ \frac{2}{5} & \frac{1}{5} \end{bmatrix} \end{aligned}$$

So, $B = (A)^{-1}$

$$= \begin{bmatrix} -\frac{3}{5} & \frac{1}{5} \\ \frac{2}{5} & \frac{1}{5} \end{bmatrix}$$

Prob. 39.

$$A = \begin{bmatrix} -1 & 1 \\ 2 & 3 \end{bmatrix} \tag{9.60}$$

From the problem no 37, we know that,

$$\begin{aligned} A^{-1} &= \begin{bmatrix} -\frac{3}{5} & \frac{1}{5} \\ \frac{2}{5} & \frac{1}{5} \end{bmatrix} So, \\ A^{-1}[A^{-1}]^{-1} &= I_2 \end{aligned}$$

Let us take $[A^{-1}]^{-1} = B$

$$B = \begin{bmatrix} b_{11} & b_{12} \\ b_{21} & b_{22} \end{bmatrix}$$

$$So,$$

$$\begin{bmatrix} -\frac{3}{5} & \frac{1}{5} \\ \frac{2}{5} & \frac{1}{5} \end{bmatrix}\begin{bmatrix} b_{11} & b_{12} \\ b_{21} & b_{22} \end{bmatrix} = \begin{bmatrix} 1 & 0 \\ 0 & 1 \end{bmatrix}$$

If we write the above matrix in the linear equations form

$$-\frac{3}{5}b_{11} + \frac{1}{5}b_{21} = 1 \quad (9.61)$$
$$\frac{2}{5}b_{11} + \frac{1}{5}b_{21} = 0 \quad (9.62)$$
$$-\frac{3}{5}b_{12} + \frac{1}{5}b_{22} = 0 \quad (9.63)$$
$$\frac{2}{5}b_{12} + \frac{1}{5}b_{22} = 1 \quad (9.64)$$

Solving eqn. (9.61) and eqn. (9.62) eqn.9.61 - eqn. (9.62)

$$-\frac{5}{5}b_{11} = 1$$
$$b_{11} = -1 \quad (9.65)$$
$$(9.66)$$

Substituting in the equation (9.61)

$$\frac{3}{5} + \frac{1}{5}b_{21} = 1$$
$$b_{21} = 2 \quad (9.67)$$

Solving eqn. (9.63) and eqn. (9.64) eqn. (9.63) - eqn. (9.64)

$$-\frac{5}{5}b_{12} = -1$$
$$-b_{12} = -1$$
$$b_{12} = 1 \quad (9.68)$$

Substituting in the equation (9.63)

$$-\frac{3}{5} + \frac{1}{5}b_{22} = 0$$
$$\frac{1}{5}b_{22} = \frac{3}{5}$$
$$b_{22} = 3$$
$$\begin{bmatrix} b_{11} & b_{12} \\ b_{21} & b_{22} \end{bmatrix} = \begin{bmatrix} -\frac{3}{5} & \frac{1}{5} \\ \frac{2}{5} & \frac{1}{5} \end{bmatrix}$$

So, $B = [A^{-1}]^{-1}$

$$= \begin{bmatrix} -1 & 1 \\ 2 & 3 \end{bmatrix}$$

Prob. 41.

$$C = \begin{bmatrix} 2 & 4 \\ 3 & 6 \end{bmatrix}$$
$$CC^{-1} = I_2$$

Let us take $A^{-1} = B$

$$B = \begin{bmatrix} b_{11} & b_{12} \\ b_{21} & b_{22} \end{bmatrix}$$

So,

$$\begin{bmatrix} 2 & 4 \\ 3 & 6 \end{bmatrix} \begin{bmatrix} b_{11} & b_{12} \\ b_{21} & b_{22} \end{bmatrix} = \begin{bmatrix} 1 & 0 \\ 0 & 1 \end{bmatrix}$$

If we write the above matrix in the linear equations form

$$2b_{11} + 4b_{21} = 1 \quad (9.69)$$
$$3b_{11} + 6b_{21} = 0 \quad (9.70)$$
$$2b_{12} + 4b_{22} = 0 \quad (9.71)$$
$$3b_{12} + 6b_{22} = 1 \quad (9.72)$$

Solving eqn. (9.69) and eqn. (9.70)
3 x eqn.9.69 - 2 x eqn. (9.70)

$$6b_{11} + 12b_{21} = 3 \quad (9.73)$$
$$6b_{11} + 12b_{21} = 0 \quad (9.74)$$

Eqn (9.73) & eqn (9.74) are contradictory. So there will not be any solution. So this matrix is not invertible.

Prob. 43.

(a) Writing the above matrix in the equation form

$$-x + 0 = -2 \quad (9.75)$$
$$2x - 3y = -5 \quad (9.76)$$

From the eqn. (9.181),
$x = 2$ Substituting the value of x in eqn. (9.182)

$$2(2) - 3y = -5$$
$$-3y = -9$$
$$y = 3$$
$$So,\ detX = \begin{bmatrix} 2 \\ 3 \end{bmatrix} \quad (9.77)$$

(b) Using inverse matrix method

$$\begin{aligned}
A &= \begin{bmatrix} -1 & 0 \\ 2 & -3 \end{bmatrix} \\
D &= \begin{bmatrix} -2 \\ -5 \end{bmatrix} \\
AX &= D \\
X &= (A^{-1})D \\
A &= \begin{bmatrix} -1 & 0 \\ 2 & -3 \end{bmatrix} \\
AA^{-1} &= I_2
\end{aligned}$$

Let us take $A^{-1} = B$

$$B = \begin{bmatrix} b_{11} & b_{12} \\ b_{21} & b_{22} \end{bmatrix}$$

$$So,$$

$$\begin{bmatrix} -1 & 0 \\ 2 & -3 \end{bmatrix} \begin{bmatrix} b_{11} & b_{12} \\ b_{21} & b_{22} \end{bmatrix} = \begin{bmatrix} 1 & 0 \\ 0 & 1 \end{bmatrix}$$

If we write the above matrix in the linear equations form

$$-b_{11} + 0 = 1 \tag{9.78}$$

$$2b_{11} - 3b_{21} = 0 \tag{9.79}$$

$$-b_{12} + 0 = 0 \tag{9.80}$$

$$2b_{12} - 3b_{22} = 1 \tag{9.81}$$

From the eqn. (9.78) & 43.b.3 we get,

$$b_{11} = -1 \tag{9.82}$$

$$b_{12} = 0 \tag{9.83}$$

$$\tag{9.84}$$

Substituting b_{11} in eqn. (9.79)

$$-2 - 3b_{21} = 0$$

$$b_{21} = -\frac{2}{3} \tag{9.85}$$

Substituting b_{12} in eqn. (9.81)

$$0 - 3b_{22} = 1$$

$$b_{22} = -\frac{1}{3} \tag{9.86}$$

So, $B = (A)^{-1}$

$$\begin{aligned}
&= \begin{bmatrix} -1 & 0 \\ -\frac{2}{3} & -\frac{1}{3} \end{bmatrix} \\
X &= (A^{-1})D \\
X &= \begin{bmatrix} -1 & 0 \\ -\frac{2}{3} & -\frac{1}{3} \end{bmatrix} \begin{bmatrix} -2 \\ -5 \end{bmatrix} \\
&= \begin{bmatrix} 2+0 \\ \frac{4}{3}+\frac{5}{3} \end{bmatrix} \\
&= \begin{bmatrix} 2 \\ \frac{9}{3} \end{bmatrix} \\
X &= \begin{bmatrix} 2 \\ 3 \end{bmatrix} \qquad (9.87)
\end{aligned}$$

Prob. 45.

$$A = \begin{bmatrix} 2 & -1 \\ 1 & 3 \end{bmatrix}$$

To check whether A is invertible, we compute the determinant of A.

$$\begin{aligned}
[A] &= (2)(3) - (1)(-1) \\
[A] &= 6+1 \\
[A] &= 7
\end{aligned}$$

As $[A] \neq 0$, A is invertible.

Prob. 47.

$$A = \begin{bmatrix} 4 & -1 \\ 8 & -2 \end{bmatrix}$$

To check whether A is invertible, we compute the determinant of A.

$$\begin{aligned}
[A] &= (4)(-2) - (8)(-1) \\
[A] &= -8+8 \\
[A] &= 0
\end{aligned}$$

As $[A] = 0$, A is not an invertible matrix.

Prob. 49.

(a)

$$A = \begin{bmatrix} 2 & 4 \\ 3 & 6 \end{bmatrix}$$

To check whether A is invertible, we compute the determinant of A.

$$\begin{aligned} [A] &= (2)(6) - (3)(4) \\ [A] &= 12 - 12 \\ [A] &= 0 \end{aligned}$$

As $[A] = 0$, A is not an invertible matrix.

(b)

$$\begin{aligned} x &= \begin{bmatrix} x \\ y \end{bmatrix} \\ B &= \begin{bmatrix} b_1 \\ b_2 \end{bmatrix} \\ AX &= B \\ \begin{bmatrix} 2 & 4 \\ 3 & 6 \end{bmatrix}\begin{bmatrix} x \\ y \end{bmatrix} &= \begin{bmatrix} b_1 \\ b_2 \end{bmatrix} \end{aligned}$$

Writing the above matrix in the form of linear equations,

$$\begin{aligned} 2x + 4y &= b_1 \\ 3x + 6y &= b_2 \end{aligned}$$

(c)

$$\begin{aligned} A &= \begin{bmatrix} 2 & 4 \\ 3 & 6 \end{bmatrix} \\ x &= \begin{bmatrix} x \\ y \end{bmatrix} \\ B &= \begin{bmatrix} 3 \\ \frac{9}{2} \end{bmatrix} \\ AX &= B \\ \begin{bmatrix} 2 & 4 \\ 3 & 6 \end{bmatrix}\begin{bmatrix} x \\ y \end{bmatrix} &= \begin{bmatrix} 3 \\ \frac{9}{2} \end{bmatrix} \end{aligned}$$

Writing the above matrix in the form of linear equations,

$$2x + 4y = 3 \tag{9.88}$$

$$3x + 6y = \frac{9}{2} \tag{9.89}$$

multiplying the eqn. (9.89) by 2

$$6x + 12y = 9 \tag{9.90}$$

$$\tag{9.91}$$

multiplying the eqn. (9.88) by 3

$$6x + 12y = 9 \tag{9.92}$$

$$\tag{9.93}$$

So, eqn. (9.90 & eqn. (9.92) are the same. So they are identical and parallel. So it will have infinite number of solutions. Graphical Solution:

for the eqn (9.88)	for the eqn (9.89)
at x = 0, y = $\frac{3}{4}$	at x = 0, y = $\frac{3}{4}$
at x = 1, y = $\frac{1}{4}$	at x = 1, y = $\frac{1}{4}$
at x = 2, y = $-\frac{3}{4}$	at x = 2, y = $-\frac{3}{4}$
at x = 3, y = $-\frac{3}{4}$	at x = 3, y = $-\frac{3}{4}$

(d) To find AX=B has no solution.

$$2x + 4y = b_1 \tag{9.94}$$

$$3x + 6y = b_2 \tag{9.95}$$

multiplying the eqn. (9.94) by 2 & multiplying the eqn. (9.95) by 2

$$6x + 12y = 3b_1 \tag{9.96}$$

$$6x + 12y = 2b_2 \tag{9.97}$$

$$\tag{9.98}$$

Equating the above two equations by subtracting,
i.e., eqn. (9.96) - eqn. (9.97) gives

$$3b_1 - 2b_2 = 0$$

$$3b_1 = 2b_2$$

$$b_1 = \frac{2}{3}b_2 \tag{9.99}$$

when the eqn. (9.99) is satisfied, then there will be infinite number of solutions. So, $b_1 \neq \frac{2}{3}b_2$ is the condition to have no solution.

Prob. 51.

$$\begin{aligned} A &= \begin{bmatrix} 1 & -1 \\ 0 & 2 \end{bmatrix} \\ AA^{-1} &= I_2 \end{aligned}$$

To check whether A is invertible, we compute the determinant of A.

$$\begin{aligned} [A] &= (1)(2) - (0)(-1) \\ [A] &= 2 + 0 \\ [A] &= 2 \end{aligned}$$

As $[A] \neq 0$, A is invertible.
Let us take $A^{-1} = B$

$$B = \begin{bmatrix} b_{11} & b_{12} \\ b_{21} & b_{22} \end{bmatrix}$$

$$So,$$

$$\begin{bmatrix} 1 & -1 \\ 0 & 2 \end{bmatrix} \begin{bmatrix} b_{11} & b_{12} \\ b_{21} & b_{22} \end{bmatrix} = \begin{bmatrix} 1 & 0 \\ 0 & 1 \end{bmatrix}$$

If we write the above matrix in the linear equations form

$$b_{11} - b_{21} = 1 \tag{9.100}$$

$$2b_{21} = 0 \tag{9.101}$$

$$b_{12} - b_{22} = 0 \tag{9.102}$$

$$2b_{22} = 1 \tag{9.103}$$

From the eqn. (9.101) & eqn. (9.103) we get

$$b_{21} = 0 \tag{9.104}$$

$$b_{22} = \frac{1}{2} \tag{9.105}$$

Substituting eqn (9.104) in eqn. (9.100)

$$\begin{aligned} b_{11} &= 1 + b_{21} \\ &= 1 \end{aligned} \tag{9.106}$$

Substituting eqn (9.105) in eqn. (9.101)

$$\begin{aligned} b_{12} &= b_{22} \\ &= \frac{1}{2} \\ So, B = (A)^{-1} &= \begin{bmatrix} 1 & 1/2 \\ 0 & 1/2 \end{bmatrix} \\ AX &= 0 \\ So,\ X &= 0xA^{-1} \\ &= 0 \\ So, X &= \begin{bmatrix} 0 \\ 0 \end{bmatrix} \end{aligned} \tag{9.107}$$

So this system has a trivial solution.

Prob. 53.

$$C = \begin{bmatrix} 1 & 3 \\ 1 & 3 \end{bmatrix}$$

To check whether C is invertible, we compute the determinant of C.

$$\begin{aligned} [C] &= (1)(3) - (1)(3) \\ [C] &= 3 - 3 \\ [C] &= 0 \end{aligned}$$

As $[C] = 0$, C is non invertible matrix.
Let us take $C^{-1} = B$

$$B = \begin{bmatrix} b_{11} & b_{12} \\ b_{21} & b_{22} \end{bmatrix}$$

$$So,$$

$$\begin{bmatrix} 1 & 3 \\ 1 & 3 \end{bmatrix} \begin{bmatrix} b_{11} & b_{12} \\ b_{21} & b_{22} \end{bmatrix} = \begin{bmatrix} 1 & 0 \\ 0 & 1 \end{bmatrix}$$

If we write the above matrix in the linear equations form

$$b_{11} + 3b_{21} = 1 \tag{9.108}$$

$$b_{11} + 3b_{21} = 0 \tag{9.109}$$

$$b_{12} + 3b_{22} = 0 \tag{9.110}$$

$$b_{12} + 3b_{22} = 1 \tag{9.111}$$

The eqns. (9.108) & (9.109) and also the eqns. (9.110)& (9.111) are contradictory. As this system is not having any solution we can't find the inverse So we can't find $CX = 0$

Prob. 55.

$$A = \begin{bmatrix} 2 & -1 & -1 \\ 2 & 1 & 1 \\ -1 & 1 & -1 \end{bmatrix} \tag{9.112}$$

To check whether A is invertible, we compute the determinant of A.

$$\begin{aligned} [A] &= 2(-1-1) + (-2+1) - (2+1) \\ [A] &= 2(-2) + (-1) - (3) \\ [A] &= -4 - 1 - 3 \\ [A] &= -8 \end{aligned}$$

As $[A] \neq 0$, A is invertible. Let us take $A^{-1} = B$

$$AA^{-1} = I_3$$

Let us take $A^{-1} = B$

$$B = \begin{bmatrix} b_{11} & b_{12} & b_{13} \\ b_{21} & b_{22} & b_{23} \\ b_{31} & b_{32} & b_{33} \end{bmatrix}$$

So,

$$\begin{bmatrix} 2 & -1 & -1 \\ 2 & 1 & 1 \\ -1 & 1 & -1 \end{bmatrix} \begin{bmatrix} b_{11} & b_{12} & b_{13} \\ b_{21} & b_{22} & b_{23} \\ b_{31} & b_{32} & b_{33} \end{bmatrix}$$

$$= \begin{bmatrix} 1 & 0 & 0 \\ 0 & 1 & 0 \\ 0 & 0 & 1 \end{bmatrix}$$

If we write the above matrix in the linear equations form

$$2b_{11} - b_{21} - b_{31} = 1 \tag{9.113}$$

$$2b_{11} + b_{21} + b_{31} = 0 \tag{9.114}$$

$$-b_{11} + b_{21} - b_{31} = 0 \quad (9.115)$$
$$2b_{12} - b_{22} - b_{32} = 0 \quad (9.116)$$
$$2b_{12} + b_{22} + b_{32} = 1 \quad (9.117)$$
$$-b_{12} + b_{22} - b_{32} = 0 \quad (9.118)$$
$$2b_{13} - b_{23} - b_{33} = 0 \quad (9.119)$$
$$2b_{13} + b_{23} + b_{33} = 0 \quad (9.120)$$
$$-b_{13} + b_{23} - b_{33} = 1 \quad (9.121)$$

Solving eqn. (9.192) and eqn. (9.193) by adding those two eqns

$$\begin{aligned} 4b_{11} &= 1 \\ b_{11} &= \frac{1}{4} \end{aligned} \quad (9.122)$$

Solving eqn. (9.193) and eqn. (9.194) by adding those two eqns. and substituting eqn (9.122)gives

$$\begin{aligned} b_{11} + 2b_{21} &= 0 \\ 2b_{21} &= -\frac{1}{4} \\ b_{21} &= -\frac{1}{8} \end{aligned} \quad (9.123)$$

substituting eqn b_{11} & b_{21} in eqn (9.192) gives

$$b_{31} = -\frac{3}{8} \quad (9.124)$$

Solving eqn. (9.116) and eqn. (9.117) by adding those two eqns

$$\begin{aligned} 4b_{12} &= 1 \\ b_{12} &= \frac{1}{4} \end{aligned} \quad (9.125)$$

Solving eqn. (9.117) and eqn. (9.118) by adding those two eqns. and substituting eqn (9.125)gives

$$\begin{aligned} b_{12} + 2b_{22} &= 1 \\ 2b_{22} &= 1 - \frac{1}{4} \\ b_{22} &= \frac{3}{8} \end{aligned} \quad (9.126)$$

substituting eqn b_{12} & b_{22} in eqn (9.118) gives

$$\begin{aligned} -\frac{1}{4}+\frac{3}{8}-b_{32} &= 0 \\ b_{32} &= \frac{-2+3}{8} \\ b_{32} &= \frac{1}{8} \end{aligned} \tag{9.127}$$

Solving eqn. (9.119) and eqn. (9.120) by adding those two eqns

$$\begin{aligned} 4b_{13} &= 0 \\ b_{13} &= 0 \end{aligned} \tag{9.128}$$

Solving eqn. (9.120) and eqn. (9.121) by adding those two eqns. and substituting eqn (9.128)gives

$$\begin{aligned} b_{13}+2b_{23} &= 1 \\ 2b_{23} &= 1 \\ b_{23} &= \frac{1}{2} \end{aligned} \tag{9.129}$$

substituting eqn b_{13} & b_{23} in eqn (9.121) gives

$$\begin{aligned} \frac{1}{2}-b_{33} &= 1 \\ b_{33} &= -\frac{1}{2} \end{aligned} \tag{9.130}$$

So, $B=(A)^{-1}$

$$= \begin{bmatrix} 1/4 & 1/4 & 0 \\ -1/8 & -3/8 & 1/2 \\ -3/8 & 1/8 & -1/2 \end{bmatrix}$$

Prob. 57.

$$A = \begin{bmatrix} -1 & 0 & -1 \\ 0 & -2 & 0 \\ -1 & 1 & 2 \end{bmatrix} \tag{9.131}$$

Let us take $A^{-1}=B$

$$AA^{-1} = I_3$$

Let us take $A^{-1} = B$

$$B = \begin{bmatrix} b_{11} & b_{12} & b_{13} \\ b_{21} & b_{22} & b_{23} \\ b_{31} & b_{32} & b_{33} \end{bmatrix}$$

So,

$$\begin{bmatrix} -1 & 0 & -1 \\ 0 & -2 & 0 \\ -1 & 1 & 2 \end{bmatrix} \begin{bmatrix} b_{11} & b_{12} & b_{13} \\ b_{21} & b_{22} & b_{23} \\ b_{31} & b_{32} & b_{33} \end{bmatrix}$$

$$= \begin{bmatrix} 1 & 0 & 0 \\ 0 & 1 & 0 \\ 0 & 0 & 1 \end{bmatrix}$$

If we write the above matrix in the linear equations form

$$-b_{11} - b_{31} = 1 \quad (9.132)$$

$$-2b_{21} = 0 \quad (9.133)$$

$$-b_{11} + b_{21} + 2b_{31} = 0 \quad (9.134)$$

$$-b_{12} - b_{32} = 0 \quad (9.135)$$

$$-2b_{22} = 1 \quad (9.136)$$

$$-b_{12} + b_{22} + 2b_{32} = 0 \quad (9.137)$$

$$-b_{13} - b_{33} = 0 \quad (9.138)$$

$$-2b_{23} = 0 \quad (9.139)$$

$$-b_{13} + b_{23} + 2b_{33} = 1 \quad (9.140)$$

From the equation (9.136)

$$b_{21} = 0 \quad (9.141)$$

Solving eqn. (9.135) and eqn. (9.136) by adding

$$3b_{31} = -1$$

$$b_{31} = -\frac{1}{3} \quad (9.142)$$

substituting the value of b_{31} in eqn. (9.135)

$$-b_{11} = \frac{2}{3}$$

$$b_{11} = -\frac{2}{3} \quad (9.143)$$

From the equation (9.136)

$$\begin{aligned} -2b_{22} &= 1 \\ b_{22} &= -\frac{1}{2} \end{aligned} \tag{9.144}$$

Solving eqn. (9.137) and eqn. (9.135) by subtracting i.e., eqn. (9.137) - eqn. (9.135)

$$\begin{aligned} b_{22} + 3b_{32} &= 0 \\ b_{32} &= \frac{1}{6} \end{aligned} \tag{9.145}$$

substituting the value of b_{32} and b_{22} in eqn. (9.197)

$$\begin{aligned} -b_{12} &= b_{32} \\ b_{12} &= \frac{1}{6} \end{aligned} \tag{9.146}$$

$$\tag{9.147}$$

From the equation (9.139)

$$b_{23} = 0 \tag{9.148}$$

Solving eqn. (9.140) and eqn. (9.138) by subtracting i.e., eqn. (9.140) - eqn. (9.138)

$$\begin{aligned} b_{23} + 3b_{33} &= 1 \\ b_{33} &= \frac{1}{3} \end{aligned} \tag{9.149}$$

substituting the value of b_{23} in eqn. (9.138)

$$\begin{aligned} -b_{13} &= b_{33} \\ b_{13} &= -\frac{1}{3} \end{aligned} \tag{9.150}$$

$$\tag{9.151}$$

So, $B = (A)^{-1}$

$$= \begin{bmatrix} -\frac{2}{3} & -\frac{1}{6} & -\frac{1}{3} \\ 0 & -\frac{1}{2} & 0 \\ -\frac{1}{3} & \frac{1}{6} & \frac{1}{3} \end{bmatrix}$$

Prob. 59.

$$\begin{aligned} N_1(t+1) &= 0.2N_0(t) \\ N_2(t+1) &= 0.7N_1(t) \end{aligned}$$

For finding the values at t = 1

$$\begin{aligned} N(1) &= \begin{bmatrix} 0 & 3.2 & 1.7 \\ 0.2 & 0 & 0 \\ 0 & 0.7 & 0 \end{bmatrix} \begin{bmatrix} 2000 \\ 800 \\ 200 \end{bmatrix} \\ &= \begin{bmatrix} (3.2)(800) + (1.7)(200) \\ (0.2)(2000) \\ (0.7)(800) \end{bmatrix} \\ &= \begin{bmatrix} 2560 + 340 \\ 400 \\ 560 \end{bmatrix} \\ &= \begin{bmatrix} 2900 \\ 400 \\ 560 \end{bmatrix} \end{aligned}$$

For finding the values at t = 2

$$\begin{aligned} N(2) &= \begin{bmatrix} 0 & 3.2 & 1.7 \\ 0.2 & 0 & 0 \\ 0 & 0.7 & 0 \end{bmatrix} \begin{bmatrix} 2900 \\ 400 \\ 560 \end{bmatrix} \\ &= \begin{bmatrix} (3.2)(400) + (1.7)(560) \\ (0.2)(2900) \\ (0.7)(400) \end{bmatrix} \\ &= \begin{bmatrix} 1280 + 952 \\ 580 \\ 280 \end{bmatrix} \\ &= \begin{bmatrix} 2232 \\ 580 \\ 280 \end{bmatrix} \end{aligned}$$

Prob. 61. Population is divided into four age classes.
For finding the values at time = 1

$$N(1) = \begin{bmatrix} 0 & 0 & 4.6 & 3.7 \\ 0.7 & 0 & 0 & 0 \\ 0 & 0.5 & 0 & 0 \\ 0 & 0 & 0.1 & 0 \end{bmatrix} \begin{bmatrix} 1500 \\ 500 \\ 250 \\ 50 \end{bmatrix}$$

$$= \begin{bmatrix} (4.6)(250) + (3.7)(50) \\ (0.7)(1500) \\ (0.5)(500) \\ (0.1)(250) \end{bmatrix}$$

$$= \begin{bmatrix} 1150 + 185 \\ 1050 \\ 250 \\ 25 \end{bmatrix}$$

$$N(1) = \begin{bmatrix} 1335 \\ 1050 \\ 250 \\ 25 \end{bmatrix}$$

For finding the values at time = 2

$$N(2) = \begin{bmatrix} 0 & 0 & 4.6 & 3.7 \\ 0.7 & 0 & 0 & 0 \\ 0 & 0.5 & 0 & 0 \\ 0 & 0 & 0.1 & 0 \end{bmatrix} \begin{bmatrix} 1335 \\ 1050 \\ 250 \\ 25 \end{bmatrix}$$

$$= \begin{bmatrix} (4.6)(250) + (3.7)(25) \\ (0.7)(1325) \\ (0.5)(1050) \\ (0.1)(250) \end{bmatrix}$$

$$= \begin{bmatrix} 1150 + 92.5 \\ 934.5 \\ 525 \\ 25 \end{bmatrix}$$

$$N(2) = \begin{bmatrix} 1242.5 \\ 934.5 \\ 525 \\ 25 \end{bmatrix}$$

Prob. 63.

$$L = \begin{bmatrix} 2 & 3 & 2 & 1 \\ 0.4 & 0 & 0 & 0 \\ 0 & 0.6 & 0 & 0 \\ 0 & 0 & 0.8 & 0 \end{bmatrix}$$

(i) There are four age classes.

(ii) 60% of the one year olds are surviving until the end of the following breeding section.

(iii) The average number of female offspring of a two-year-old is 2.

Prob. 65.

$$L = \begin{bmatrix} 1 & 2.5 & 3 & 1.5 \\ 0.9 & 0 & 0 & 0 \\ 0 & 0.3 & 0 & 0 \\ 0 & 0 & 0.2 & 0 \end{bmatrix}$$

(i) 20% of the one year olds are surviving until the end of the following breeding section.

(ii) The average number of female offspring of one-year-old female is 3.

Prob. 67.

$$\begin{aligned} L &= \begin{bmatrix} 1.2 & 3.2 \\ 0.8 & 0 \end{bmatrix} \\ N_0(0) &= 100 \\ N_1(0) &= 0 \end{aligned}$$

From the above matrix

$$\begin{aligned} N_0(t+1) &= 1.2N_0(t) + 3.2N_1(t) \\ N_1(t+1) &= 0.8N_0(t) \\ \begin{bmatrix} N_0(1) \\ N_1(1) \end{bmatrix} &= \begin{bmatrix} 1.2 & 3.2 \\ 0.8 & 0 \end{bmatrix} \begin{bmatrix} 100 \\ 0 \end{bmatrix} \\ N(1) &= \begin{bmatrix} (1.2)(100) + (3.2)(0) \\ (0.8)(100) + 0 \end{bmatrix} \\ N(1) &= \begin{bmatrix} 120 \\ 80 \end{bmatrix} \end{aligned}$$

At t = 2

$$\begin{aligned} N(2) &= \begin{bmatrix} (1.2)(120) + (3.2)(80) \\ (0.8)(120) + 0 \end{bmatrix} \\ N(2) &= \begin{bmatrix} 400 \\ 96 \end{bmatrix} \end{aligned}$$

At t = 3

$$N(3) = \begin{bmatrix} (1.2)(400) + (3.2)(96) \\ (0.8)(400) + 0 \end{bmatrix}$$
$$N(3) = \begin{bmatrix} 787.2 \\ 320 \end{bmatrix}$$

At t = 4

$$N(4) = \begin{bmatrix} (1.2)(787.2) + (3.2)(320) \\ (0.8)(787.2) + 0 \end{bmatrix}$$
$$N(4) = \begin{bmatrix} 1968 \\ 630 \end{bmatrix}$$

At t = 5

$$N(5) = \begin{bmatrix} (1.2)(1968) + (3.2)(630) \\ (0.8)(1968) + 0 \end{bmatrix}$$
$$N(5) = \begin{bmatrix} 4377 \\ 1574 \end{bmatrix}$$

At t = 6

$$N(6) = \begin{bmatrix} (1.2)(4377) + (3.2)(1574) \\ (0.8)(4377) + 0 \end{bmatrix}$$
$$N(6) = \begin{bmatrix} 10288 \\ 3501 \end{bmatrix}$$

At t = 7

$$N(7) = \begin{bmatrix} (1.2)(10288) + (3.2)(3501) \\ (0.8)(10288) + 0 \end{bmatrix}$$
$$N(7) = \begin{bmatrix} 23548 \\ 8230 \end{bmatrix}$$

At t = 8

$$N(8) = \begin{bmatrix} (1.2)(23548) + (3.2)(8230) \\ (0.8)(23548) + 0 \end{bmatrix}$$
$$N(8) = \begin{bmatrix} 54593 \\ 18838 \end{bmatrix}$$

At t = 9

$$N(9) = \begin{bmatrix} (1.2)(54593) + (3.2)(18838) \\ (0.8)(54593) + 0 \end{bmatrix}$$

$$N(9) = \begin{bmatrix} 125792 \\ 43674 \end{bmatrix}$$

At t = 10

$$N(10) = \begin{bmatrix} (1.2)(125792) + (3.2)(43674) \\ (0.8)(125792) + 0 \end{bmatrix}$$

$$N(10) = \begin{bmatrix} 290706 \\ 100633 \end{bmatrix}$$

t	1	2	3	4	5	6	7	8	9
$q_0(t)$	3.3	1.9	2.5	2.2	2.3	2.3	2.3	2.3	2.3

t	1	2	3	4	5	6	7	8	9
$q_1(t)$	1.2	3.3	2	2.5	2.2	2.3	2.3	2.3	2.3

The stable age distribution is $q_0(t) = q_1(t) = 2.3$

9.3 Linear Maps, Eigenvectors and Eigenvalues

Prob. 1.

(a)

$$A = \begin{bmatrix} 2 & 1 \\ 3 & 4 \end{bmatrix}$$

$$x = \begin{bmatrix} x_1 \\ x_2 \end{bmatrix}$$

$$y = \begin{bmatrix} y_1 \\ y_2 \end{bmatrix}$$

To Prove $A(x + y) = Ax + Ay$

LHS

$$x + y = \begin{bmatrix} x_1 + y_1 \\ x_2 + y_2 \end{bmatrix}$$

$$\begin{aligned}
A(x+y) &= \begin{bmatrix} 2 & 1 \\ 3 & 4 \end{bmatrix} \begin{bmatrix} x_1 + y_1 \\ x_2 + y_2 \end{bmatrix} \\
&= \begin{bmatrix} 2(x_1 + y_1) + 1(x_2 + y_2) \\ 3(x_1 + y_1) + 4(x_2 + y_2) \end{bmatrix} \\
&= \begin{bmatrix} 2x_1 + 2y_1 + x_2 + y_2 \\ 3x_1 + 3y_1 + 4x_2 + 4y_2) \end{bmatrix}
\end{aligned}$$

RHS

$$\begin{aligned}
Ax &= \begin{bmatrix} 2 & 1 \\ 3 & 4 \end{bmatrix} \begin{bmatrix} x_1 \\ x_2 \end{bmatrix} \\
&= \begin{bmatrix} 2x_1 + x_2 \\ 3x_1 + 4x_2 \end{bmatrix} \\
Ay &= \begin{bmatrix} 2 & 1 \\ 3 & 4 \end{bmatrix} \begin{bmatrix} y_1 \\ y_2 \end{bmatrix} \\
&= \begin{bmatrix} 2y_1 + y_2 \\ 3y_1 + 4y_2 \end{bmatrix}
\end{aligned}$$

Adding the above and rearranging

$$Ax + Ay = \begin{bmatrix} 2x_1 + 2y_1 + x_2 + y_2 \\ 3x_1 + 3y_1 + 4x_2 + 4y_2 \end{bmatrix}$$

so,

$$\text{LHS} = \text{RHS}$$

(b)

$$\begin{aligned}
A &= \begin{bmatrix} 2 & 1 \\ 3 & 4 \end{bmatrix} \\
x &= \begin{bmatrix} x_1 \\ x_2 \end{bmatrix} \\
y &= \begin{bmatrix} y_1 \\ y_2 \end{bmatrix}
\end{aligned}$$

To Prove $A(\lambda x) = \lambda(Ax)$

LHS

$$\begin{aligned}
\lambda x &= \begin{bmatrix} \lambda x_1 \\ \lambda x_2 \end{bmatrix} \\
A\lambda x &= \begin{bmatrix} 2 & 1 \\ 3 & 4 \end{bmatrix} \begin{bmatrix} \lambda x_1 \\ \lambda x_2 \end{bmatrix} \\
&= \begin{bmatrix} 2\lambda x_1 + \lambda x_2 \\ 3\lambda x_1 + 4\lambda x_2 \end{bmatrix}
\end{aligned}$$

LHS

$$\begin{aligned}
Ax &= \begin{bmatrix} 2 & 1 \\ 3 & 4 \end{bmatrix} \begin{bmatrix} x_1 \\ x_2 \end{bmatrix} \\
&= \begin{bmatrix} 2x_1 + x_2 \\ 3x_1 + 4x_2 \end{bmatrix} \\
\lambda(Ax) &= \begin{bmatrix} 2\lambda x_1 + \lambda x_2 \\ 3\lambda x_1 + 4\lambda x_2 \end{bmatrix}
\end{aligned}$$

so,

$$\text{LHS} = \text{RHS}$$

Prob. 3. Length: $|x| = \sqrt{2^2 + 2^2} = \sqrt{8} = 2\sqrt{2}$

Angle: $\tan\alpha = \frac{x_2}{x_1} = \frac{2}{2} = 1$ so $\alpha = \tan^{-1}$, that is $\alpha = 45^o$

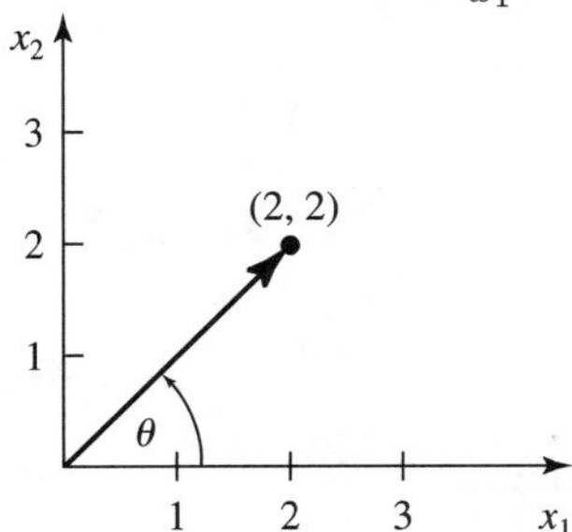

Prob. 5. Length:$|x| = \sqrt{(0)^2 + 3^2} = 3$.

Angle: tan $\alpha = \frac{x_2}{x_1} = \frac{3}{(0)}$ which is not defined, so $\alpha = 90^o$.

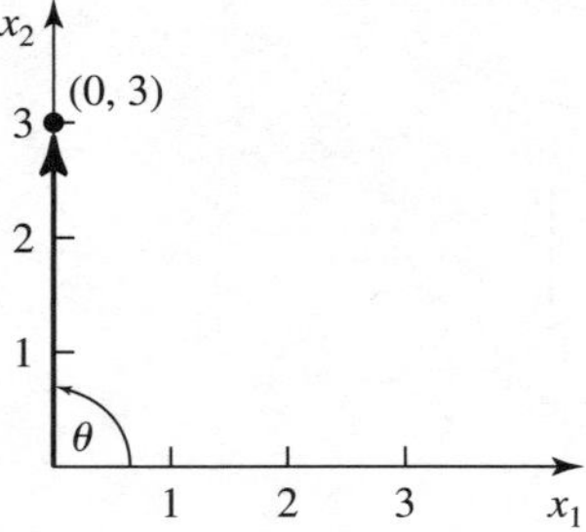

Prob. 7. Length: $|x| = \sqrt{(\sqrt{3})^2 + (1)^2} = \sqrt{4} = 2$.

Angle: tan $\alpha = \frac{x_2}{x_1} = \frac{-1}{\sqrt{3}}$, so $\alpha = \tan^{-1}(\frac{-1}{\sqrt{3}}) = 150^o$

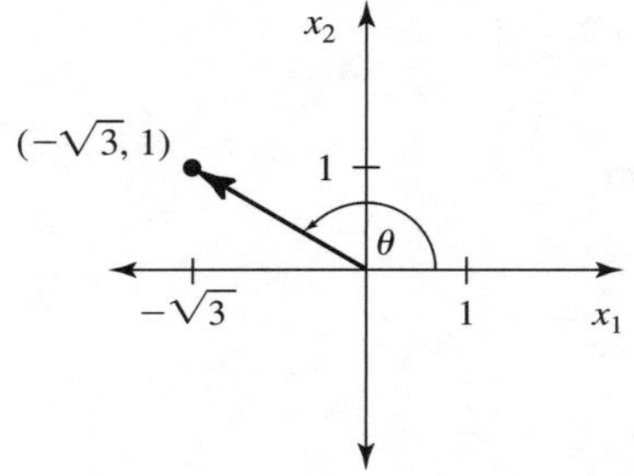

Prob. 9.

$$
\begin{aligned}
x_1 &= 2\cos(30) = 2\frac{\sqrt{3}}{2} = \sqrt{3} \\
x_2 &= 2\sin(30) = 2\frac{1}{2} = 1 \\
\begin{bmatrix} x_1 \\ x_2 \end{bmatrix} &= \begin{bmatrix} \sqrt{3} \\ 1 \end{bmatrix}
\end{aligned}
$$

Prob. 11.

$$
\begin{aligned}
x_1 &= 1\cos(70) = 1(0.342) = 0.342 \\
x_2 &= 1\sin(70) = 1(0.9397) = 0.9397 \\
\begin{bmatrix} x_1 \\ x_2 \end{bmatrix} &= \begin{bmatrix} 0.342 \\ 0.939 \end{bmatrix}
\end{aligned}
$$

Prob. 13. $\alpha = 360 - 15^o = 345^o$ so, $x_1 = 3\cos(345) = 3(0.966) = 2.898$ and $x_2 = 3\sin(345) = 3(-0.2588) = -0.7764$.

$$\begin{bmatrix} x_1 \\ x_2 \end{bmatrix} = \begin{bmatrix} 2.898 \\ -0.7764 \end{bmatrix}$$

Prob. 15. With $r = 5$ and $\alpha = 90^o + 25^o = 115^o$,

$$\begin{aligned} x_1 &= 5\cos(115^o) = 5(-0.4226) = -2.113 \\ x_2 &= 5\sin(115^o) = 5(0.9063) = 4.53 \\ \begin{bmatrix} x_1 \\ x_2 \end{bmatrix} &= \begin{bmatrix} -2.113 \\ 4.53 \end{bmatrix} \end{aligned}$$

Prob. 17.

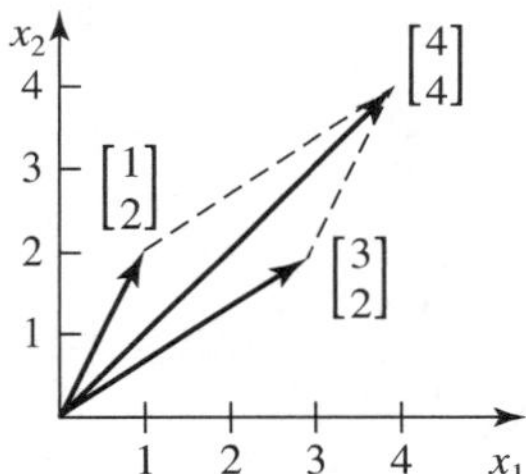

To add $\begin{bmatrix} 1 \\ 2 \end{bmatrix}$ to $\begin{bmatrix} 3 \\ 2 \end{bmatrix}$, move the vector $\begin{bmatrix} 1 \\ 2 \end{bmatrix}$ to the tip of $\begin{bmatrix} 3 \\ 2 \end{bmatrix}$ without changing the length of the former. The sum of two vectors start at (0,0). The sum is the diagonal in the parallelogram formed by these two vectors i.e., $\begin{bmatrix} 4 \\ 4 \end{bmatrix}$

Prob. 19.

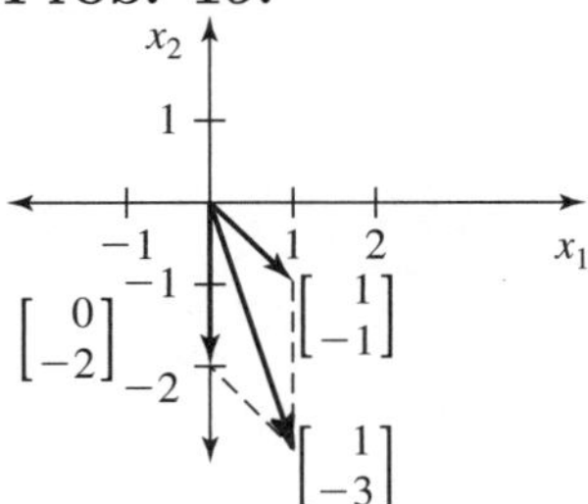

To add $\begin{bmatrix} 0 \\ -2 \end{bmatrix}$ to $\begin{bmatrix} 1 \\ -1 \end{bmatrix}$, move the vector $\begin{bmatrix} 0 \\ -2 \end{bmatrix}$ to the tip of $\begin{bmatrix} 1 \\ -1 \end{bmatrix}$ without changing the length of the former. The sum of two vectors start at (0,0). The sum is the diagonal in the parallelogram formed by these two vectors i.e.,$\begin{bmatrix} 1 \\ -3 \end{bmatrix}$.

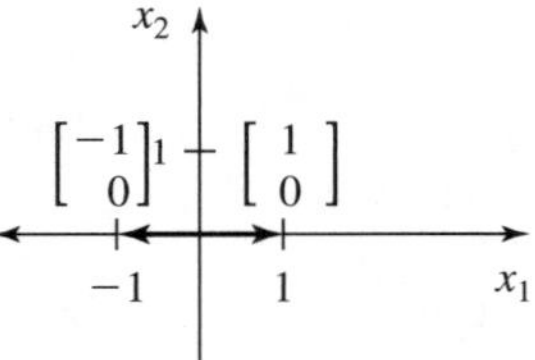

Prob. 21.

To add $\begin{bmatrix} 1 \\ 0 \end{bmatrix}$ to $\begin{bmatrix} -1 \\ 0 \end{bmatrix}$, move the vector $\begin{bmatrix} 1 \\ 0 \end{bmatrix}$ to the tip of $\begin{bmatrix} -1 \\ 0 \end{bmatrix}$ without changing the length of the former. The sum of two vectors start at (0,0). The sum is the diagonal in the parallelogram formed by these two vectors i.e.,$\begin{bmatrix} 0 \\ 0 \end{bmatrix}$.

Prob. 23.

$$\begin{aligned} a &= 2 \\ x &= \begin{bmatrix} -2 \\ 1 \end{bmatrix} \\ ax &= \begin{bmatrix} -4 \\ 2 \end{bmatrix} \end{aligned}$$

If we multiply the vector with the scalar, there will not be any change in direction but in the length. The length will be increased by a factor 'a' (here it is 2). Since 'a' is positive here, the resulting vector points in the same direction as of $\begin{bmatrix} -2 \\ 1 \end{bmatrix}$.

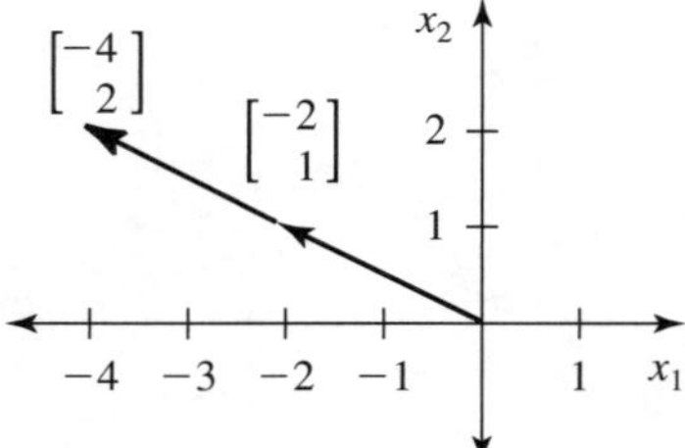

Prob. 25.

$$\begin{aligned} a &= 0.5 \\ x &= \begin{bmatrix} 0 \\ -2 \end{bmatrix} \\ ax &= \begin{bmatrix} 0 \\ -1 \end{bmatrix} \end{aligned}$$

If we multiply the vector with the scalar, there will not be any change in direction but in the length. The length will be increased by a factor 'a' (here it is 0.5, so the length will be less than the original). Since 'a' is positive here, the resulting vector points in the same direction as of $\begin{bmatrix} 0 \\ -2 \end{bmatrix}$.

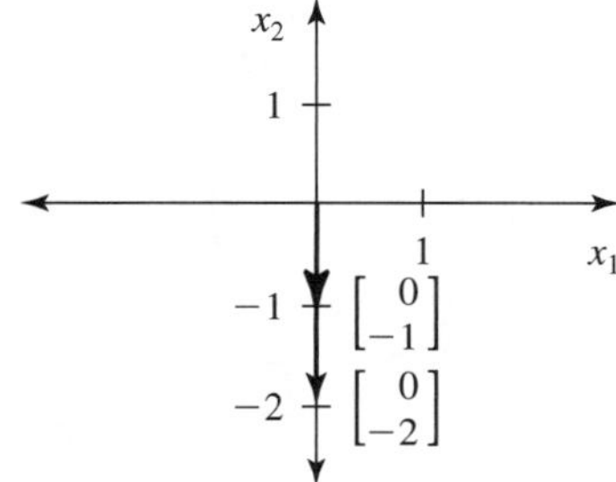

Prob. 27.

$$\begin{aligned} a &= \frac{1}{4} \\ x &= \begin{bmatrix} -4 \\ 1 \end{bmatrix} \\ ax &= \begin{bmatrix} -1 \\ \frac{1}{4} \end{bmatrix} \end{aligned}$$

If we multiply the vector with the scalar, there will not be any change in direction but in the length. The length will be increased by a factor 'a' (here it is $\frac{1}{4}$, so the length will be less than the original). Since 'a' is positive here, the resulting vector points in the same direction as of $\begin{bmatrix} -4 \\ 1 \end{bmatrix}$.

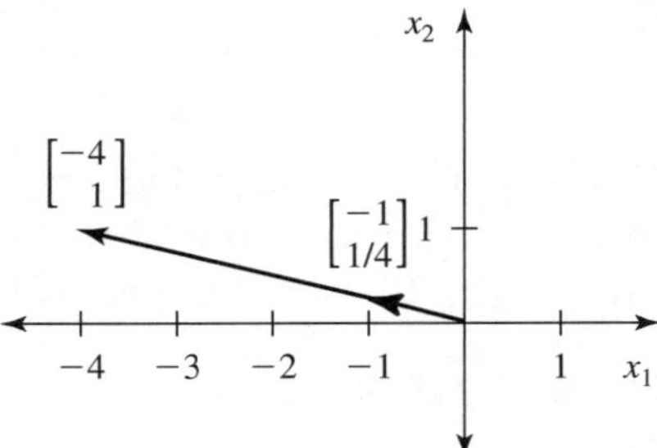

Prob. 29.

$$\begin{aligned} u &= \begin{bmatrix} 3 \\ 4 \end{bmatrix} \\ v &= \begin{bmatrix} -1 \\ -2 \end{bmatrix} \\ u + v &= \begin{bmatrix} 2 \\ 2 \end{bmatrix} \end{aligned}$$

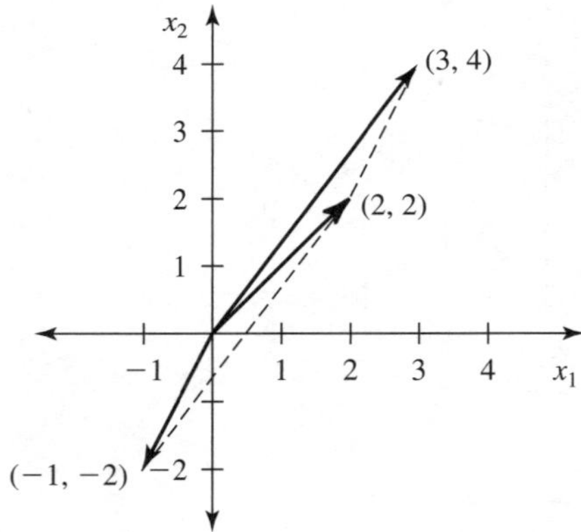

To add $\begin{bmatrix} 3 \\ 4 \end{bmatrix}$ to $\begin{bmatrix} -1 \\ -2 \end{bmatrix}$, move the vector $\begin{bmatrix} 3 \\ 4 \end{bmatrix}$ to the tip of $\begin{bmatrix} -1 \\ -2 \end{bmatrix}$ without changing the length of the former. The sum of two vectors start at (0,0). The sum is the diagonal in the parallelogram formed by these two vectors i.e., $\begin{bmatrix} 2 \\ 2 \end{bmatrix}$

Prob. 31.

$$\begin{aligned} w &= \begin{bmatrix} 1 \\ -2 \end{bmatrix} \\ u &= \begin{bmatrix} 3 \\ 4 \end{bmatrix} \\ w-u &= \begin{bmatrix} -2 \\ -6 \end{bmatrix} \end{aligned}$$

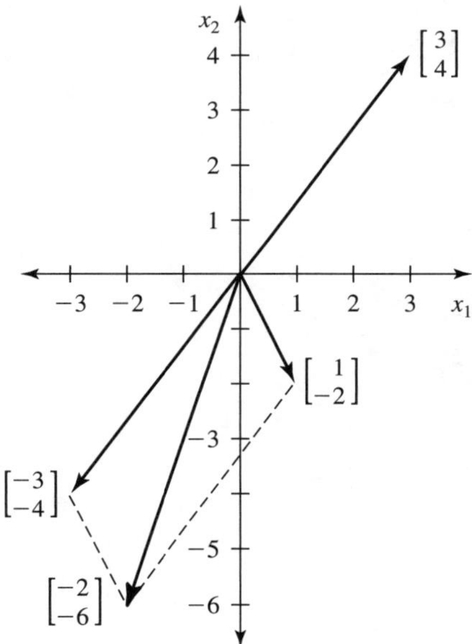

Prob. 33.

$$\begin{aligned} u &= \begin{bmatrix} 3 \\ 4 \end{bmatrix} \\ v &= \begin{bmatrix} -1 \\ -2 \end{bmatrix} \\ w &= \begin{bmatrix} 1 \\ -2 \end{bmatrix} \\ u + v + w &= \begin{bmatrix} 3 \\ 0 \end{bmatrix} \end{aligned}$$

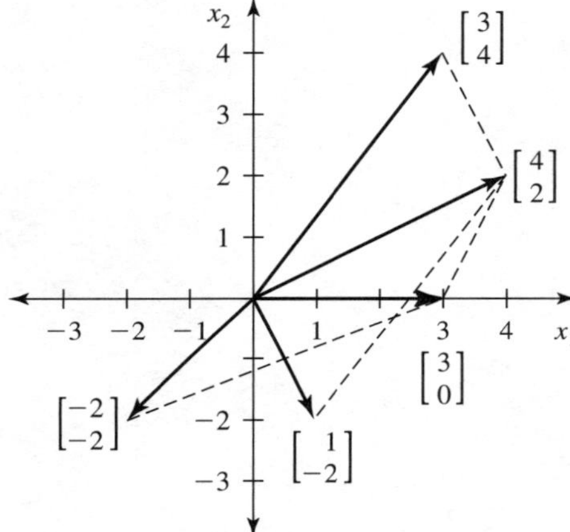

Prob. 35.

$$A = \begin{bmatrix} 1 & 0 \\ 0 & 1 \end{bmatrix}$$

$$x \rightarrow Ax$$

$$\begin{aligned} \begin{bmatrix} x_1 \; x_2 \end{bmatrix} &\rightarrow \begin{bmatrix} 1 & 0 \\ 0 & 1 \end{bmatrix} \begin{bmatrix} x_1 \\ x_2 \end{bmatrix} \\ &= \begin{bmatrix} x_1 + 0 \\ 0 + x_2 \end{bmatrix} \\ &= \begin{bmatrix} x_1 \\ x_2 \end{bmatrix} \end{aligned}$$

So the vector is unchanged.

Prob. 37.

$$A = \begin{bmatrix} 0 & -1 \\ 1 & 0 \end{bmatrix}$$

$$x \to Ax$$

$$\begin{aligned} \begin{bmatrix} x_1 \\ x_2 \end{bmatrix} &\to \begin{bmatrix} 0 & -1 \\ 1 & 0 \end{bmatrix} \begin{bmatrix} x_1 \\ x_2 \end{bmatrix} \\ &= \begin{bmatrix} 0 - x_2 \\ x_1 \end{bmatrix} \\ &= \begin{bmatrix} -x_2 \\ x_1 \end{bmatrix} \end{aligned}$$

Counterclockwise rotation by $\theta = \pi/2$.

Prob. 39.

$$A = \frac{1}{2} \begin{bmatrix} \sqrt{3} & -1 \\ 1 & \sqrt{3} \end{bmatrix}$$

$$x \to Ax$$

$$\begin{aligned} \begin{bmatrix} x_1 \\ x_2 \end{bmatrix} &\to \frac{1}{2} \begin{bmatrix} \sqrt{3} & -1 \\ 1 & \sqrt{3} \end{bmatrix} \begin{bmatrix} x_1 \\ x_2 \end{bmatrix} \\ &= \begin{bmatrix} \frac{\sqrt{3}}{2} x_1 - \frac{1}{2} x_2 \\ \frac{1}{2} x_1 + \frac{\sqrt{3}}{2} x_2 \end{bmatrix} \\ &= \begin{bmatrix} \frac{\sqrt{3}}{2} x_1 - \frac{1}{2} x_2 \\ \frac{1}{2} x_1 + \frac{\sqrt{3}}{2} x_2 \end{bmatrix} \end{aligned}$$

Counterclockwise rotation by $\theta = \pi/6$.

Prob. 41.

$$\begin{aligned} R_{\pi/6} &= \begin{bmatrix} \cos(\pi/6) & -\sin(\pi/6) \\ \sin(\pi/6) & \cos(\pi/6) \end{bmatrix} \\ &= \begin{bmatrix} \cos(30) & -\sin(30) \\ \sin(30) & \cos(30) \end{bmatrix} \begin{bmatrix} -1 \\ 2 \end{bmatrix} \\ &= \begin{bmatrix} \sqrt{3}/2 & 1/2 \\ 1/2 & \sqrt{3}/2 \end{bmatrix} \begin{bmatrix} -1 \\ 2 \end{bmatrix} \end{aligned}$$

$$
\begin{aligned}
&= \begin{bmatrix} -\sqrt{3}/2 - 1 \\ -1/2 + \sqrt{3} \end{bmatrix} \\
&= 1/2 \begin{bmatrix} -\sqrt{3} - 2 \\ -1 + 2\sqrt{3} \end{bmatrix}
\end{aligned}
$$

Prob. 43.

$$
\begin{aligned}
R_{\pi/12} &= \begin{bmatrix} \cos(\pi/12) & -\sin(\pi/12) \\ \sin(\pi/12) & \cos(\pi/12) \end{bmatrix} \begin{bmatrix} 5 \\ 2 \end{bmatrix} \\
&= \begin{bmatrix} 5\cos\frac{\pi}{12} & -2\sin\frac{\pi}{2} \\ 5\sin\frac{\pi}{2} & 2\cos\frac{\pi}{12} \end{bmatrix}
\end{aligned}
$$

Prob. 45.

$$
\begin{aligned}
R_{\pi/4} &= \begin{bmatrix} \cos(-\frac{\pi}{4}) & -\sin(-\frac{\pi}{4}) \\ \sin(-\pi/4) & \cos(-\pi/4) \end{bmatrix} \begin{bmatrix} 2 \\ 1 \end{bmatrix} \\
&= \begin{bmatrix} 1/\sqrt{2} & 1/\sqrt{2} \\ -1/\sqrt{2} & 1/\sqrt{2} \end{bmatrix} \begin{bmatrix} 2 \\ 1 \end{bmatrix} \\
&= \begin{bmatrix} \sqrt{2} + 1/\sqrt{2} \\ -2 + 1/\sqrt{2} \end{bmatrix}
\end{aligned}
$$

Prob. 47.

$$
\begin{aligned}
R_{\pi/7} &= \begin{bmatrix} \cos(\frac{\pi}{7}) & -\sin(\frac{\pi}{7}) \\ \sin(\frac{\pi}{7}) & \cos(\frac{\pi}{7}) \end{bmatrix} \begin{bmatrix} 5 \\ -3 \end{bmatrix} \\
&= \begin{bmatrix} 5\cos(\frac{\pi}{7}) & -3\sin(\frac{\pi}{7} \\ -5\sin(\frac{\pi}{7}) & -3\cos(\frac{\pi}{7}) \end{bmatrix}
\end{aligned}
$$

Prob. 49.

$$
\begin{aligned}
A &= \begin{bmatrix} 2 & 3 \\ 0 & -1 \end{bmatrix} \\
Ax = \lambda I x & \\
(A - \lambda I)x &= 0 \\
[A - \lambda I] &= 0
\end{aligned}
$$

To get the above,

$$
\begin{aligned}
[A-\lambda I] &= \begin{bmatrix} 2 & 3 \\ 0 & -1 \end{bmatrix} - \lambda \begin{bmatrix} 1 & 0 \\ 0 & 1 \end{bmatrix} \\
&= \begin{bmatrix} 2-\lambda & 0 \\ 0 & -1-\lambda \end{bmatrix} \\
(2-\lambda)(-1-\lambda) &= 0 \\
So, & \\
\lambda &= 2 \;\&\; -1
\end{aligned}
$$

The eigenvalues are 2 & -1.
To find the eigenvector associated with the eigenvalue $\lambda = 2$, we must determine x_1 and x_2 (not both equal to 0), so that

$$
\begin{bmatrix} 2 & 3 \\ 0 & -1 \end{bmatrix} \begin{bmatrix} x_1 \\ x_2 \end{bmatrix} = 2 \begin{bmatrix} x_1 \\ x_2 \end{bmatrix}
$$

Writing this as a system of linear equations,

$$
\begin{aligned}
2x_1 + 3x_2 &= 2x_1 \qquad (9.152) \\
-2x_2 &= 2x_2 \\
So, & \\
x_2 &= 0
\end{aligned}
$$

If we apply $x_2 = 0$, in the eqn (9.189), will give $x_1 = x_1$. x_1 can take any value but not zero. So $x_1 = 0$
The eigenvector is (1,0)

To find the eigenvector associated with the eigenvalue $\lambda = -1$, we must determine x_1 and x_2 (not both equal to 0), so that

$$
\begin{bmatrix} 2 & 3 \\ 0 & -1 \end{bmatrix} \begin{bmatrix} x_1 \\ x_2 \end{bmatrix} = -1 \begin{bmatrix} x_1 \\ x_2 \end{bmatrix}
$$

Writing this as a system of linear equations,

$$
\begin{aligned}
2x_1 + 3x_2 &= -x_1 \qquad (9.153) \\
-2x_2 &= -x_2
\end{aligned}
$$

Eqn (9.190), gives $3x_1+3x_2 = 0$ i.e., in the eqn (9.189), will give $3x_1+3x_2 = 0$ so, $x_1 = -x_2$. If $x_1 = 1$ then $x_2 = -1$ The eigenvector is (1,-1)

$$Av_1 = 2\begin{bmatrix} 1 \\ 0 \end{bmatrix} = \begin{bmatrix} 2 \\ 0 \end{bmatrix}$$

$$Av_2 = -1\begin{bmatrix} 1 \\ -1 \end{bmatrix} = \begin{bmatrix} -1 \\ 1 \end{bmatrix}$$

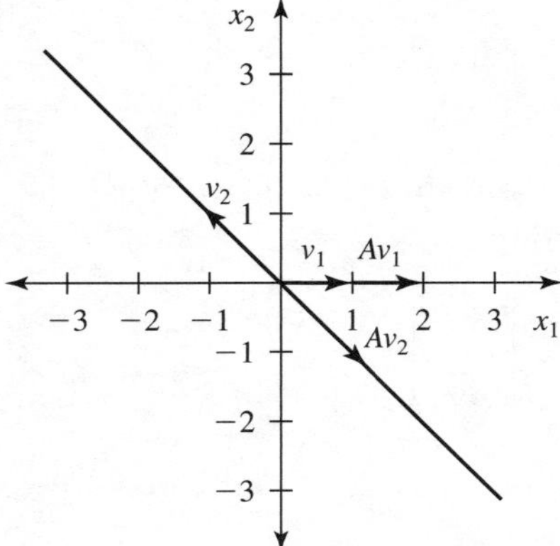

Prob. 51.

$$\begin{aligned} A &= \begin{bmatrix} 1 & 0 \\ 1 & -1 \end{bmatrix} \\ Ax = \lambda Ix & \\ (A - \lambda I)x &= 0 \\ [A - \lambda I] &= 0 \end{aligned}$$

To get the above,

$$\begin{aligned} [A - \lambda I] &= \begin{bmatrix} 1 & 0 \\ 1 & -1 \end{bmatrix} - \lambda \begin{bmatrix} 1 & 0 \\ 0 & 1 \end{bmatrix} \\ &= \begin{bmatrix} 1-\lambda & 0 \\ 0 & -1-\lambda \end{bmatrix} \\ (1-\lambda)(-1-\lambda) &= 0 \\ So, & \\ \lambda &= 1 \,\&\, -1 \end{aligned}$$

The eigenvalues are 1 & -1.
To find the eigenvector associated with the eigenvalue $\lambda = 1$, we must determine x_1 and x_2 (not both equal to 0), so that

$$\begin{bmatrix} 1 & 0 \\ 1 & -1 \end{bmatrix} \begin{bmatrix} x_1 \\ x_2 \end{bmatrix} = 1 \begin{bmatrix} x_1 \\ x_2 \end{bmatrix}$$

Writing this as a system of linear equations,

$$\begin{aligned} x_1 &= x_1 \\ -x_2 &= x_2 \\ So, & \\ x_2 &= 0 \end{aligned}$$

From $x_1 = x_1$ we can conclude that x_1 can take any value other than zero. The eigenvector is (1,0)

To find the eigenvector associated with the eigenvalue $\lambda = -1$, we must determine x_1 and x_2 (not both equal to 0), so that

$$\begin{bmatrix} 1 & 0 \\ 1 & -1 \end{bmatrix} \begin{bmatrix} x_1 \\ x_2 \end{bmatrix} = -1 \begin{bmatrix} x_1 \\ x_2 \end{bmatrix}$$

Writing this as a system of linear equations,

$$\begin{aligned} x_1 &= -x_1 \\ x_1 - x_2 &= -x_2 \\ So, & \\ x_1 &= 0 \end{aligned}$$

From $x_2 = x_2$ we can conclude that x_2 can take any value other than zero.

The eigenvector is (0,1)
Av = eigenvalue x corresponding eigenvector

$$Av_1 = 1\begin{bmatrix}1\\0\end{bmatrix} = \begin{bmatrix}1\\0\end{bmatrix}$$
$$Av_2 = -1\begin{bmatrix}0\\1\end{bmatrix} = \begin{bmatrix}0\\-1\end{bmatrix}$$

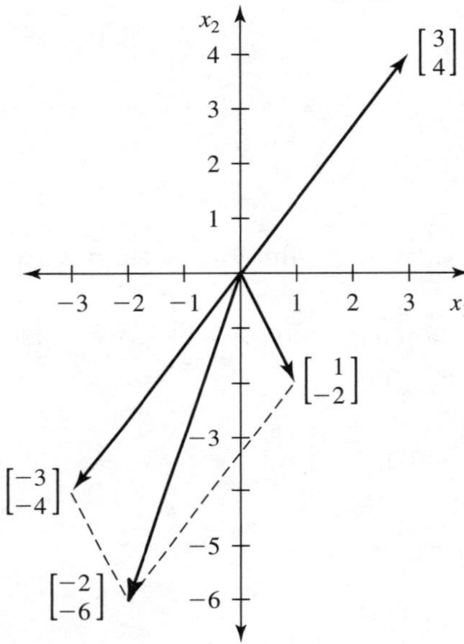

Prob. 53.

$$A = \begin{bmatrix}-4 & 2\\-3 & 1\end{bmatrix}$$
$$Ax = \lambda I x$$
$$(A - \lambda I)x = 0$$
$$[A - \lambda I] = 0$$

To get the above,

$$[A - \lambda I] = \begin{bmatrix}-4 & 2\\-3 & 1\end{bmatrix} - \lambda\begin{bmatrix}1 & 0\\0 & 1\end{bmatrix}$$
$$= \begin{bmatrix}-4-\lambda & 2\\-3 & 1-\lambda\end{bmatrix}$$
$$[A - \lambda I] = (-4-\lambda)(1-\lambda) + 6$$

$$\begin{aligned}
(-4-\lambda)(1-\lambda)+6 &= 0 \\
\lambda^2+4\lambda+-\lambda+6-4 &= 0 \\
\lambda^2+3\lambda+2 &= 0 \\
(1+\lambda)(2+\lambda) &= 0 \\
So, & \\
\lambda &= -1 \ \& \ -2
\end{aligned}$$

The eigenvalues are -1 & -2. To find the eigenvector associated with the eigenvalue $\lambda = -1$, we must determine x_1 and x_2 (not both equal to 0), so that

$$\begin{bmatrix} -4 & 2 \\ -3 & 1 \end{bmatrix} \begin{bmatrix} x_1 \\ x_2 \end{bmatrix} = -1 \begin{bmatrix} x_1 \\ x_2 \end{bmatrix}$$

Writing this as a system of linear equations,

$$\begin{aligned}
-4x_1+2x_2 &= -x_1 \\
-3x_1+x_2 &= -x_2 \\
3x_1+2x_2 &= 0 \\
3x_1 &= 2x_2 \\
So, & \\
x_2 &= \frac{3}{2}
\end{aligned}$$

If $x_1 = 1$, then $x_2 = \frac{3}{2}$.
The eigenvector is $(1,\frac{3}{2})$

To find the eigenvector associated with the eigenvalue $\lambda = -2$, we must determine x_1 and x_2 (not both equal to 0), so that

$$\begin{bmatrix} -4 & 2 \\ -3 & 1 \end{bmatrix} \begin{bmatrix} x_1 \\ x_2 \end{bmatrix} = -2 \begin{bmatrix} x_1 \\ x_2 \end{bmatrix}$$

Writing this as a system of linear equations,

$$\begin{aligned}
-4x_1+2x_2 &= -2x_1 \\
-3x_1+x_2 &= -2x_2 \\
-x_1+x_2 &= 0 \\
So, & \\
x_1 &= x_2
\end{aligned}$$

If $x_1 = 1$, then $x_2 = -1$.
The eigenvector is (1,1)
Av = eigenvalue x corresponding eigenvector

$$Av_1 = -2\begin{bmatrix} 1 \\ 1 \end{bmatrix} = \begin{bmatrix} -2 \\ -2 \end{bmatrix}$$

$$Av_2 = -1\begin{bmatrix} 1 \\ \frac{3}{2} \end{bmatrix} = \begin{bmatrix} -1 \\ -\frac{3}{2} \end{bmatrix}$$

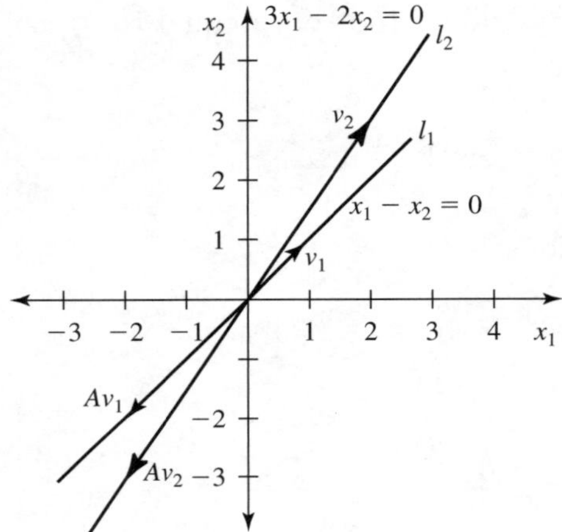

Prob. 55.

$$\begin{aligned} A &= \begin{bmatrix} 2 & 1 \\ 2 & 3 \end{bmatrix} \\ Ax = \lambda I x & \\ (A - \lambda I)x &= 0 \\ [A - \lambda I] &= 0 \end{aligned}$$

To get the above,

$$\begin{aligned} [A - \lambda I] &= \begin{bmatrix} 2 & 1 \\ 2 & 3 \end{bmatrix} - \lambda \begin{bmatrix} 1 & 0 \\ 0 & 1 \end{bmatrix} \\ &= \begin{bmatrix} 2-\lambda & 1 \\ 2 & 3-\lambda \end{bmatrix} \\ [A - \lambda I] &= (2-\lambda)(3-\lambda) - 2 \\ (2-\lambda)(3-\lambda) - 2 &= 0 \\ \lambda^2 - 2\lambda + -3\lambda + 6 - 2 &= 0 \end{aligned}$$

$$\begin{aligned}
\lambda^2 - 5\lambda + 4 &= 0 \\
(\lambda - 1)(\lambda - 4) &= 0 \\
So, & \\
\lambda &= 1 \ \& \ 4
\end{aligned}$$

The eigenvalues are 1 & 4. To find the eigenvector associated with the eigenvalue $\lambda = 1$, we must determine x_1 and x_2 (not both equal to 0), so that

$$\begin{bmatrix} 2 & 1 \\ 2 & 3 \end{bmatrix} \begin{bmatrix} x_1 \\ x_2 \end{bmatrix} = 1 \begin{bmatrix} x_1 \\ x_2 \end{bmatrix}$$

Writing this as a system of linear equations,

$$\begin{aligned}
2x_1 + x_2 &= x_1 \\
2x_1 + 3x_2 &= x_2
\end{aligned}$$

rearranging,

$$\begin{aligned}
x_1 + x_2 &= 0 \\
x_1 &= -x_2
\end{aligned}$$

If $x_1 = 1$, then $x_2 = -1$. The eigenvector is (1,-1)

To find the eigenvector associated with the eigenvalue $\lambda = 4$, we must determine x_1 and x_2 (not both equal to 0), so that

$$\begin{bmatrix} 2 & 1 \\ 2 & 3 \end{bmatrix} \begin{bmatrix} x_1 \\ x_2 \end{bmatrix} = 4 \begin{bmatrix} x_1 \\ x_2 \end{bmatrix}$$

Writing this as a system of linear equations,

$$\begin{aligned}
2x_1 + x_2 &= 4x_1 \\
2x_1 + 3x_2 &= 4x_2
\end{aligned}$$

rearranging,

$$\begin{aligned}
-2x_1 + x_2 &= 0 \\
2x_1 - x_2 &= 0 \\
So, & \\
2x_1 &= x_2
\end{aligned}$$

If $x_1 = 1$, then $x_2 = 2$.
The eigenvector is (1,2)
Av = eigenvalue x corresponding eigenvector

$$Av_1 = 1\begin{bmatrix} 1 \\ -1 \end{bmatrix} = \begin{bmatrix} 1 \\ -1 \end{bmatrix}$$

$$Av_2 = 4\begin{bmatrix} 1 \\ 2 \end{bmatrix} = \begin{bmatrix} 4 \\ 8 \end{bmatrix}$$

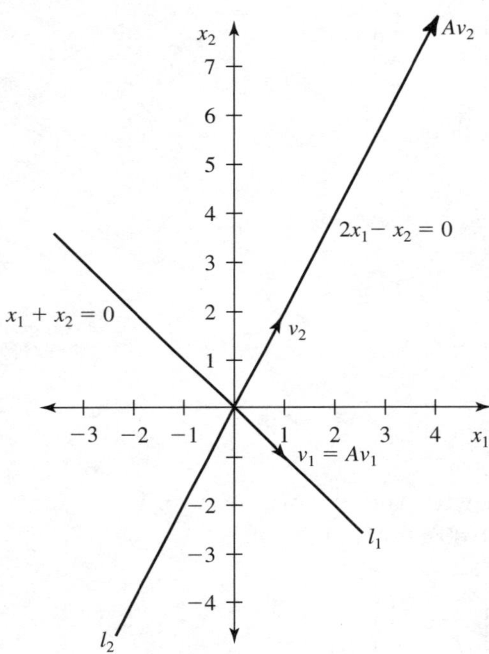

Prob. 57.

$$A = \begin{bmatrix} 4 & 0 \\ 0 & 3 \end{bmatrix}$$

$$Ax = \lambda I x$$

$$(A - \lambda I)x = 0$$

$$[A - \lambda I] = 0$$

To get the above,

$$[A - \lambda I] = \begin{bmatrix} 4 & 0 \\ 0 & 3 \end{bmatrix} - \lambda \begin{bmatrix} 1 & 0 \\ 0 & 1 \end{bmatrix}$$

$$\begin{aligned} &= \begin{bmatrix} 4-\lambda & 0 \\ 0 & 3-\lambda \end{bmatrix} \\ [A-\lambda I] &= (4-\lambda)(3-\lambda) \\ \lambda &= 4 \;\&\; 3 \end{aligned}$$

The eigenvalues are 3 & 4.

Prob. 59.

$$A = \begin{bmatrix} 1 & -3 \\ 0 & 2 \end{bmatrix}$$

$$\begin{aligned} Ax &= \lambda I x \\ (A-\lambda I)x &= 0 \\ [A-\lambda I] &= 0 \end{aligned}$$

To get the above,

$$\begin{aligned} [A-\lambda I] &= \begin{bmatrix} 1 & -3 \\ 0 & 2 \end{bmatrix} - \lambda \begin{bmatrix} 1 & 0 \\ 0 & 1 \end{bmatrix} \\ &= \begin{bmatrix} 1-\lambda & -3 \\ 0 & 2-\lambda \end{bmatrix} \\ [A-\lambda I] &= (1-\lambda)(2-\lambda) \\ \lambda &= 1 \;\&\; 2 \end{aligned}$$

The eigenvalues are 1 & 2.

Prob. 61.

$$A = \begin{bmatrix} -\frac{3}{2} & 0 \\ -1 & \frac{1}{2} \end{bmatrix}$$

$$\begin{aligned} Ax &= \lambda I x \\ (A-\lambda I)x &= 0 \\ [A-\lambda I] &= 0 \end{aligned}$$

To get the above,

$$\begin{aligned} [A-\lambda I] &= \begin{bmatrix} -\frac{3}{2} & 0 \\ -1 & \frac{1}{2} \end{bmatrix} - \lambda \begin{bmatrix} 1 & 0 \\ 0 & 1 \end{bmatrix} \\ &= \begin{bmatrix} -\frac{3}{2}-\lambda & 0 \\ -1 & \frac{1}{2}-\lambda \end{bmatrix} \\ [A-\lambda I] &= (-\frac{3}{2}-\lambda)(\frac{1}{2}-\lambda) \\ \lambda &= -\frac{3}{2} \;\&\; \frac{1}{2} \end{aligned}$$

The eigenvalues are $-\frac{3}{2}$ & $\frac{1}{2}$.

If one or both off-diagonal elements of 2x2 matrix are equal to zero, then the eigenvalues are that matrix is equal to the diagonal elements.

Prob. 63. The real part of λ_1 and λ_2 will be negative, if and only if, trace $A < 0$ and determinant $A > 0$

$$\begin{aligned} tr\ A &= 2 - 3 = -1 < 0 \\ \det A &= (2)(-3) - (4)(-2) = 2 > 0 \end{aligned}$$

So, the real parts of both eigenvalues are negative.

Prob. 65. The real part of λ_1 & λ_2 will be negative, if and only if, trace $A < 0$ and determinant $A > 0$

$$\begin{aligned} tr\ A &= 4 - 3 = 1 > 0 \\ \det A &= (4)(-3) - (4)(-4) = 4 > 0 \end{aligned}$$

Though determinant A is positive, the trace of A is not negative. As it is not satisfying the lemma, the real parts of both eigenvalues are not negative.

Prob. 67. The real part of λ_1 & λ_2 will be negative, if and only if, trace $A < 0$ and determinant $A > 0$

$$\begin{aligned} tr\ A &= 2 - 3 = -1 < 0 \\ \det A &= (2)(-3) - (2)(-5) = 4 > 0 \end{aligned}$$

So, the real parts of both eigenvalues are negative.

Prob. 69.

(a) To be a linearly independent, if $\lambda_1 \neq \lambda_2$, then the eigenvector associated with λ_1 and the eigenvector associated with λ_2 are linearly independent.

$$\begin{aligned} A &= \begin{bmatrix} -1 & 1 \\ 0 & 2 \end{bmatrix} \\ Ax = \lambda I x & \\ (A - \lambda I)x &= 0 \\ [A - \lambda I] &= 0 \end{aligned}$$

To get the above,

$$[A - \lambda I] = \begin{bmatrix} -1 & 1 \\ 0 & 2 \end{bmatrix} - \lambda \begin{bmatrix} 1 & 0 \\ 0 & 1 \end{bmatrix}$$

$$\begin{aligned}
&= \begin{bmatrix} -1-\lambda & 1 \\ 0 & 2-\lambda \end{bmatrix} \\
(-1-\lambda)(2-\lambda) &= 0 \\
So, & \\
\lambda &= 2 \;\&\; -1
\end{aligned}$$

The eigenvalues are -1 and 2. To find the eigenvector associated with the eigenvalue $\lambda = -1$, we must determine x_1 and x_2 (not both equal to 0), so that

$$\begin{bmatrix} -1 & 1 \\ 0 & 2 \end{bmatrix} \begin{bmatrix} x_1 \\ x_2 \end{bmatrix} = -1 \begin{bmatrix} x_1 \\ x_2 \end{bmatrix}$$

Writing this as a system of linear equations,

$$\begin{aligned}
-x_1 + x_2 &= -x_1 \\
x_2 &= 0 \\
then & \\
x_1 &= x_1 \\
So, & \\
x_2 &= 0
\end{aligned}$$

From $x_1 = x_1$ we can conclude that x_1 can take any value other than zero. The eigenvector is (1,0)

To find the eigenvector associated with the eigenvalue $\lambda = 2$, we must determine x_1 and x_2 (not both equal to 0), so that

$$\begin{bmatrix} -1 & 1 \\ 0 & 2 \end{bmatrix} \begin{bmatrix} x_1 \\ x_2 \end{bmatrix} = 2 \begin{bmatrix} x_1 \\ x_2 \end{bmatrix}$$

Writing this as a system of linear equations,

$$\begin{aligned}
-x_1 + x_2 &= 2x_1 \\
-3x_1 + x_2 &= 0 \\
x_2 &= 3x_1
\end{aligned}$$

So, if $x_1 = x_1$, x_2 will be equal to 3. The eigenvector is (1,3).

(b) Representing $\begin{bmatrix} 1 \\ -3 \end{bmatrix}$ as a linear combination of u_1 and u_2. We get:

$$\begin{bmatrix} 1 \\ -3 \end{bmatrix} = 2 \begin{bmatrix} 1 \\ 0 \end{bmatrix} - \begin{bmatrix} 1 \\ 3 \end{bmatrix}$$

(c) To compute A^{20}
$\lambda = -1,\ 2$

$$\begin{aligned}
A^{20} &= A^{20}\begin{bmatrix} 1 \\ -3 \end{bmatrix} \\
&= A^{20}\left[2\begin{pmatrix} 1 \\ 0 \end{pmatrix} - \begin{pmatrix} 1 \\ 3 \end{pmatrix}\right] \\
&= 2A^{20}\begin{pmatrix} 1 \\ 0 \end{pmatrix} - A^{20}\begin{pmatrix} 1 \\ 3 \end{pmatrix} \\
&= 2(-1)^{20}\begin{pmatrix} 1 \\ 0 \end{pmatrix} - (2)^{20}\begin{pmatrix} 1 \\ 3 \end{pmatrix} \\
&= \begin{pmatrix} 2 \\ 0 \end{pmatrix} - 1048576\begin{pmatrix} 1 \\ 3 \end{pmatrix} \\
&= \begin{pmatrix} 2 \\ 0 \end{pmatrix} - \begin{pmatrix} 1048576 \\ 3145728 \end{pmatrix} \\
&= \begin{pmatrix} -1048574 \\ 3145728 \end{pmatrix}
\end{aligned}$$

Prob. 71.

$$A = \begin{bmatrix} -1 & 0 \\ 3 & 1 \end{bmatrix}$$

To get the eigenvalues,

$$\begin{aligned}
Ax &= \lambda I x \\
(A - \lambda I)x &= 0 \\
[A - \lambda I] &= 0
\end{aligned}$$

To get the above,

$$\begin{aligned}
[A - \lambda I] &= \begin{bmatrix} -1 & 0 \\ 3 & 1 \end{bmatrix} - \lambda\begin{bmatrix} 1 & 0 \\ 0 & 1 \end{bmatrix} \\
&= \begin{bmatrix} -1-\lambda & 0 \\ 3 & 1-\lambda \end{bmatrix} \\
[A - \lambda I] &= (-1-\lambda)(1-\lambda) \\
\lambda &= 1 \;\&\; -1
\end{aligned}$$

The eigenvalues are 1 & -1.

To find the eigenvector associated with the eigenvalue $\lambda = 1$, we must determine x_1 and x_2 (not both equal to 0), so that

$$\begin{bmatrix} -1 & 0 \\ 3 & 1 \end{bmatrix}\begin{bmatrix} x_1 \\ x_2 \end{bmatrix} = 1\begin{bmatrix} x_1 \\ x_2 \end{bmatrix}$$

Writing this as a system of linear equations,

$$\begin{aligned} -x_1 + 0 &= -x_1 \\ 3x_1 + x_2 &= -x_2 \end{aligned}$$

rearranging,

$$\begin{aligned} x_1 &= x_1 \\ 3x_1 &= -2x_2 \\ So, & \\ x_2 &= -\frac{3}{2}x_1 \end{aligned}$$

If $x_1 = 1$, then $x_2 = -\frac{3}{2}$.
The eigenvector is $(1,-\frac{3}{2})$

To compute $A^{15}\begin{bmatrix} 2 \\ 0 \end{bmatrix}$, By substituting u_1 & u_2 in the eqn $x = a_1u_1 + a_2u_2$
Using the previous problem's method, we can find out the value for a_1 and a_2 and applying the λ_1 and λ_2 we get,

$$A^{15}\begin{bmatrix} 2 \\ 0 \end{bmatrix} = \begin{bmatrix} -2 \\ 6 \end{bmatrix}$$

Prob. 73.

$$A = \begin{bmatrix} 5 & 7 \\ -2 & -4 \end{bmatrix}$$

To get the eigenvalues,

$$\begin{aligned} Ax = \lambda Ix & \\ (A - \lambda I)x &= 0 \\ [A - \lambda I] &= 0 \end{aligned}$$

To get the above,

$$\begin{aligned}
[A-\lambda I] &= \begin{bmatrix} 5 & 7 \\ -2 & -4 \end{bmatrix} - \lambda \begin{bmatrix} 1 & 0 \\ 0 & 1 \end{bmatrix} \\
&= \begin{bmatrix} 5-\lambda & 7 \\ -2 & -4-\lambda \end{bmatrix} \\
[A-\lambda I] &= (5-\lambda)(-4-\lambda)+14 \\
-\lambda+\lambda^2-6 &= 0 \\
(\lambda-3)(\lambda+2) &= 0 \\
\lambda &= 3 \ \& \ -2
\end{aligned}$$

The eigenvalues are 3 & -2.

To find the eigenvector associated with the eigenvalue $\lambda = 3$, we must determine x_1 and x_2 (not both equal to 0), so that

$$\begin{bmatrix} 5 & 7 \\ -2 & -4 \end{bmatrix} \begin{bmatrix} x_1 \\ x_2 \end{bmatrix} = 3 \begin{bmatrix} x_1 \\ x_2 \end{bmatrix}$$

Writing this as a system of linear equations,

$$\begin{aligned}
5x_1 + 7x_2 &= 3x_1 \\
-2x_1 - 4x_2 &= 3x_2
\end{aligned}$$

rearranging,

$$\begin{aligned}
7x_2 &= -2x_1 \\
x_2 &= -\frac{2}{7}x_1
\end{aligned}$$

If $x_1 = 7$, then $x_2 = -2$. The eigenvector is $(7, -2))$

To find the eigenvector associated with the eigenvalue $\lambda = -2$, we must determine x_1 and x_2 (not both equal to 0), so that

$$\begin{bmatrix} 5 & 7 \\ -2 & -4 \end{bmatrix} \begin{bmatrix} x_1 \\ x_2 \end{bmatrix} = -2 \begin{bmatrix} x_1 \\ x_2 \end{bmatrix}$$

Writing this as a system of linear equations,

$$\begin{aligned}
5x_1 + 7x_2 &= -2x_1 \\
7x_1 &= -7x_2
\end{aligned}$$

rearranging,

$$x_1 = -x_2$$

If $x_1 = 1$, then $x_2 = -1$. The eigenvector is (1, -1) To compute $A^{20}\begin{bmatrix} -3 \\ -2 \end{bmatrix}$, use $x = a_1u_1 + a_2u_2$ Using the previous problem's method, we can find out the value for a_1 and a_2 and applying the λ_1 and λ_2 we get,

$$A^{20}\begin{bmatrix} -3 \\ -2 \end{bmatrix} = \begin{bmatrix} (-7)(3)^{20} + (4)(-2)^{20} \\ (2)(3)^{20} - (4)(-2)^{20} \end{bmatrix}$$

Prob. 75.

(a)

$$\begin{aligned} L &= \begin{bmatrix} 2 & 4 \\ 0.3 & 0 \end{bmatrix} \\ Lx = \lambda I x & \\ (L - \lambda I)x &= 0 \\ [L - \lambda I] &= 0 \end{aligned}$$

To get the above,

$$\begin{aligned} [L - \lambda I] &= \begin{bmatrix} 2 & 4 \\ 0.3 & 0 \end{bmatrix} - \lambda \begin{bmatrix} 1 & 0 \\ 0 & 1 \end{bmatrix} \\ &= \begin{bmatrix} 2-\lambda & 4 \\ 0.3 & -\lambda \end{bmatrix} \\ (2-\lambda)(-\lambda) - 1.2 &= 0 \\ \lambda^2 - 2\lambda &= 1.2 \\ i.e., & \\ \lambda_1 &= 1 + \sqrt{2.2} \\ \lambda_2 &= 1 - \sqrt{2.2} \end{aligned}$$

The eigenvalues are $1 + \sqrt{2.2}$ and $1 - \sqrt{2.2}$

(b) Biological Interpretation:

Among these two eigenvalues, the larger one determines the growth rate of the population. Here the larger one is $1 + \sqrt{2.2}$.

(c) Stable age distribution:

The eigenvector corresponding to the larger eigenvalue is a stable age distribution. To find the eigenvector associated with the larger eigenvalue $\lambda = 1 + \sqrt{2.2}$,

$$\begin{bmatrix} 2 & 4 \\ 0.3 & 0 \end{bmatrix} \begin{bmatrix} x_1 \\ x_2 \end{bmatrix} = (1 + \sqrt{2.2}) \begin{bmatrix} x_1 \\ x_2 \end{bmatrix}$$

Writing this as a system of linear equations,

$$\begin{aligned} 2x_1 + 4x_2 &= (1 + \sqrt{2.2})x_1 \\ 0.3x_1 &= (1 + \sqrt{2.2})x_2 \\ x_2 &= \frac{0.3}{1 + \sqrt{2.2}} x_1 \end{aligned}$$

If $x_1 = 1$ then $x_2 = \frac{0.3}{1+\sqrt{2.2}}$

So the stable age distribution is approximately $\begin{pmatrix} 0.893 \\ 0.107 \end{pmatrix}$

Prob. 77.

(a)

$$\begin{aligned} L &= \begin{bmatrix} 7 & 3 \\ 0.1 & 0 \end{bmatrix} \\ Lx = \lambda I x & \\ (L - \lambda I)x &= 0 \\ [L - \lambda I] &= 0 \end{aligned}$$

To get the above,

$$\begin{aligned} [L - \lambda I] &= \begin{bmatrix} 7 & 3 \\ 0.1 & 0 \end{bmatrix} - \lambda \begin{bmatrix} 1 & 0 \\ 0 & 1 \end{bmatrix} \\ &= \begin{bmatrix} 7 - \lambda & 3 \\ 0.1 & -\lambda \end{bmatrix} \\ (7 - \lambda)(-\lambda) - 0.3 &= 0 \\ \lambda^2 - 7\lambda &= 0.3 \\ \lambda &= \pm\frac{1}{10}(\sqrt{1255} + 35) \\ i.e., & \\ \lambda &= \frac{1}{10}(\sqrt{1255} + 35) \\ \lambda &= -\frac{1}{10}(\sqrt{1255} + 35) \end{aligned}$$

The eigenvalues are 7.042 & -0.0426

(b) Biological Interpretation:

Among these two eigenvalues, the larger one determines the growth rate of the population. Here the larger one is 7.042.

(c) Stable age distribution:

The eigenvector corresponding to the larger eigenvalue is a stable age distribution. To find the eigenvector associated with the larger eigenvalue $\lambda = 7.042$,

$$\begin{bmatrix} 7 & 3 \\ 0.1 & 0 \end{bmatrix} \begin{bmatrix} x_1 \\ x_2 \end{bmatrix} = 2.033 \begin{bmatrix} x_1 \\ x_2 \end{bmatrix}$$

Writing this as a system of linear equations,

$$\begin{aligned} 7x_1 + 3x_2 &= 7.042x_1 \\ 0.1x_1 &= 7.042x_2 \\ x_2 &= \frac{0.1}{7.042}x_1 \\ x_2 &= 0.0142x_1 \end{aligned}$$

If $x_1 = 1$ then $x_2 = 0.0142$ The eigenvector is $(1,\ 0.0142)$. So the stable age distribution is $\begin{pmatrix} 0.986 \\ 0.014 \end{pmatrix}$

Prob. 79.

(a)

$$\begin{aligned} L &= \begin{bmatrix} 0 & 5 \\ 0.09 & 0 \end{bmatrix} \\ Lx = \lambda I x & \\ (L - \lambda I)x &= 0 \\ [L - \lambda I] &= 0 \end{aligned}$$

To get the above,

$$\begin{aligned} [L - \lambda I] &= \begin{bmatrix} 0 & 5 \\ 0.09 & 0 \end{bmatrix} - \lambda \begin{bmatrix} 1 & 0 \\ 0 & 1 \end{bmatrix} \\ &= \begin{bmatrix} -\lambda & 5 \\ 0.09 & -\lambda \end{bmatrix} \\ (-\lambda)(-\lambda) - 0.45 &= 0 \end{aligned}$$

$$\begin{aligned} \lambda^2 &= 0.45 \\ \lambda &= \pm 0.67 \end{aligned}$$

i.e.,

$$\lambda = 0.67, \ -0.67$$

The eigenvalues are 0.67 & -0.67

(b) Biological Interpretation:

Among these two eigenvalues, the larger one determines the growth rate of the population. Here the larger one is 0.67.

(c) Stable age distribution:

The eigenvector corresponding to the larger eigenvalue is a stable age distribution.
To find the eigenvector associated with the larger eigenvalue $\lambda = 0.67$,

$$\begin{bmatrix} 0 & 5 \\ 0.09 & 0 \end{bmatrix} \begin{bmatrix} x_1 \\ x_2 \end{bmatrix} = 0.67 \begin{bmatrix} x_1 \\ x_2 \end{bmatrix}$$

Writing this as a system of linear equations,

$$\begin{aligned} 5x_2 &= 0.67x_1 \\ 0.09x_1 &= 0.67x_2 \\ x_2 &= \frac{0.09}{0.67}x_1 \\ x_2 &= 0.134x_1 \end{aligned}$$

If $x_1 = 1$ then $x_2 = 0.134$ So the stable age distribution is $\begin{pmatrix} 0.882 \\ 0.118 \end{pmatrix}$

9.4 Analytic Geometry

Prob. 1.

$$\begin{aligned} x &= \begin{bmatrix} 1 \\ 4 \\ -1 \end{bmatrix} \\ y &= \begin{bmatrix} -2 \\ 1 \\ 0 \end{bmatrix} \end{aligned}$$

(a)

$$x + y = \begin{bmatrix} 1 - 2 \\ 4 + 1 \quad -1 + 0 \end{bmatrix} = \begin{bmatrix} -1 \\ 5 \\ -1 \end{bmatrix}$$

(b)

$$2x = \begin{bmatrix} (2)(1) \\ (2)(4) \\ (2)(-1) \end{bmatrix} = \begin{bmatrix} 2 \\ 8 \\ -2 \end{bmatrix}$$

(c)

$$-3y = \begin{bmatrix} (-3)(-2) \\ (-3)(1) \\ (-3)(0) \end{bmatrix} = \begin{bmatrix} -6 \\ -3 \\ 0 \end{bmatrix}$$

Prob. 3.

$$\begin{aligned} A &= (2, 3) \\ B &= (4, 1) \\ \overrightarrow{AB} &= \begin{bmatrix} b_1 - a_1 \\ b_2 - a_2 \end{bmatrix} \\ &= \begin{bmatrix} 4 - 2 \\ 1 - 3 \end{bmatrix} \\ &= \begin{bmatrix} 2 \\ -2 \end{bmatrix} \end{aligned}$$

Prob. 5.

$$\begin{aligned} A &= (0,1,-3) \\ B &= (-1,-1,2) \\ \overrightarrow{AB} &= \begin{bmatrix} b_1 - a_1 \\ b_2 - a_2 \\ b_3 - a_3 \end{bmatrix} = \begin{bmatrix} -1-0 \\ 1-(-1) \\ 2-(-3) \end{bmatrix} = \begin{bmatrix} -1 \\ 2 \\ 5 \end{bmatrix} \end{aligned} \tag{9.154}$$

Prob. 7. Length of x

$$\begin{aligned} A &= [1,3]' \\ i.e., & \\ A &= \begin{bmatrix} 1 \\ 3 \end{bmatrix} \\ |x| &= \sqrt{(1)^2 + (3)^2} \\ &= \sqrt{1+9} \\ &= \sqrt{10} \end{aligned}$$

Prob. 9. Length of x

$$\begin{aligned} A &= [0,1,5]' \\ i.e., & \\ A &= \begin{bmatrix} 0 \\ 1 \\ 5 \end{bmatrix} \\ |x| &= \sqrt{(0)^2 + (1)^2 + (5)^2} \\ &= \sqrt{0+1+25} \\ &= \sqrt{26} \end{aligned}$$

Prob. 11. Normalizing the vector

$$\begin{aligned} vector &= [1,3,-1]' \\ i.e., & \\ A &= \begin{bmatrix} 1 \\ 3 \\ -1 \end{bmatrix} \end{aligned}$$

First calculate the length of the given vector

$$\begin{aligned} |x| &= \sqrt{(1)^2 + (3)^2 + (-1)^2} \\ &= \sqrt{1 + 9 + 1} \\ &= \sqrt{11} \end{aligned}$$

Normalizing the vector is finding out the unit vector.

$$\begin{aligned} \hat{x} &= \frac{x}{|x|} \\ &= \frac{1}{\sqrt{11}} \begin{bmatrix} 1 \\ 3 \\ -1 \end{bmatrix} \\ &= \begin{bmatrix} \frac{1}{\sqrt{11}} \\ \frac{3}{\sqrt{11}}3 \\ -\frac{1}{\sqrt{11}} \end{bmatrix} \end{aligned}$$

Prob. 13. Normalizing the vector

$$\begin{aligned} vector &= [1, 0, 0]' \\ i.e., & \\ A &= \begin{bmatrix} 1 \\ 0 \\ 0 \end{bmatrix} \end{aligned}$$

First calculate the length of the given vector

$$\begin{aligned} |x| &= \sqrt{(1)^2 + (0)^2 + (0)^2} \\ &= \sqrt{1 + 0 + 0} \\ &= \sqrt{1} \\ &= 1 \end{aligned}$$

Normalizing the vector is finding out the unit vector.

$$\begin{aligned} \hat{x} &= \frac{x}{|x|} \\ &= \frac{1}{1} \begin{bmatrix} 1 \\ 0 \\ 0 \end{bmatrix} \\ &= \begin{bmatrix} 1 \\ 0 \\ 0 \end{bmatrix} \end{aligned}$$

Prob. 15. Dot Product of $x = [1, 2]'$ and $y = [3, -1]'$

$$\begin{aligned} x &= \begin{bmatrix} 1 \\ 2 \end{bmatrix} \\ y &= \begin{bmatrix} 3 \\ -1 \end{bmatrix} \\ x.y &= \begin{bmatrix} 1 & 2 \end{bmatrix} \begin{bmatrix} 3 \\ -1 \end{bmatrix} \\ &= \begin{bmatrix} 3 - 2 \end{bmatrix} \\ &= 1 \end{aligned}$$

Prob. 17. Dot Product of $x = [0, -1, 3]'$ and $y = [-3, 1, 1]'$

$$\begin{aligned} x &= \begin{bmatrix} 0 \\ -1 \\ 3 \end{bmatrix} \\ y &= \begin{bmatrix} -3 \\ 1 \\ 1 \end{bmatrix} \\ x.y &= \begin{bmatrix} 0 & -1 & 3 \end{bmatrix} \begin{bmatrix} -3 \\ 1 \\ 1 \end{bmatrix} \\ &= \begin{bmatrix} 0 - 1 + 3 \end{bmatrix} \\ &= 2 \end{aligned}$$

Prob. 19. Using the dot product to compute the length of $[0, -1, 2]'$

$$|x| = \sqrt{x \cdot x} = \sqrt{(0)^2 + (-1)^2 + (2)^2} = \sqrt{5}$$

Prob. 21. Using the dot product to compute the length of $[1, 2, 3, 4]'$

$$|x| = \sqrt{x.x} = \sqrt{(1)^2 + (2)^2 + (3)^2 + (4)^2} = \sqrt{30}$$

Prob. 23.

$$\begin{aligned} x &= \begin{bmatrix} 1 \\ 2 \end{bmatrix} \\ y &= \begin{bmatrix} 3 \\ -1 \end{bmatrix} \end{aligned}$$

To determine the angle θ between x & y

$$\begin{aligned}
x.y &= |x||y|\cos\theta && (9.155)\\
|x| &= \sqrt{(1)^2+(2)^2} \\
&= \sqrt{1+4} \\
&= \sqrt{5} && (9.156)\\
|y| &= \sqrt{(3)^2+(-1)^2} \\
&= \sqrt{9+1} \\
&= \sqrt{10} && (9.157)\\
x.y &= \begin{bmatrix} 1 & 2 \end{bmatrix}\begin{bmatrix} 3 \\ -1 \end{bmatrix} \\
&= [3-2] \\
&= 1 && (9.158)
\end{aligned}$$

Substituting eqn (9.156), (9.157) and (9.158) in the eqn (9.155)

$$\begin{aligned}
cos\theta &= \frac{1}{\sqrt{5}\sqrt{10}} \\
&= \frac{1}{5\sqrt{2}} \\
\theta &= \cos^{-1}\frac{1}{5\sqrt{2}}
\end{aligned}$$

Prob. 25.

$$\begin{aligned}
x &= \begin{bmatrix} 0 \\ -1 \\ 3 \end{bmatrix} \\
y &= \begin{bmatrix} -3 \\ 1 \\ 1 \end{bmatrix}
\end{aligned}$$

To determine the angle θ between x & y

$$\begin{aligned}
x.y &= |x||y|\cos\theta && (9.159)\\
|x| &= \sqrt{(0)^2+(-1)^2+(3)^2} \\
&= \sqrt{1+9} \\
&= \sqrt{10} && (9.160)
\end{aligned}$$

$$\begin{aligned}
|y| &= \sqrt{(-3)^2+(1)^2+(1)^2} \\
&= \sqrt{9+1+1} \\
&= \sqrt{11} \\
&= \sqrt{11} && (9.161) \\
x.y &= \begin{bmatrix} 0 & -1 & 3 \end{bmatrix} \begin{bmatrix} -3 \\ 1 \\ 1 \end{bmatrix} \\
&= [0-1+3] \\
&= 2 && (9.162)
\end{aligned}$$

Substituting eqn (9.160), (9.161) and (9.162) in the eqn (9.159)

$$\begin{aligned}
cos\theta &= \frac{2}{\sqrt{10}\sqrt{11}} \\
&= \frac{2}{\sqrt{110}} \\
\theta &= \cos^{-1}\frac{2}{\sqrt{110}}
\end{aligned}$$

Prob. 27. The vectors x & y are perpendicular if $x.y = 0$ $x = [1, -1]'$ Let us take

$$\begin{aligned}
y &= \begin{bmatrix} y_1 \\ y_2 \end{bmatrix} \\
x.y &= \begin{bmatrix} 1 & -1 \end{bmatrix} \begin{bmatrix} y_1 \\ y_2 \end{bmatrix} \\
y_1 - y_2 &= 0 \\
So, & \\
y_1 &= y_2
\end{aligned}$$

From the above, we can assure that, any choice of numbers (y_1, y_2) that satisfies this equation would give a vector which will be perpendicular to x. So, if we give $y_1 = 1$ then $y_2 = 1$

$$y = \begin{bmatrix} 1 \\ 1 \end{bmatrix}$$

Prob. 29. The vectors x & y are perpendicular if $x.y = 0$ $x = [1, -2, 4]'$
Let us take

$$\begin{aligned} y &= \begin{bmatrix} y_1 \\ y_2 \\ y_3 \end{bmatrix} \\ x.y &= \begin{bmatrix} 1 & -2 & 4 \end{bmatrix} \begin{bmatrix} y_1 \\ y_2 \\ y_3 \end{bmatrix} \\ y_1 - 2y_2 + 4y_3 &= 0 \end{aligned}$$

From the above, we can assure that, any choice of numbers (y_1, y_2, y_3) that satisfies the above equation would give a vector which will be perpendicular to x. So, if we give $y_1 = 2$, then

$$2 - 2y_2 + 4y_3 = 0$$

if we give $y_3 = -1$, then

$$\begin{aligned} 2 - 2y_2 - 4 &= 0 \\ y_2 &= -1 \end{aligned}$$

Substituting all the values,

$$y = \begin{bmatrix} 2 \\ -1 \\ -1 \end{bmatrix}$$

Prob. 31. The coordinates are $P = (0, 0)$, $Q = (4, 0)$ and $R = (4, 3)$

$$\begin{aligned} \overrightarrow{PQ} &= \begin{bmatrix} q_1 - p_1 \\ q_2 - p_2 \end{bmatrix} \\ &= \begin{bmatrix} 4 - 0 \\ 0 - 0 \end{bmatrix} \\ &= \begin{bmatrix} 4 \\ 0 \end{bmatrix} \\ \overrightarrow{QR} &= \begin{bmatrix} r_1 - q_1 \\ r_2 - q_2 \end{bmatrix} \\ &= \begin{bmatrix} 4 - 4 \\ 3 - 0 \end{bmatrix} \end{aligned}$$

$$
\begin{aligned}
&= \begin{bmatrix} 0 \\ 3 \end{bmatrix} \\
\overrightarrow{PR} &= \begin{bmatrix} r_1 - p_1 \\ r_2 - p_2 \end{bmatrix} \\
&= \begin{bmatrix} 4 - 0 \\ 3 - 0 \end{bmatrix} \\
&= \begin{bmatrix} 4 \\ 3 \end{bmatrix}
\end{aligned}
$$

To find out the length of the vector $\overrightarrow{PQ}$

$$
\begin{aligned}
|\overrightarrow{PQ}| &= \sqrt{(4)^2 + (0)^2} \\
&= \sqrt{16 + 0} \\
&= 4
\end{aligned} \tag{9.163}
$$

To find out the length of the vector $\overrightarrow{QR}$

$$
\begin{aligned}
|\overrightarrow{QR}| &= \sqrt{(0)^2 + (3)^2} \\
&= \sqrt{0 + 9} \\
&= 3
\end{aligned} \tag{9.164}
$$

To find out the length of the vector $\overrightarrow{PR}$

$$
\begin{aligned}
|\overrightarrow{PR}| &= \sqrt{(4)^2 + (3)^2} \\
&= \sqrt{16 + 9} \\
&= 5
\end{aligned} \tag{9.165}
$$

Calculating the dot products

$$
\begin{aligned}
\overrightarrow{PQ}.\overrightarrow{QR} &= \begin{bmatrix} 4 & 0 \end{bmatrix} \begin{bmatrix} 0 \\ 3 \end{bmatrix} \\
&= 0
\end{aligned} \tag{9.166}
$$

$$
\begin{aligned}
\overrightarrow{QR}.\overrightarrow{PR} &= \begin{bmatrix} 0 & 3 \end{bmatrix} \begin{bmatrix} 4 \\ 3 \end{bmatrix} \\
&= 9
\end{aligned} \tag{9.167}
$$

$$
\begin{aligned}
\overrightarrow{PQ}.\overrightarrow{PR} &= \begin{bmatrix} 4 & 0 \end{bmatrix} \begin{bmatrix} 4 \\ 3 \end{bmatrix} \\
&= 16 + 0 \\
&= 0
\end{aligned} \tag{9.168}
$$

To determine the angle θ between $\overrightarrow{PQ}$ & $\overrightarrow{QR}$

$$\begin{aligned}
\overrightarrow{PQ}.\overrightarrow{QR} &= |\overrightarrow{PQ}||\overrightarrow{PQ}|\cos\theta \\
cos\theta &= \frac{\overrightarrow{PQ}.\overrightarrow{QR}}{|\overrightarrow{PQ}||\overrightarrow{PQ}|}
\end{aligned}$$

Substituting the values from the eqn (9.163), (9.164) & (9.166) in the above equation,

$$\begin{aligned}
cos\theta &= \frac{0}{(4)(3)} \\
&= 0 \\
\theta &= \cos^{-1}(0) \qquad (9.169)
\end{aligned}$$

To determine the angle θ between $\overrightarrow{QR}$ & $\overrightarrow{PR}$

$$\begin{aligned}
\overrightarrow{QR}.\overrightarrow{PR} &= |\overrightarrow{QR}||\overrightarrow{PR}|\cos\theta \\
cos\theta &= \frac{\overrightarrow{QR}.\overrightarrow{PR}}{|\overrightarrow{QR}||\overrightarrow{PR}|}
\end{aligned}$$

Substituting the values from the eqn (9.164), (9.165) & (9.167) in the above equation,

$$\begin{aligned}
cos\theta &= \frac{9}{(3)(5)} \\
\theta &= \cos^{-1}\left(\frac{3}{5}\right) \qquad (9.170)
\end{aligned}$$

To determine the angle θ between $\overrightarrow{PQ}$ & $\overrightarrow{PR}$

$$\begin{aligned}
\overrightarrow{PQ}.\overrightarrow{PR} &= |\overrightarrow{PQ}||\overrightarrow{PR}|\cos\theta \\
cos\theta &= \frac{\overrightarrow{PQ}.\overrightarrow{PR}}{|\overrightarrow{PQ}||\overrightarrow{PR}|}
\end{aligned}$$

Substituting the values from the eqn (9.163), (9.164) & (9.168) in the above equation,

$$\begin{aligned}
cos\theta &= \frac{16}{(4)(5)} \\
&= \frac{4}{5} \\
\theta &= \cos^{-1}\left(\frac{4}{5}\right) \qquad (9.171)
\end{aligned}$$

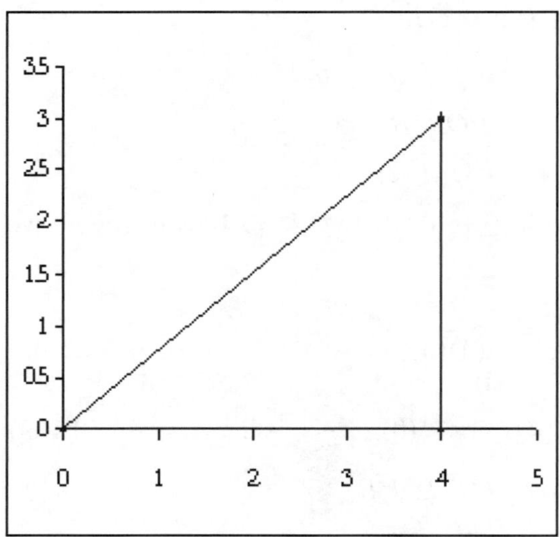

Prob. 33. The coordinates are $P = (1, 2, 3)$, $Q = (1, 5, 2)$ and $R = (2, 4, 2)$

$$\overrightarrow{PQ} = \begin{bmatrix} q_1 - p_1 \\ q_2 - p_2 \\ q_3 - p_3 \end{bmatrix} = \begin{bmatrix} 1-1 \\ 5-2 \\ 2-3 \end{bmatrix} = \begin{bmatrix} 0 \\ 3 \\ -1 \end{bmatrix}$$

$$\overrightarrow{QR} = \begin{bmatrix} r_1 - q_1 \\ r_2 - q_2 \\ r_3 - q_3 \end{bmatrix} = \begin{bmatrix} 2-1 \\ 4-5 \\ 2-2 \end{bmatrix} = \begin{bmatrix} 1 \\ -1 \\ 0 \end{bmatrix}$$

$$\overrightarrow{RP} = \begin{bmatrix} p_1 - r_1 \\ p_2 - r_2 \\ p_3 - r_3 \end{bmatrix} = \begin{bmatrix} 1-2 \\ 2-4 \\ 3-2 \end{bmatrix} = \begin{bmatrix} -1 \\ -2 \\ 1 \end{bmatrix}$$

To find out the length of the vector $\overrightarrow{PQ}$

$$|\overrightarrow{PQ}| = \sqrt{(0)^2 + (3)^2 + (-1)^2} = \sqrt{0 + 9 + 1} = \sqrt{10} \quad (9.172)$$

To find out the length of the vector $\overrightarrow{QR}$

$$|\overrightarrow{QR}| = \sqrt{(1)^2 + (-1)^2 + (0)^2} = \sqrt{1 + 1} = \sqrt{2} \quad (9.173)$$

To find out the length of the vector $\overrightarrow{RP}$

$$|\overrightarrow{RP}| = \sqrt{(-1)^2 + (-2)^2 + (1)^2} = \sqrt{1 + 4 + 1} = \sqrt{6} \quad (9.174)$$

Calculating the dot products

$$\overrightarrow{PQ}.\overrightarrow{QR} = \begin{bmatrix} 0 & 3 & -1 \end{bmatrix} \begin{bmatrix} 1 \\ -1 \\ 0 \end{bmatrix} = 0 - 3 + 0 = -3 \quad (9.175)$$

$$\overrightarrow{QR}.\overrightarrow{RP} = \begin{bmatrix} 1 & -1 & 0 \end{bmatrix} \begin{bmatrix} -1 \\ -2 \\ 1 \end{bmatrix} = -1 + 2 + 0 = 1 \quad (9.176)$$

$$\overrightarrow{PQ}.\overrightarrow{RP} = \begin{bmatrix} -1 & -2 & 1 \end{bmatrix} \begin{bmatrix} 0 \\ 3 \\ -1 \end{bmatrix} = 0 - 6 - 1 = -7 \quad (9.177)$$

To determine the angle θ between $\overrightarrow{QP}$ & $\overrightarrow{QR}$

$$\begin{aligned} \overrightarrow{PQ}.\overrightarrow{QR} &= |\overrightarrow{PQ}||\overrightarrow{QR}| \cos\theta \\ cos\theta &= \frac{\overrightarrow{PQ}.\overrightarrow{QR}}{|\overrightarrow{PQ}||\overrightarrow{QR}|} \end{aligned}$$

Substituting the values from the eqn (9.172), (9.173) & (9.175) in the above equation,

$$\begin{aligned} cos\theta &= \frac{3}{(\sqrt{10})(\sqrt{2})} = cos\theta = \frac{3}{2\sqrt{5}} \\ \theta &= \cos^{-1}\left(\frac{3}{2\sqrt{5}}\right) \end{aligned} \quad (9.178)$$

To determine the angle θ between $\overrightarrow{RQ}$ & $\overrightarrow{RP}$

$$\begin{aligned} \overrightarrow{RQ}.\overrightarrow{RP} &= |\overrightarrow{QR}||\overrightarrow{RP}| \cos\theta \\ cos\theta &= \frac{-\overrightarrow{QR}.\overrightarrow{RP}}{|\overrightarrow{QR}||\overrightarrow{RP}|} \end{aligned}$$

Substituting the values from the eqn (9.173), (9.174) & (9.176) in the above equation,

$$\begin{aligned} cos\theta &= \frac{-1}{(\sqrt{2})(\sqrt{6})} \\ &= \frac{-1}{2\sqrt{3}} \\ \theta &= \cos^{-1}\left(\frac{-1}{2\sqrt{3}}\right) \end{aligned} \quad (9.179)$$

To determine the angle θ between $\overrightarrow{PQ}$ & $\overrightarrow{PR}$

$$\begin{aligned} \overrightarrow{PQ}.\overrightarrow{PR} &= |\overrightarrow{PQ}||\overrightarrow{RP}|\cos\theta \\ cos\theta &= -\frac{\overrightarrow{PQ}.\overrightarrow{RP}}{|\overrightarrow{PQ}||\overrightarrow{RP}|} \end{aligned}$$

Substituting the values from the eqn (9.172), (9.173) & (9.177) in the above equation,

$$\begin{aligned} cos\theta &= \frac{7}{(\sqrt{6})(\sqrt{10})} \\ &= \frac{7}{2\sqrt{15}} \\ \theta &= \cos^{-1}\left(\frac{7}{2\sqrt{15}}\right) \end{aligned} \tag{9.180}$$

Prob. 35.

$$\begin{aligned} x_o &= (2,1) \\ A &= \begin{bmatrix} 1 \\ 2 \end{bmatrix} \end{aligned}$$

To find the eqn. of the line

$$\begin{aligned} a(x-x_o)+b(y-y_o) &= 0 \\ 1(x-2)+2(y-1) &= 0 \\ x-2+2y-2 &= 0 \\ x+2y-4 &= 0 \end{aligned}$$

Eqn. of the line is $x+2y-4=0$

Prob. 37.

$$x_o = (1,-2)$$
$$A = \begin{bmatrix} 4 \\ 1 \end{bmatrix}$$

To find the eqn. of the line

$$\begin{aligned} a(x-x_o)+b(y-y_o) &= 0 \\ (4)(x-1)+1(y+2) &= 0 \\ 4x-4+y+2 &= 0 \\ 4x+y-2 &= 0 \end{aligned}$$

Eqn. of the line is $4x + y - 2 = 0$

Prob. 39.

$$\begin{aligned} x_o &= (-1, 2, 3) \\ A &= \begin{bmatrix} 0 \\ -1 \\ 1 \end{bmatrix} \end{aligned}$$

To find the eqn. of the line

$$\begin{aligned} a(x - x_o) + b(y - y_o) + c(z - z_o) &= 0 \\ 0(x - 1) - 1(y - 2) + 1(z - 3) &= 0 \\ 0 - 1(y - 2) + z - 3 &= 0 \\ -y + 2 + z - 3 &= 0 \\ -y + z - 1 &= 0 \\ y - z + 1 &= 0 \end{aligned}$$

Eqn. of the line is:
$y - z + 1 = 0$

Prob. 41.

$$\begin{aligned} x_o &= (0, 0, 0) \\ A &= \begin{bmatrix} 1 \\ 0 \\ 0 \end{bmatrix} \end{aligned}$$

To find the eqn. of the line

$$\begin{aligned} a(x - x_o) + b(y - y_o) + c(z - z_o) &= 0 \\ 1(x + 0) - 0(y - 0) + 0(z - 0) &= 0 \\ x &= 0 \\ x &= 0 \end{aligned}$$

Eqn. of the line is: $x = 0$

Prob. 43. Parametric equation of the line in the x-y plane that goes through the point (1,-1) in the direction of $\begin{bmatrix} 2 \\ 1 \end{bmatrix}$ To find the eqn. of the line

$$\begin{bmatrix} x \\ y \end{bmatrix} = \begin{bmatrix} 1 \\ -1 \end{bmatrix} + t \begin{bmatrix} 2 \\ 1 \end{bmatrix}$$

Writing the above matrix in the form of linear equations,

$$x = 1 + 2t \tag{9.181}$$
$$y = -1 + t \tag{9.182}$$

Prob. 45. Parametric equation of the line in the x-y plane that goes through the point (3,-4) in the direction of $\begin{bmatrix} -1 \\ 2 \end{bmatrix}$ To find the eqn. of the line

$$\begin{bmatrix} x \\ y \end{bmatrix} = \begin{bmatrix} -1 \\ -2 \end{bmatrix} + t \begin{bmatrix} 1 \\ -3 \end{bmatrix} \tag{9.183}$$

Writing the above matrix in the form of linear equations,

$$x = -1 + t \tag{9.184}$$
$$y = -2 - 3t \tag{9.185}$$

Prob. 47. Let us consider one of the two points say (-1,2) as the point P_o. Then the direction of the indicated vector

$$\begin{aligned} \vec{U} &= \begin{bmatrix} q_1 - p_1 \\ q_2 - p_2 \end{bmatrix} \\ &= \begin{bmatrix} 3 - (-1) \\ 4 - 2 \end{bmatrix} \\ &= \begin{bmatrix} 3 + 1 \\ 4 - 2 \end{bmatrix} \\ &= \begin{bmatrix} 4 \\ 2 \end{bmatrix} \end{aligned}$$

$$\begin{bmatrix} x \\ y \end{bmatrix} = \begin{bmatrix} -1 \\ 2 \end{bmatrix} + t \begin{bmatrix} 4 \\ 2 \end{bmatrix}$$

Writing the above matrix in the form of linear equations,

$$x = -1 + 4t \tag{9.186}$$
$$y = 2 + 2t \tag{9.187}$$

for eliminating t,

$$\begin{aligned} 4t &= x+1 \\ t &= \frac{x+1}{4} \end{aligned} \tag{9.188}$$

Substituting eqn (9.188) in eqn. (9.187)

$$\begin{aligned} y &= 2+2\left(\frac{x+1}{4}\right) \\ &= 2+\frac{x}{2}+\frac{1}{2} \end{aligned}$$

multiplying by 2 on both sides

$$\begin{aligned} 2y &= x+5 \\ x-2y+5 &= 0 \end{aligned}$$

$x-2y+5=0$ is the standard form of the equation of the line.

Prob. 49. Let us consider one of the two points say (1, -3) as the point P_o. Then the direction of the indicated vector

$$\begin{aligned} \vec{U} &= \begin{bmatrix} q_1-p_1 \\ q_2-p_2 \end{bmatrix} \\ &= \begin{bmatrix} 4-1 \\ 0-(-3) \end{bmatrix} \\ &= \begin{bmatrix} 3 \\ 3 \end{bmatrix} \end{aligned}$$

$$\begin{bmatrix} x \\ y \end{bmatrix} = \begin{bmatrix} 1 \\ -3 \end{bmatrix} + t\begin{bmatrix} 3 \\ 3 \end{bmatrix}$$

Writing the above matrix in the form of linear equations,

$$x = 1+3t \tag{9.189}$$

$$y = 3+3t \tag{9.190}$$

for eliminating t,

$$\begin{aligned} 3t &= x-1 \\ t &= \frac{x-1}{3} \end{aligned} \tag{9.191}$$

Substituting eqn (9.191) in eqn. (9.190)

$$\begin{aligned} y &= -3+3\left(\frac{x-1}{3}\right) \\ &= -3+x-1 \\ x-y-4 &= 0 \end{aligned}$$

$x-y-4=0$ is the standard form of the equation of the line.

Prob. 51.

$$3x+4y-1 = 0$$

Eliminating the above equation

$$\begin{aligned} 4y &= -3x+1 \\ y &= -\frac{3}{4}x+\frac{1}{4} \end{aligned}$$

with $x=t$ we can write the parametric equation as

$$\begin{aligned} x &= t \\ y &= -\frac{3}{4}t+\frac{1}{4} \end{aligned}$$

Prob. 53.

$$2x+y-3 = 0$$

Eliminating the above equation

$$y = -2x+3$$

with $x=t$ we can write the parametric equation as

$$\begin{aligned} x &= t \\ y &= -2t+3 \end{aligned}$$

Prob. 55.

$$\begin{bmatrix} x \\ y \\ z \end{bmatrix} = \begin{bmatrix} 1 \\ -1 \\ 2 \end{bmatrix} + t\begin{bmatrix} 1 \\ -2 \\ 1 \end{bmatrix}$$

Writing the above matrix in the form of linear equations,

$$x = 1 + t \tag{9.192}$$
$$y = -1 - 2t \tag{9.193}$$
$$z = 2 + t \tag{9.194}$$

Prob. 57.

$$\begin{bmatrix} x \\ y \\ z \end{bmatrix} = \begin{bmatrix} -1 \\ 3 \\ -2 \end{bmatrix} + t \begin{bmatrix} -1 \\ -2 \\ 4 \end{bmatrix}$$

Writing the above matrix in the form of linear equations,

$$x = -1 - t \tag{9.195}$$
$$y = 3 - 2t \tag{9.196}$$
$$z = -2 + 4t \tag{9.197}$$

The eqn (9.195), eqn (9.196) & eqn (9.197) are the parametric equations of the line in the given $x - y - z$ space

Prob. 59. Let us consider one of the two points say (5, 4, -1) as the point P_o.
Then the direction of the indicated vector

$$\begin{aligned} \overrightarrow{U} &= \begin{bmatrix} q_1 - p_1 \\ q_2 - p_2 \\ q_3 - p_3 \end{bmatrix} \\ &= \begin{bmatrix} 2 - 5 \\ 0 - 4 \\ 3 - (-1) \end{bmatrix} \\ &= \begin{bmatrix} -3 \\ -4 \\ 4 \end{bmatrix} \end{aligned}$$

$$\begin{bmatrix} x \\ y \\ z \end{bmatrix} = \begin{bmatrix} 5 \\ 4 \\ -1 \end{bmatrix} + t \begin{bmatrix} -3 \\ -4 \\ 4 \end{bmatrix}$$

Writing the above matrix in the form of linear equations,

$$x = 5 - 3t \tag{9.198}$$
$$y = 4 - 4t \tag{9.199}$$
$$y = -1 + 4t \tag{9.200}$$

The eqn (9.198), eqn (9.199) & eqn (9.200) are the parametric equations of the line in the given $x - y - z$ space.

Prob. 61. The two points are (2, -3, 1) & (-5, 2, 1) Let us consider one of the two points say (2, -3, 1) as the point P_o. Then the direction of the indicated vector

$$\begin{aligned}\vec{U} &= \begin{bmatrix} q_1 - p_1 \\ q_2 - p_2 \\ q_3 - p_3 \end{bmatrix} \\ &= \begin{bmatrix} -5 - 2 \\ 2 - (-3) \\ 1 - 1 \end{bmatrix} \\ &= \begin{bmatrix} -7 \\ 2 + 3 \\ 0) \end{bmatrix} \\ &= \begin{bmatrix} -7 \\ 5 \\ 0 \end{bmatrix}\end{aligned}$$

$$\begin{bmatrix} x \\ y \\ z \end{bmatrix} = \begin{bmatrix} 2 \\ -3 \\ 1 \end{bmatrix} + t \begin{bmatrix} -7 \\ 5 \\ 0 \end{bmatrix}$$

Writing the above matrix in the form of linear equations,

$$x = 2 - 7t \tag{9.201}$$
$$y = -3 + 5t \tag{9.202}$$
$$z = 1 \tag{9.203}$$

The eqn (9.201), eqn (9.202) & eqn (9.203) are the parametric equations of the line in the given $x - y - z$ space.

Prob. 63. The given plane is (1, -1, 2) The perpendicular line to the above mentioned plane is $\begin{bmatrix} 1 \\ 2 \\ 1 \end{bmatrix}$. So, the equation of the plane in the three dimensional space is

$$\begin{aligned} x_o &= (1, -1, 2) \\ A &= \begin{bmatrix} 1 \\ 2 \\ 1 \end{bmatrix} \end{aligned}$$

To find the eqn. of the line

$$\begin{aligned} a(x - x_o) + b(y - y_o) + c(z - z_o) &= 0 \\ 1(x - 1) + 2(y + 1) + 1(z - 2) &= 0 \\ x - 1 + 2y + 2 + z - 2 &= 0 \\ x + 2y + z - 1 &= 0 \end{aligned}$$

The line through the given points (0, -3, 2) and (-1, -2, 3)

$$\begin{aligned} \vec{U} &= \begin{bmatrix} q_1 - p_1 \\ q_2 - p_2 \\ q_3 - p_3 \end{bmatrix} \\ &= \begin{bmatrix} -1 - 0 \\ -2 + 3 \\ 3 - 2 \end{bmatrix} \\ &= \begin{bmatrix} -1 \\ 1 \\ 1 \end{bmatrix} \quad \text{and hence} \quad \begin{bmatrix} x \\ y \\ z \end{bmatrix} = \begin{bmatrix} 0 \\ -3 \\ 2 \end{bmatrix} + t \begin{bmatrix} -1 \\ 1 \\ 1 \end{bmatrix} \end{aligned}$$

To find out the point of intersection of the plane and the line: $-t + 2(-3 + t) + (2 + t) = 1$ which yields $t = 5/2$ and hence the point of intersection is $1 - 5/2, -1/2, 9/2)$.

Prob. 65. The given plane is (0, -2, 1) The perpendicular line to the above mentioned plane is $\begin{bmatrix} -1 \\ 1 \\ -1 \end{bmatrix}$. So, the equation of the plane in the three dimensional space is

$$x_o = (0, -2, 1)$$

$$A = \begin{bmatrix} -1 \\ 1 \\ -1 \end{bmatrix}$$

To find the eqn. of the line

$$\begin{aligned} a(x - x_o) + b(y - y_o) + c(z - z_o) &= 0 \\ -1(x - 0) + 1(y + 2) - 1(z - 1) &= 0 \\ -x + 0 + y + 2 - z + 1 &= 0 \\ x - y + z - 3 &= 0 \end{aligned}$$

The line through the point (5, -1, 0) and parallel to the given plane.

Parallel to the plane indicates, perpendicular to the given vector A. A vector that is perpendicular to A is $\begin{bmatrix} 1 \\ 1 \\ 0 \end{bmatrix}$

Let us take the points of the unknown equation is (x, y, z)

$$\begin{aligned} \overrightarrow{U} &= \begin{bmatrix} q_1 - p_1 \\ q_2 - p_2 \\ q_3 - p_3 \end{bmatrix} \\ &= \begin{bmatrix} x - 5 \\ y + 1 \\ z + 0 \end{bmatrix} \end{aligned}$$

If the line is parallel to the plane i.e., perpendicular to the given line, then

$$\begin{bmatrix} x \\ y \\ z \end{bmatrix} = \begin{bmatrix} 5 \\ -1 \\ 0 \end{bmatrix} + t \begin{bmatrix} 1 \\ 1 \\ 0 \end{bmatrix}$$

the equation of the line which is parallel to the given plane is $x - y + z - 6 = 0$

9.6 Review Problems

Prob. 1.

(a)

$$\begin{aligned} A &= \begin{bmatrix} -1 & 1 \\ 0 & 2 \end{bmatrix} \\ x &= \begin{bmatrix} 1 \\ 1 \end{bmatrix} \end{aligned}$$

$$x \to Ax$$

$$\begin{aligned}
\begin{bmatrix} x_1 \\ x_2 \end{bmatrix} &\to \begin{bmatrix} -1 & 1 \\ 0 & 2 \end{bmatrix}\begin{bmatrix} 1 \\ 1 \end{bmatrix} \\
&= \begin{bmatrix} -1+1 \\ 0+2 \end{bmatrix} \\
&= \begin{bmatrix} 0 \\ 2 \end{bmatrix}
\end{aligned}$$

So this map stretches or contracts each coordinate separately. They are in the same quadrant.

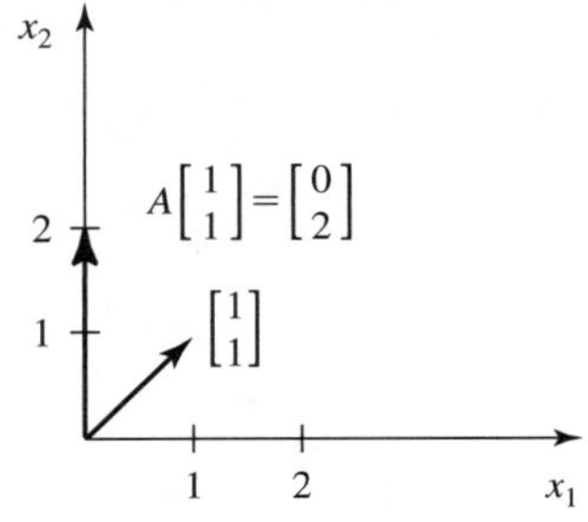

Length: $|x| = \sqrt{1^2+1^2} = \sqrt{1+1} = \sqrt{2}$

Angle: $\tan\alpha = \frac{x_2}{x_1} = \frac{1}{1} = 1$, so $\alpha = 45^o$.

(b) Eigenvalue & Eigenvector

$$\begin{aligned}
A &= \begin{bmatrix} -1 & 1 \\ 0 & 2 \end{bmatrix} \\
Ax &= \lambda I x \\
(A - \lambda I)x &= 0 \\
[A - \lambda I] &= 0
\end{aligned}$$

To get the above,

$$\begin{aligned}
[A - \lambda I] &= \begin{bmatrix} -1 & 1 \\ 0 & 2 \end{bmatrix} - \lambda \begin{bmatrix} 1 & 0 \\ 0 & 1 \end{bmatrix} \\
&= \begin{bmatrix} -1-\lambda & 1 \\ 0 & 2-\lambda \end{bmatrix} \\
(-1-\lambda)(2-\lambda) &= 0 \\
So, & \\
\lambda &= -1 \,\&\, 2
\end{aligned}$$

The eigenvalues are 2 & -1. To find the eigenvector associated with the eigenvalue $\lambda = -1$, we must determine x_1 and x_2 (not both equal to 0), so that

$$\begin{bmatrix} -1 & 1 \\ 0 & 2 \end{bmatrix} \begin{bmatrix} x_1 \; x_2 \end{bmatrix} = -1 \begin{bmatrix} x_1 \\ x_2 \end{bmatrix}$$

Writing this as a system of linear equations,

$$\begin{aligned} -x_1 + x_2 &= -x_1 \\ -2x_2 &= -x_2 \end{aligned}$$

$3x_2 = 0$ $x_2 = 0$ So substituting $x_2 = 0$ in the eqn $-x_1 + x_2 = -x_1$, $-x_1 = -x_1$ so, x_1 can take any value other than zero. So, let us take $x_2 = 1$ The eigenvector is (1, 0)

To find the eigenvector associated with the eigenvalue $\lambda = 2$, we must determine x_1 and x_2 (not both equal to 0), so that

$$\begin{bmatrix} -1 & 1 \\ 0 & 2 \end{bmatrix} \begin{bmatrix} x_1 \; x_2 \end{bmatrix} = 2 \begin{bmatrix} x_1 \\ x_2 \end{bmatrix}$$

Writing this as a system of linear equations,

$$\begin{aligned} -x_1 + x_2 &= 2x_1 \\ 2x_2 &= 2x_2 \end{aligned}$$

rearranging the above equations,

$$\begin{aligned} x_2 &= 3x_1 \\ x_1 &= x_2 \end{aligned}$$

If we apply $x_1 = 1$, in the equation $x_2 = 3$. The eigenvector is (1, 3)

(c)

$$\begin{aligned} A &= \begin{bmatrix} -1 & 1 \\ 0 & 2 \end{bmatrix} \\ u_i &= \begin{bmatrix} 1 \\ 0 \end{bmatrix} \\ Au_i &= \begin{bmatrix} -1 & 1 \\ 0 & 2 \end{bmatrix} \begin{bmatrix} -1+0 \\ 0+0 \end{bmatrix} \end{aligned}$$

$$
\begin{aligned}
&= \begin{bmatrix} -1+0 \\ 0+0 \end{bmatrix} \\
&= \begin{bmatrix} -1 \\ 0 \end{bmatrix} \\
Au_{ii} &= \begin{bmatrix} -1 & 1 \\ 0 & 2 \end{bmatrix} \begin{bmatrix} 1 \\ 3 \end{bmatrix} \\
&= \begin{bmatrix} -1+3 \\ 0+6 \end{bmatrix} \\
&= \begin{bmatrix} 2 \\ 6 \end{bmatrix}
\end{aligned}
$$

That will lie on the same eigenvector.

$$
\begin{aligned}
x &= a_1u_1 + a_2u_2 \\
\begin{bmatrix} 1 \\ 1 \end{bmatrix} &= a_1 \begin{bmatrix} 1 \\ 0 \end{bmatrix} + a_2 \begin{bmatrix} 1 \\ 1 \end{bmatrix} \\
a_1 + a_2 &= 1 \qquad (9.204) \\
3a_2 &= 1 \\
a_2 &= \frac{1}{3} \qquad (9.205)
\end{aligned}
$$

Substituting eqn (9.205) in (9.204)

$$
\begin{aligned}
a_1 &= 1 - \frac{1}{3} \\
&= \frac{2}{3}
\end{aligned}
$$

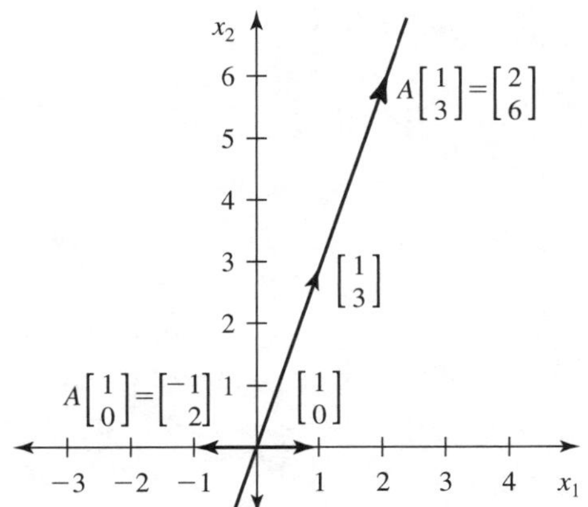

(d) u_1 and u_2 are both eigenvectors corresponding to A.

$$\begin{aligned} Au_1 &= \lambda_1 u_1 \\ Au_2 &= \lambda_2 u_2 \\ Ax &= a_1\lambda_1 u_1 + a_2\lambda_2 u_2 \\ &= \frac{2}{3}(-1)\begin{bmatrix} 1 \\ 0 \end{bmatrix} + \frac{1}{3}(2)\begin{bmatrix} 1 \\ 3 \end{bmatrix} \\ &= \begin{bmatrix} -\frac{2}{3} \\ 0 \end{bmatrix} + \begin{bmatrix} \frac{2}{3} \\ 2 \end{bmatrix} \\ &= \begin{bmatrix} 0 \\ 2 \end{bmatrix} \end{aligned}$$

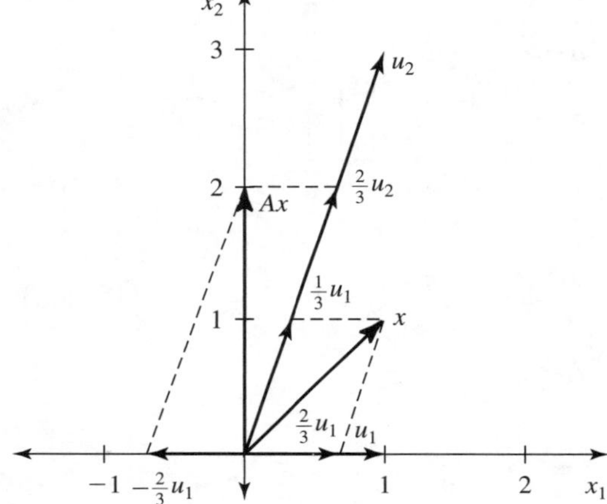

Prob. 3.

$$L = \begin{bmatrix} 1.5 & 0.875 \\ 0.5 & 0 \end{bmatrix}$$

$$\begin{aligned} Lx &= \lambda I x \\ (L - \lambda I)x &= 0 \\ [L - \lambda I] &= 0 \end{aligned}$$

To get the above,

$$\begin{aligned} [L - \lambda I] &= \begin{bmatrix} 1.5 & 0.875 \\ 0.5 & 0 \end{bmatrix} - \lambda \begin{bmatrix} 1 & 0 \\ 0 & 1 \end{bmatrix} \\ &= \begin{bmatrix} 1.5 - \lambda & 0.875 \\ 0.5 & -\lambda \end{bmatrix} \end{aligned}$$

$$\begin{aligned}(1.5-\lambda)(-\lambda)-(0.875)(0.5) &= 0\\ \lambda^2-1.5\lambda-0.4375 &= 0\\ (\lambda-1.75)(\lambda+0.25) &= 0\\ \lambda &= 1.75,\ -0.25\end{aligned}$$

The eigenvalues are 1.75 & -0.25 Among these two eigenvalues, the larger one determines the growth rate of the population. Here the larger one is 1.75.

Stable age distribution: The eigenvector corresponding to the larger eigenvalue is a stable age distribution. To find the eigenvector associated with the larger eigenvalue $\lambda = 1.75$,

$$\begin{bmatrix} 1.5 & 0.875 \\ 0.5 & 0 \end{bmatrix}\begin{bmatrix} x_1 \\ x_2 \end{bmatrix} = 1.75\begin{bmatrix} x_1 \\ x_2 \end{bmatrix}$$

Writing this as a system of linear equations,

$$\begin{aligned}1.5x_1+0.875x_2 &= 1.75x_1\\ 0.5x_1 &= 1.75x_2\\ \frac{1}{2}x_1 &= \frac{7}{4}x_2\\ x_2 &= \frac{2}{7}x_1\end{aligned}$$

If $x_1 = 1$ then $x_2 = \frac{2}{7}$ The eigenvector is $(1, \frac{2}{7})$. So the stable age distribution is $\begin{pmatrix} 1 \\ \frac{2}{7} \end{pmatrix}$

Prob. 5.

$$\begin{aligned}AB &= \begin{bmatrix} 0 & 1 \\ 2 & -1 \end{bmatrix}\\ A^{-1} &= \begin{bmatrix} 4 & -1 \\ 8 & -1 \end{bmatrix}\\ B &= A^{-1}AB\\ So&\\ A^{-1}AB &= \begin{bmatrix} 4 & -1 \\ 8 & -1 \end{bmatrix}\begin{bmatrix} 0 & 1 \\ 2 & -1 \end{bmatrix}\\ &= \begin{bmatrix} 0-2 & 4+1 \\ 0-2 & 8+1 \end{bmatrix}\\ &= \begin{bmatrix} -2 & 5 \\ -2 & 9 \end{bmatrix}\end{aligned}$$

Prob. 7. The two different ways to solve the system of equations of the form

I **Graphical Solution**

Recall that the standard form of a linear equation in two variables is $Ax + By = C$ where A,B, and C are constants, A and B are not both equal to 0, and x and y are the two variables; its graph is a straight line. Any point (x, y) on this straight line satisfies (or solves) the equation Ax + By = C. We can extend this to more than one equation, to get the following system

$$\begin{aligned} a_{11}x_1 + a_{12}x_2 &= b_1 \\ a_{21}x_1 + a_{22}x2 &= b_2 \end{aligned}$$

where $a_{11},\ a_{12},\ a_{21},\ a_{22}\ b_1,\ and\ b_2$ are constants, x and y are the two variables. Finding an ordered pair (x, y) that satisfies each equation of the system given above. Because each equation in the system describes a straight line, we are therefore finding out the the point of intersection of these two lines. The following three cases are possible:

1. The two lines have exactly one point of intersection. In this case, the system has exactly one solution.
2. The two lines are parallel and do not intersect. In this case, the system has no solution.
3. The two lines are parallel and intersect (that is, they are identical). In this case, the system has infinitely many solutions, namely each point on the line.

II **Solution Method**

$$\begin{aligned} a_{11}x_1 + a_{12}x_2 &= b_1 \\ a_{21}x_1 + a_{22}x2 &= b_2 \end{aligned}$$

We will transform this system into an equivalent system in upper triangular form. To do so we will use the following three basic operations:

i. Multiplying an equation by a nonzero constant
ii. Adding one equation to another
iii. Rearranging the order of the equations

This method is named as Gaussian elimination. The above mentioned three cases are identified such that:

1. If $a_{11}a_{22} - a_{12}a_{22} \neq 0$ Then the two lines have exactly one point of intersection. In this case, the system has exactly one solution.
2. If $a_{11}a_{22} - a_{12}a_{22} = 0$ and $b_1 \neq b_2$ then the two lines are parallel and do not intersect. In this case, the system has no solution.
3. If $a_{11}a_{22} - a_{12}a_{22} = 0$ and $b_1 = b_2$ then the two lines are parallel and intersect (that is, they are identical). In this case, the system has infinitely many solutions, namely each point on the line.

Prob. 9.

$$\begin{aligned} a_{11}x_1 + a_{12}x_2 &= b_1 \\ a_{21}x_1 + a_{22}x2 &= b_2 \end{aligned}$$

The system will have infinitely many solutions if and only if $a_{11}a_{12} - a_{12}a_{22} = 0$ and $b_1 = b_2$ The given system is

$$\begin{aligned} ax + 3y &= 0 \\ x - y &= 0 \end{aligned}$$

If we write it in the form of matrix (equivalent system)

$$\begin{bmatrix} a & 3 \\ 1 & -1 \end{bmatrix} \begin{bmatrix} x \\ y \end{bmatrix} = \begin{bmatrix} 0 \\ 0 \end{bmatrix}$$

Here $b_1 = b_2$ i.e., $0, -a - 3 = 0$ So, if $a = -3$ then the system will have infinitely many solutions. i.e., the two lines are parallel and intersect (that is, they are identical).

Prob. 11.

$$L = \begin{bmatrix} 0.5 & 2.3 \\ a & 0 \end{bmatrix} \tag{9.206}$$

Finding the eigenvalue

$$\begin{aligned} L &= \begin{bmatrix} 0.5 & 2.3 \\ a & 0 \end{bmatrix} \\ Lx = \lambda I x & \\ (L - \lambda I)x &= 0 \\ [L - \lambda I] &= 0 \end{aligned}$$

To get the above,

$$\begin{aligned}
[L - \lambda I] &= \begin{bmatrix} 0.5 & 2.3 \\ a & 0 \end{bmatrix} - \lambda \begin{bmatrix} 1 & 0 \\ 0 & 1 \end{bmatrix} \\
&= \begin{bmatrix} 0.5 - \lambda & 2.3 \\ a & 0 - \lambda \end{bmatrix} \\
(0.5 - \lambda)(-\lambda) - 2.3a &= 0
\end{aligned}$$

$$\lambda^2 - 0.5\lambda - 2.3a = 0$$

$$\lambda_{12} = \frac{0.5 \pm \sqrt{0.5^2 + (4)(2.3)a}}{2}$$

$$\lambda_1 = \frac{0.5 + \sqrt{0.5^2 + (4)(2.3)a}}{a} > 1$$

if

$$a > \frac{5}{23}$$

Since $a \leq 1$, the population will be growing for $\frac{5}{23} < a \leq 1$.

10 Multivariable Calculus

10.1 Functions of Two or More Independent Variables

Prob. 1.

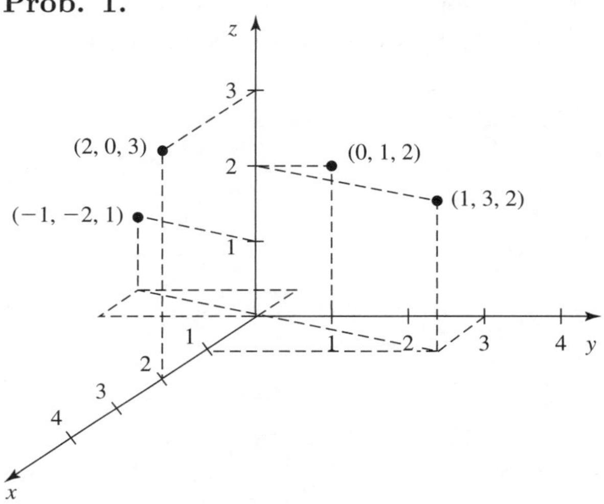

Prob. 3. $\frac{4}{13}$

Prob. 5. 0.904837

Prob. 7. Domain is $\mathbb{R}^2$. Range is $\mathbb{R}_+$. The level curves are circles $x^2+y^2 = c$ for $c \geq 0$ (for $c = 0$ the level curve is just a point).

Prob. 9. Domain is the set $\{(x, y) \in \mathbb{R}^2 : y > x^2\}$. Range is $\mathbb{R}$. Level curves are parabolas $y - x^2 = e^c$, for every real c.

Prob. 11. 10.23

Prob. 13. 10.24

Prob. 15.

(a) The level curves are circles.

Let $z = f_1(x, y)$. We have $f_1(x, 0) = x^2$ on the intersection of the surface with the plane xOz, therefore the curve has equation $z = x^2$. Similarly, $z = f_1(0, y) = y^2$ is the intersection with the plane yOz.

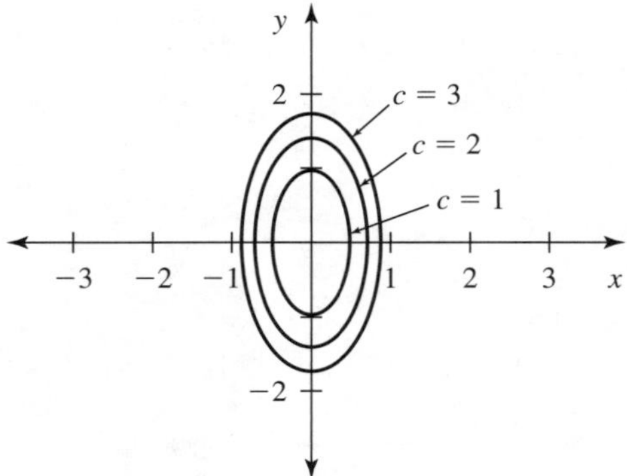

(b) For $z = f_4(x, y)$, we have $f_4(x, 0) = 4x^2$ on the intersection of the surface with the plane xOz, therefore the curve has equation $z = 4x^2$. Similarly, the surface intersects the plane yOz along the curve $z = y^2$.

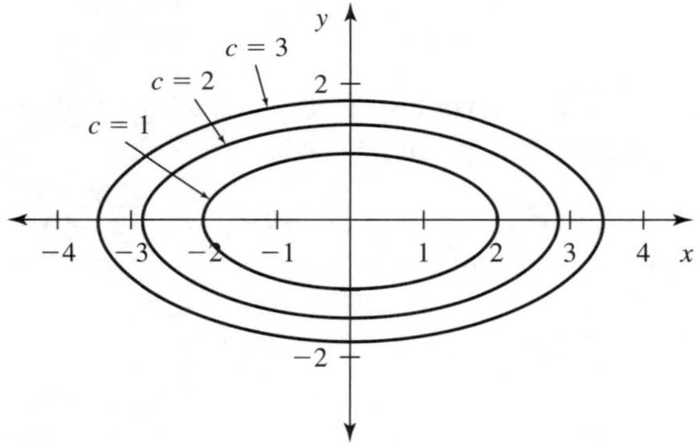

(c) For $z = f_{1/4}(x, y)$, we have $f_{1/4}(x, 0) = 0.25x^2$ on the intersection of the surface with the plane xOz, therefore the curve has equation $z = 0.25x^2$. Similarly, the surface intersects the plane yOz along the curve $z = y^2$.

Prob. 17. $23, 18, 15$.

10.2 Limits and Continuity

Prob. 1. 1.

Prob. 3. $-\frac{1}{2}$.

Prob. 5. $-\frac{3}{2}$.

Prob. 7. $\frac{2}{3}$.

Prob. 9. We have $\lim_{x\to 0^+} \frac{x^2-2\times 0^2}{x^2+0^2} = 1 \neq \lim_{y\to 0^+} \frac{0^2-2y^2}{0^2+y^2} = -2$.

Prob. 11. We have $\lim_{x\to 0} \frac{4x\times 0}{x^2+0^2} = 0 = \lim_{y\to 0} \frac{4\times 0\times y}{0^2+y^2}$, but $\lim_{x\to 0} \frac{4x^2}{2x^2} = 2$, so the limit does not exist.

Prob. 13. For $y = mx$ we get $\lim_{x\to 0} \frac{2mx^2}{x^3+mx^2} = 2$. For $y = x^2$ we get $\lim_{x\to 0} \frac{2x^3}{2x^3} = 1$, so the given limit does not exist.

Prob. 15. $(x-1)^2 + (y+1)^2 < 4$.

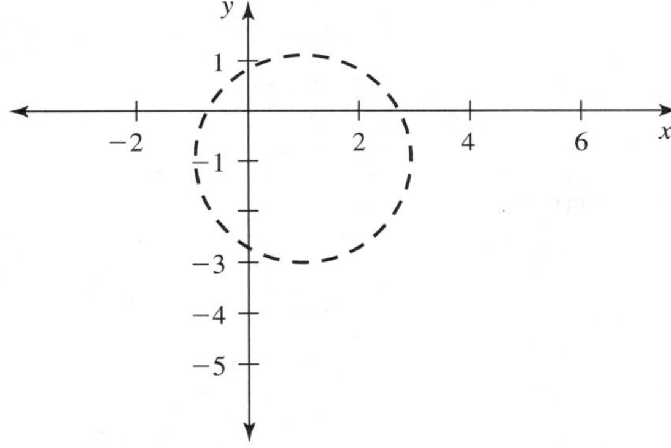

Prob. 17. Open disk centered at $(0, 2)$ with radius 3.

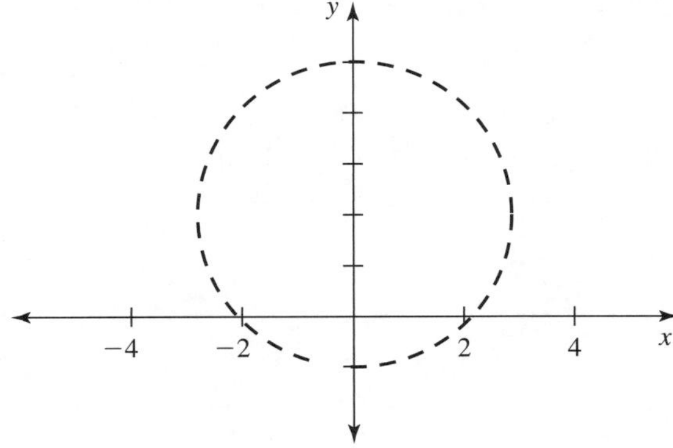

Prob. 19. As $2x^2+y^2 \le 2(x^2+y^2)$, given $\varepsilon > 0$ we can take $\delta^2 = \varepsilon/2$. We have $0 < \sqrt{x^2+y^2} < \delta \Rightarrow x^2+y^2 < \varepsilon/2 \Rightarrow 2(x^2+y^2) < \varepsilon \Rightarrow |f(x,y)| < \varepsilon$.

Prob. 21. It is easy to see that $\lim_{(x,y)\to(0,0)}(x^2+y^2)$ exists and is equal to 0. We have also $f(0,0)=0$, therefore f is continuous at $(0,0)$.

Prob. 23. By Problem 11 f has no limit at $(0,0)$.

Prob. 25. By Problem 13 f has no limit at $(0,0)$.

Prob. 27.

(a) $h = f \circ g$ where $g(x,y) = x^2+y^2$ and $f(z) = \sin z$.

(b) All $\mathbb{R}^2$.

Prob. 29.

(a) $h = f \circ g$ where $g(x,y) = xy$ and $f(z) = e^z$.

(b) All $\mathbb{R}^2$.

10.3 Partial Derivatives

Prob. 1. $\frac{\partial f}{\partial x} = 2xy + y^2$; $\frac{\partial f}{\partial y} = x^2 + 2xy$.

Prob. 3. $\frac{\partial f}{\partial x} = \frac{3\sqrt{xy^3}}{2} - \frac{2\sqrt[3]{y^2}}{3\sqrt[3]{x}}$;
$\frac{\partial f}{\partial y} = \frac{3\sqrt{yx^3}}{2} - \frac{2\sqrt[3]{x^2}}{3\sqrt[3]{y}}$.

Prob. 5. $\frac{\partial f}{\partial x} = \frac{\partial f}{\partial y} = \cos(x+y)$.

Prob. 7. $\frac{\partial f}{\partial x} = -4x\cos(x^2-2y)\sin(x^2-2y)$; $\frac{\partial f}{\partial y} = 4\cos(x^2-2y)\sin(x^2-2y)$.

Prob. 9. $\frac{\partial f}{\partial x} = \frac{\partial f}{\partial y} = \frac{e^{\sqrt{x+y}}}{2\sqrt{x+y}}$.

Prob. 11. $\frac{\partial f}{\partial x} = e^x\sin(xy) + e^x y\cos(xy)$; $\frac{\partial f}{\partial y} = e^x x\cos(xy)$.

Prob. 13. $\frac{\partial f}{\partial x} = \frac{2}{(2x+y)}$; $\frac{\partial f}{\partial y} = \frac{1}{2x+y}$.

Prob. 15. $\frac{\partial f}{\partial x} = \frac{-2x}{(y^2-x^2)\ln 3}$; $\frac{\partial f}{\partial y} = \frac{2y}{(y^2-x^2)\ln 3}$.

Prob. 17. 6.

Prob. 19. $3e^5$.

Prob. 21. $\frac{\partial f}{\partial z} = \frac{1}{xz}x$, so $f_z(e, 1) = 1$.

Prob. 23. $\frac{2}{9}$.

Prob. 25. $\frac{\partial f}{\partial x}(1, 1) = \frac{\partial f}{\partial y}(1, 1) = -2$. The fact that $\frac{\partial f}{\partial y}(1, 1) = -2$ means that the slope of the tangent to the graph of the function $y \mapsto f(1, y)$ at the point $y = 1$ is -2. The fact that $\frac{\partial f}{\partial x}(1, 1) = -2$ means that the slope of the tangent to the graph of the function $x \mapsto f(x, 1)$ at the point $x = 1$ is -2.

Prob. 27. $\frac{\partial f}{\partial x}(-2, 1) = -4$; $\frac{\partial f}{\partial y}(-2, 1) = 4$. The fact that $\frac{\partial f}{\partial x}(-2, 1) = -4$ means that the slope of the tangent to the graph of the function $x \mapsto f(x, 1)$ at the point $x = -2$ is -4. The fact that $\frac{\partial f}{\partial y}(-2, 1) = 4$ means that the slope of the tangent to the graph of the function $y \mapsto f(-2, y)$ at the point $y = 1$ is 4.

Prob. 29.

(a) $\frac{\partial P_e}{\partial a} = \frac{NT}{(1+aT_h N)^2} > 0$, therefore P_e increases when a increases.

(b) $\frac{\partial P_e}{\partial T} = \frac{aN}{1+T_h N} > 0$, therefore P_e increases when N increases.

Prob. 31. $\frac{\partial f}{\partial x} = 2xz - y$; $\frac{\partial f}{\partial y} = z^2 - x$; $\frac{\partial f}{\partial z} = x^2 + 2yz$.

Prob. 33. $\frac{\partial f}{\partial x} = 3x^2y^2z + \frac{1}{yz}$; $\frac{\partial f}{\partial y} = 2x^3yz - \frac{x}{y^2z}$; $\frac{\partial f}{\partial z} = x^3y^2 - \frac{x}{yz^2}$.

Prob. 35. $\frac{\partial f}{\partial x} = e^{x+y+z}$; $\frac{\partial f}{\partial y} = e^{x+y+z}$; $\frac{\partial f}{\partial z} = e^{x+y+z}$.

Prob. 37. $\frac{\partial f}{\partial x} = \frac{1}{x+y+z}$; $\frac{\partial f}{\partial y} = \frac{1}{x+y+z}$; $\frac{\partial f}{\partial z} = \frac{1}{x+y+z}$.

Prob. 39. $\frac{\partial^2 f}{\partial x^2} = 2y$.

Prob. 41. $\frac{\partial^2 f}{\partial x \partial y} = e^y$.

Prob. 43. $\frac{\partial^2 f}{\partial u^2} = 2\sec^2(u + w)\tan(u + w)$.

Prob. 45. $\frac{\partial^3 f}{\partial x^2 \partial y} = -6x \sin y$.

Prob. 47. $\frac{\partial^3 f}{\partial x^3} = \frac{2}{(x+y)^3}$.

Prob. 49.

(a) We have $\frac{\partial f}{\partial N} = \frac{2b^2NT+cb^2N^2T}{(1+cN+bT_hN^2)^2} > 0$, therefore an increase in N brings an increase in f.

(b) We have $\frac{\partial f}{\partial T} = \frac{b^2N^2}{1+cN+bT_hN^2} > 0$, therefore an increase in T brings an increase in f.

(c) We have $\frac{\partial f}{\partial T_h} = \frac{-bN^2}{(1+cN+bT_hN^2)^2} < 0$, therefore when T_h increases f decreases.

(d)

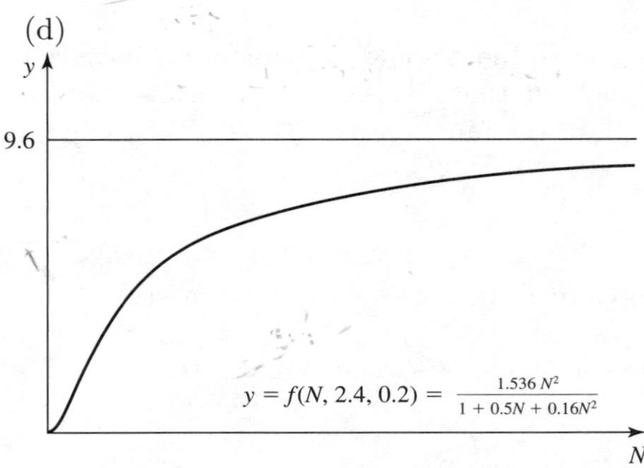

Prob. 51.

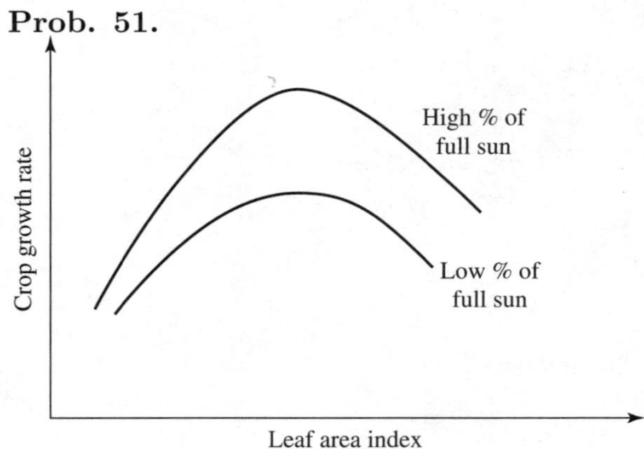

10.4 Tangent Planes, Differentiability, and Linearization

Prob. 1. $z = 6x + 4y - 8$.

Prob. 3. $z = 2ex - e$.

Prob. 5. $z = y$.

Prob. 7. $\frac{\partial f}{\partial x} = y^2 + 2xy$, $\frac{\partial f}{\partial y} = 2xy + x^2$, as these are polynomials they are continuous everywhere, therefore f is differentiable at any point.

Prob. 9. $\frac{\partial f}{\partial x} = -\sin(x+y) = \frac{\partial f}{\partial y}$, as this function is continuous everywhere, f is differentiable at any point.

Prob. 11. $\frac{\partial f}{\partial x} = 1 - 2y$, $\frac{\partial f}{\partial y} = 2y - 2x$, as these are polynomials they are continuous everywhere, therefore f is differentiable at any point.

Prob. 13. $L(x, y) = \frac{1}{2}x + 2y + \frac{1}{2}$.

Prob. 15. $L(x, y) = x + y$.

Prob. 17. $L(x, y) = x + \frac{1}{2}y + \ln 2 - \frac{3}{2}$.

Prob. 19. $L(x, y) = 1 + x + y$, $L(0.1, 0.05) = 1.15$, $f(0.1, 0.05) = 1.1618$.

Prob. 21. $L(x, y) = 2x - 3y - 2$, $L(1.1, 0.1) = -0.1$, $f(1.1, 0.1) = -0.0943$.

Prob. 23. $\begin{bmatrix} 1 & 1 \\ 2x^{-}2y & \end{bmatrix}$.

Prob. 25. $\begin{bmatrix} e^{x-y} & -e^{x-y} \\ e^{x+y} & e^{x+y} \end{bmatrix}$.

Prob. 27. $\begin{bmatrix} -\sin(x-y)\sin(x-y) \\ -\sin(x+y) - \sin(x+y \end{bmatrix}$.

Prob. 29. $\begin{bmatrix} 4xy + 1\ 2x^2 - 3 \\ e^x \sin\ y\ e^x \cos\ y \end{bmatrix}$.

Prob. 31. $L(x, y) = \begin{bmatrix} 4x + 2y - 4 \\ -x - y + 3 \end{bmatrix}$.

Prob. 33. $L(x, y) = \begin{bmatrix} e^{2x-y} \\ 2x - y - 1 \end{bmatrix}$.

Prob. 35. $L(x, y) = \begin{bmatrix} x - y + 1 \\ y - x + 1 \end{bmatrix}$.

Prob. 37. $L(x,y) = \begin{bmatrix} 1-y \\ 12y-13 \end{bmatrix}$. $L(1.1,1.9) = \begin{bmatrix} -0.9 \\ 9.8 \end{bmatrix}$, $f(1.1,1.9) = \begin{bmatrix} -0.88 \\ 9.83 \end{bmatrix}$.

Prob. 39. $L(x,y) = \begin{bmatrix} 10x-10y-25 \\ -24x+8y+48 \end{bmatrix}$, $L(1.9,-3.1) = \begin{bmatrix} 25 \\ -22.4 \end{bmatrix}$, $f(1.9,-3.1) = \begin{bmatrix} 25 \\ -22.382 \end{bmatrix}$.

10.5 More About Derivatives

Prob. 1. $18\ln 2 + 8$.

Prob. 3. $\frac{4\pi+3\sqrt{3}}{2\sqrt{4\pi^2+27}}$.

Prob. 5. 0.

Prob. 7. $\frac{\partial f}{\partial x}\frac{du}{dt} + \frac{\partial f}{\partial y}\frac{dv}{dt}$.

Prob. 9. $\frac{-2x}{x^2+2y+y^2}$.

Prob. 11. $\frac{3x^2y+3y^3-2x}{2y-3x^3-3xy^2}$.

Prob. 13. $\frac{-1}{\sqrt{1-x^2}}$, for $-1 \le x \le 1$.

Prob. 15. We have $\frac{\partial r}{\partial t} = \frac{\partial r}{\partial F}\frac{\partial F}{\partial t} + \frac{\partial r}{\partial N}\frac{\partial N}{\partial t} < 0$, therefore the growth rate is negative.

Prob. 17. $\begin{bmatrix} 3x^2y^2 \\ 2x^3y \end{bmatrix}$.

Prob. 19. $\frac{[\ \frac{3x^2-3y}{2\sqrt{x^3-3xy}}\ ]\ -3x}{2\sqrt{x^3-3xy}}$.

Prob. 21. $\frac{[\ \frac{xe^{\sqrt{x^2+y^2}}}{\sqrt{x^2+y^2}}\ ]\ ye^{\sqrt{x^2+y^2}}}{\sqrt{x^2+y^2}}$.

Prob. 23. $\begin{bmatrix} \frac{x^2-y^2}{x(x^2+y^2)} \\ \frac{y^2-x^2}{y(x^2+y^2)} \end{bmatrix}$.

Prob. 25. $\frac{2}{\sqrt{3}}$.

Prob. 27. $-\sqrt{2}$.

Prob. 29. $\frac{15}{\sqrt{10}}$.

Prob. 31. $\frac{13}{\sqrt{2}}$.

Prob. 33. $\frac{-1}{4\sqrt{29}}$.

Prob. 35. $\begin{bmatrix} 5 \\ -3 \end{bmatrix}$.

Prob. 37. $\begin{bmatrix} 5/4 \\ -3/4 \end{bmatrix}$.

Prob. 39. $\begin{bmatrix} 3/5 \\ 4/5 \end{bmatrix}$.

Prob. 41. $\frac{1}{\sqrt{733}}\begin{bmatrix} 2 \\ -27 \end{bmatrix}$.

Prob. 43. $\begin{bmatrix} -\frac{4}{5} \\ -\frac{4}{5} \end{bmatrix}$.

10.6 Applications

Prob. 1. Minimum at $(1, 0)$.

Prob. 3. Saddle at $(\pm 2, 4)$.

Prob. 5. Saddle at $(0, 3)$.

Prob. 7. Maximum at $(0, 0)$.

Prob. 9. Saddle at $(0, (2k+1)\frac{\pi}{2})$, $k \in \mathbb{Z}$.

Prob. 11.

(b) $0, 2$.

(c) f_1 has a minimum at $(0,0)$ but there is a direction along which f_1 is constant; f_2 has a saddle at $(0,0)$; f_3 has a minimum at $(0,0)$.

Prob. 13. The absolute maximum is at (1,-1) and $f(1,-1) = 3$. The absolute minimum is at (-1,1) and $f(-1,1) = -3$.

Prob. 15. The absolute maxima are at (1,0) and (-1,0) and $f(1,0) = f(-1,0) = 1$. The absolute minima are at (0,1) and (0,-1) and $f(0,1) = f(0,-1) = -1$.

Prob. 17. We find that

$$\nabla f(x,y) = \begin{bmatrix} 2x-1 \\ 2y+2 \end{bmatrix}$$

Absolute maxima occur on the boundary at (0,1), (1,0), (1,-2), (0,-2) where the value of f is equal to 0. The absolute minimum is at $(1/2,-1)$ where the value of f is $-5/4$.

Prob. 19. Absolute maximum at $(2/3, 2/3)$; absolute minima occur at all points along the boundary of the domain.

Prob. 21. Absolute minimum at $(-2,0)$ and absolute maximum at $(3,0)$.

Prob. 23. Absolute minimum at $(-1/2, 1/2)$ and absolute maximum at $(1/\sqrt{2}, -1/\sqrt{2})$.

Prob. 25. Yes.

Prob. 27. $(1,1)$.

Prob. 29. The volume is $V = xyz$ and the surface is $S = 2xy+2xz+2yz = 48m^2$. We can substitute

$$z = \frac{24-xy}{x+y}$$

Then,

$$V = xy\frac{24-xy}{x+y} \quad \text{with} \quad x,y>0$$

We find

$$\frac{\partial V}{\partial x} = \frac{(24y-2xy^2)(x+y)-24xy+x^2y^2}{(x+y)^2}$$

and

$$\frac{\partial V}{\partial y} = \frac{(24x-2yx^2)(x+y)-24xy+x^2y^2}{(x+y)^2}$$

Setting the partial derivatives equal to 0 yields $x = y = 0$ and $x = y = 2\sqrt{2}$. Since x and y are both positive, only the second solution is in the domain of the function. We find $z = 2\sqrt{2}$ and $V = (2/\sqrt{2})^3$, which is the maximum volume since the volume tends to 0 when any of the three sides tend to 0.

Prob. 31. Since $V = xyz = 216m^3$, we find for the surface area

$$S(x, y) = 2xy + \frac{432}{y} + \frac{432}{x} \quad \text{with} \quad x, y > 0$$

Hence,

$$\nabla S(x, y) = \begin{bmatrix} 2y - 432/x^2 \\ 2x - 432/y^2 \end{bmatrix}$$

Setting the gradient equal to 0, we obtain $x = y = 6$ and hence $z = 6$. This is where the function attains a minimum since it goes to infinity for x or y close to 0. The minimum surface area is therefore $216m^2$.

Prob. 33. The square of the distance between the plane and the origin is

$$f(x, y) = x^2 + y^2 + (1 - x - y)^2$$

where x and y are both real numbers. We find

$$\nabla f(x, y) = \begin{bmatrix} 2x + 2(1 - x - y)(-1) \\ 2y + 2(1 - x - y)(-1) \end{bmatrix}$$

Setting the gradient equal to 0, we find $x = y = 1/3$. Since $f(x, y)$ can take on arbitrarily large values, this is where the minimum occurs. Hence the minimum distance is $1/\sqrt{3}$.

Prob. 35.

(a) We have $H = p_1 \ln p_1 + p_2 \ln p_2 + p_3 \ln p_3 = p_1 \ln p_1 + p_2 \ln p_2 + (1 - p_1 - p_2) \ln(1 - p_1 - p_2)$. As the function $\ln x$ is only defined for $x > 0$, we must have $p_1 > 0$, $p_2 > 0$, $1 > p_1 + p_2$.

(b) H is differentiable on its domain. The limit on the boundary of the triangular domain is 0. H is positive, so it attains a maximum in the interior of the triangle. The only critical point is $(p_1, p_2) = (\frac{1}{3}, \frac{1}{3})$.

Prob. 37. We need to solve $6x = 2x\lambda$, $1 = 2y\lambda$, and $x^2 + y^2 = 1$. The maxima are at $(\pm\sqrt{35}/6, 1/6)$ and hence the maximum of f is $37/12$. The minimum is at (0, -1) and the minimum of f is -1.

Prob. 39. We need to solve $y = 2\lambda$, $x = -4\lambda$, and $2x - 4y = 1$. The minimum is at $(1/4, -1/8)$ and the minimum of f is $-1/32$.

Prob. 41. We need to solve $2x = 3\lambda$, $2y = -2\lambda$, and $3x - 2y = 4$. The minimum is at $(12/13, -8/13)$ and the minimum of f is $208/169$.

Prob. 43. We need to solve $2xy = 2x\lambda$, $x^2 = 3\lambda$, and $x^2 + 3y = 1$. A local minimum is at $(0, 1/3)$ where the value of f is 0. The absolute maxima are at $(\pm\sqrt{1/2}, 1/6)$ where the value of f is $1/12$.

Prob. 45. We need to solve $2xy^2 = 2x\lambda$, $2yx^2 = -2y\lambda$, and $x^2 - y^2 = 1$. The minimum is at $(\pm 1, 0)$ and the minimum of f is 0.

Prob. 47. The perimeter is $P = 2x + 2y$. The area is $A = xy$. We thus have to maximize $f(x, y) = xy$, $0 \leq x, y \leq P/2$, under the constraing $g(x, y) = 2x + 2y - P = 0$. We have

$$\nabla f(x, y) = \begin{bmatrix} y \\ x \end{bmatrix} \quad \text{and} \quad \nabla g(x, y) = \begin{bmatrix} 2 \\ 2 \end{bmatrix}.$$

We thus need to solve: $y = 2\lambda$, $x = 2\lambda$, $2x + 2y = P$. This yields $x = y = P/4$. Checking the boundary of the domain, either x or y is equal to 0; the area is then equal to 0. Thus among all rectangles with a given perimeter, the one with equal sides (i.e., the square) has the largest area.

Prob. 49. We need to minimize $f(x, y) = 3x + 2y$, $x, y \geq 0$, under the constraint $g(x, y) = xy - 384 = 0$. This yields $3 = \lambda y$, $2 = \lambda x$, $xy = 384$. Solving this yields $x = 16$, $y = 24$. Since the function f tends to infinity along the boundary of the domain because of the constraint, the candidate for the extremum is a minimum. The amount of fence that is required is 96 ft.

Prob. 51. We need to maximize $f(x, y) = xy/2$, $0 \leq x \leq 4$, $y = \sqrt{16 - x^2}$, under the constraint $g(x, y) = x^2 + y^2 - 16 = 0$. This yields $y/2 = 2x\lambda$, $x/2 = 2y\lambda$, $x^2 + y^2 = 16$. Solving this yields $x = y = 2\sqrt{2}$. Since the function f is equal to 0 along the boundary of the domain, the candidate for the extremum is a maximum. The largest possible area is thus 4.

Prob. 53. This is Problem 9 of Section 5.4.

Prob. 55. We need to minimize $f(r, \theta) = r\theta + 2r$, $0 \leq \theta \leq 2\pi$, $r^2\theta/2 = A$, under the constraint $g(r, k\theta) = r^2\theta/2 - A = 0$. This yields $\theta + 2 = r\theta\lambda$, $r = r^2\lambda/2$, $r^2\theta/2 = A$. Solving this yields $\theta = 2$, $r = \sqrt{A}$. The area is thus equal to $r\sqrt{A}$. When $\theta = 0$, then the perimeter is infinite. When $\theta = 2\pi$, it follows that $r = \sqrt{A/\pi}$, which yields $P = 4\sqrt{A\pi}$. Hence the smallest perimeter for a given area is obtained with $\theta = 2$, $r = \sqrt{A}$.

Prob. 57.

(a) We need to optimize $f(x, y) = x + y$, $x, y \neq 0$, under the constraint $g(x, y) = 1/x + 1/y - 1 = 0$. We need to solve $x^2 = -\lambda$, $y^2 = -\lambda 1/x + 1/y = 1$. This yields $x = y = 2$, and $f(2, 2) = 4$.

(b) There are no global extrema since the function f can take on arbitrarily large values under the constraint.

Prob. 61.

(a) With $R = 10$, $n = 3$, the constraint function is $3x_1 + 3x_2 = 10$.

(b) The fitness function is given by

$$f(x, y) = \frac{3x_1}{t + x_1} + \frac{3x_2}{10 + 2x_2}$$

We have

$$\nabla f(x_1, x_2) = \begin{bmatrix} 15/(5 + x_2)^2 \\ 30/(10 + 2x_2)^2 \end{bmatrix} \quad \text{and} \quad \nabla g(x_1, x_2) = \begin{bmatrix} 3 \\ 3 \end{bmatrix}$$

Hence, we need to solve

$$\frac{15}{(5 + x_1)^2} = 3\lambda, \quad \frac{30}{(10 + 2x_2)^2} = 3\lambda, \quad 3x_1 + 3x_2 = 10$$

This yields

$$(x_1, x_2) = \left(\frac{65 - 40\sqrt{2}}{3}, \frac{40\sqrt{2} - 55}{3}\right)$$

Other candidates are the boundary points $(x_1, x_2) = (10/3, 0)$ and $(x_1, x_2) = (0, 10/3)$. The absolute maximum is obtained for $(x_1, x_2) = \left(\frac{65-40\sqrt{2}}{3}, \frac{40\sqrt{2}-55}{3}\right)$.

Prob. 65.

(a) ii) We have $\frac{\partial c}{\partial x} = \frac{-x}{2Dt}c(x,t)$, therefore c increases when x is negative and decreases when x is positive.

iv) We have $\frac{\partial^2 c}{\partial x^2} = \frac{\exp\left[-\frac{x^2}{4Dt}\right]}{2Dt\sqrt{4\pi Dt}}\left[\frac{x^2}{2Dt} - 1\right]$, therefore $\frac{\partial^2 c}{\partial x^2} = 0$ when $|x| = \sqrt{2Dt}$, $\frac{\partial^2 c}{\partial x^2} > 0$ when $|x| > \sqrt{2Dt}$ and $\frac{\partial^2 c}{\partial x^2} < 0$ when $|x| < \sqrt{2Dt}$.

(b)

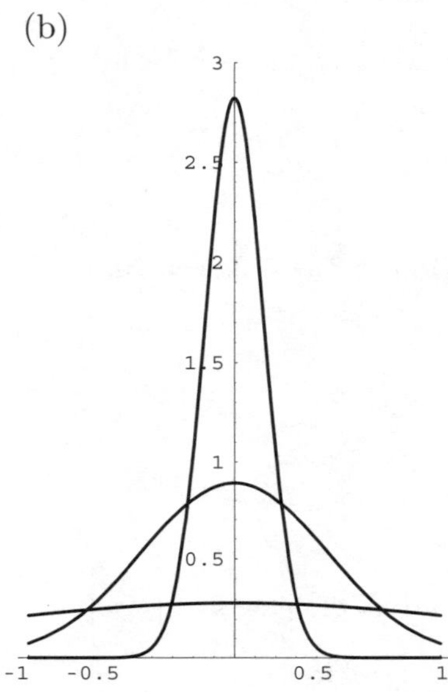

10.7 Systems of Difference Equations

Prob. 1. Refer to the table below:

t	N_t	P_t
0	5	0
1	7.5	0
2	11.25	0
3	16.87	0
4	25.31	0
5	37.96	0
6	56.95	0
7	85.42	0
8	128.14	0
9	192.21	0
10	288.32	0

Prob. 3. For any t we have $P_{t+1} = cN_t \times 0 = 0$, since $1 - \exp 0 = 0$. We have also $N_{t+1} = b^{t+1}N_0$.

Prob. 5. Refer to the table below:

t	N_t	P_t
0	5	5
1	6.78	1.42
2	9.89	0.57
3	14.67	0.33
4	21.85	0.29
5	32.59	0.38
6	48.50	0.75
7	71.67	2.18
8	102.91	9.18
9	128.47	51.81
10	68.36	248.67
11	0.70	203.69
12	0.01	2.09
13	0.02	0.002
14	0.03	3.48×10^{-6}
15	0.05	8.15×10^{-9}

Prob. 7. Refer to the table below:

t	N_t	P_t
0	5	0
1	7.5	0
2	11.25	0
3	16.87	0
4	25.31	0
5	37.96	0
6	56.95	0
7	85.42	0
8	128.14	0
9	192.21	0
10	288.32	0

Prob. 9. For any t we have $P_{t+1} = cN_t \times 0 = 0$. We have also $N_{t+1} = b^{t+1}N_0$.

Prob. 11. Refer to the table below:

t	N_t	P_t
0	100	50
1	79.45	141.09
2	36.96	164.42
3	15.68	79.52
4	10.02	27.01
5	10.00	10.05
6	12.56	4.89
7	17.18	3.31
8	24.19	3.17
9	34.14	4.28
10	47.21	7.98
11	61.27	19.10
12	67.49	48.83
13	54.16	94.14
14	31.68	99.14
15	18.01	59.00

Prob. 13.

(a)

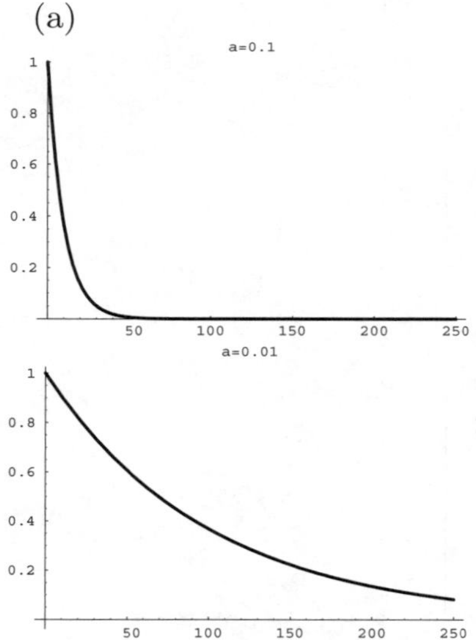

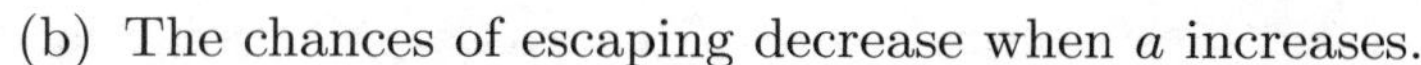
(b) The chances of escaping decrease when a increases.

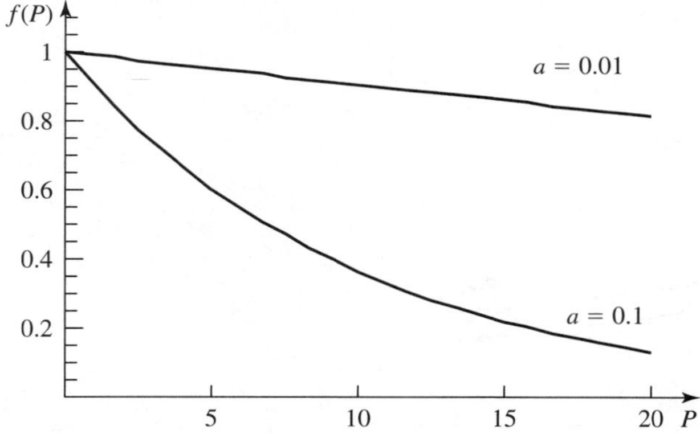

Prob. 15.

(a)

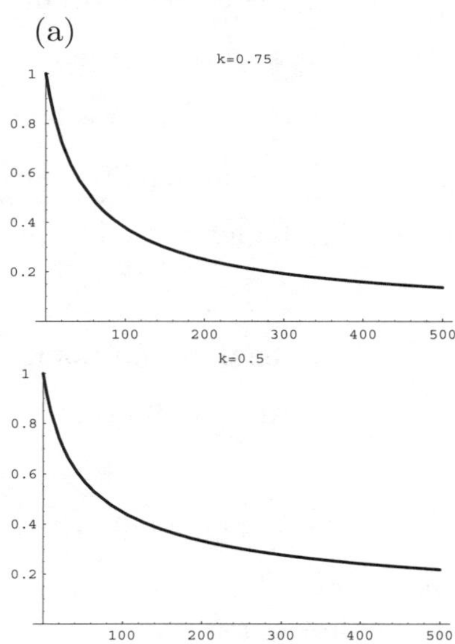

(b) The chances of escaping decrease when k increases. Maximum is 2, minimum is 0.

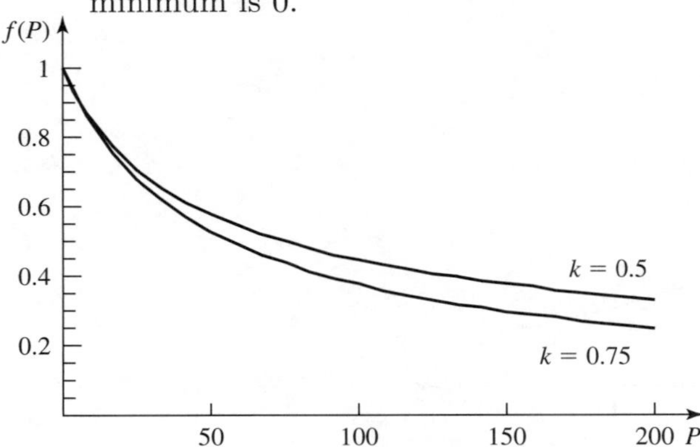

Prob. 17. We have $\begin{bmatrix} -0.7 & 0 \\ -0.3 & 0.2 \end{bmatrix}\begin{bmatrix} 0 \\ 0 \end{bmatrix} = \begin{bmatrix} 0 \\ 0 \end{bmatrix}$, so $\begin{bmatrix} 0 \\ 0 \end{bmatrix}$ is an equilibrium point. As the eigenvalues are -0.7 and 0.2, both with modulus less than 1, this equilibrium is stable.

Prob. 19. We have $\begin{bmatrix} -1.4 & 0 \\ -0.5 & 0.1 \end{bmatrix}\begin{bmatrix} 0 \\ 0 \end{bmatrix} = \begin{bmatrix} 0 \\ 0 \end{bmatrix}$, so $\begin{bmatrix} 0 \\ 0 \end{bmatrix}$ is an equilibrium point. As there is an eigenvalue, -1.4, with modulus larger than 1, this equilibrium is unstable.

Prob. 21. We have $\begin{bmatrix} 1 & 2 \\ 3 & 2 \end{bmatrix}\begin{bmatrix} 0 \\ 0 \end{bmatrix} = \begin{bmatrix} 0 \\ 0 \end{bmatrix}$, so $\begin{bmatrix} 0 \\ 0 \end{bmatrix}$ is an equilibrium point. As there is an eigenvalue, 4, with modulus larger than 1, this equilibrium is unstable.

Prob. 23. We have $\begin{bmatrix} -0.2 & -0.4 \\ 0.6 & 0.1 \end{bmatrix}\begin{bmatrix} 0 \\ 0 \end{bmatrix} = \begin{bmatrix} 0 \\ 0 \end{bmatrix}$, so $\begin{bmatrix} 0 \\ 0 \end{bmatrix}$ is an equilibrium point. As the eigenvalues are complex and the square of the modulus of each equals the determinant, which is 0.22, this equilibrium is stable.

Prob. 25. We have $\begin{bmatrix} 4.2 & -3.4 \\ 2.4 & -1.1 \end{bmatrix}\begin{bmatrix} 0 \\ 0 \end{bmatrix} = \begin{bmatrix} 0 \\ 0 \end{bmatrix}$, so $\begin{bmatrix} 0 \\ 0 \end{bmatrix}$ is an equilibrium point. As the eigenvalues are complex and the square of the modulus of each equals the determinant, which is 3.54, this equilibrium is unstable.

Prob. 27. The corresponding Jacobi matrix is $\begin{bmatrix} 0 & 0.25 \\ 2 & 0 \end{bmatrix}$, which has eigenvalues $\pm 1/\sqrt{2}$, both with modulus less than 1, so the equilibrium is stable.

Prob. 29. The corresponding Jacobi matrix is $\begin{bmatrix} 0 & 1 \\ -0.5 & 1 \end{bmatrix}$, which has complex eigenvalues. As its determinant is $0.5 < 1$ the equilibrium is stable.

Prob. 31. We have $\begin{bmatrix} 0 \\ 0 \end{bmatrix} = \begin{bmatrix} a \times 0 \\ 2 \times 0 - \cos 0 + 1 \end{bmatrix}$, so $\begin{bmatrix} 0 \\ 0 \end{bmatrix}$ is an equilibrium point. The Jacobi matrix for the corresponding linarization is $\begin{bmatrix} 0 & a \\ 2 & 0 \end{bmatrix}$ which has eigenvalues $\pm\sqrt{2a}$, so the equilibrium is stable when $a < \frac{1}{2}$.

Prob. 33. The nonnegative equilibrium is $\begin{bmatrix} 0 \\ 0 \end{bmatrix}$. The Jacobi matrix of the corresponding linearization is $\begin{bmatrix} 0 & 1 \\ \frac{1}{2} & \frac{2}{3} - 2y \end{bmatrix}$. One of the eigenvalues of the Jacobi matrix at $\begin{bmatrix} 0 \\ 0 \end{bmatrix}$ is 1.115, which has modulus larger than 1, so this equilibrium is unstable.

Prob. 35. The corresponding Jacobi matrix is $\begin{bmatrix} 0 & a \\ 1 & 0 \end{bmatrix}$. The corresponding eigenvalues are the solutions of the equation $\lambda^2 = a$, therefore the equilibrium is stable if $-1 < a < 1$.

Prob. 37.

(a) $\begin{bmatrix} r - \frac{1}{2} \\ r - \frac{1}{2} \end{bmatrix}$ is an equilibrium point, so if $r > 1/2$, then $(r - 1/2, r - 1/2)$ is an equilibrium.

(b) The Jacobi matrix of the corresponding linearization is $\begin{bmatrix} 0 & 1 \\ \frac{1}{2} & r - 2y \end{bmatrix}$.
At the equilibrium the eigenvalues are given by $\frac{1-r\pm\sqrt{r^2-2r+3}}{2}$
The quantity inside the square root $(r^2 - 2r + 3)$ is always positive. To see that observer that $r^2 - 2r + 3 = (r - 1)^2 + 1$, so both eigenvalues are real.

Forcing the modulus of the eigenvalues to be less than 1, we see that the numerator should be bound by:

$$-2 < 1 - r \pm \sqrt{r^2 - 2r + 3} < 2$$

Solving the inequality on the right we obtain $r < 1/2$, which was already known and the one on the left produces $r > 3/2$, so for values of $1/2 < r < 3/2$, the modulus of the eigenvalues are all less then 1 and the equilibrium is stable.

Prob. 39. There is a stable equilibrium at $\begin{bmatrix} \frac{\log 4}{0.1} \\ \frac{4 \log 4}{0.3} \end{bmatrix}$.

Prob. 41. There is a stable equilibrium at $\begin{bmatrix} 1000 \\ 750 \end{bmatrix}$.

10.9 Review Problems

Prob. 1. See graphs below:

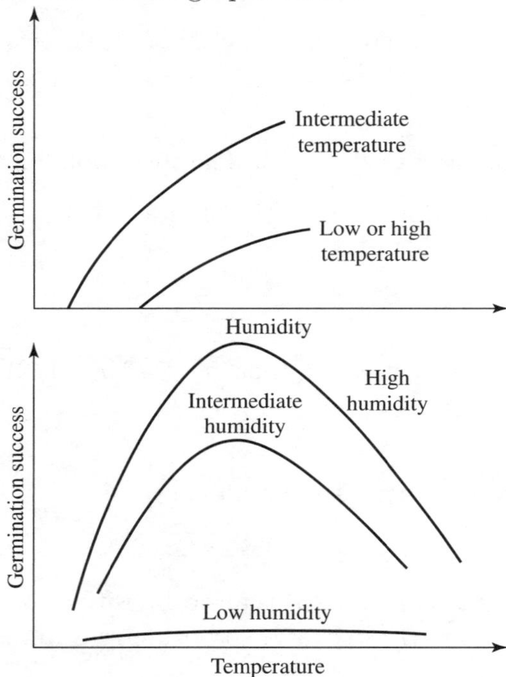

Prob. 3.

(a) $\frac{\partial A_i}{\partial F} > 0\,, \frac{\partial A_i}{\partial D} < 0.$

(b) The area covered by the introduced species increases with the amount of fertilizer used and also with the disturbance intensity.

(c) Both have areas that increase with the amount of fertilizer used, but the indigenous species area decreases when the disturbance intensity increases, while the introduced species area increases when the disturbance intensity increases.

Prob. 5. $\begin{bmatrix} 2x & -1 \\ 3x^2 & -2y \end{bmatrix}$.

11 Systems of Differential Equations

11.1 Linear Systems: Theory

Prob. 1. $\frac{dx}{dt} = \begin{bmatrix} 2 & 3 \\ -4 & 1 \end{bmatrix} x(t)$

Prob. 3. $\frac{dx}{dt} = \begin{bmatrix} 0 & 0 & 1 \\ -1 & 0 & 0 \\ 1 & 1 & 0 \end{bmatrix} x(t)$

Prob. 5. $(1,0) \to \frac{dx}{dt} = \begin{bmatrix} -1 \\ 1 \end{bmatrix}$

$(0,1) \to \frac{dx}{dt} = \begin{bmatrix} 2 \\ 0 \end{bmatrix}$

$(-1,0) \to \frac{dx}{dt} = \begin{bmatrix} 1 \\ -1 \end{bmatrix}$

$(0,-1) \to \frac{dx}{dt} = \begin{pmatrix} -2 \\ 0 \end{pmatrix}$

$(1,1) \to \frac{dx}{dt} = \begin{pmatrix} +1 \\ 1 \end{pmatrix}$

$(0,0) \to \frac{dx}{dt} = \begin{pmatrix} 0 \\ 0 \end{pmatrix}$

$(-2,1) \to \frac{dx}{dt} = \begin{pmatrix} 4 \\ -2 \end{pmatrix}$

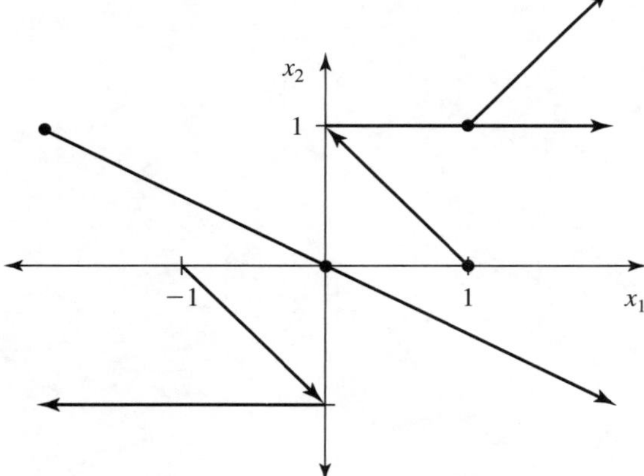

Prob. 7. $(1,0) \to \frac{dx}{dt} = \begin{pmatrix} 1 \\ -1 \end{pmatrix}$

$(0,1) \to \frac{dx}{dt} = \begin{pmatrix} 3 \\ 2 \end{pmatrix}$

$(-1,1) \to \frac{dx}{dt} = \begin{pmatrix} 2 \\ 3 \end{pmatrix}$

$(0,-1) \to \frac{dx}{dt} = \begin{pmatrix} -3 \\ -2 \end{pmatrix}$

$(-3,1) \to \frac{dx}{dt} = \begin{pmatrix} 0 \\ 5 \end{pmatrix}$

$(0,0) \to \frac{dx}{dt} = \begin{pmatrix} 0 \\ 0 \end{pmatrix}$

$(-2,1) \to \frac{dx}{dt} = \begin{pmatrix} 1 \\ 4 \end{pmatrix}$

Prob. 9.

(a) Fig. 11.18

(b) Fig. 11.19

(c) Fig. 11.20

(d) Fig. 11.21

Prob. 11.

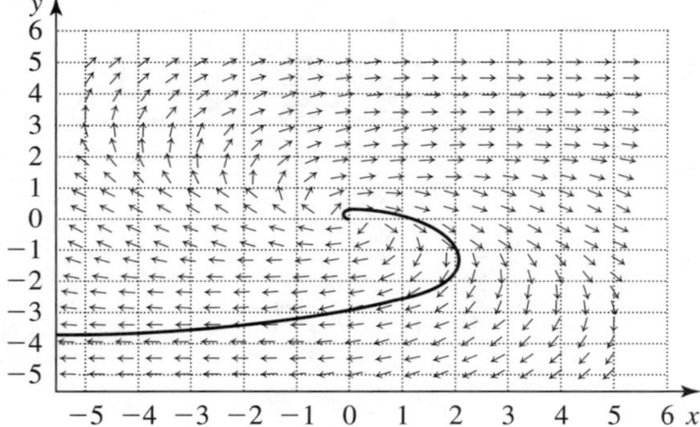

Prob. 13. $\frac{dy}{dt} = \begin{pmatrix} 1 & 3 \\ 5 & 3 \end{pmatrix} x(t)$

$A = \begin{pmatrix} 1 & 3 \\ 5 & 3 \end{pmatrix}$ and the characteristic equation is $\det(A - \lambda I) = 0$, or

$$\begin{vmatrix} 1-\lambda & 3 \\ 5 & 3-\lambda \end{vmatrix} = 0$$

$$(1-\lambda)(3-\lambda) - 15 = 0$$

$$\lambda^2 - 4\lambda - 12 = 0$$

and the eigenvalues are $\lambda = 6$ and $\lambda = -2$ Computating the eigenvector we have: $\lambda = G$; the system of equations reduce to

$$\begin{pmatrix} -5 & 3 \\ 5 & -3 \end{pmatrix} \begin{pmatrix} u_1 \\ u_2 \end{pmatrix} = \begin{pmatrix} 0 \\ 0 \end{pmatrix}$$

or $5u_1 = 3u_2$, and one eigenvalue is the vector $\begin{pmatrix} 3 \\ 5 \end{pmatrix}$.

$\lambda = -2$, the equation reduces to

$$\begin{pmatrix} 3 & 3 \\ 5 & 5 \end{pmatrix} \begin{pmatrix} u_1 \\ u_2 \end{pmatrix} = \begin{pmatrix} 0 \\ 0 \end{pmatrix}$$

and the eigenvalue will be $\begin{pmatrix} 1 \\ -1 \end{pmatrix}$, so the general solution is:

$$x(t) = c_1 e^{6t} \begin{pmatrix} 3 \\ 5 \end{pmatrix} + c_2 e^{-2t} \begin{pmatrix} 1 \\ -1 \end{pmatrix}$$

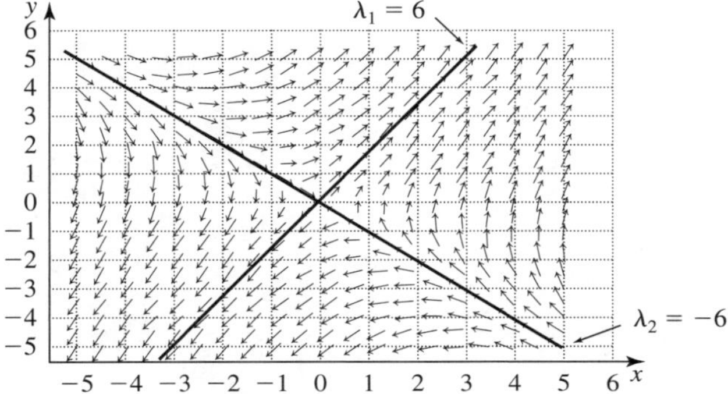

Prob. 15. $\frac{dx}{dt} = \begin{pmatrix} -3 & 3 \\ 6 & 4 \end{pmatrix} x(t)$

$A = \begin{pmatrix} -3 & 3 \\ 6 & 4 \end{pmatrix}$ and the characteristic equation is $\det(A - \lambda I) = 0$, or

$$\begin{vmatrix} -3-\lambda & 3 \\ 6 & 4-\lambda \end{vmatrix} = 0$$

$$(-3-\lambda)(4-\lambda) - 18 = 0$$

and the eigenvalues are $\lambda = 6$ and $\lambda = -5$. Computing the eigenvector we have

$\lambda = 6$;

$$\begin{pmatrix} -9 & 3 \\ 6 & -2 \end{pmatrix} \begin{pmatrix} u_1 \\ u_2 \end{pmatrix} = \begin{pmatrix} 0 \\ 0 \end{pmatrix}$$

or $6u_1 = 2u_2$ or $3u_1 = u_2$ and the eigenvector is $\begin{pmatrix} 1 \\ 3 \end{pmatrix}$.

Similarly for $\lambda = -5$

$$\begin{pmatrix} 2 & 3 \\ 6 & 9 \end{pmatrix} \begin{pmatrix} u_1 \\ u_2 \end{pmatrix} = \begin{pmatrix} 0 \\ 0 \end{pmatrix}$$

or $2u_1 + 3u_2 = 0$ and the eigenvector is $\begin{pmatrix} 3 \\ -2 \end{pmatrix}$ and the general solution is:

$$x(t) = c_1 e^{6t} \begin{pmatrix} 1 \\ 3 \end{pmatrix} + c_2 e^{-5t} \begin{pmatrix} 3 \\ -2 \end{pmatrix}$$

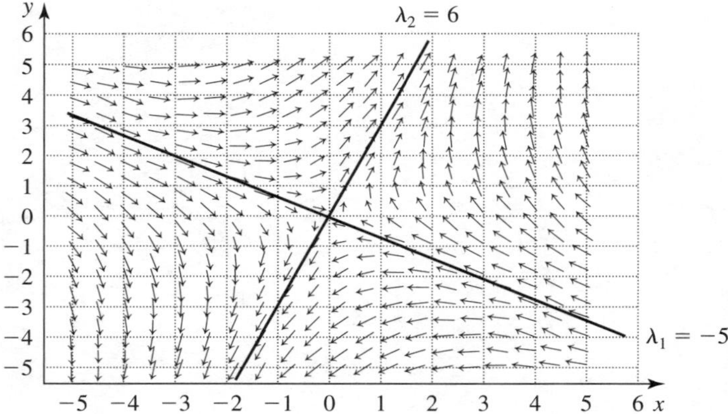

Prob. 17. $\frac{dx}{dt} = \begin{pmatrix} -2 & 0 \\ -3 & 1 \end{pmatrix} x(t)$

In this case $A = \begin{pmatrix} -2 & 0 \\ -3 & 1 \end{pmatrix}$, and the characteristic equation is

$$\begin{vmatrix} -2-\lambda & 0 \\ -3 & 1-\lambda \end{vmatrix} = 0$$

$$(-2-\lambda)(1-\lambda) = 0$$

and the characteristic values can be read straight out of the equation and are $\lambda = -2$ and $\lambda = 1$. Computing the eigenvector

$\lambda = -2$;

$$\begin{pmatrix} 0 & 0 \\ -3 & 3 \end{pmatrix} \begin{pmatrix} u_1 \\ u_2 \end{pmatrix} = \begin{pmatrix} 0 \\ 0 \end{pmatrix}$$

and $u_1 = u_2$, that is, the eigenvector is $\begin{pmatrix} 1 \\ 1 \end{pmatrix}$

$\lambda = 1$;

$$\begin{pmatrix} -3 & 0 \\ -3 & 0 \end{pmatrix} \begin{pmatrix} u_1 \\ u_2 \end{pmatrix} = \begin{pmatrix} 0 \\ 0 \end{pmatrix}$$

and $3u_1 = 0$, or $u_1 = 0$ and an eigenvector is $\begin{pmatrix} 0 \\ 1 \end{pmatrix}$, so the general solution is

$$x(t) = c_1 e^{-2t} \begin{pmatrix} 1 \\ 1 \end{pmatrix} + c_2 e^{t} \begin{pmatrix} 0 \\ 1 \end{pmatrix}$$

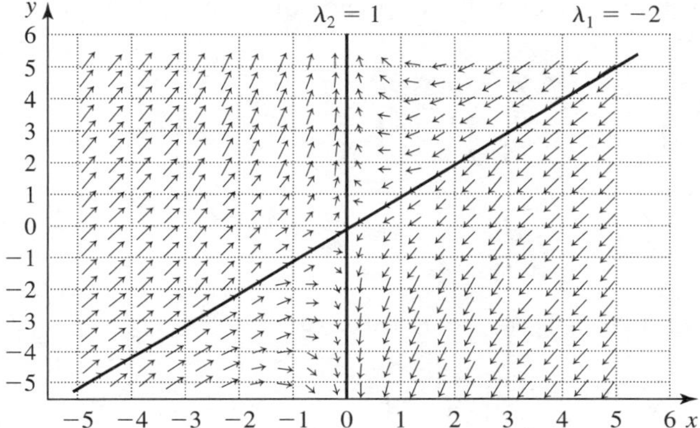

Prob. 19. $\frac{dx}{dt} = \begin{pmatrix} -3 & 0 \\ 4 & 2 \end{pmatrix} x(t)$

$A = \begin{pmatrix} -3 & 0 \\ 4 & 2 \end{pmatrix}$, and the characteristic equation is

$$\begin{vmatrix} -3-\lambda & 0 \\ 4 & 2-\lambda \end{vmatrix} = 0$$

or $(-3-\lambda)(2-\lambda) = 0$ which was roots or characteristic values $\lambda = -3$ and $\lambda = 2$. Computing the eigenvectors, we have:

$\lambda = -3$;

$$\begin{pmatrix} 0 & 0 \\ 4 & 5 \end{pmatrix} \begin{pmatrix} u_1 \\ u_2 \end{pmatrix} = \begin{pmatrix} 0 \\ 0 \end{pmatrix}$$

$4u_1 = -5u_2$, and the eigenvector is $\begin{pmatrix} 5 \\ -4 \end{pmatrix}$

$\lambda = 2$;

$$\begin{pmatrix} -5 & 0 \\ 4 & 0 \end{pmatrix} \begin{pmatrix} u_1 \\ u_2 \end{pmatrix} = \begin{pmatrix} 0 \\ 0 \end{pmatrix}$$

and $u_1 = 0$, in this case, and the eigenvector is $\begin{pmatrix} 0 \\ 1 \end{pmatrix}$. So the general solution is

$$x(t) = c_1 e^{-3t} \begin{pmatrix} 5 \\ -4 \end{pmatrix} + c_2 e^{2t} \begin{pmatrix} 0 \\ 1 \end{pmatrix}$$

Since $x_1(0) = -5$ and $x_2(0) = 5$. We end up with the equation, substituting for $t = 0$.

$$\begin{pmatrix} -5 \\ 5 \end{pmatrix} = c_1 \begin{pmatrix} 5 \\ -4 \end{pmatrix} + c_2 \begin{pmatrix} 0 \\ 1 \end{pmatrix}$$

or

$$\begin{cases} 5c_1 = -5 \\ -4c_1 + c_2 = 5 \end{cases}$$

whose solution is $c_1 = -1$ and $c_2 = 1$, so the solution is:

$$x(t) = -e^{-3t} \begin{pmatrix} 5 \\ -4 \end{pmatrix} + e^{2t} \begin{pmatrix} 0 \\ 1 \end{pmatrix}.$$

Prob. 21. $A = \begin{pmatrix} 3 & -2 \\ 0 & 1 \end{pmatrix}$, so the characteristic equation is:

$$\begin{vmatrix} 3-\lambda & -2 \\ 0 & 1-\lambda \end{vmatrix} = 0$$

or $(5 - \lambda)(6 - \lambda) = 0$ with eigenvectors $\lambda = 1$ and $\lambda = 3$. Computing the eigenvalues, we have:

$\lambda = 1$;

$$\begin{pmatrix} 2 & -2 \\ 0 & 0 \end{pmatrix} \begin{pmatrix} u_1 \\ u_2 \end{pmatrix} = \begin{pmatrix} 0 \\ 0 \end{pmatrix}$$

$2u_1 = u_2$, and the eigenvector is $\begin{pmatrix} 1 \\ 1 \end{pmatrix}$

$\lambda = 3$;

$$\begin{pmatrix} 0 & -2 \\ 0 & -2 \end{pmatrix} \begin{pmatrix} u_1 \\ u_2 \end{pmatrix} = \begin{pmatrix} 0 \\ 0 \end{pmatrix}$$

and $u_2 = 0$, giving us an eigenvector $\begin{pmatrix} 1 \\ 0 \end{pmatrix}$, so the general solution is

$$x(t) = c_1 e^t \begin{pmatrix} 1 \\ 1 \end{pmatrix} + c_2 e^{3t} \begin{pmatrix} 1 \\ 0 \end{pmatrix}.$$

with the initial conditions, we obtain the equations

$$\begin{pmatrix} 1 \\ 1 \end{pmatrix} = c_1 \begin{pmatrix} 1 \\ 1 \end{pmatrix} + c_2 \begin{pmatrix} 1 \\ 0 \end{pmatrix}$$

whose solution is $c_1 = 1$ and $c_2 = 0$, so the solution is

$$x(t) = e^t \begin{pmatrix} 1 \\ 1 \end{pmatrix}$$

or better

$$x_1(t) = e^t = x_2(t)$$

Prob. 23. $A = \begin{pmatrix} 4 & 7 \\ 1 & -2 \end{pmatrix}$, and the characteristic equation is:

$$\begin{vmatrix} 4-\lambda & 7 \\ 1 & -2-\lambda \end{vmatrix} = 0$$

$$(4-\lambda)(-2-\lambda) - 7 = 0$$

$$\lambda^2 - 2\lambda - 15 = 0$$

and the eigenvalues are $\lambda = -3$ and $\lambda = 5$, so computing the eigenvectors we have:

$\lambda = -3$;

$$\begin{pmatrix} 7 & 7 \\ 1 & 1 \end{pmatrix} \begin{pmatrix} u_1 \\ u_2 \end{pmatrix} = \begin{pmatrix} 0 \\ 0 \end{pmatrix}$$

or $u_2 = -u_1$, and the eigenvector is $\begin{pmatrix} 1 \\ -1 \end{pmatrix}$

$\lambda = 5$;

$$\begin{pmatrix} -1 & 7 \\ 1 & -7 \end{pmatrix} \begin{pmatrix} u_1 \\ u_2 \end{pmatrix} = \begin{pmatrix} 0 \\ 0 \end{pmatrix}$$

or $u_1 = 7u_2$ and the eigenvector is $\begin{pmatrix} 7 \\ 1 \end{pmatrix}$ and the general solution is

$$x(t) = c_1 e^{-3t} \begin{pmatrix} 1 \\ -1 \end{pmatrix} + c_2 e^{5t} \begin{pmatrix} 7 \\ 1 \end{pmatrix}.$$

Using the initial conditions we end up with

$$\begin{pmatrix} -1 \\ -2 \end{pmatrix} = c_1 \begin{pmatrix} 1 \\ -1 \end{pmatrix} + c_2 \begin{pmatrix} 7 \\ 1 \end{pmatrix}$$

or

$$\begin{cases} c_1 + 7c_2 = -1 \\ -c_1 + c_2 = -2 \end{cases}$$

whose solution is $c_1 = \frac{13}{8}$ and $c_2 = \frac{3}{8}$ so the solution is

$$x(t) = \frac{13}{8} e^{-3t} \begin{pmatrix} 1 \\ -1 \end{pmatrix} - \frac{3}{8} e^{5t} \begin{pmatrix} 7 \\ 1 \end{pmatrix}.$$

Prob. 25.

(a) $A = \begin{pmatrix} 1 & 0 \\ 0 & 1 \end{pmatrix}$, and a straight computation shows that

$$\begin{vmatrix} 1-\lambda & 0 \\ 0 & 1-\lambda \end{vmatrix} = 0$$

that is, $(1-\lambda)^2 = 0$ whose only root is $\lambda = 1$, so A has repested eigenvalues.

(b) Let's call the two vectors e_1 and e_2, respectively, then it is a trivial confirmation that

$$Ae_1 = e_1 \quad \text{and} \quad Ae_2 = e_2$$

so e_1 and e_2 are eigenvectors of $A = Id$.

Now any vector $\begin{pmatrix} c_1 \\ c_2 \end{pmatrix}$ can be written as $c_1 e_1 + c_2 e_2$, just use the two coordenates of the vector as coeficients of the decomposition

(c) Now let $x(t) = c_1 e^t \begin{pmatrix} 1 \\ 0 \end{pmatrix} + c_2 e^t \begin{pmatrix} 0 \\ 1 \end{pmatrix}$, by substituting $t = 0$ we can easily see that it satisfies the initial conditions $x_1(0) = c_1$ and $x_2(0) = c_2$, let's now differentiate it individually

$$x_1 = c_1 e^t$$

$$x_2 = c_2 e^t$$

we get

$$\frac{dx_1}{dt} = c_1e^t = x_1$$

$$\frac{dx_2}{dt} = c_2e^t = x_2,$$

so

$$\begin{pmatrix} \frac{dx_1}{dt} \\ \frac{dx_2}{dt} \end{pmatrix} = \begin{pmatrix} 1 & 0 \\ 0 & 1 \end{pmatrix} \begin{pmatrix} x_1(t) \\ x_2(t) \end{pmatrix}$$

satisfying the equation.

Prob. 27.

$$\det(A - \lambda I) = \begin{vmatrix} 1-\lambda & 0 \\ 1 & -2-\lambda \end{vmatrix} = (1-\lambda)(-2-\lambda) = 0,$$

so the eigenvalues are $\lambda = 1$ and $\lambda = -2$, a saddle.

Prob. 29.

$$\begin{aligned} \det(A - \lambda I) &= \begin{vmatrix} 6-\lambda & -4 \\ -3 & 5-\lambda \end{vmatrix} = (6-\lambda)(5-\lambda) - 12 \\ &= \lambda^2 - 11\lambda + 18 = 0 \end{aligned}$$

and the roots are 9 and 2, both positive, so a source.

Prob. 31.

$$\begin{aligned} \det(A - \lambda I) &= \begin{vmatrix} -3-\lambda & -1 \\ 1 & -6-\lambda \end{vmatrix} = (3+\lambda)(6+\lambda) + 1 \\ &= \lambda^2 + 9\lambda + 19 = 0 \end{aligned}$$

and the roots are $\frac{-9\pm\sqrt{5}}{2}$, both negative, so a sink.

Prob. 33.

$$\begin{aligned} \det(A - \lambda I) &= \begin{vmatrix} -\lambda & -2 \\ -1 & 3-\lambda \end{vmatrix} = (-\lambda)(3-\lambda) - 2 \\ &= \lambda^2 - 3\lambda - 2 \end{aligned}$$

and the roots are $\frac{3\pm\sqrt{17}}{2}$, one positive, one negative, a saddle.

Prob. 35.

$$\begin{aligned} \det(A - \lambda I) &= \begin{vmatrix} -2-\lambda & -3 \\ 1 & 3-\lambda \end{vmatrix} = (-2-\lambda)(3-\lambda) + 3 \\ &= \lambda^2 - \lambda - 3 = 0 \end{aligned}$$

and the roots are $\frac{1 \pm \sqrt{13}}{2}$, one positive, one negative, so a saddle.

Prob. 37.

$$\begin{aligned} \det(A - \lambda I) &= \begin{vmatrix} 1-\lambda & 3 \\ -2 & -2-\lambda \end{vmatrix} = (1-\lambda)(-2-\lambda) + 6 \\ &= \lambda^2 + \lambda + 4 \end{aligned}$$

and the roots are $\frac{-1 \pm \sqrt{-15}}{2}$, or $-\frac{1}{2} \pm \frac{\sqrt{15}}{2} i$ which have negative real part, so it is a stable spiral. Observe that all our equations are of the type

$$\lambda^2 + b\lambda + c = 0$$

with complex roots, so the real part will always be $-b/2$, and so the sign will always be the apposite sign of b, and *ih* the case $b = 0$ we have a center.

In fact, we can go further and observe that the term b in the second degree equation of these matrices is equal to minus the sum of the two diagonal elements. So, for these matrices that have imaginary roots, if the sum of two diagonal elements is positive it will be a unstable spiral, if the sum is zero a center, and if the sum is negative a stable spiral.

Prob. 39. Positive diagonal sum, by the Solution to Problem 37, an unstable spiral.

Prob. 41. Diagonal sum zero, by the Solution of Problem 37, a center.

Prob. 43. Diagonal sum is zero, so by the Solution to Problem 37, a center.

Prob. 45. Negative diagonal sum, so by the Solution to Problem 37, a stable spiral.

Prob. 47.

$$\begin{aligned} \det(A - \lambda I) &= \begin{vmatrix} -1-\lambda & -2 \\ 1 & 3-\lambda \end{vmatrix} = (\lambda - 3)(\lambda + 1) + 2 \\ &= \lambda^2 - 2\lambda - 1 = 0 \end{aligned}$$

and the roots are $\lambda_{1,2} = \frac{2 \pm \sqrt{4+t}}{2}$, so it is a saddle.

Prob. 49.

$$\begin{aligned} \det(A - \lambda I) &= \begin{vmatrix} -1-\lambda & -1 \\ 5 & -3-\lambda \end{vmatrix} = (\lambda+1)(\lambda+3)+5 \\ &= \lambda^2 + 4\lambda + 8 = 0 \end{aligned}$$

and the roots are $\lambda = \frac{-41 \pm \sqrt{-16}}{2} = -2 \pm 2i$, so a stable spiral.

Prob. 51.

$$\begin{aligned} \det(A - \lambda I) &= \begin{vmatrix} 1-\lambda & 3 \\ 2 & 3-\lambda \end{vmatrix} = (\lambda-1)(\lambda-3)-6 \\ &= \lambda^2 - 4\lambda - 3 = 0 \end{aligned}$$

and the roots are $\lambda_{112} = \frac{4 \pm \sqrt{16+12}}{2}$, so a saddle.

Prob. 53.

$$\begin{aligned} \det(A - \lambda I) &= \begin{vmatrix} -2-\lambda & 3 \\ 1 & -4-\lambda \end{vmatrix} = (\lambda+4)(\lambda+2)-3 \\ &= \lambda^2 + 6\lambda + 5 = 0 \end{aligned}$$

with roots -1 and -5, so a sink.

Prob. 55.

$$\begin{aligned} \det(A - \lambda I) &= \begin{vmatrix} 3-\lambda & -5 \\ 2 & -1-\lambda \end{vmatrix} = (\lambda-3)(\lambda+1)+10 \\ &= \lambda^2 - 2\lambda + 7 = 0 \end{aligned}$$

with roots $1 \pm \sqrt{-6}$, which are complex imaginary with positive real part, so an unstable spiral.

Prob. 57.

(a) $\det(A - \lambda I) = 0$ produces the equation

$$\begin{vmatrix} 4-\lambda & 8 \\ 1 & 2-\lambda \end{vmatrix} = 0$$

$$(\lambda-2)(\lambda-4) - 8 = 0$$

$$\lambda^2 - 6\lambda = 0$$

so the eigenvalues are $\lambda = 0$ and $\lambda = 6$

$\lambda = 0$;

$$\begin{pmatrix} 4 & 8 \\ 1 & 2 \end{pmatrix} \begin{pmatrix} u_1 \\ u_2 \end{pmatrix} = \begin{pmatrix} 0 \\ 0 \end{pmatrix}$$

which reduces to $u_1 + 2u_2 = 0$ and the associated eigenvector $\begin{pmatrix} -2 \\ 1 \end{pmatrix}$.

$\lambda = 6$;

$$\begin{pmatrix} -2 & 8 \\ 1 & -4 \end{pmatrix} \begin{pmatrix} u_1 \\ u_2 \end{pmatrix} = \begin{pmatrix} 0 \\ 0 \end{pmatrix}$$

which reduces to $u_1 = 4u_2$ and the associated eigenvector $\begin{pmatrix} 4 \\ 1 \end{pmatrix}$. The general solution is given by

$$x(t) = c_1 e^{0t} \begin{pmatrix} -2 \\ 1 \end{pmatrix} + c_2 e^{6t} \begin{pmatrix} 4 \\ 1 \end{pmatrix}.$$

or

$$x(t) = \begin{pmatrix} -2c_1 \\ c_1 \end{pmatrix} + c_2 e^{6t} \begin{pmatrix} 4 \\ 1 \end{pmatrix}.$$

and the individual solutions are:

$$\begin{aligned} x_1(t) &= 4c_2 e^{6t} - 2c_1 \\ x_2(t) &= c_2 e^{6t} + c_1 \end{aligned}$$

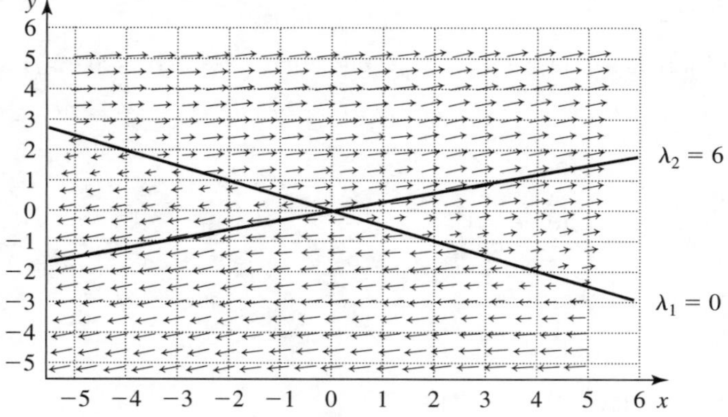

(c)

$$\begin{aligned}\frac{dx_1}{dt} &= 4c_2 6e^{6t}\\ \frac{dx_2}{dt} &= 6c_2 e^{6t}\end{aligned}$$

so $\frac{dx_2}{dx_1} = \frac{1}{4}$ which is the direction parallel to eigenvector associated with the non-zero eigenvalue.

Solutions diverge always from the direction of the eigenvector associate with the zero eigenvalue.

11.2 Linear Systems: Applications

Prob. 1. The system is

$$\begin{aligned}\frac{dx_1}{dt} &= -(0.5 + 0.05)x_1 + 0.1x_2\\ \frac{dx_2}{dt} &= 0.5x_1 - (0.1 + 0.02)x_2\end{aligned}$$

The determinant in this case is

$$\Delta = (0.12)(0.55) - (0.5)(0.1) > 0$$

which is strictly positive, then both eigenvalues are strictly negative and the equilibrium (0,0) is a stable node.

Prob. 3. The system is

$$\begin{aligned}\frac{dx_1}{dt} &= -(2.5)x_1 + 0.7x_2\\ \frac{dx_2}{dt} &= 2.5x_2 - (0.7 + 0.1)x_2\end{aligned}$$

The determinant is

$$\Delta = (2.5)(0.8) - (2.5)(0.7) > 0$$

which is strictly positive, then the eigenvalues are strictly negative and the equilibrium (0,0) is a stable node.

Prob. 5. The system is

$$\begin{aligned}\frac{dx_1}{dt} &= -0.1x_1 + 0x_2\\ \frac{dx_2}{dt} &= 0x_1 - 0.3x_2\end{aligned}$$

The determinant is $\Delta = (0.1)(0.3) = 0.3 > 0$ which is strictly positive, so the eigenvalues are strictly negative and the equilibrium is a stable node.

Prob. 7. Setting the system to find the values of a, b, c and d we have

$$\begin{aligned}a + c &= 0.4\\ b &= 0.3\\ a &= 0.1\\ b + d &= 0.5\end{aligned}$$

so $a = 0.1$, $b = 0.3$, $c = 0.3$ and $d = 0.2$.

Prob. 9. Set up the system to find the values of a, b, c and d yields $a = 0.3$, $b = 0$, $c = 0.9$, $d = 0.2$

Prob. 11. Set up the system to find the values of a, b, c and d yields $a = 0$, $b = 0$, $c = 0.2$, $d = 0.3$

Prob. 13. Since $\frac{dx_1}{dt} = -0.3x_1(t)$ we can integrate this equation to obtain

$$x_1(t) = x(0)e^{-0.3t}$$

for the expression for $x_1(t)$, or better

$$x_1(t) = 4e^{-0.3t}$$

since the amount in the drive will be the complement of what is in the tissue

$$\begin{aligned}x_2(t) &= 4 - x_1(t)\\ &= 4 - 4e^{-03t}\end{aligned}$$

Another Solution: We can also look at the problem as a system of two linear differential equations, the first one being given by

$$\frac{dx_1}{dt} = -0.3x_1(t)$$

and the second being exactly the same equation with the opposite sign; because the total amount of drug in the body is constant, so

$$\frac{dx_2}{dt} = 0.3x_1(t)$$

which gives us the system $\frac{dx}{dt} = Ax(t)$, where

$$A = \begin{pmatrix} -0.3 & 0 \\ 0.3 & 0 \end{pmatrix}.$$

Computing the eigenvalues we have

$$\det(A - \lambda I) = \begin{vmatrix} -0.3 - \lambda & 0 \\ 0.3 & -\lambda \end{vmatrix} = \lambda(\lambda + 0.3) = 0$$

which are $\lambda = 0$ and $\lambda = -0.3$, then proceeding to compute the eigenvectors, we have

$\lambda = 0$;

$$\begin{pmatrix} -0.3 & 0 \\ 0.3 & 0 \end{pmatrix} \begin{pmatrix} u_1 \\ u_2 \end{pmatrix} = \begin{pmatrix} 0 \\ 0 \end{pmatrix}$$

and it is easy to see that $\begin{pmatrix} 0 \\ 1 \end{pmatrix}$ is an eigenvector $\lambda = 0$;

$$\begin{pmatrix} 0 & 0 \\ 0.3 & +0.3 \end{pmatrix} \begin{pmatrix} u_1 \\ u_2 \end{pmatrix} = \begin{pmatrix} 0 \\ 0 \end{pmatrix}$$

which gives $\begin{pmatrix} 1 \\ -1 \end{pmatrix}$ is an eigenvector, so the general solution is

$$x(t) = c_1 \begin{pmatrix} 0 \\ 1 \end{pmatrix} + c_2 e^{-0.3t} \begin{pmatrix} 1 \\ -1 \end{pmatrix}$$

and if we use the initial conditions

$$\begin{cases} 4 = c_2 \\ 0 = c_1 - c_2 \end{cases}$$

so the solution to the system is $c_1 = 4$, $c_2 = 4$, and the final equations for $x_1(t)$ and $x_2(t)$ are:

$$x_1(t) = 4e^{-0.3t}$$

$$x_2(t) = 4 - 4e^{-0.3t}$$

Prob. 15. (a) The system that gives each component of the diagram is

$$\begin{aligned} a + c &= 0.2 \\ b &= 0.1 \\ a &= 0.2 \\ b + d &= 0.1 \end{aligned}$$

so separating each component we have $a = 0.2$, $b = 0.1$, $c = 0$ and $d = 0$.

(b)

$$\begin{aligned} \frac{d}{dt}(x_1 + x_2) &= \frac{dx_1}{dt} + \frac{dx_2}{dt} \\ &= -0.2x_1 + 0.1x_2 + 0.2x_1 - 0.1x_2 \\ &= 0 \end{aligned}$$

so the total quantity $x_1 + x_2$ is constant, $x_1 + x_2 = A$. A here denotes the total Area, the one occupied by adult trees plus the ones occupied by the gaps.

(c) The function $A(t) = x_1(t) + x_2(t)$, is constant, because the derivative is zero, so if $A(0) = 20$, so it is for all values of $t > 0$.

(d) Since $x_1(t) + x_2(t) = 20$, we can write

$$x_2(t) = 20 - x_1(t)$$

and substituting in the first equation of the system

$$\frac{dx_1(t)}{dt} = -0.2x_1 + 0.1(20 - x_1(t))$$

$$\frac{dx_1}{dt} = 2 - 0.3x_1$$

which is what we needed.

(e) To solve the system we first compute the eigenvalues of

$$A = \begin{pmatrix} -0.2 & 0.1 \\ 0.2 & -0.1 \end{pmatrix}$$

$$\begin{aligned} \det(A - \lambda I) &= \begin{vmatrix} -0.2 - \lambda & 0.1 \\ 0.2 & -0.1 - \lambda \end{vmatrix} = (\lambda + 0.1)(\lambda + 0.2) - 0.02 \\ &= \lambda^2 + 0.3\lambda = \lambda(\lambda + 0.3) = 0 \end{aligned}$$

so the eigenvalues are $\lambda = 0$ and $\lambda = -0.3$. Computing the eigenvectors we have

$\lambda = 0$;

$$\begin{pmatrix} -0.2 & 0.1 \\ 0.2 & -0.1 \end{pmatrix} \begin{pmatrix} u_1 \\ u_2 \end{pmatrix} = \begin{pmatrix} 0 \\ 0 \end{pmatrix}$$

which reduces to $u_2 = 2u_1$ and the eigenvector is $\begin{pmatrix} 1 \\ 2 \end{pmatrix}$ is an eigenvector

$\lambda = 0.3$;

$$\begin{pmatrix} 0.1 & 0.1 \\ 0.2 & 0.2 \end{pmatrix} \begin{pmatrix} u_1 \\ u_2 \end{pmatrix} = \begin{pmatrix} 0 \\ 0 \end{pmatrix}$$

which reduces to $u_2 = -u_1$ and the eigenvector is $\begin{pmatrix} 1 \\ -1 \end{pmatrix}$ so the general solution is

$$x(t) = c_1 \begin{pmatrix} 1 \\ 2 \end{pmatrix} + c_2 e^{-0.3t} \begin{pmatrix} 1 \\ -1 \end{pmatrix}$$

Using the initial conditions, we have

$$\begin{cases} 2 = c_1 + c_2 \\ 18 = 2c_1 - c_2 \end{cases}$$

and the solution is $c_1 = \frac{20}{3}$ and $c_2 = \frac{-14}{3}$, that is

$$x_1(t) = \frac{20}{3} - \frac{14}{3} e^{-0.3t}$$

$$x_2(t) = \frac{40}{3} + \frac{14}{3} e^{-0.3t}.$$

So the percentage of the forest occupied by adult trees is $x_2(t)/A$, that is

$$\left(\frac{2}{3} + \frac{7}{30} e^{-0.3t} \right) \times 100$$

or

$$\frac{200}{3} + \frac{70}{3} e^{-0.3t}$$

when $t \to \infty$, the second term of the sum will approach 0 and the amount covered by adult trees will approach $(200/3)\%$, or approximately 66.6%

Prob. 17. The general solution is

$$x(t) = c_1 \sin 2t + c_2 \cos 2t$$

using the initial condition $x(0) = 0$, we get

$$0 = x(0) = c_2 \cos 0 = c_2$$

which implies that $c_2 = 0$, reducing the solution to

$$x(t) = c_1 \sin 2t$$

derivating once and using the initial condition on the derivative, we get:

$$G = \frac{dx}{dt}(0) = c_1 (\cos 2t) 2|_{t=0}$$

implying that $c_1 = 3$ and the solution is

$$x(t) = 3 \sin 2t$$

Prob. 19. Let $y = \frac{dx}{dt}$, then $\frac{dy}{dt} = \frac{d^2x}{dt^2} = 3x$, so the system becomes

$$\begin{cases} \frac{dx}{dt} = y \\ \frac{dy}{dt} = 3x \end{cases}$$

Prob. 21. Let $y = \frac{dx}{dt}$, then $\frac{dy}{dt} = \frac{d^2x}{dt^2} = x - \frac{dx}{dt} = x - y$, and the system is:

$$\begin{cases} \frac{dx}{dt} = y \\ \frac{dy}{dt} = x - y \end{cases}$$

11.3 Nonlinear Autonomous Systems: Theory

Prob. 1. The Jacobian matrix will be given by

$$Df = \begin{pmatrix} 1 + x_2 & -2 + x_1 \\ -1 & 1 \end{pmatrix}$$

and at the origin it is

$$Df(0) = \begin{pmatrix} 1 & -2 \\ -1 & 1 \end{pmatrix}$$

computing the trace and determinant we have, $\tau = 2$ and $\Delta = -1$, so it is a saddle.

Prob. 3. The Jacobian matrix will be given by

$$Df = \begin{pmatrix} 1 + 2x_1 - 2x_2 & -2x_1 + 1 \\ 1 & 0 \end{pmatrix}$$

at the origin it is,

$$Df(0) = \begin{pmatrix} 1 & 1 \\ 1 & 0 \end{pmatrix}$$

computing the trace and determinant we have, $\tau = 1$ and $\Delta = -1$, so it is a saddle.

Prob. 5. First let's find the equilibrium points, which are given by the solutions to the system:

$$\begin{cases} -x_1 + 2x_1(1 - x_1) = 0 \\ -x_2 + 5x_2(1 - x_1 - x_2) = 0 \end{cases}$$

or rewriting it:

$$\begin{cases} x_1 = 2x_1(1 - x_1) \\ x_2 = 5x_2(1 - x_1 - x_2) \end{cases}$$

so one obvious solution is $x_1 = x_2 = 0$. To find the other ones let's assume $x_1 \neq 0$, then solving the first equation we get $1 = 2(1-x_1)$, that is, $x_1 = 1/2$. Substituting that on the second equation we have $x_2 = 5x_2(\frac{1}{2} - x_2)$, whose solutions are $x_2 = 0$, or $x_2 = \frac{3}{10}$, giving two more solutions.

Now let's assume that $x_1 = 0$, then the first equation is easily verified and the second one reduces to the second degree $x_2 = 5x_2(1 - x_2)$ whose solutions $x_2 = 0$ and $x_2 = 4/5$. So the equilibrium points are $(0, 0)$, $(1/2, 0)$, $(1/2, 3/10)$ and $0, 4/5)$. Now the Jacobian is given by the expression

$$Df = \begin{pmatrix} 1 - 4x_1 & 0 \\ -5x_2 & 4 - 5x_1 - 10x_2 \end{pmatrix}$$

and the determinant is $\Delta = (1 - 4x_1)(4 - 5x_1 - 10x_2)$ and the trace is $\tau = 5 - 9x_1 - 10x_2$, so analyzing it on each of the equilibrium points we have:

$(0, 0) \to \Delta = 4$, $\tau = 5 \to$ unstable node $(\tau^2 > 4\Delta)$
$(1/2, 0) \to \Delta = -3/2$, $\tau = 1/2 \to$ saddle $(\Delta < 0)$
$(1/2, 3/10) \to \Delta = 3/2$, $\tau = -5/2 \to$ stable node $(\tau^2 > 4\Delta)$
$(0, 4/5) \to \Delta = -4$, $\tau = -3 \to$ saddle $(\Delta < 0)$

Prob. 7. The equilibrium points are given by the equations:

$$\begin{cases} 4x_1(1 - x_1) - 2x_1x_2 = 0 \\ x_2(2 - x_2) - x_2 = 0 \end{cases}$$

or other

$$\begin{cases} 2x_1(1 - x_1) = x_1x_2 \\ x_2(2 - x_2) = x_2 \end{cases}$$

The second equation has degree 2 and the roots are $x_2 = 0$ or $x_2 = 1$. In the frist case if we substitute on the first equation we get

$$x_1(1 - x_1) = 0$$

which has solutions $x_1 = 0$ and $x_1 = 1$, generating two equilibrium points $(0,0)$ and $(0,1)$. Now if we take $x_2 = 1$ on the first equation we end up with

$$2x_1(1 - x_1) = x_1$$

which has the obvious solution $x_1 = 0$ which is already noted as one of the equilibrium points, and the other solution (simplyfing x_1) is

$$2(1 - x_1) = 1$$

or $x_1 = 1/2$, which gives the last equilibrium point as $(1/2, 1)$.

The Jacobian is given by the expression

$$Df = \begin{pmatrix} 4 - 8x_1 - 2x_2 & -2x_1 \\ 0 & 1 - 2x_2 \end{pmatrix}$$

and the determinant and trace are given by:

$$\begin{aligned} \Delta &= (4 - 8x_1 - 2x_1)(1 - 2x_2) \\ \tau &= 5 - 8x_1 - 4x_2 \end{aligned}$$

So at the equilibrium points we have:

$(0,0) \to \Delta = 4$, $\tau = 5 \to \tau^2 > 4\Delta \to$ unstable node
$(0,1) \to \Delta = -4$, $\tau = 1 \to \Delta < 0 \to$ saddle
$(1/2,1) \to \Delta = 2$, $\tau = -3 \to \tau^2 > 4\Delta \to$ stable node.

Prob. 9. The equilibrium points are given by the equations:

$$\begin{cases} x_1 - x_2 = 0 \\ x_1 x_2 - x_2 = 0. \end{cases}$$

So from the first equation we know that $x_1 = x_2$, and substituting on the second, we get

$$x_2^2 - x_2 = 0$$

or

$$x_2(x_2 - 1) = 0$$

with solutions $x_2 = 0$ and $x_2 = 1$, so the two equilibrium points are $(0, 0)$ and $(1, 1)$. Looking at the Jacobian at these points we have

$$Df = \begin{pmatrix} 1 & -1 \\ x_2 & x_1 - 1 \end{pmatrix}$$

so determinant is $\Delta = x_1 + x_2 - 1$ and the trace $\tau = x_1$, so at the equilibrium points

$(0, 0) \to \Delta = -1 \to$ saddle

$(1, 1) \to \Delta = 1$, $\tau = 1 \to \tau^2 < 4\Delta \to$ unstable spiral

Prob. 11.

(a) The isoclines are the curves

$$x_1(10 - 2x_1 - x_2) = 0$$

and

$$x_2(10 - x_1 - 2x_2) = 0$$

so solving each one of the equations we get

$$10x_1 - 2x_1^2 - x_1x_2 = 0$$

$$x_1x_2 = 10x_1 - 2x_1^2$$

so for $x_1 \neq 0$, we have $x_2 = 10 - 2x_1$, for the first curve and solving the second we have

$$10x_2 - x_1x_2 - 2x_2^2 = 0$$

$$x_1x_2 = 10x_2 - 2x_2^2$$

and in a similar way, for $x_2 \neq 0$

$$x_1 = 10 - 2x_2$$

so the two graphs are

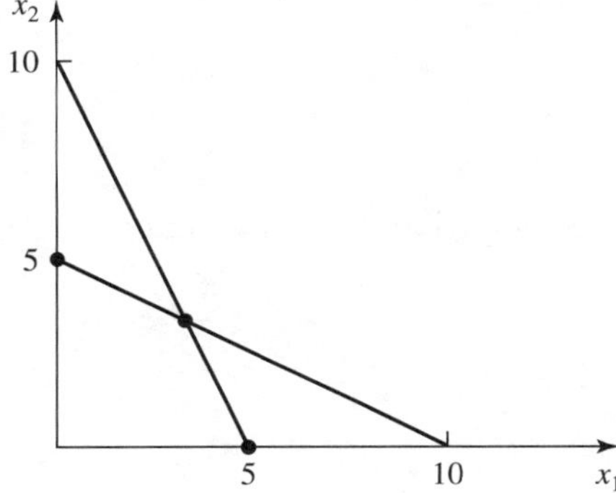

(b) Looking for the intersection point of two isoclines is the same as solving the system

$$\begin{cases} x_1 = 10 - 2x_2 \\ x_2 = 10 - 2x_1 \end{cases}$$

and substituting one equation into another we get

$$x_1 = 10 - 2(10 - x_1)$$

whose solution is $x_1 = 10/3$, and substituting back on the second equation we find $x_2 = 10/3$, so $\left(\frac{10}{3}, \frac{10}{3}\right)$ is an equilibrium.

The Jacobian of the system is given by

$$Df = \begin{pmatrix} 10 - 4x_1 - x_2 & -x_1 \\ -x_2 & 10 - x_1 - 4x_2 \end{pmatrix}$$

so at the equilibrium point

$$Df = \begin{pmatrix} 220/3 & -10/3 \\ -10/3 & -20/3 \end{pmatrix}$$

so $\Delta = 100/3$ and $\tau = -40/3$, since $\tau^2 > 4\Delta$, the equilibrium is a stable node.

Prob. 13. The matrix of the signs of elements of the Jacobian is

$$\begin{bmatrix} - & - \\ - & - \end{bmatrix}$$

so the determinant is unknown and the trace has negative sign, so the equilibrium can be one of three, a stable spiral, a stable node or a saddle.

Prob. 15. The matrix of signs of the Jacobian is

$$\begin{bmatrix} + & - \\ - & - \end{bmatrix}$$

so the trace is unknown and determinant has negative sign, so the equilibrium is a saddle.

Prob. 17. The matrix of signs of the Jacobian

$$\begin{bmatrix} - & 0 \\ - & - \end{bmatrix}$$

so determinant is positive and trace negative and the equilibrium is a stable node or a stable spiral.

Prob. 19.

(a) The first isocline will be given by the equation

$$x_1(2 - x_1) - x_1 x_2 = 0$$

so for $x_1 \neq 0$, it solves to $x_2 = 2 - x_1$.

The second isocline is given by

$$x_1 x_2 - x_2 = 0$$

and for $x_2 \neq 0$ it is the vertical line $x_1 = 1$, so the graph looks like:

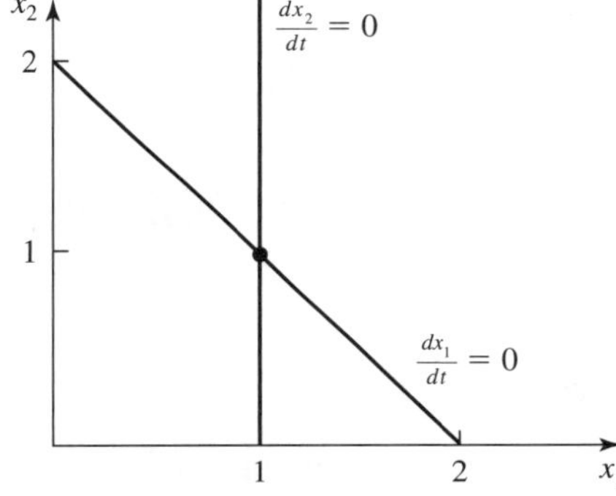

(b) The two lines intersect at $(1, 1)$ in the first quadrant, by trivial substitutive and this is the equilibrium point.

Looking at the signs of both functions on the side of each line we have and the matrix of signs of the Jacobian is

$$\begin{bmatrix} - & - \\ + & 0 \end{bmatrix}$$

so the determinant is positive and the trace negative, making it a stable node or stable spiral.

11.4 Nonlinear Systems: Applications

Prob. 1. $N_2 = 20$ and $\alpha_{12} N_2 = 4$ so $\alpha_{12} = 1/5$ and by the same token $N_1 = 30$ and $\alpha_{21} N_1 = 6$ so $\alpha_{21} = 1/5$, so the equation are

$$\frac{dN_1}{dt} = 2N_1 \left(1 - \frac{N_1}{20} - \frac{N_2}{100}\right)$$

$$\frac{dN_2}{dt} = 3N_2\left(1 - \frac{N_2}{15} - \frac{N_1}{75}\right)$$

Prob. 3. In this case $K_1 = 10$, $K_2 = 15$, $\alpha_{12} = 0.7$ and $\alpha_{21} = 0.3$, and we clearly see that

$$K_1 = 10 < 10.5 = \alpha_{12}K_2$$

and

$$K_2 = 15 > 3 = \alpha_{21}K_1$$

and species 2 excludes species 1.

Prob. 5. In this case $K_1 = 20$, $K_2 = 15$, $\alpha_{12} = 4$ and $\alpha_{21} = 5$, so we can clearly see that

$$K_1 = 20 < 60 = \alpha_{12}K_2$$

$$K_2 = 15 < 100 = \alpha_{21}K_1$$

and this is a case of founder control.

Prob. 7. The four equilibrium points will be $(0,0)$, $(K_1,0)$, $(0,K_2)$ and the positive solutions to the system of equations

$$\begin{cases} N_1 + 1.3N_2 = 18 \\ 0.6N_1 + N_2 = 20 \end{cases}$$

a simple substitution argument shows that the systems solution is negative, so we only have to analyze the other three equilibrium points.

The equilibrium at $(0,0)$ has eigenvalues $\lambda_1 = r_1 = 3$ and $\lambda_2 = r_2 = 2$, making it unstable.

The equilibrium at $(k_1,0) = (18,0)$ has eigenvalues $\lambda_1 = -r_1 = -3$ and

$$\lambda_2 = r_2\left(1 - \alpha_{21}\frac{k_1}{k_2}\right) = 2\left(1 - 0.6\frac{18}{20}\right) > 0$$

making it unstable.

The equilibrium at $(0,k_2) = (0,20)$ has eigenvalues $\lambda_2 = -r_1 = -2$ and

$$\lambda_1 = r_1\left(1 - \alpha_{12}\frac{k_2}{k_1}\right) = 3\left(1 - 1.3\frac{20}{18}\right) < 0,$$

so it is locally stable.

Prob. 9. The equilibrium points are $(0,0)$, $(k_1,0)$, $(0,k_2)$ and the positive solutions to the system of equations

$$\begin{cases} N_1 + 3N_2 = 35 \\ 4N_2 + N_1 = 40 \end{cases}$$

multiplying the second by 3 and subtracting out of the first equation we get

$$-11N_1 = -85$$

guaranteeing a positive solution for N_1; and with substitution back into the first equation we see the positive solution for N_2 as well.

The equilibrium at $(0,0)$ is always unstable with eigenvalues $\lambda_1 = r_1 = 1$ and $\lambda_2 = r_2 = 3$.

The equilibrium at $(k_1, 0) = (35, 0)$ has eigenvalues $\lambda_1 = -r_1 = -1$ and

$$\lambda_2 = r_2\left(1 - \alpha_{21}\frac{k_1}{k_2}\right) = 3\left(1 - 4.\frac{35}{40}\right) < 0$$

so the equilibrium is stable.

The equilibrium at $(0, K_2) = (0, 40)$ has eigenvalues $\lambda_2 = -r_2 = -3$ and

$$\lambda_1 = r_1\left(1 - \alpha_{12}\frac{k_2}{k_1}\right) = 1\left(1 - 3.\frac{40}{35}\right) < 0$$

so the equilibrium is stable.

Now, for the non-trivial point, $\alpha_{12}\alpha_{21} > 1$ and the equilibrium is unstable.

Prob. 11. The difference among the two situations for the N_1 population is 20, so

$$\alpha_{12} = \frac{20}{N_2} = \frac{20}{80} = \frac{1}{4}$$

The difference among the two situations for the N_2 population is 70, so

$$\alpha_{21} = \frac{70}{N_1} = \frac{70}{180} = \frac{7}{18}$$

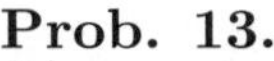

Prob. 13.

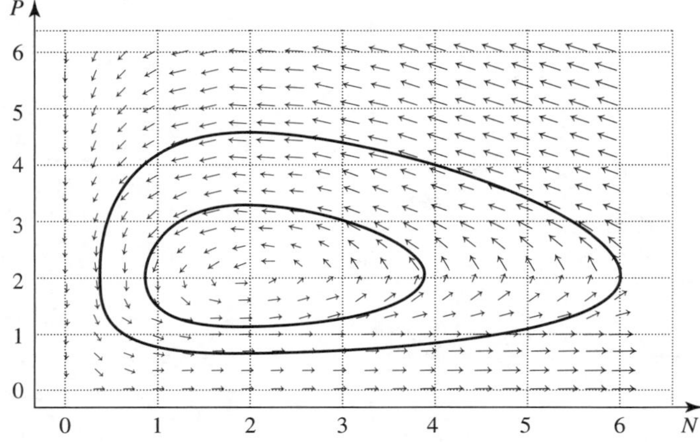

Prob. 15.

(a) The two isoclines are $N - 4PN = 0$ and $2PN - 3P = 0$ so with the first equation we have

$$N(1 - 4P) = 0$$

with solutions $N = 0$ and $P = 1/4$. As for the second isocline we have

$$2PN - 3P = 0$$

or factoring $P(2N - 3) = 0$ with solutions $P = 0$, and $N = 3/2$, so there are two points of equilibrium $(0, 0)$ and $(3/2, 1/4)$.

(b) The Jacobi matrix at $(0, 0)$ is given by

$$Df(0,0) = \begin{bmatrix} 1 & 0 \\ 0 & -3 \end{bmatrix}$$

a saddle, unstable equilibrium.

(c) At the non-trivial equilibrium point

$$Df(3/2, 1/4) = \begin{bmatrix} 0 & -1 \\ 5 & 0 \end{bmatrix}$$

and the eigenvalues can be computed and are $\lambda_1 = i\sqrt{5}$ and $\lambda_2 = -i\sqrt{5}$, which does not allows us to infer anything about the stability of the equilibrium.

(d)

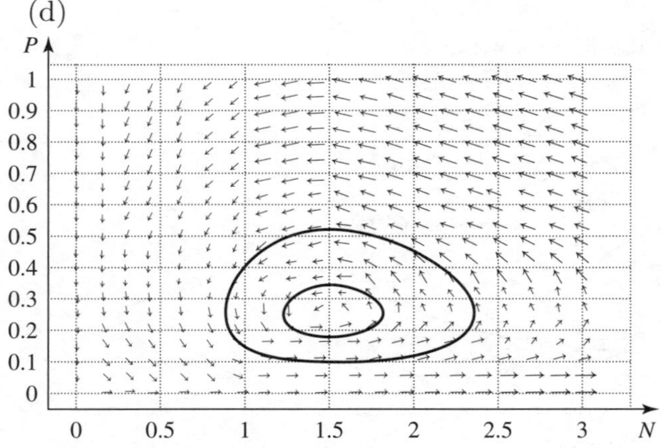

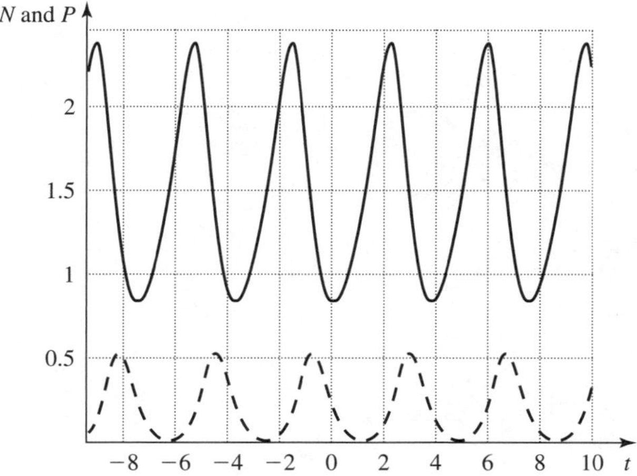

Prob. 17.

(a) If the predator is absent $P = 0$ and the equation simply becomes

$$\frac{dN}{dt} = 5N$$

Separating the variables we have

$$\frac{dN}{N} = 5dt$$

and integrating both sides

$$\int \frac{dN}{N} = \int 5dt$$

or

$$\ln N = 5t + c$$

or

$$N(t) = ce^{5t}$$

so in the absence of a predator the insect population will increase exponentially.

(b) When the predator is introduced the growth rate for $N(t)$ decreases, according to the equation

$$\frac{dN}{dt} = 5N - 3PN$$

which will help control the density of the insect.

(c)

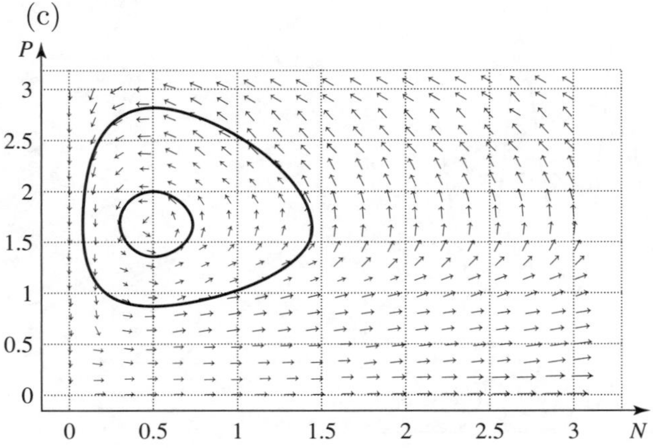

Prob. 19.

(a) In the absence of a predator $P = 0$ and the first equation becomes

$$\frac{dN}{dt} = 3N\left(1 - \frac{N}{10}\right)$$

(b) The isoclines are

$$3N\left(1 - \frac{N}{10}\right) - 2PN = 0$$

and

$$PN - 4P = 0$$

The first one solves to $N = 0$ and $P = \frac{3}{2}\left(1 - \frac{N}{10}\right)$ and the second to $P = 0$ and $N = 4$, so the equilibrium points are $(0, 0)$ and $(4, 9/10)$.

The Jacobian matrix is given by

$$Df(N, P) = \begin{pmatrix} 3 - \frac{3N}{5} - 2P & -2N \\ P & N - 4 \end{pmatrix}.$$

So at $(0, 0)$ it is

$$Df(0, 0) = \begin{pmatrix} 3 & 0 \\ 0 & -4 \end{pmatrix}$$

which is a diagonal matrix with eigenvalues 3 and -4 so the equilibrium is unstable.

For the other point

$$Df(4, 9/10) = \begin{pmatrix} -6/5 & -8 \\ 9/10 & 0 \end{pmatrix}$$

so with $\Delta = 36/5$ and $\tau = -6/5$ it is a stable spiral.

(c)

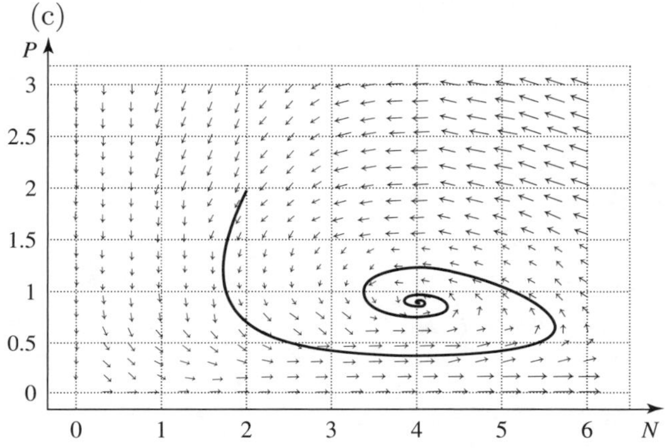

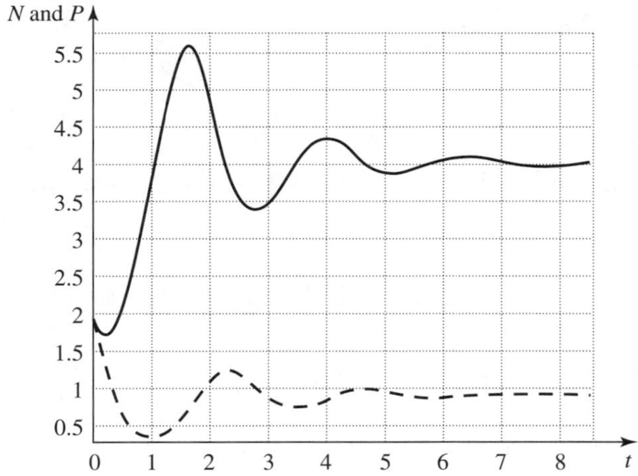

Prob. 21.

(a) The isoclines will be given by the equations

$$N\left(1-\frac{N}{20}\right)=5PN$$

$$PN=4P$$

so they are $N=0$ and $P=\frac{1}{5}-\frac{N}{100}$ for the first equation and $P=0$ and $N=4$ for the second.

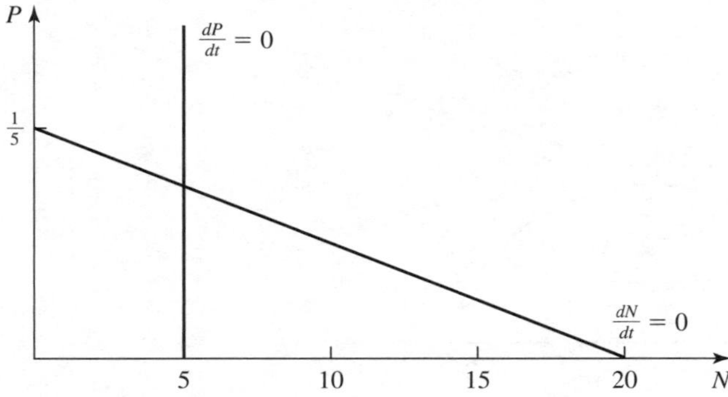

(b) The signs of the Jacobian matrix are

$$Df=\begin{bmatrix} - & - \\ + & 0 \end{bmatrix}$$

with positive determinant and negative trace, so it is a stable equilibrium.

Prob. 23. The intersection of th zero-isoclines will happen at the point

$$N=\frac{d}{c} \quad \text{and} \quad P=\frac{a}{b}\left(1-\frac{d}{ck}\right)$$

so the density of the prey does not depend on the value of "a" and will increase the density of the predator.

Prob. 25. The equilibrium point of Problem 22 is given by

$$N=\frac{d}{c} \quad \text{and} \quad P=\frac{a}{b}\left(1-\frac{d}{ck}\right)$$

so an increase in the value of "c" will increase the denominator of $N=\frac{d}{c}$ and then reduce the abundance of the prey. At the same time increasing the value of "c" will decrease the amount that gets subtracted in P, increasing the abundance of the predator at the equilibrium point.

Prob. 27. It is a $(+-)$ case of *Predation*, the determinant is positive, trace negative, so stable.

Prob. 29. It is a $(++)$ case of *Mutualism*, or *Symbiosis*, the determinant $\Delta > 0$ and trace negative, so stable.

Prob. 31. It is a $(++)$ case of *Mutualism* or *Symbiosis*, the determinant $\Delta = 1.5 - 2.6 = -1.1 < 0$, so unstable.

Prob. 33. It is a $(--)$ case of *Competition*, the determinant negative, so unstable.

Prob. 35. The signs of the Jacobian matrix are

$$\begin{bmatrix} - & - \\ - & - \end{bmatrix}$$

so the determinant is unknown and the equilibrium cannot be determined from the data.

Prob. 37. The signs of the Jacobian matrix are

$$\begin{bmatrix} - & - \\ - & - \end{bmatrix}$$

so the determinant is unknown and the equilibrium cannot be determined.

Prob. 39. The signs of the Jacobian matrix are

$$\begin{bmatrix} - & - \\ - & - \end{bmatrix}$$

so the determinant is unknown and the type of equilibrium cannot de determined.

Prob. 41. The diagonal element a_{ii} measures the effect species i has on itself, and the value of the derivative

$$a_{ii} = \frac{af_i(N_1, N_2)}{aN_i}$$

so if they are negative the growth of the number of individuals will imply in a lower growth rate.

Prob. 43.

(a) The zero-isoclines are given by the equations

$$aN = bNP$$

$$cNP = dP$$

with solutions $N = 0$ and $P = a/b$ for the first and $P = 0$ and $N = d/c$ for the second, so the non-trivial equilibrium happens at the point:

$$\left(\frac{d}{c}, \frac{a}{b}\right)$$

(b) The community matrix is given by

$$A = \begin{bmatrix} \frac{af_1}{aN} & \frac{af_1}{aP} \\ \frac{af_2}{aN} & \frac{af_2}{aP} \end{bmatrix} = \begin{bmatrix} a - bP & -bN \\ cP & cN - d \end{bmatrix}$$

(c) $a_{12} = -bN < 0$, so the growth rate of the prey is decreased if the abundance of the predator increase.

$a_{21} = cP > 0$, so the grouwth rate of the predator increases if the abundance of the prey increase.

Prob. 45.

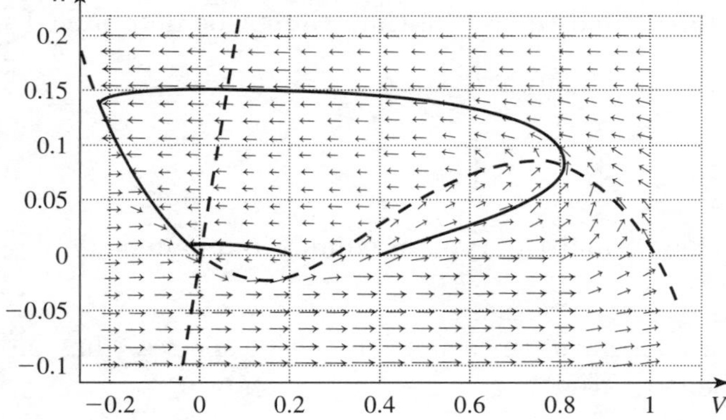

Prob. 47. We have first to verify both eigenvalues are real and negative, for one to observe an action potential. In this particular case the Jacobian

$$Df = \begin{bmatrix} -3V^2 + 2.6V - 0.3 & -1 \\ 0.01 & -0.004 \end{bmatrix}$$

so at $(0,0)$ it reduces to

$$Df = \begin{bmatrix} -0.3 & -1 \\ 0.01 & -0.004 \end{bmatrix}$$

so computing the eigenvalues

$$\begin{aligned} \det(A - \lambda I) &= \begin{vmatrix} -0.3 - \lambda & -1 \\ 0.01 & -0.004 - \lambda \end{vmatrix} \\ &= (\lambda + 0.004)(\lambda + 0.3) + 0.01 \\ &= \lambda^2 + 0.304\lambda + 0.0012 + 0.01 \\ &= \lambda^2 + 0.304\lambda + 0.0112 \end{aligned}$$

which can easily verified to have roots which are real and negative, so we will observe an action potential for any value of $V \in (0.3, 1)$.

Prob. 49.

$$\begin{aligned} \frac{db}{dt} &= -kab \\ \frac{dc}{dt} &= kab \end{aligned}$$

Prob. 51.

$$\begin{aligned} \frac{ds}{dt} &= -k_1 se \\ \frac{de}{dt} &= k_2 c - k_1 se \\ \frac{dc}{dt} &= k_1 se - k_2 c \\ \frac{dp}{dt} &= k_2 c \end{aligned}$$

Prob. 53. The conserved quantity is $x + y$, because

$$\frac{d}{dt}(x + y) = \frac{dx}{dt} + \frac{dy}{dt} = 2x - 3y + 3y - 2x = 0$$

Prob. 55. The conserved quantity is $x + y + z$, because

$$\begin{aligned} \frac{d}{dt}(x + y + z) &= \frac{dx}{dt} + \frac{dy}{dt} + \frac{dz}{dt} \\ &= (-x + 2y + z) + (-2xy) + (x - z) = 0 \end{aligned}$$

Prob. 57.

(a)

$$\begin{aligned}\lim_{s\to\infty} f(s) &= \lim_{s\to\infty} \frac{\vartheta_m s}{k_m + s} = \lim_{s\to\infty} \frac{\frac{\vartheta_m s}{s}}{\frac{k_m+s}{s}} \\ &= \lim_{s\to\infty} \frac{\vartheta_m}{\frac{k_m}{s} + 1} \\ &= \vartheta_m\end{aligned}$$

because the denominator tends to 1.

(b)

$$f(k_m) = \frac{\vartheta_m \cdot k_m}{k_m + k_m} = \frac{\vartheta_m \cdot k_m}{2k_m} = \frac{\vartheta_m}{2}$$

(c) (i) Both numerator and denominator of $f(s)$ are positive, so $f(s) > 0$

(ii) Assume that s_1 and s_2 are on the line and that $s_1 < s_2$, then taking the inverses and multiplying by k_m (positive) we get

$$\frac{k_m}{s_1} > \frac{k_m}{s_2}$$

adding 1 and taking inverses again changes the direction of the inequality

$$\frac{1}{\frac{k_m}{s_1} + 1} < \frac{1}{\frac{k_m}{s_2} + 1}$$

so multiplying both sides ϑ_m we get

$$\frac{\vartheta_m}{\frac{k_m}{s_1} + 1} < \frac{\vartheta_m}{\frac{k_m}{s_2} + 1}$$

multiplying both numerator and denominator on the left by s_1 and on the right by s_2, should not change the value of the fractions and

$$\frac{\vartheta_m}{k_m + s_1} < \frac{\vartheta_m}{k_m + s_2}$$

showing that $f(s_1) < f(s_2)$, that is, $f(s)$ is increasing.

Another way to see this result is to compute the derivative of $f(s)$ and show that it is positive everywhere

$$f'(s) = \frac{(k_m + s)(\vartheta_m) - \vartheta_m s}{(k_m + s)^2} = \frac{k_m \vartheta_m}{(k_m + s)^2} > 0$$

so $f(s)$ is increasing.

(iii) To show that $f(s)$ is concave down we compute the second derivative

$$f''(s) = -\frac{k_m \vartheta_m}{(k_m + s)^3} < 0$$

so $f(s)$ is concave down

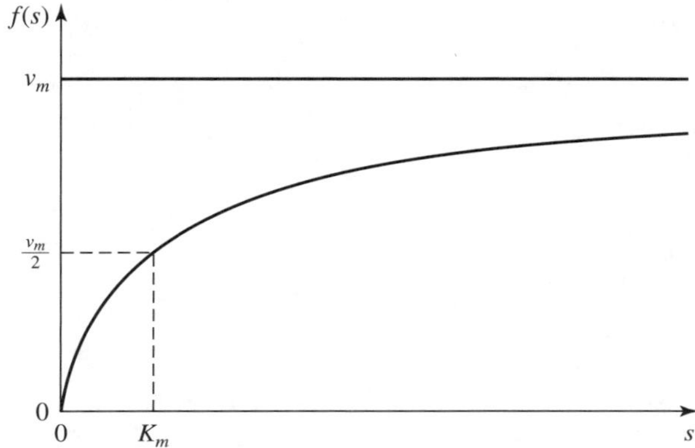

(d) The reaction rate $\frac{dp}{dt} = \frac{\vartheta_m s}{k_m + s}$ is limited above by ϑ_m and according to the graph depends on s, the total amount of substrate.

Prob. 59.

(a) Setting the equations equal to zero and adding algebraically we see that the equilibrium point is

$$\hat{s} = \frac{Dk_m}{y\vartheta_m - D}$$

which is an increasing function of D, to see this just compute the derivative of

$$f(x) = \frac{xk_m}{y\vartheta_m - x}$$

with respect to the variable x, or observe the fact that $1/f(x)$ can be easily seen as a decreasing function of x.

(b) To find $\hat{s}$, just mark $\frac{D}{y}$ on the q-axis of the graph and find the inverse point (mark the horizontal until the graph and then the vertical until the s-axis). The following calculation show why this is true

$$q\left(\frac{Dk_m}{y\vartheta_m - D}\right) = \frac{\vartheta_m \frac{Dk_m}{y\vartheta_m - D}}{k_m + \frac{Dk_m}{y\vartheta_m - D}} = \frac{D}{y}$$

As it can be seen from the graph the higher value of D, the higher will be the value of $\hat{s}$

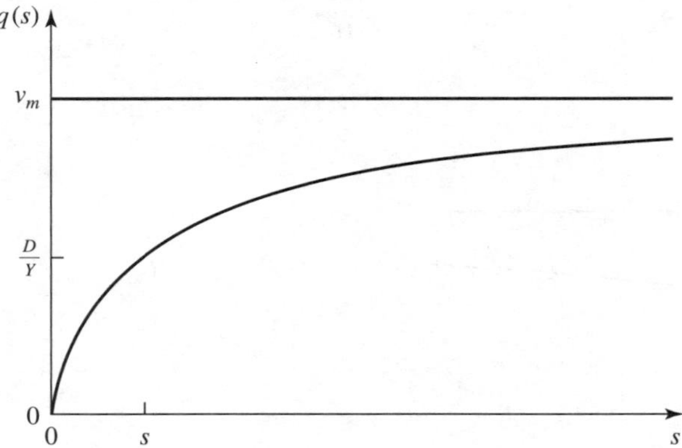

Prob. 61. In this case $D = 2$, $s_0 = 4$, $\vartheta_m = 3$, $k_m = 1$ and $Y = 1$. We can easily see that the inequality

$$0 < D/Y < \vartheta_m$$

is verified and we will have two equilibrium points

$$(s_0, 0) = (4, 0)$$

and

$$\left(\frac{Dk_m}{y\vartheta_m - D}, y\left(s_0 - \frac{Dk_m}{y\vartheta_m - D}\right)\right) = (2, 2)$$

At the first point the Jacobi matrix is

$$Df(s_0, 0) = \begin{bmatrix} -D & -q(s_0) \\ 0 & yq(s_0) - D \end{bmatrix} = \begin{bmatrix} -2 & -12/5 \\ 0 & 2/5 \end{bmatrix}$$

with determinant $\Delta = -4/5$ negative making it an unstable point.

From the text we see that when the nontrivial equilibrium exists (our case here) it is always stable.

11.6 Review Problems

Prob. 1. Since $Z(t) = N_1(t)/N_2(t)$, we can apply logarithms

$$\ln Z(t) = \ln N_1(t) - \ln N_2(t)$$

and computing the derivative

$$\begin{aligned}\frac{d}{dt}\ln Z(t) &= \frac{d}{dt}\ln N_1(t) - \frac{d}{dt}\ln N_2(t)\\ &= \frac{N_1'(t)}{N_1(t)} - \frac{N_2'(t)}{N_2(t)}\\ &= r_1 - r_2\end{aligned}$$

Now solving the differential equation

$$\frac{d}{dt}\ln Z(t) = r_1 - r_2$$

we get

$$\ln Z(t) = (r_1 - r_2)t + c$$

and taking exponentiation on both sides

$$Z(t) = Z_0 e^{(r_1 - r_2)t}$$

so if $r_1 > r_2$, the coefficient for t is positive and $\lim_{t\to\infty} Z(t) = \infty$, so if $r_1 > r_2$, population 1 becomes numerically dominant.

Prob. 3.

(a) The zero isoclines are given by the equations

$$2N(1 - \frac{N}{10}) = 3PN$$

and

$$PM = 3P$$

with solutions $N = 0$ and $P = \frac{2}{3}(1 - \frac{N}{10})$ for the first and $P = 0$ and $N = 3$ for the second

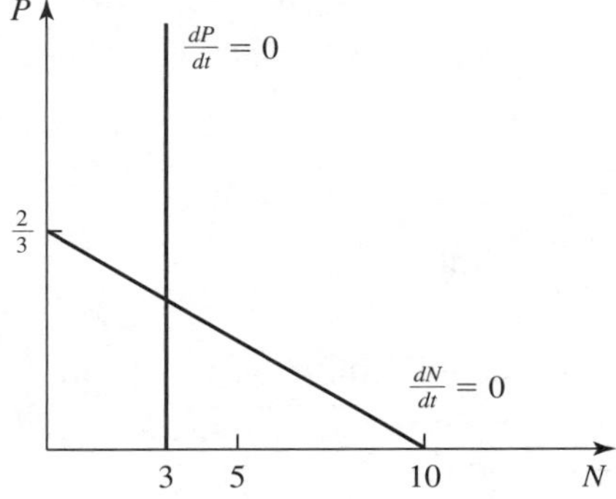

(b) The matrix of signs of the Jacobian are

$$\begin{bmatrix} + & 0 \\ - & - \end{bmatrix}$$

so the determinant is negative and the equilibrium unstable.

Prob. 5.

(a) In the absence of species 2, $p_2 = 0$, so the two equations reduceto

$$\frac{dp_1}{dt} = c_1 p_1 (1 - p_1) - m_1 p_1$$

and the equilibrium points are given by

$$c_1 p_1 (1 - p_1) = m_1 p_1$$

with solutions $p_1 = 0$ and $p_1 = \frac{m_1}{c_1} - 1$, so a nontrivial equilibrium will happen with $o < p_1 \leq 1$, if and only if $1 < \frac{m_1}{c_1} \leq 2$, that is

$$c_1 < m_1 \leq 2c_1$$

(b) We can re-write the equations in the Lotka-Volterra format to obtain

$$\begin{aligned}
\frac{dp_1}{dt} &= c_1 p_1 (1 - p_1 - p_2) - c_1 p_1 \frac{m_1}{c_1} \\
&= c_1 p_1 (1 - \frac{m_1}{c_1} - p_1 - p_2) \\
&= c_1 p_1 \left(\frac{c_1 - m_1}{c_1} - p_1 - p_2 \right) \\
&= p_1 (c_1 - m_1 - c_1 p_1 - c_1 p_2) \\
&= (c_1 - m_1) p_1 \left(1 - \frac{c_1}{c_1 - m_1} p_1 - \frac{c_1}{c_1 - m_1} p_2 \right)
\end{aligned}$$

which is our standard equation with

$$r_1 = c_1 - m_1, \quad K_1 = 1 - \frac{m_1}{c_1} \quad \text{and} \quad \alpha_{12} = 1$$

and in the same fashion we rearrange the second equation to obtain

$$r_2 = c_2 - m_2, \quad K_2 = 1 - \frac{m_2}{c_2} \quad \text{and} \quad \alpha_{21} = 1.$$

Assuming that $\frac{c_1}{m_1} > \frac{c_2}{m_2}$ we can invert to obtain

$$\frac{m_1}{c_1} < \frac{m_2}{c_2}$$

and

$$1 - \frac{m_1}{c_2} > 1 - \frac{m_2}{c_2}$$

that is $K_1 > K_2$. Since $\alpha_{12} = \alpha_{21} = 1$, we have both $K_1 > \alpha_{12}K_2$ and $K_2 < \alpha_{21}K_1$, and species 1 will exclude species 2, if they start on a positive parte of the plane.

Prob. 7.

(a) The zero-isoclines of the system is given by the equations

$$D(s_0 -) = q(s)x$$

and

$$Yq(s)x = Dx$$

so the nontrivial solutions are $q(\hat{x}) = \frac{D}{Y}$, and using this on the first equation we get

$$D(s_0 - \hat{s}) = \frac{D}{Y}\hat{x}$$

so at the equilibrium point: $\hat{x} = Y(s_0 - \hat{s})$, which is positive when $\hat{s} < s_0$.

(b) Since $q(\hat{s}) = \frac{D}{Y}$, we can write out the equation

$$\frac{v_m \hat{s}}{K_m + \hat{s}} = \frac{D}{Y}$$

and solving for $\hat{s}$ we get: $\hat{s} = \frac{DK_m}{Yv_m - D}$ so the equilibrium abundance of the microbe is

$$\hat{x} = Y(x_0 - \hat{s}) = \frac{YDK_m}{Yv_m - D}$$

Consider now the ivnerse of $\hat{x}$, that is

$$\frac{1}{\hat{x}} = \frac{Yv_m - D}{YDK_m} = \frac{Yv_m}{YDK_m} - \frac{D}{YDK_m} = \frac{v_m}{DK_m} - \frac{1}{YK_m}$$

so we can clearly see that $\frac{1}{\hat{x}}$ is a decreasing function of D and an increasing function of Y, so $\hat{x}$ will be increasing with D and decreasing with Y.

12 Probability and Statistics

12.1 Counting

Prob. 1. One experiment (trial) consists of using one fertilizer with one light level. There are 5 different fertilizers and each is to be tested with every one of the two light levels. Thus one has to conduct $5 \cdot 2 = 10$ trials. But each trial is required to have 4 replicates. Hence the total number of replicates is given by $5 \cdot 2 \cdot 4$, which is equal to 40.

Prob. 3. The beetle is to be tested with 2 satiation levels. For each satiation level, it is to be given two (out of three) items of food. Since there are 3 ways in which two food items can be selected out of 3 and since 2 satiation levels are to be tested, six experiments have to be conducted on the beetle to complete all possibilities. However, we are told that to ensure accuracy of results, each experiment is to be repeated 20 times. Thus there are a total number of 120 $(20 \cdot 6)$ replicates.

Prob. 5. For a breakfast a person takes, as per the problem, one beverage, one cereal and one fruit. Since there is a choice of 3 beverages, 7 cereals and 4 fruits, a person has a choice of $(3 \cdot 7 \cdot 4 = 84)$ different breakfasts.

Prob. 7. This problem can best be tackled by imagining that there are only 3 cities, A, B and C to be toured. If City A is chosen first, there are 2 alternatives, B and C or C and B. The other options are B,A,C and B,C,A; C,A,B and C,B,A. (We are actually using enumeration here to understand the problem). Thus there are a total of 6 options, which is really 3!. Thus for 5 cities, the number of different routes are 5!. This is equal to 120.

Prob. 9. Seven books are to be arranged in a shelf. Remember that the order is important. Thus, the seven books can be arranged in 7! different ways, which works out to 5040.

Prob. 11. Out of 26 letters, four are to be chosen at a time. Remember that the order is very important, since each order stands for a different word. This is a clear case for using permutation. Hence the number of words that can formed in this case is $P(26, 4)$. This is equal to $26!/22!$, which is $23 \cdot 24 \cdot 25 \cdot 26$. So the number of words is a phenomenal 358,800.

Prob. 13. Three awards are to be distributed. Since the awards are not identical, but different, this is a clear case in which order is important. Using principles of permutation illustrated in Section 12.1.2, the number of ways in which three awards can be presented to the class of 15 students is $P(15, 3)$. The result is $(13 \cdot 14 \cdot 15)$ which is 2730.

Prob. 15. The order in which the candies are chosen is not important. Hence this is a case for using Combination, as explained in Section 12.1.3. The answer is $C(10, 3)$, which is 120.

Prob. 17. Since no specific tasks are assigned to the members of the committee, this is a case for using Combinations. The three people can be selected out of ten in $C(10, 3)$ ways. This works out to $10! \div (3! \cdot 7!) = 8 \cdot 9 \cdot 10 \div (2 \cdot 3) = 120$ ways.

Prob. 19. Four balls can be selected out of fifteen balls in $C(15, 4)$ ways. Applying the formula for combinations, $15! \div (4! \cdot 11!) = 12 \cdot 13 \cdot 14 \cdot 15 \div (2 \cdot 3 \cdot 4)$ $= 1365$ ways.

Prob. 21. If all the letters were different, they could have been arranged in 9! ways. Since there are four A's and five B's some of the words would be indistinguishable. The number of different words that can be formed is $9! \div (4! \cdot 5!) = 6 \cdot 7 \cdot 8 \cdot 9 \div (2 \cdot 3 \cdot 4) = 126$ ways.

Prob. 23.

(a) The two red balls can be picked out of the five red balls in $C(5, 2)$ ways. This works out to $5! \div (3! \cdot 2!) = 10$ ways. The two blue balls can be picked up from the four blue balls in $C(4, 2)$ ways. This works out to $4! \div (2! \cdot 2!) = 6$ ways. One red ball and one blue ball can be picked in $C(5, 1) \cdot C(4, 1)$ ways. The first term works out to $5! \div 4! = 5$ ways. The second term works out to $4!/3! = 4$ ways. Hence the product is 20 ways.

(b) The totals of all three combinations works out to $10 + 6 + 20 = 36$ ways. Now the way in which two balls can be picked from nine balls in $C(9, 2)$ ways. This works out $9! \div (2! \cdot 7!) = 8 \cdot 9 \div 2 = 36$ ways. This total agrees with the sum obtained in part A.

Prob. 25. If all the letters were different, they could have been arranged in 14! ways. But some of the words formed will be indistinguishable from each other since the letters repeat. Hence the total number of different words that can be formed is $14! \div (5! \cdot 3! \cdot 6!) = 7 \cdot 8 \cdot 9 \cdot 10 \cdot 11 \cdot 12 \cdot 13 \cdot 14/(2 \cdot 3 \cdot 4 \cdot 5 \cdot 2 \cdot 3) =$ $168,168$ ways.

Prob. 27. S= a,b,c. The subsets of this given set are

1) ϕ the null set. 2) a 3) b 4) c 5) a,b 6) b,c 7) a,c 8) a,b,c

The total number of subsets is therefore given by $2^3 = 8$.

Another way to see this is that for each element we need to decide whether or not it becomes a member of the subset (2 ways). Since there are 3 elements, there are $2^2 = 8$ different subsets.

Prob. 29. Let the positions on the bench be numbered 1,2,3,4. If Peter sits on the extreme ends ie. 1 or 4, Melissa would have to be on positions 2 and 3 respectively. Brian and Hillary can sit on any of the remaining two positions. So the number of ways this is possible is $2 \cdot 1 \cdot 2 \cdot 1 = 4$ ways. If Peter sits on positions 2 or 3, Melissa can sit on either side of him. Brian and Hillary can sit on the other two benches. So the number of ways is $2 \cdot 2 \cdot 2 \cdot 1 = 8$ ways, so the total is $8 + 4 = 12$ ways.

Prob. 31. The total number of ways in which 3 people can be selected from a group of 7 is $C(7, 3) = 7! \div (4! \cdot 3!) = 5 \cdot 6 \cdot 7 \div (2 \cdot 3) = 35$ ways. Now let us count the number of committees in which the two people who do not want to be together are included. If the two people appear together in a committee, the third person can be selected in $C(5, 1) = 5$ ways. Hence the two people who do not want to serve together will appear in 5 ways.

Subtracting 5 from 35 = 30 ways is the total number of ways in which 3 people can be selected out of 7 if two of them do not want to serve together.

Prob. 33. There are three ways to have at least one perennial plant in the selection:

1. two annual plants and one perennial plant: $\binom{4}{2}\binom{3}{1} = 18$
2. one annual plants and two perennial plants: $\binom{4}{1}\binom{3}{2} = 12$
3. no annual plants and three perennial plants: $\binom{4}{0}\binom{3}{3} = 1$

Hence, the total number of ways is $18 + 12 + 1 = 31$.

Prob. 35. $(x + y)^4 = (x + y)^2(x + y)^2 = (x^2 + 2xy + y^2)(x^2 + 2xy + y^2) = x^4 + 4(x^3)y + 6(x^2)(y^2) + 4x(y^3) + y^4$

Prob. 37. The four red cards can be drawn in $C(26, 4) = 14950$ ways. The five black cards can be drawn in $C(26, 5) = 65780$ ways. Therefore the total number of ways $= C(26, 4) \cdot C(26, 5) = 983,411,000$ ways.

Prob. 39. We first choose the value of the two pairs ($C(13,2)$). For each value, their suits can be chosen in $C(4,2)$ ways. The one remaining card needs to be chosen from the remaining 44 cards ($C(44,1)$ ways). Hence, the number of ways of picking exactly two pairs are

$$\binom{13}{2}\binom{4}{2}\binom{4}{2}\binom{44}{1} = 123,552$$

Prob. 41. There are 13 ways in which 4 cards of the same value can be picked, since there are 13 values attributed to the 13 cards of each suit. The one remaining card of the *hand* can be picked in $C(48,1)$ ways. Hence the total number of ways *four of a kind* can be picked is 13 * 48 which is 624.

Prob. 43. There are $3! = 6$ ways to arrange the three voices.

12.2 What is Probability?

Prob. 1. The sample space for the experiment of tossing a coin three times is:

$$\Omega = \{HHH, HHT, HTH, HTT, THH, THT, TTH, TTT\}$$

where H stands for Heads, T stands for tails. The Sample Space consists of a total of 8 elements.

Prob. 3. If the balls are selected simultaneously, then the Sample Space of selecting two balls without replacement is Ω = {1,2; 1,3; 1,4; 1,5; 2,3; 2,4; 2,5; 3,4; 3,5; 4,5; 10} where 1,2,3,4,5 are the numbers on the balls. The Sample Space thus contains 10 elements.

Prob. 5. $\Omega = \{1, 2, 3, 4, 5, 6\}$ A= {1,3,5}, B= {1,2,3} A$\cup$ B = {1,2,3,5} A$\cap$B = {1,3}

Prob. 7. (A$\cup$B)^c = {1,2,3,5}^{c} = {4,6}

Prob. 9. Since Pr(1) + Pr(2) + Pr(3) + Pr(4) + Pr(5) = 1,

$$\text{Pr}(5) = 1 - \Sigma_{i=1}^{i=4} Pr(i) = 1 - 0.1 - 0.2 - 0.05 - 0.05 = 0.6$$

Prob. 11. $\text{Pr}(A)^c = \text{Pr}(2) + \text{Pr}(4) = 0.25$

Prob. 13. Pr(A$\cap$B) = 0.1; Pr(A) = 0.4; $\text{Pr}(A^c \cap B^c) = 0.2$. According to De Morgan's law, (A^c $\cap$ B^c)= (A$\cup$B)^c Therefore (A$\cup$B)^c = 0.2. Therefore

(A∪B) = 1 - 0.2 = 0.8 Pr(A∪B)= Pr(A) + Pr(B) - Pr(A∩B) Therefore $0.8 = 0.4 + \Pr(B) - 0.1$ $\Pr(B) = 0.5$

Prob. 15. The Sample Space Ω = (HH, HT, TH, TT). Since all events are equally possible (the coins are fair coins) and three of the events are favorable to obtaining at least one head, the probability of obtaining at least one head = 3/4

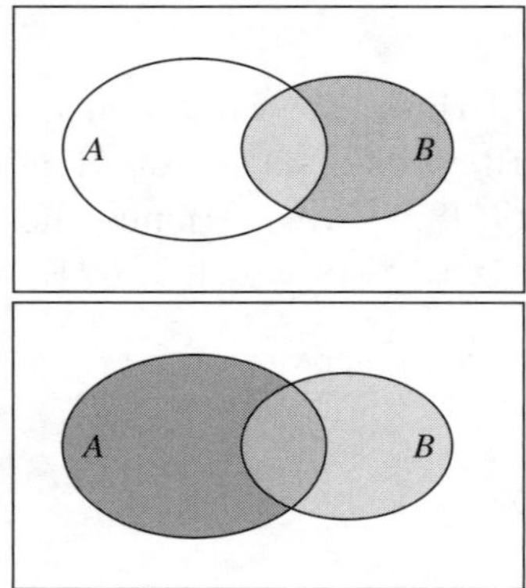

Prob. 17. P(at least one head)
$= 1 - P$ (no heads)
$= 1 - P(TT) = 1 - 0.25$
$= 0.75$

Prob. 19. The Probability of obtaining at least one four in a throw = 1 - Probability of obtaining no 4's on either throw is: $\frac{6+6-1}{36} = \frac{11}{36}$.

Prob. 21. The Sample Space in the experiment in which two fair dice are thrown one after the other is identically the same as that in which one fair die is thrown twice, as in Problem 2. Thus the Sample Space has 36 elements which are enumerated in Problem 2(and also in Problem 22 below). All outcomes are equally probable. Out of those outcomes, {2,1; 3,1; 3,2; 4,1; 4,2; 4,3; 5,1; 5,2; 5,3; 5,4; 6,1; 6,2; 6,3; 6,4; 6,5} = 15 elements are favorable to the probability that the first number is greater than the second number. Hence the answer is $15 \div 36 = 5/12$

Prob. 23. The four possible outcomes of a cross between Cc and Cc is Ω = (CC, Cc, cC and cc). Each has probability of $1 \div 4$ since all outcomes are equally probable. Only the last case produces white flowers. Hence the probability of obtaining white flowers is 1/4.

Prob. 25. The possible outcomes of a cross between Aa and Aa are Ω $= \{$AA, Aa, aA, aa$\}$. Since Aa = aA, the outcomes are = $\{$AA, Aa, Aa, aa$\}$ all outcomes are equally probable. So the probability of obtaining Aa is $(1 + 1) \div 4 = 0.5$

Prob. 27. Let a boy be represented by B and girl as G. The possible outcomes are $\Omega = \{$BBB, BBG, BGB, BGG, GBB, GBG, GGB, GGG$\}$ All outcomes are equally probable since the sex ratio is 1:1. So the probability of obtaining all girls is $1 \div 8$

Prob. 29.

(a) The total number of fish is denoted by N. We capture 100 fish. So $K = 100$. Later we catch 10 fish. So n = 10. 3 of them are marked ie k=3. The total number of ways of capturing 10 fish out of N fish is $C(N, 10)$ ways. 3 of 10 fish will be marked if if select 3 marked fish out of 100 marked fish and 7 unmarked fish from $(N - 100)$ unmarked fish. So the probability that exactly 3 fish will be marked is $P(N) = (C(100, 3) \cdot C(N - 100, 7)) \div C(N, 10)$

(b) We now investigate when the value arrived in a) is maximized. This value is maximized when $P(N) \div P(N-1)$ is greater than or equal to 1. $P(N-1) = (C(100, 3) \cdot C(N-101, 7)) \div C(N-1, 10)$ Therefore $P(N) \div P(N-1) = C(N-100, 7) \cdot C(N-1, 10) \div (C(N, 10) \cdot C(N-101, 7))$ The above my be written as: $((N - 100!) \cdot (N - 1!) \cdot (N - 10!) \cdot (N - 108!)) \div ((N - 107!) \cdot (N - 11!) \div (N!) \cdot (N - 101))$ We use the following facts to simplify this equation $1)(N - 107!) \div (N - 108!) = (N - 107)$; $2)(N - 10!) \div (N - 11!) = N - 10$; $3)(N - 100!) \div (N - 101!) = N - 100$; $4)(N!) \div (N - 1!) = N$;

Using the above four facts, we obtain $P(N) \div P(N-1) = (N-100) \cdot (N-10) \div ((N) \cdot (N-107))$ This is maximized when $P(N) \div P(N-1) \geq 1$ $(N - 100) \cdot (N - 10) \div (N \exp 2 - 107 \cdot N) \geq 1$ $N^2 - 10 \cdot N - 100 \cdot N + 1000 \geq (N^2 - 107 \cdot N)$ $1000 \geq 3N$ $N \leq (1000 \div 3)$ N is approximately 333. Now by example 13, N is approximately equal to $Kn \div k$ Substituting values we obtain N is approximately equal to $100 \cdot (10 \div 3) = 1000 \div 3 = 333$ Therefore the two values agree.

Prob. 31. The probability that both cards are spades is $\dfrac{\binom{13}{2}\binom{39}{0}}{\binom{52}{2}} = \frac{1}{17}$

Prob. 33. The total number of ways in which 3 balls can be picked from 12 balls is $C(12, 4) = 220$ ways. Out of these ways, let us count the number of favorable outcomes. If the balls are of different color, 1 ball has to be green, 1 ball has to be blue and 1 ball has to be red. 1 green ball can be picked from 4 green balls in $C(4, 1) = 4$ ways. 1 blue ball can be picked from 6 blue balls in $C(6, 1) = 6$ ways. 1 red ball can be picked from 2 red balls in $C(2, 1) = 2$ ways. Hence the total number of ways of picking 1 green, 1 blue and 1 red balls $= 4 \cdot 6 \cdot 2 = 48$ ways. So the probability of picking all balls of different colors is $48 \div 220 = 0.218181818$

Prob. 35. The probability of obtaining at least one ace = 1 - Probability of obtaining no aces at all.

$$\text{So the probability of obtaining at least one ace} = 1 - \frac{\binom{48}{4}}{\binom{52}{4}}.$$

Prob. 37. So the probability of all cards being red $\binom{26}{13} / \binom{52}{13}$.

Prob. 39. We first choose a value for the two pairs. There are $\binom{13}{2}$ ways. Each pair can be picked in $C(4, 2)$ ways. The remaining card can be picked in $C(44, 1)$ or $4 \cdot C(11, 1)$ ways. The five cards can be selected in $C(52, 5)$ ways.

Combining the above steps, the probability of obtaining exactly two pairs is: $(C(13, 2) \cdot C(4, 2) \cdot C(12, 2) \cdot C(44, 1)) \div C(52, 5)$.

12.3 Conditional Probability and Independence

Prob. 1. $P(B/A) = 13/51$ (since there are 13 spades out of the remaining 51 cards)

Prob. 3. We define two events

A = (probability that first ball is blue)

B = (probability that the second ball is green)

$P(A) = 5/11$ (there are 5 blue balls out of 11 balls) It is given that the event A has occurred. We have to compute $P(B/A)$ Now $P(B/A) = 6/10 = 3/5$ (since there are 6 green balls out of the remaining 10 balls)

Prob. 5. We define two events

A = (the younger child is a girl)

B = (the older child is a girl)

$P(A) = 1/2$ (assuming a 1:1 sex ratio)

It is given that the younger child is a girl.

$P(B|A) = P(A \cap B)/P(A) = (1/4)/(1/2) = 1/2$ since the probability that both children are girls in 1/4.

Prob. 7. The sample space of numbers add up to 7 is $\Omega = (1, 6; 2, 5; 3, 4; 4, 3; 5, 2; 6, 1)$ and the probability that the first number is a 4 is therefore 1/6

Prob. 9. The sample space $\Omega = (HHH, HHT, HTH, HTT, THH, THT, TTH)$ since at least one head has occurred. (The only element left out is TTT). Hence the probability that the first coin is heads = 4/7.

Prob. 11. Let A = (test result is negative)

Let B1 = person is infected

Let B2 = person is not infected

$P(B1) = 1/100 = 0.01$

$P(B2) = 99/100 = 0.99$

$P(A/B1) = 1 - 0.9 = 0.1$

$P(A/B2) = 1 - 0.15 = 0.85$

$P(A) = P(A/B1) \cdot P(B1) + P(A/B2) \cdot P(B2)$

$P(A) = 0.1 \cdot 0.01 + 0.85 \cdot 0.99 = 0.001 + 0.8415 = 0.8425$

Prob. 13. Bag 1 contains 2 blue balls

Bag 2 contains 2 green balls

Bag 3 contains 1 blue and 1 green

Let $P(A)$ = probability of picking a blue ball

Let P(Bi) = probability of picking i-th bag, i = 1 to 3

$P(A/B1) = 1$ since bag 1 has only blue balls

$P(A/B2) = 0$ since bag 2 contains only green balls

$P(A/B3) = 1/2$ since bag 3 has one blue and 1 green ball

Also $P(B1) = P(B2) = P(B3) = 1/3$ since there is equal chance of picking a bag.

We have $P(A) = P(A/B1) \cdot P(B1) + P(A/B2) \cdot P(B2) + P(A/B3) \cdot P(B3)$

$P(A) = 1 \cdot 1/3 + 0 + 1/2 \cdot 1/3 = 1/3 + 1/6 = 1/2$

Prob. 15. Let $P(A)$ = probability that the first card is an ace.

Let $P(B)$ = probability that the second card is an ace.

$P(A) = 4/52 = 1/13$ since there are 4 aces out of 52 cards.

Now we compute $P(B)$.

(Now the second card can be an ace if the first card is not an ace and the second card is an ace OR if both the cards are aces)

Therefore $P(B) = 48/52 \cdot 4/51 + 4/52 \cdot 3/51 = 1/13$

Hence the two probabilities are equal.

Prob. 17. 40 percent are of genotype CC

60 percent are of genotype Cc

White flowering tea plant is of genotype cc

Let $P(A)$ =probability of obtaining white flowers

Let $P(Bi)$ = probability of picking red flowering pea plant

Therefore $P(B1) = 0.4$and $P(B2) = 0.6$

Now cross between CC and cc results in red plants because the only possible result of such a cross is Cc.

Therefore $P(A/B1) = 0$

Cross between Cc and cc results in either (Cc, cc)

Therefore $P(A/B2) = 1/2$

$P(A) = P(A/B1) \cdot P(B1) + P(A/B2) \cdot P(B2)$

$P(A) = 0 + (1/2 \cdot 0.6) = 0.3$

Prob. 19. Let $P(A)$ be probability of picking heads

Let (PBi) be probability of picking i-th coin, $i = 1, 2$

$P(B1) = P(B2) = \frac{1}{2}$ since there are equal chances of picking either coin

$P(A/B1) = 1/2$ since the first coin is fair

$P(A/B2) = 1$ since the second coin has two heads.

$P(A) = P(A/B1) \cdot P(B1) + P(A/B2) \cdot P(B2) = 1/2 \cdot 1/2$
$+1 \cdot 1/2 = 1/4 + 1/2 = 3/4$

Prob. 21. Let $P(A)$ denote probability that the card is a spade

$P(A) = 13/52$

Let $P(B)$ denote probability that the card is an ace $P(B) = 4/52$ Let $P(A \cap B) =$ P(card is ace of spade)$= 1/52$ Since $P(A) \cdot P(B) = 13/52 \cdot 4/52 = P(A \cap B)$ it follows that A and B are independent.

Prob. 23. 5 green and 6 blue balls. $P(A) = 5/11$ (since there are 5 green out of 11 balls); $P(B/A) = 4/10$ (since there will be 4 green balls left out of 10 balls (if the first ball picked is green) which is $P(A \cap B)/P(A)$

Therefore $P(A \cap B) = 4/10 \cdot 5/11 = 20/110 = 2/11$ Now the second ball can be green if the both the balls are green or if the first is blue and the second is green. $P(B) = 5/11 \cdot 4/10 + 6/11 \cdot 5/10 = 20/110 + 30/110 = 50/110$
$P(A) \cdot P(B) = 5/11 \cdot 50/110 = 250/1210 = 25/121$

Since $P(A) \cdot P(B)$ is not equal to $P(A \cap B)$, the events are not independent.

Prob. 25. The sample space is $\Omega = (BBB, BBG, BGB, BGG, GBB, GBG, GGB, GGG)$ where B denotes a boy and G denotes a girl.

(a) A = (all children are girls)= 1/8 (element 8)

(b) B = (at least one boy) s 7/8(elements 1 through 7)

(c) C = (at least two girls) is 4/8 (elements 4,6,7,8) = 1 ÷ 2

(d) D = (at most two boys) = 7/8 (elements 2 through 8)

Prob. 27. Let $P(A)$ = probability that the student will pass. Let $P(Bi)$ = probability that the student guesses the i-th question correctly. P(Bi) = 1/4 for i = 1 to 10 Since the events are independent, $P(A) = P(B1) \cdot P(B2) \cdot P(B3) \cdots P(Bn) = (1/4)^{10} = 1/1048576$ This means that, if you guess the answers, the chance of passing is less than one million!

Prob. 29. In this problem, it is more convenient to think in terms of complement of an event. Probability that an insect lives more than 5 days = 0.1 Probability that at least one insect will be alive after 5 days = 1 - (Probability that no insects will be alive after 5 days) = $1 - 0.9^{10} = 1 - 0.348678 = 0.651321$

Prob. 31. Let A = test result is positive, B1 be the probability of infection, and B2 be the probability of non infection then B1 = 1/50, B2 = 49/50 , and $P(A/B1) = 0.95$, $P(A/B2) = 0.1$.

We are interested in $P(B1/A)$ that is the probability that the person is infected given that the result is positive. $P(Bi/A) = X/Y$ where $X = P(A/Bi) \cdot P(Bi)$ and $Y = \Sigma_{j=1}^{j=n} P(A/Bj) \cdot P(Bj)$ Hence the required probability is: 1/50 · 0.95 / (1/50 · 0.95 + 49/50 · 0.10) = 0.019/ (0.019 + 0.098) = 0.019/0.117 = 0.16239

Prob. 33. Let the probability that the fair coin was picked be P(A), P(B) denote probability that heads was got.

So P(B) = 1/2·1/2 + 1/2·1 = 3/4 and P(A|B) = P(A∩B)/P(B) = (1/2·1/2) / (3/4) = 1/3

Prob. 35. Let E denote the probability that the woman is a carrier, then $P(E) = 1/2$. If F denotes the event that her son is healthy, $P(F/E) = 1/2$.

Hence $P(E/F) = P(E \cap F)/P(F)$ and $P(F/E)P(E)/P(F) = (1/2)(1/2)/P(F)$ and $P(F) = (1/2)(1/2) + (1/2)1 = 3/4$ and $P(E/F) = (1/4)/(3/4) = 1/3$

Prob. 37.

(a) We have no other information about II-3 other than she may have inherited it from her mother. Therefore, the probability is 1/2 that II-3 is a carrier.

(b) We have no other information about III 2 other than she may have inherited it from her mother. Therefore the probability is 1/2 that II 2 is a carrier.

(c) Here the problem is different. We have some more information about II 2 in that she has a healthy son. Let P(E) be the probability that II 2 is a carrier. P(E) = 1/2 Let P(F) denote the event that III 1 is healthy. We have to compute P(E/F) $P(E/F) = P(E \cap F)/P(F) = P(F/E) \cdot P(E)/P(F)$

Now $P(F/E) = 1/2$ We need to compute P(F) as $P(F) = 1/2 \cdot 1/2 + 1/2 \cdot 1 = 3/4$ $P(E/F) = (0.5 \cdot 0.5)/(0.75) = 1/3$

12.4 Discrete Random Variables and Discrete Distributions

Prob. 1. The sample space $\Omega = (HH, HT, TH, TT)$

Outcome	Value	Outcome	Value	Outcome	Value	Outcome	Value
X(HH)	0	X(HT)	1	X(TH)	1	X(TT)	2

Probability Table					
Probability	Value	Probability	Value	Probability	Value
$P(X = 0)$	1/4	$P(X = 1)$	2/4	$P(X = 2)$	1/4

Prob. 3. 3 Green and 2 blue

Probability	Computation	Value
$P(X = 0)$	$(2/5) \cdot (1/4)$	2/20
$P(X = 1)$	$(3/5) \cdot (2/4) + (2/5) \cdot (3/4)$	12/20
$P(X = 2)$	$(3/5) \cdot (2/4)$	6/20

Prob. 5. Values of distributions function F(x) for given values of mass function and $P(X = x)$ are tabulated below. Graph for $F(x)$ follows.

x	P(X=x)
−3	0.2
−1	0.3
1.5	0.4
2	0.1

$$F(x) = \begin{cases} 0 & \text{for} \quad x < -3 \\ 0.2 & \text{for} \quad -3 \leq x < -1 \\ 0.5 & \text{for} \quad -1 \leq x < 1.5 \\ 0.9 & \text{for} \quad 1.5 \leq x < 2 \\ 1.0 & \text{for} \quad x \geq 2 \end{cases}$$

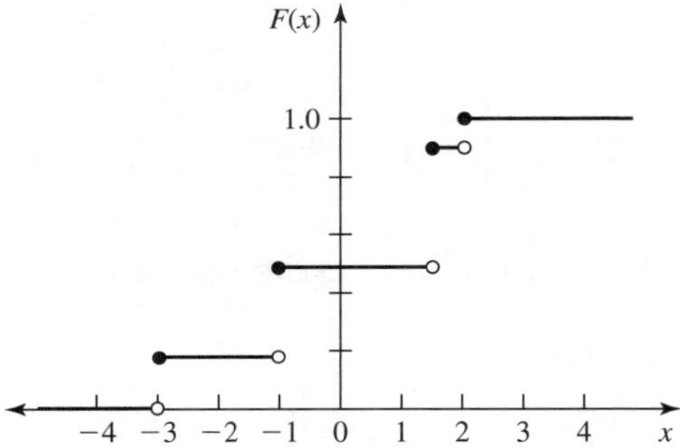

Prob. 7. This exercise is the 'converse' of the previous exercise, in the sense that, given $F(x)$, we have to find P. The values are tabulated below.

Range of x	< -2	$-2 \leq x < 0$	$0 \leq x < 1$	$1 \leq x < 2$	$x >= 2$
F(x)	0	0.2	0.3	0.7	1

x	$P(X = x)$
-2	0.2
0	0.1
1	0.4
2	0.3

Prob. 9. $S = (1, 2, 3, 4, 5, 6, 7, 8, 9, 10)$ and $p(k) = k/N$

(a) By the definition of probability,

$$\sum_{k=1}^{10} p(k) = \sum_{k=1}^{10} \frac{k}{N} = \frac{1}{N}\frac{10 \cdot 11}{2} = \frac{55}{N} = 1$$

therefore $N = 55$

(b) The probability that $X < 8$ is given by $(1+2+3+4+5+6+7)/55 = 28/55$

Prob. 11.

(a)

No. of leaves	19	21	20	13	18	14	17	15	12	16
Relative frequency	1/25	1/25	2/25	4/25	2/25	5/25	5/25	3/25	1/25	1/25

(b) 1) Direct average can be found by first evaluating the numerator.

$$\begin{bmatrix} 19 + 21 + 20 + 13 + 18+ \\ 14 + 17 + 14 + 17 + 17+ \\ 13 + 15 + 12 + 15 + 17+ \\ 15 + 16 + 18 + 17 + 14+ \\ 14 + 14 + 13 + 20 + 13 \end{bmatrix}$$

The numerator is thus 396. Hence the direct average is $396/25 = 15.84$

2) By using relative frequency $(19 \cdot 1 + 21 \cdot 1 + 20 \cdot 2 + 13 \cdot 4 + 18 \cdot 2 + 14 \cdot 5 + 17 \cdot 5 + 15 \cdot 3 + 12 \cdot 1 + 16 \cdot 1)/25 = 396/25 = 15.84$

Prob. 13.

(a) Forming the relevant table, we have:

No. of leaves	7	8	3	2	5	6	9	10	4
Relative frequency	2/25	6/25	1/25	1/25	3/25	4/25	3/25	3/25	2/25

(b) 1) Direct average can be found by first finding the numerator.

$$\left[\begin{array}{l}(7+8+8+3+2+5+6+9+10+6+8+8+7+ \\ 6+6+9+10+4+4+8+6+9+10+5+5+8)\end{array}\right]$$

The total is 171. Hence the direct average is $171/25 = 6.84$

2) By using relative frequency

$$\left[\begin{array}{l}(7\cdot 2+8\cdot 6+3\cdot 1+2\cdot 1+5\cdot 3+ \\ 6\cdot 4+9\cdot 3+10\cdot 3+4\cdot 2)\end{array}\right]$$

The above totals to 171. Hence the direct average is: $171/25 = 6.84$

Prob. 15. The table to be used is as follows:

x	-2	-1	0	1
P(X=x)	0.1	0.4	0.3	0.2

(a) $EX = -2\cdot 0.1 + -1\cdot 0.4 + 0\cdot 0.3 + 1\cdot 0.2 = -0.4$

(b) $EX^2 = (-2)\cdot(-2)\cdot 0.1 + (-1)\cdot(-1)\cdot 0.4 + (0)\cdot(0)\cdot 0.3 + (1)\cdot(1)\cdot 0.2 = 1$

(c) $E(X(X-1)) = (-2)\cdot(-3)\cdot 0.1 + (-1)\cdot(-2)\cdot 0.4 + (0) + 1\cdot(0)\cdot 0.2 = 1.4$

Prob. 17. The mean μ is obtained by totaling the entries in the third row, so it is -0.1

x	-3	-1	1.5	2
$P(x = X)$	0.2	0.3	0.4	0.1
$x \cdot P(x = X)$	-0.6	-0.3	0.6	0.2

Variance $= (-3-(-0.1))^2\cdot(0.2) + (-1-(-0.1))^2\cdot(0.3) + (1.5-(-0.1))^2\cdot(0.4) + (2-(-0.1))^2\cdot(0.1) = 3.39$

Standard Deviation $= \sqrt{\text{Variance}} = 1.841195$

Prob. 19. In this instance, all cases are equiprobable. Hence for each $x_i, i = 1, 2, 3 \cdots 10$, $P(x_i = X) = 1/10 = 0.1$

(a) $EX = \Sigma_{i=1}^{i=10} x_i \cdot P(X) = 55(0.1) = 5.5$

(b) $var(X) = E(X - \mu)^2$ For clarity, the terms that are to be summed up to find variance are tabulated below.

x	$(x-\mu)^2 \cdot P$	Value
1	$(1-5.5)^2) \cdot (0.1)$	2.025
2	$(2-5.5)^2 \cdot (0.1)$	1.225
3	$(3-5.5)^2 \cdot (0.1)$	0.625
4	$(4-5.5)^2 \cdot (0.1)$	0.225
5	$(5-5.5)^2 \cdot (0.1)$	0.025
6	$(6-5.5)^2 \cdot (0.1)$	0.025
7	$(7-5.5)^2 \cdot (0.1)$	0.225
8	$(8-5.5)^2 \cdot (0.1)$	0.625
9	$(9-5.5)^2 \cdot (0.1)$	1.225
10	$(10-5.5)^2 \cdot (0.1)$	2.025
VARIANCE = 8.25		

Prob. 21.

(a) By definition of expectation, $E(x) = \Sigma_x x \cdot p(x)$, we have $E(aX + b) = \Sigma_x (ax + b) \cdot p(x)$

(b) $E(aX + b) = \Sigma_x (ax + b) \cdot p(x) = \Sigma_x (ax) \cdot p(x) + \Sigma_x b \cdot p(x) = a \cdot \Sigma_x x \cdot p(x) + b \cdot 1 = aE(x) + b$ (since $\Sigma_x p(x) = 1$)

Prob. 23.

(a) $P(X = 1, Y = 0) = 0.1$

(b) $P(X = 1) = 0.1 + 0.4 = 0.5$

(c) $P(Y = 0) = 0.3 + 0.1 = 0.4$

(d) $P(Y = 0/X = 1) = (0.1)/(0.1 + 0.4) = 0.2$

Prob. 25.

(a) $E(X) = -2 \cdot 0.1 + -1 \cdot 0 + 0 \cdot 0.3 + 1 \cdot 0.4 + 2 \cdot 0.05 + 3 \cdot 0.15 = 0.75.$

$E(Y) = -2 \cdot 0.2 + -1 \cdot 0.2 + 0 + 1 \cdot 0.3 + 2 \cdot 0 + 3 \cdot 0.2 = 0.3$

(b) Since X and Y are independent, $E(X+Y) = EX + EY = 0.75 + 0.3 = 1.05$.

(c) $var(X) = E(X^2) - (EX)^2$ Let us compute $E(X^2)$: $E(X^2) = (-2) \cdot (-2) \cdot (0.1) + 0 + 0 + (1) \cdot (1) \cdot (0.4) + 2 \cdot 2 \cdot (0.05) + 3 \cdot 3 \cdot (0.15) = 0.4 + 0.4 + 0.2 + 1.35 = 2.35$

$var(X) = 2.35 - 0.75 \cdot (0.75) = 1.7875$

$var(Y) = E(Y^2) - (EY)^2$

Let us compute now $E(Y2) = (-2) \cdot (-2) \cdot (0.2) + (-1) \cdot (-1) \cdot (0.2) + 0 + 1 \cdot 1 \cdot (0.3) + 0 + 3 \cdot 3 \cdot (0.2) = 0.8 + 0.2 + 0.3 + 1.8 = 3.1$

$var(Y) = 3.1 - 0.3 \cdot (0.3) = 3.01$

(d) Since X and Y are independent $var(X+Y) = var(X) + var(Y) = 1.7875 + 3.01 = 4.7975$

Prob. 27.

(a) $var(X) = E(X - EX)^2$. Lets look at $X - EX$. This value may be positive or negative. However, $(X - EX)^2$ is always positive because $(+ve) \cdot (+ve) = (+ve)$ and $(-ve) \cdot (-ve)$ is also positive, therefore var(X) is always ≥ 0

(b) Now $var(X) = EX^2 - (EX)^2$, since $var(X) \geq 0, [EX^2 - (EX)^2] \geq 0$, $EX^2 \geq (EX)^2$

Prob. 29. The total number of outcomes is $2^{10} = 1024$ If X is the number of heads, lets calculate the probability distribution of X.

$$\begin{bmatrix} X & P(x=X) & \\ 0 & C(10,0)/1024 & 1/1024 \\ 1 & C(10,1)/1024 & 10/1024 \\ 2 & C(10,2)/1024 & 45/1024 \\ 3 & C(10,3)/1024 & 120/1024 \\ 4 & C(10,4)/1024 & 210/1024 \\ 5 & C(10,5)/1024 & 252/1024 \\ 6 & C(10,6)/1024 & 210/1024 \\ 7 & C(10,7)/1024 & 120/1024 \\ 8 & C(10,8)/1024 & 45/1024 \\ 9 & C(10,9)/1024 & 10/1024 \\ 10 & C(10,10)/1024 & 1/1024 \end{bmatrix}$$

$P(X = 5) = 252/1024 = 0.24609375$
$P(X \geq 8) = (45 + 10 + 1)/1024 = 0.0546875$
$P(X \leq 9) = (1+10+45+120+210+252+210+120+45+10)/1024 = 1023/1024 = 0.999023$

Prob. 31. Toss a fair die 6 times. The probability of success is 1/6. Hence, $P(X = k) = \binom{6}{k}(1/6)^k(5/6)^{6-k}$, $k = 0, 1, 2, \ldots, 6$.

Prob. 33. 20% of all plants are infected, therefore $(100 - 20) = 80\%$ are not infected. If 20 plants are picked, Since the events are independent, the probability that none of them carried aphids is $(0.8)^{20} = 0.011529215$.

Prob. 35. In a box contains 10 apples the probability that an apple is spoiled is 0.1

(a) The expected number of spoiled apples per box is $EX = (10)(0.1) = 1$.

(b) The probability that a box picked at random from the shipment will contain no spoiled apples is $(0.9)^{10} = 0.3486$. Hence the expected number of boxes that contain no spoiled apples is 3.486

Prob. 37. The probability of guessing a question right is 1/4, so $E(X) = (1/4) \cdot 50 = 12.5$

Prob. 39. With 12 green and 24 blue balls

(a) If ten balls are taken out and six of the ten are blue. The total number of ways that ten balls can be picked from 36 balls is $C(36, 10) = 254186856$. The number of ways that 6 blue balls can be picked from 24 blue balls is $C(24, 6) = 134596$. The number of ways that 4 green balls can be picked from 12 green balls is is $C(12, 4) = 495$, therefore the probability of picking 6 blue balls $134596 \cdot 495/(254186856) = 0.26211$

(b) The probability distribution is binomial with $n = 10$, and success probability is 24/36.

The probability $P(S10 = 6)$ is: $C(10, 6) \cdot (24/36)^6 \cdot (12/36)^4 = (210) \cdot (2/3)^6 \cdot (1/3)^4 = 13440/59049 = 0.2276$

Prob. 41. $N1 = 10, N2 = 14, N3 = 6$. so $p1 = 0.2, p2 = 0.35, p3 = 0.45$ and $P(N1 = 10, N2 = 14, N3 = 6) = ((30!)/(10! \cdot 14! \cdot 6!)) \cdot (0.2)^{10} \cdot (0.35)^{14} \cdot (0.45)^6$

Prob. 43. Let's go through Example 28 where Mendel's experiments are described,

Result of Mendel's experiment		
Total No. of seeds is 40		
Characteristic of seed	Quantity	Probability of outcome shown in Col.2
Round Yellow	20	9/16
Round Green	10	3/16
Wrinkled Yellow	8	3/16
Wrinkled Green	2	1/16

This is obviously an exercise in multinomial distribution. The number of ways in which the 40 seeds can belong to the four categories mentioned is:

$$\frac{40!}{(20! \cdot 10! \cdot 8! \cdot 2!)}$$

Now the probability of each phenotype occurring from among the 16 possible genotypes is given in the Table above. Taking this into account, the probability of the outcome given in the above Table is:

$$(9/16)^{20} \cdot (3/16)^{10} \cdot (3/16)^8 \cdot (1/16)^2$$

So, finally the probability of the 'successful' outcome as desired by Mendel can be found by multiplying the two long preceding expressions with factorials and powers of numbers and so on.

Prob. 45. This is a multinomial distribution

Colour and Number of balls sampled		Probability of picking	
Color	Number	Symbol	Value
Green	N1	p1	6/24
Blue	N2	p2	8/24
Red	N3	p3	10/24

We want to find $P(N1 = 2, N2 = 2, N3 = 2) = \frac{6!}{2!2!2!}(\frac{6}{24})^2(\frac{8}{24})^2(\frac{10}{24})^2$.

Prob. 47. For convinience, let tabulate the data:

Genotype and Number			Probability	
Genotype	Symbol	Value	Symbol	Value
CC	N1	5	p1	1/4
Cc	N2	12	p2	2/4
cc	N3	6	p3	1/4

We want to find $P(N1, N2, N3)$. This is a multinomial distribution The number of offspring is $5 + 12 + 6 = 23$ and the required probability is equal to

$$\frac{23!}{5!12!6!}(\frac{1}{4})^5(\frac{1}{2})^{12}(\frac{1}{4})^6$$

Prob. 49. For $k = 1$, the probability of heads appearing on the first throw is 1/2 For $k = 2$, the probability of tails and then heads is $(1/2)^2 = 1/4$ For $k = 3$, the probability of tails, tails and then heads is $(1/2)^3 = 1/8$

Prob. 51.

$$\begin{bmatrix} k & P(X = k) \\ 1 & 1/2 = 0.5 \\ 2 & 1/2 \cdot 1/2 = 0.25 \\ 3 & (1/2)^3 = 0.125 \end{bmatrix}$$

Therefore $P(X > 3) = (1 - 0.5 - 0.25 - 0.125) = (1 - 0.875) = 0.125$

Prob. 53.

$$\begin{bmatrix} x & Computing P(X=x) & Final value of P \\ 1 & 1/2 & 0.5 \\ 2 & (1/2)^2 & 0.25 \\ 3 & (1/2)^3 & 0.125 \\ 4 & (1/2)^4 & 0.0625 \end{bmatrix}$$

Therefore $P(X \leq 4) = 0.5 + 0.25 + 0.125 + 0.0625 = 0.9375$

Prob. 55. The probability that at least 20 draw are needed is equal to $(14/15)^{19}$ since the probability of success is $1/15$.

Prob. 57. The distribution is geometric with $p = 5/30$, so $ET = 1/p = 30/5 = 6$ and $var(T) = (1-p)/p^2 = (1-5/30)/(5/30)^2 = 30$

Prob. 59.

(a) The probability that exactly 6 balls are needed if the balls are replaced is $(9/10)^5 \cdot (1/10) = 0.059049$

(b) If the balls are not replaced, the probability is $(9/10) \cdot (8/9) \cdot (7/8) \cdot (6/7) \cdot (5/6) \cdot (1/5) = 0.1$

Prob. 61.

(a) The probability that the first success occurs on the k-th trial is $(1-p)^{(k-1)} \cdot p$

(b) $\binom{k-1}{1} p^2 (1-p)^{k-2}$

Prob. 63. We can see that it is Poisson distributed with $\lambda = 2$ $P(X = k) = e^{(-\lambda)} \cdot \lambda^k / k!$, so

$$\begin{bmatrix} x & P(x) \\ 0 & e^{-2} \\ 1 & 2e^{-2} \\ 2 & 2e^{-2} \\ 3 & (4/3)e^{-2} \end{bmatrix}$$

Prob. 65. X is Poisson distributed with $\lambda = 1$

(a) $P(X \geq 2) = 1 - P(X = 0) - P(X = 1) = 1 - e^{-1}(1 + 1) \approx 0.2642$

(b) $P(1 \leq X \leq 3) = P(1)+P(2)+P(3)$, $P(1 \leq X \leq 3) = e^{-1}(1+\frac{1}{2}+\frac{1}{6}) \approx 0.6131$.

Prob. 67. X is Poisson distributed with $\lambda = 1.5$
$P(X = k) = e^{-\lambda} \cdot \lambda^k/k!$
$P(X = 0) = 0.2231$
$P(X = 1) = 0.3346$
$P(X = 2) = 0.2510$
$P(X = 3) = 0.1255$
$P(X > 3) = 1 - P(X = 0) - P(X = 1) - P(X = 2) - P(X = 3) = 1 - e^{-1.5}(1 + 1.5 + \frac{(1.5)^2}{2} + \frac{(1.5)^3}{6})$.

Prob. 69. X is Poisson distributed with $\lambda = 2$
$P(X = k) = e^{-\lambda} \cdot \lambda^k/k!$
$P(X = 0) = 0.1353 = e^{-2}$
$P(X = 1) = 0.2706 = 2e^{-2}$
The probability that X is at least $2 = 1 - P(X = 0) - P(X = 1) = 1 - e^{-2}(1 + 2) \approx 1 - 0.1353 - 0.2706 = 0.5940$

Prob. 71. X is Poisson distributed with $\lambda = 7$, $P(X = k) = e^{-\lambda} \cdot \lambda^k/k!$ and $P(X = 0) = e^{-7}$

Prob. 73. X is Poisson distributed with $\lambda = 0.5$ $P(X = k) = e^{-\lambda} \cdot \lambda^k/k!$ Probability that there is at least one typo in a page = 1 - Probability that there are no typos $= 1 - P(X = 0) = 1 - e^{-0.5} \approx 0.3934$

Prob. 75. X is Poisson distributed with $\lambda = 3$, $P(X = k) = e^{-\lambda} \cdot \lambda^k/k!$, $P(X = 0) = 0.0497$ $P(X = 1) = 0.1493$. The probability of at least two substitutions is $1 - P(0) - P(1) = 1 - 0.0497 - 0.1493 = 0.801$

Prob. 77. Number of customers arriving at the post office is assumed to be Poisson distributed with a mean λ of 4 customers per hour.

(a) If no one arrives during an hour, the value of k in the Poisson formula (see the earlier exercise) is zero. $P(X = 0) = e^{-4} \cdot (-4)^0/0! = 0.01831564$ If only one person arrives during the hour(we will need this result in subdivision (c) below), all we need to do is to use the Poisson formula with k = 1 (of course λ is still 4) $P(X = 1) = (e^{-4} \cdot -4^1)/1! = 0.07326255$

(b) Two customers arrive during an hour. So $\lambda = 4$ and $k = 2$ $P(X = 2) = (e^{-4} \cdot 4^2)/2! = 0.14652511$

(c) Exactly two customers arrive between 2 PM and 4 PM. Note that the duration is *two successive hours* and that the arrivals in these two one hour periods are *two independent events*. Note that there are *three* different ways in which the *two* independent events could have taken place. $\lambda = 4$ for all cases.

Customers		Probability	k	Probability Value
Number	Hour	Symbol		using Poisson formula
Zero	First	$P(X = 0)$	0	0.01831564
One	First	$P(X = 1)$	1	0.07326255
Two	First	$P(X = 2)$	2	0.14652511

A Table similar to the above can be drawn up for the second independent event, namely $P(Y = k), k = 0, 1, 2$. These will naturally have the same values as the respective $P(X = k)$. Now we come to the important task of finding the probability that two persons visited during two successive hours. Note that there are *three* alternative ways in which the TWO independent events could have taken place.

Customers				Probability	Probability
Hour	Number	Hour	Number	of first event	of second event
First	0	Second	2	$P(X = 0)$	$P(Y = 2)$
First	1	Second	1	$P(X = 1)$	$P(Y = 1)$
First	2	Second	0	$P(X = 2)$	$P(Y = 0)$

Note that we have listed above the three alternative ways in which the visit of the three customers could have taken place. The respective probabilities are:

$$\begin{bmatrix} \textbf{Alternative} & ProbabilityExpression & Calculation & Value \\ 1 & P(X=0)\cdot P(Y=2) & (0.01831564)\cdot(0.14652511) & 0.00268370 \\ 2 & P(X=1)\cdot P(Y=1) & (0.07326255)^2 & 0.00536740 \\ 3 & P(X=2)\cdot P(Y=0) & (0.14652511)\cdot(0.01831564) & 0.00268370 \end{bmatrix}$$

Since there are three ways in which the desired event (of three customers visiting in two successive hours) could have taken place, the probability of the desired event taking place is the sum of the three probabilities listed in the Table above. Some simplification is possible since numerically $P(X = k) = P(Y = k), k = 0, 1, 2$ The final answer (at last!) is $2 \cdot (P(X = 0) \cdot P(X = 2)) + P(X = 1)^2 = 0.0107348$

By the way, do you see that the probability of the first alternative is equal to that of the third alternative? What is more surprising is that the probability of the first (or the third) alternative is half that of the second alternative! Can you guess the reason?

(d) We have

$$P(Y = 2|X + Y = 2) = \frac{P(Y = 2; X + Y = 2)}{P(X + Y = 2)} = \frac{P(Y = 2; X = 0)}{P(X + Y = 2)}$$

Using independence in the numberator and the result in (c) for the denominator, we find that this probability is equal to

$$\frac{P(Y = 2)P(X = 0)}{P(X + Y = 2)} = \frac{e^{-4}\frac{4^2}{2}e^{-4}}{e^{-8}32} = \frac{1}{4}$$

Prob. 79.

(a) Given that $\lambda = 3$, we are asked to find the probability $P(X + Y) = 2$. Let us first tabulate the probability $P(X = k)$ for k = 0,1,2 using Poisson equation and making the appropriate substitutions.

k	$e^{-\lambda}$	λ^k	$k!$	$P(X = k) or P(Y = k)$
0	0.049787068	1	1	0.049787068
1	0.049787068	3	1	0.149361205
2	0.049787068	9	2	0.224041807

$X + Y$ will have the value 2 under anyone of the following conditions.

$(X = 0)$ AND $(Y = 2)$ OR

$(X = 1)$ AND $(Y = 1)$ OR

$(X = 2)$ AND $(Y = 0)$

It follows therefore that $P(X+Y = 2) = P(X = 0) \cdot P(Y = 2) + P(X = 1) \cdot P(Y = 1)$

$+P(X = 2) \cdot P(Y = 0) = 2 \cdot (P(X = 0) \cdot P(X = 2) + (P(X = 1))^2 = 2 \cdot 0.049787068 \cdot 0.224041807 + (0.149361205)^2 = 0.044617539$

(b) Given that $X + Y = 2$, we are asked to find the probability that $X = kfork = 0, 1, 2$

Case1: Let $k = 0$. This means that $X = 0$ and $Y = 2$ The required probability for case 1 is:

Case 1 $P(X = 0), (Y = 2))/(P(X+Y) = 2)$ which is: $0.011154385/0.044617539 = 0.2500$

Case 2: $(P(X = 1) \cdot P(Y = 1))/(P(X+Y) = 2) = 0.022308770/0.044617539 = 0.5000$

Case 3: is identical to case 1. Notice that the probabilities of the above three cases add up to exactly unity.

Prob. 81. Since only 1 in 1000 experiences side effect $p = 1/1000$ Since 500 is the strength of the test group, $n = 500$. We shall use Poisson approximation approach. $\lambda = n \cdot p = 500 \cdot 1/1000 = 0.5$ $P(X = k) = e^{-\lambda} \cdot \lambda^k/k!$ $P(X = 0) = 0.60653$ is the probability that none of the 500 persons experiences side effect

Prob. 83.

(a) The probability that there are no cases of Down's syndrome $P(X = 0) = (699/700)^{1000} = 0.2394$

The probability that there is exactly one case of Down's syndrome $P(1) = C(1000, 1) \cdot (699/700)^{999} \cdot (1/700) = (1000) \cdot (0.239748) \cdot (1/700) = 0.342498$. Hence the probability that there is at most one case is: $P(0) + P(1) = 0.5818$

(b) We are using a Poisson approximation $\lambda = n \cdot p = 1000 \cdot 1/700 = 1.42857$ $P(X = k) = e^{-\lambda} \cdot \lambda^k/k!$ $P(X = 0) = 0.2396$ $P(X = 1) = 0.3423$. Hence the probability that there is at most one case is $P(X = 0) + P(X = 1) = 0.5819$

Prob. 85. There are $P = 50$ parasitoids, vying with one another to find a host. The probability of any one of them encountering a host is $a = 0.03$ Hence, the probability that the host escapes detection is (for such cases, we take the only the first term of Poisson distribution) is $e^{-0.3 \cdot 50} = e^{-1.5} = 0.22313016$

12.5 Continuous Distributions

Prob. 1. As per Equation 12.28 in Example 1 in Section 12.5.1, any non-negative function which satisfies the condition

$$\int_{-\infty}^{\infty} f(x)dx = 1$$

is a candidate for being a density function. Carrying out the indicated non-negative test, we see that $f(x) \geq 0$. Furthermore,

$$\int_{-\infty}^{\infty} (3e^{-3x})dx = \int_{0}^{\infty} (3e^{-3x})dx = \frac{3e^{-3x}}{-3}\Big|_{0}^{\infty} = 1$$

Thus $f(x)$ is indeed a probability density function. The corresponding distribution function can be seen, from above, to be:

$$F(x) = \begin{cases} 1 - e^{-3x} & \text{for } x \geq 0 \\ 0 & \text{for } x < 0 \end{cases}$$

Prob. 3. Given that $f(x) = c/(1 + x^2)$, it is necessary to determine c so that $f(x)$ is a probability density function. All we need to do is to find the area under the curve for over all real numbers. Let $x = \tan\theta$; then $dx = \sec^2\theta d\theta$

$$\int_{-\infty}^{\infty} c\frac{dx}{1+x^2} = \int_{-\pi/2}^{\pi/2} c(\sec^2\theta)\frac{d\theta}{(1+\tan^2\theta)} = c\pi$$

Hence, we conclude that $c = \frac{1}{\pi}$

Prob. 5.

$$\text{Given density function} \quad f(x) = \begin{cases} 2e^{-2x} & \text{for } x > 0 \\ 0 & \text{for } x \leq 0 \end{cases}$$

It is required to find Expectation and Variance. Using integration by parts,

$$EX = \int_{0}^{\infty} x2e^{-2x}dx = -xe^{-2x} - (0.5)e^{-2x}\Big|_{0}^{\infty} = \frac{1}{2}$$

The variance is found from the usual expression $(EX^2 - (EX)^2)$

$$EX^2 = \int_{0}^{\infty} (x^2)2e^{-2x})dx = -x^2e^{-2x}\Big|_{0}^{\infty} + \int_{0}^{\infty} 2xe^{-2x}dx = \frac{1}{2}$$

Hence the variance is $\frac{1}{2} - (\frac{1}{2})^2 = \frac{1}{4}$

Prob. 7. In this problem we are given the distribution function $F(x)$ to begin with.

$$F(x) = (1 - 1/x^3), x > 1$$

The density function is hence:

$$f(x) = \frac{3}{x^4}, x > 1$$

EX can be found by a simple integration, to be

$$\int_1^\infty x\frac{3}{x^4}dx = -\frac{3}{2}\frac{1}{x^2}\Big|_1^\infty = \frac{3}{2}$$

EX^2 can also be found, by another integration to be

$$\int_1^\infty x^2\frac{3}{x^4}dx = -\frac{3}{x}\Big|_1^\infty = 3$$

Hence Var(x) is $3 - (3/2)^2 = 3/4$.

Prob. 9.

$$\text{Given density function} \quad f(x) = \begin{cases} (a-1)x^{-a} & \text{for } x > 1 \\ 0 & \text{for } x \le 1 \end{cases}$$

(a) For $a > 2$,

$$\begin{aligned} EX &= \int_1^\infty (a-1)x^{1-a}dx \\ &= ((a-1)/(2-a))x^{2-a}\Big|_1^\infty \\ &= \text{unbounded} \end{aligned}$$

For $a = 2$, $EX = \int_1^\infty \frac{1}{x}dx$ is undefined

If $a \le 2$ then EX is unbounded, since x^{2-a} increases without limit.

(b)

$$EX = (a-1)/(a-2) \quad \text{if} \quad a > 2$$

Prob. 11.

(a) To prove that the curve is symmetric about $x = \mu$, we need to show that $f(x+\mu) = f(x-\mu)$. Define $K = 1/(\sigma\sqrt{2\pi})$, $f(x) = Ke^{-(x-\mu)^2/2\sigma^2}$, then we need to show that $f(\mu+x) = f(\mu-x)$. Obviously $e^{-(\mu+x-\mu)^2} = e^{-(\mu-x-\mu)^2}$ thus, it is clear that the normal distribution curve is symmetric about $x = \mu$

(b) To prove that the maximum of the curve is at $x = \mu$, we need to show that the first derivative is zero at that point. Let us also make the substitution $v = \sigma^2$

Let $f(x) = e^{-(x-\mu)^2/2v}$, we only need to prove that $df(x)/dx = 0$ and $f'(x) = -\frac{1}{v}(x-\mu)f(x)$, it is clear that at $x = \mu$, $f'(x) = 0$. Furthermore, $f'(x) > 0$ for $x < \mu$ and < 0 for $x > \mu$. Hence $f(x)$ has a maximum at $x = \mu$.

Note that we have omitted K in the above expression, as it plays no role in differentiation (or in the conclusion)

(c) We need to examine the second derivative at such a point. In order to reduce clutter in long algebraic expressions, let us first simplify the distribution function by making appropriate substitutions. Let us also omit the constant K as it plays no role here.

$$
\begin{aligned}
\text{Normal Curve} f(x) &= Ke^{-(x-\mu)^2/2\sigma^2} \\
\text{Put} \quad (x-\mu) &= z \\
\text{Then} \quad dx &= dz \\
\text{Variance} \quad v &= \sigma^2 \\
\text{K can be omitted for simplicity} & \\
\text{Incorporating above changes} \quad f(z) &= e^{-z^2/2v} \\
\text{Differentiating once} \quad f'(z) &= (-z/v)f(z) \\
\text{Differentiating again} \quad f''(z) &= (-1/v)(zf'(z) + f(z)) \\
\text{Simplifying} \quad f''(z) &= -(f(z)/v)(1 - z^2/v) \\
\text{It is clear that} \quad f''(z) &= 0 \quad \text{at} \quad z = \pm\sigma \\
\text{Furthermore} \quad f''(z) \quad \text{changes sign at} \quad z = \pm 5 & \\
\text{Thus, inflection points occur at} \quad x &= \mu \pm \sigma
\end{aligned}
$$

Note that f(z) is obtainable from f(x) by a mere translation by μ on the x- axis. Hence the shape of the curve is preserved

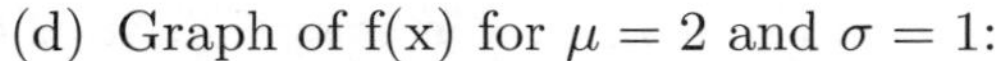
(d) Graph of f(x) for $\mu = 2$ and $\sigma = 1$:

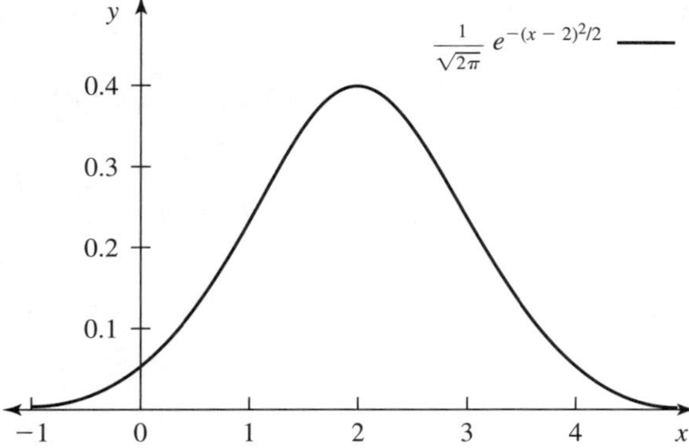

Prob. 13. Given that

$$\mu = 12.8 \quad \text{and} \quad \sigma = 2.7$$

we can evaluate the following:

$$\mu - 2\sigma = 7.4 \quad \text{and} \quad \mu + 2\sigma = 18.2$$

From the graph of normal distribution, it is known that 95 percent of the area falls between

$$\mu - 2\sigma \quad \text{and } \mu + 2\sigma$$

i.e. between 7.4 and 18.2 respectively. Hence

$$P(X \in [7.4, 18.2] = 0.95)$$

In a similar manner, it is possible to find an interval such that 99 percent falls within this interval. From the graph of normal distribution it is known that 99 percent of the area falls between

$$\mu - 3\sigma \quad \text{and} \quad \mu + 3\sigma$$

This works out to 4.7 and 20.9 respectively. Hence $P(X \in [4.7, 20.9] = 0.99)$.

Prob. 15. In this problem, we are asked to determine the fraction of population that will fall in the interval $P(\mu \leq X < \infty)$. Since the normal distribution curve is symmetric about μ, the area to the right of μ is one-half the total area. Hence the answer to the question is $\frac{1}{2}$.

Prob. 17. The range mentioned in this problem, which is similar to the previous one is: $-\infty$ and $(\mu + 3\sigma)$.

$$Area \Big|_{-\infty}^{\mu} = \frac{1}{2} \quad \text{Known fact} \tag{12.1}$$

$$Area \Big|_{\mu-3\sigma}^{\mu+3\sigma} = 0.995 \quad \text{Known fact} \tag{12.2}$$

$$\text{Hence} \quad Area \Big|_{\mu}^{\mu+3\sigma} = (0.995/2) = 0.495 \tag{12.3}$$

$$\text{Adding (1) and (3) above} \quad Area \Big|_{-\infty}^{\mu+3\sigma} = (0.5 + 0.495) = 0.995 \tag{12.4}$$

$$\tag{12.5}$$

Hence the fraction of population that will fall in the range given in this problem is 0.995

Prob. 19. In this problem, both the lower and upper limits of X are to the left of μ The range given is $(-\infty)$ to $\mu - 2\sigma$. The fraction of population in this range is: $(1.0 - 0.95)/2 = 0.025$

Prob. 21. For $\mu = 3$ and $\sigma = 2$

(a) The probability $P(X \leq 4)$ is to be determined. The normalized random variable $Z = (X - \mu)/\sigma$ is $(4 - 3)/2 = 0.5$. From Appendix B, the area under the standard normalised curve to the left of $Z = 0.5$ is 0.6915. Hence $P(X \leq 4) = 0.6915$. In other words, there is a 69.15 percent probability that the random variable is less than or equal to 4. Since parts (b), (c), (c) of this Problem are similar to (a) let us consider tabulating the key results.

(b)

$$\text{Range of X} = P(2 \leq X \leq 4)$$

$$\text{Corresponding range of Z} = P(-0.5 \leq Z \leq 0.5)$$

$$\text{Normalized curve} \quad Area \Big|_{-\infty}^{0.5} = 0.6915$$

$$\text{Normalized curve} \quad Area \Big|_{-\infty}^{0} = 0.5$$

$$\text{Hence} \quad Area \Big|_{0}^{0.5} = 0.1915$$

$$\text{Hence} \quad Area\Big|_{-0.5}^{0.5} = 2(0.1915) = 0.383$$

In other words the probability of the random variable being in the range (2 to 4) is 38.3 percent.

(c) Though this is similar to (b) above, the important difference is that the range in question is above a certain value.

$$\begin{aligned}
\text{Range of X} &= P(X > 5) \\
\text{Corresponding range of Z} &= P(Z > 1) \\
\text{Normalized curve} \quad Area\Big|_{-\infty}^{1} &= 0.8413 \\
\text{Normalized curve} \quad Area\Big|_{-\infty}^{\infty} &= 1.0 \\
\text{Hence} \quad Area\Big|_{1}^{\infty} &= 1.0 - 0.8413 = 0.1587 \\
\text{Area above given X} \quad Area\Big|_{5}^{\infty} &= 0.1587
\end{aligned}$$

To put the final result in words, the probability of the random variable X being more than 5 is 15.87 percent.

(d) In this case also, area determination has to done in a range which is negative. So the Table in Appendix B has to be used with caution!

Description	Equation
Range of X	$P(X \leq 0)$
Corresponding Range of Z	$P(Z \leq -1.5)$
Area in Normalized Curve	$Area\Big\|_{-\infty}^{1.5} = 0.9332$
Using symmetry	$Area\Big\|_{-1.5}^{\infty} = 0.9332$
So, area we are looking for	$Area\Big\|_{-\infty}^{-1.5} = 1 - 0.9332 = 0.0668$

Prob. 23. Here, we are asked to find the value of x, given its probability. So, we will have to find the range of absissa of normal curve given the value of the enclosed area. For (a) to (d) below, $\mu = 1$ and $\sigma = 2$

(a) Given that $P(X \leq x) = 0.9$, we are asked to find value of x.

$$\text{From Appendix B} \quad Area\Big|_{-\infty}^{1.28} = 0.8997 \tag{12.6}$$

$$\text{Similarly } Area\Big|_{-\infty}^{1.29} = 0.9015 \tag{12.7}$$

$$\text{Find } X = \mu + Z\sigma = 1 + (1.28)2 = 3.56 \tag{12.8}$$

$$\tag{12.9}$$

The value of Z that will give the required area of 0.9 (instead of 0.8997) is taken to be 1.28 Hence $P(X \leq 3.56) = 0.9$

(b)

$$\text{Given that} \quad P(X > x) = 0.4$$

$$\text{It follows that} \quad P(X \leq x) = 0.6$$

$$\text{from Appendix B} \quad Area\Big|_{-\infty}^{0.25} = 0.5987$$

$$\text{from Appendix B} \quad Area\Big|_{-\infty}^{0.26} = 0.6026$$

Hence, we can interpolate and say that the area to the left of the vertical line at $z = 0.255$ will be 0.6. $X = \mu + Z\sigma = 1 + 2(0.255) = 1.51$ and $P(X > 1.51) = 0.4$

(c) Note that, in working out this part, we can benefit from the result obtained above in part (b). The area in the range $1.51 < x < \infty$ has been derived to be 0.4. The area in the range $1 < x < \infty$ is known to be 0.5. Hence the area in the range $1 < x < 1.51$ is 0.1 By symmetry, the area in the range 0.49 to 1 is also 0.1. Thus the area in the range $-\infty$ to 0.49 is 0.4. Thus, $P(X \leq 0.49) = 0.4$.

(d) A new element added in this question is modulus.

$$P(|X - 1| < x) = 0.5$$

As before, $\mu = 1$ and $\sigma = 2$ First, let us take a close look at the standard normalised distribution curve. The mean is zero. The value of Z increases from zero to infinity in the positive direction and decreases from zero to infinity in the negative direction. If you consider $|Z|$, it increases in both directions.

$$\begin{aligned}
\text{Area in the region} \quad -\infty \leq z < 0.675 &= 0.75 \\
\text{Area in the region} \quad -\infty \leq z < 0 &= 0.5 \\
\text{Hence, area in the interval} \quad 0 \leq z < 0.675 &= 0.25 \\
\text{By symmetry, area in} \quad 0 \leq z < -0.675 &= 0.25 \\
\text{Thus, area in the interval} \quad 0.675 \leq z \leq 0.675 &= 0.5 \\
\text{So we can conclude that} \quad |z| \leq 0.675 &= 0.5
\end{aligned}$$

From the Z world, we can go over to the X world by a linear transformation. The shape of the curve, the area under given intervals, etc., will not change. However, the numerical values of X range will be different from those of Z.

$$\begin{aligned}
\text{Converting z1} \quad X_1 &= 1 - 2 * 0.675 = -0.3 \\
\text{Converting z2} \quad X_2 &= 1 + 2 * 0.675 = 2.3
\end{aligned}$$

it is easily verified that $P(|X - 1| \leq 1.3) = 0.5$.

Prob. 25.

(a)

Given data	$\mu = 500$; $\sigma = 100$
Question	Find $P(X > 700)$
Let us normalise x to z	$z = (700 - 500)/100 = 2.0$
Area under $-\infty < z < 2.0$	found from Tables as 0.9772
Area beyond 2.0	$(1.0 - 0.9772) = 0.0228$
Probability that score exceeds 700	Answer is 2.28 percent

(b) To find the answer, we first note, from the standard normal curve, that the area under $-\infty < z < 1.28$ is 0.8997 which is almost equal to 0.9. This means that 10 percent of students scored higher than what corresponds to a value of z of 1.28. Using the conversion formula, $x = \mu + z\sigma = 500 + (1.28)(100) = 628$ To conclude, 10 percent of students obtained a score of 628 or higher.

Prob. 27. Tabulating the data and computations as usual

Value of z corresponding to 5000 gm	$(5000 - 3720)/527 = 2.49$
Area for range $-\infty < z < 2.49$	from std. tables 0.9925
Area for range $z > 2.49$	$(1 - 0.9925) = 0.0075$
Percentage of animals having a weight exceeding 5000 gm	0.75

Prob. 29.

$$\begin{aligned} \text{Given data} &= \mu = 2 \\ \text{Given data} &= \sigma = 1 \\ P(0 \leq X \leq 3) = P(-2 \leq Z \leq 1) &= F(1) - [1 - F(2)) \\ = 0.8413 - (1 - 0.9772) &= 0.8185 \end{aligned}$$

Prob. 31.

$$\begin{aligned} f(z) &= (1/\sqrt{2\pi})e^{-z^2/2} \\ \int_{-\infty}^{\infty} f(z)dz &= 1 \\ \int_{-\infty}^{\infty} zf(z)dz &= 0 \\ \int_{-\infty}^{\infty} z^2 f(z)dz &= 1 \end{aligned}$$

Now we proceed to find $E(X)$. In all equations below, the limits of integration are $-\infty$ to ∞

$$\begin{aligned} \text{Define } K &= (1/\sqrt{2\pi\sigma^2}) \\ \int xf(x)dx &= K \int x \exp[-(x-\mu)^2/2\sigma^2]dx \\ \text{Substitute } x &= \mu + \sigma z \end{aligned}$$

$$\begin{aligned}
\text{Hence } dx &= \sigma dz \\
\int xf(x)dx &= K\mu \int e^{-z^2/2}dz + K\sigma \int ze^{-z^2/2}dz \\
\text{since } \int zf(z)dz &= 0 \\
\int xf(x)dx &= K\mu \int e^{-z^2/2}\sigma dz = \mu
\end{aligned}$$

Thus EX value is proved to be equal to μ

Prob. 33. It is required to find $E|X|$, given that X is standard normally distributed. It means that mean is zero and standard deviation is one.

Let us split the interval of integration into two parts: $-\infty$ to zero and zero to $+\infty$. Thus $EX = (EX1 + EX2)$, where EX1 and EX2 are defined and derived below. Also note that in the first interval, where x is negative, $|x|$ can be replaced by -x.

std. normal fn $\quad f(x) = (1/\sqrt{2\pi})e^{-x^2/2}$

$$\begin{aligned}
\text{substitute} \quad K &= (1/\sqrt{2\pi}) \\
\text{split integration into two parts} \quad EX &= EX1 + EX2 \\
EX1 &= \int_{-\infty}^{0} |x|e^{-x^2/2}dx \\
\text{When x is } <0, \quad |x| &= -x \\
\text{Hence} \quad EX1 &= K\int_{0}^{-\infty} xe^{-x^2/2}dx \\
\text{Due to symmetry} \quad EX1 &= K\int_{0}^{\infty} xe^{-x^2/2}dx \\
\text{Using substitution } u = x^2/2\text{, we find} \quad EX1 &= K \\
\text{By virtue of symmetry} \quad EX2 &= K \\
\text{Final answer} \quad EX &= 2K = \sqrt{2/\pi}
\end{aligned}$$

Prob. 35. Maximum score in the examination is given to be 100 marks. $\mu = 74$and $\sigma = 11$

(a) Finding the percentage of students scoring above 90:

$$\text{Converting to Z} \quad Z = (90 - 74)/11 \approx 1.45$$

$$Area\Big|_{-\infty}^{1.45} = 0.9251$$

$$Area\Big|_{-\infty}^{\infty} = 1.0 \quad \text{Known Property}$$

$$Area\Big|_{1.45}^{\infty} = 0.0749$$

So, the percentage of students scoring above 90 is 7.28%

(b) It is required to find the percentage of students who score in the range $60 \leq X \leq 80$

$$Z1 = (60 - 74)/11 = -1.273$$

$$Z2 = (80 - 74)/11 = 0.5455$$

$$Area\Big|_{-\infty}^{0.5455} = 0.7071 \quad \text{from Appendix B}$$

$$Area\Big|_{-\infty}^{1.273} = 0.898 \quad \text{from Appendix B}$$

$$Area\Big|_{-\infty}^{-1.273} = 0.102$$

$$Area\Big|_{-1.273}^{0.5455} = (0.707 - 0.102) = 0.605$$

(c) Finding the minimum score of highest 10 percent of class.

$$Area\Big|_{-\infty}^{?} = 0.9 \quad \text{to find the upper limit}$$

$$Area\Big|_{-\infty}^{1.285} = 0.9 \quad \text{from Appendix B}$$

$$X = \mu + \sigma Z$$

$$X = 74 + (11)(1.285)$$

$$= 88.14 \quad \text{lowest mark of the upper 10\% students}$$

(d) Finding the maximum score of lowest 5% of class: Area $\Big|_{-\infty}^{-1.65} = 0.05$.

Hence $X = 74 + (11)(-1.65) = 55.85$.

Prob. 37. The occurrences with first digit after decimal being 3 are: 0.31 to 0.39; 1.31 to 1.39 2.31 to 2.39 3.31 to 3.39 There are four occurrences each of width 0.1. Thus the total is 0.4. It is one-tenth of the total width of four. Hence the probability of the first digit after decimal being 3 is $(0.4/4) = 10$ percent.

Prob. 39. For a uniform distribution, the expressions for mean and variance are:

$$\begin{aligned} \text{Mean of uniform distribution} \quad \mu &= (a+b)/2 \\ \text{Variance} \quad \sigma^2 &= (b-a)^2/12 \\ \text{Given that} \quad \mu &= 4; \sigma^2 = 3 \\ \text{It can be easily shown that} \quad a &= 1; b = 7 \end{aligned}$$

Prob. 43. Computer programs are available for conducting simulation of probability trials. For example, the famous Bernoulli trial can be conducted for the case where $P(X = H) \leq 0.6$. In the 10 trials below, the computer-generated random numbers are normally distributed. The outcomes of the trials in terms of Heads or Tails are listed below:

- 0.1905 H
- 0.4285 H
- 0.9963 T
- 0.1666 H
- 0.2223 H
- 0.6885 T
- 0.0489 H
- 0.3567 H
- 0.0719 H
- 0.8661 T

Prob. 45. $X = \min(X_1, X_2, \ldots, X_n)$

(a)

$$P(X > x) = P(\min(X_1, X_2, \ldots, X_n) > x) = P(X_1 > x, X_2 > x, \ldots, X_n > x) = \prod_{i=1}^{n} P(X_i >$$

$$= \prod_{i=1}^{n}(1-x) = (1-x)^n$$

(b) $P(X > x) = (1-x)''$. Take logarithms:

$$\lim_{n\to\infty} \ln P(X > \frac{x}{n}) = \lim_{n\to\infty} (1-x/n)^n = \lim_{n\to\infty} \left[n \ln\left(1 - \frac{x}{n}\right)\right] = \lim_{n\to\infty} \frac{\ln(1-x/n)}{1/n}$$

We define $h = 1/n$. Then this limit is equal to

$$\lim_{h\to 0} \frac{\ln(1-hx)}{h} = \lim_{h\to 0} \frac{\frac{-x}{1-hx}}{1} = -x$$

where we used L'Hospital's rule in the penultimate step. Hence,

$$\lim_{n\to\infty} \ln P(X > \frac{x}{n}) = e^{-x}$$

Prob. 47. A probability density function f(x) is given:

$$f(x) = \begin{cases} \lambda e^{-\lambda x} & \text{for } x > 0 \\ 0 & \text{for } x \leq 0 \end{cases}$$

If you want to verify that f(x) given here is indeed a proper candidate for a distribution function, go to Problem 1.

$$\begin{aligned}
\text{By definition} \quad EX &= \int_0^\infty x\lambda e^{-\lambda x} dx \\
\text{Substitute} \quad u &= x \\
\text{Also substitute} \quad dv &= e^{-\lambda x} dx \\
\text{Hence} \quad v &= e^{-\lambda x}/(-\lambda) \\
\text{Substituting} \quad EX &= xe^{-\lambda x}\Big|_0^\infty + e^{(-\lambda)x}/(-\lambda)\Big|_0^\infty \\
\text{Thus} \quad EX &= (1/(\lambda)
\end{aligned}$$

Prob. 49. The average life of a battery is given to be three months. This means that $\lambda = 1/3$.

$$\text{Given density function} \quad f(x) = \begin{cases} (1/3)e^{-x/3} & \text{for } x > 0 \\ 0 & \text{for } x \leq 0 \end{cases}$$

The probability that the battery will last for more than four months can be found by calculating the area under the distribution curve beyond the value of four.

$$\begin{aligned} \text{Area under curve} \quad Area &= \int_4^\infty \frac{1}{3}e^{-x/3}dx \\ &= -e^{-x/3}\Big|_4^\infty = e^{-4/3} \\ &= 0.2636 \end{aligned}$$

So, the probability is 26% percent that the battery will last longer than four months.

Prob. 51. The lifetime of a radioactive atom is assumed to be exponentially distributed with an average lifetime of 27 days.

(a) It is required to find the probability that the atom will not decay during the first 20 days of starting to observe the atom.

$$\text{Given density function} \quad f(x) = \begin{cases} (1/27)e^{-x/27} & \text{for } x > 0 \\ 0 & \text{for } x \leq 0 \end{cases}$$

The probability that the atom will not decay during the first twenty days can be found by calculating the area under the distribution curve beyond the value of twenty

$$\begin{aligned} \text{Area under curve} \quad Area &= \int_{20}^\infty \frac{1}{27}e^{-x/27}dx \\ &= -e^{-x/27}\Big|_{20}^\infty = e^{-20/27} \\ &= 0.47676 \end{aligned}$$

So, the probability is 48 percent that the atom will not decay during the first twenty days.

(b) If the atom is to last for the second twenty days, it should definitely not have decayed during the first twenty days. Also the value of λ is constant and such a case is called a non-aging process. Thus the answer to part (b) is the same as that of (a), that is, the probability is about 48 percent that the atom will not decay during the second twenty days.

Prob. 53. The lifetime of a technical device is known to be exponentially distributed with a mean life of five years.

(a) It is required to find the probability that the device fails after three years.

$$\text{Given density function} \quad f(x) = \begin{cases} (1/5)e^{-x/5} & \text{for } x > 0 \\ 0 & \text{for } x \leq 0 \end{cases}$$

From the density function, we evaluate thd distribution function by integration. As the result has been discussed earlier, we write:

$$\text{Distribution function} \quad F(x) = \begin{cases} 1 - e^{-x/5} & \text{for } x > 0 \\ 0 & \text{for } x \leq 0 \end{cases}$$

Knowing the distribution function, it is easy to evaluate failure data.

$$\begin{aligned} \text{Distribution function} \quad F(x) &= 1 - e^{-0.2x} \\ \text{P(Device fails within 3 years)} &= F(3) = 1 - e^{-0.6} \approx 0.451 \end{aligned}$$

Thus there is a 45 percent chance that the device will fail within 3 years.

(b) Because of the non-aging property, the probability in question is $1 - F(1) = e^{-1/5}$.

Prob. 55. A technical device, the lifetime of which follows exponential distribution, has the parameter $\lambda = 0.2$ per year.

(a) Expected life time of the above device is five years.

(b)

$$\text{Median lifetime} \quad x_m \stackrel{\text{def}}{=} P(X > x_m)$$

The median lifetime is defined as age x_m, at which P(not having failed by age x_m) is 0.5.

$$\begin{aligned} \text{By definition} \quad S(x_m) &= e^{-\lambda x_m} = 0.5 \\ \text{From given data} \quad -\frac{x_m}{5} &= ln\frac{1}{2} \\ \text{Hence,} \quad x_m &= 5ln2 \end{aligned}$$

Prob. 57. In this problem, the hazard rate function is not a constant as in the previous two examples, but is itself a function of time. This is the key difference between non-ageing entities and ageing ones.

(a) In this part, it is required to find survival chance for the specified period.

$$\begin{aligned} \text{Given hazard rate function} \quad \lambda(x) &= 0.3 + 0.1e^{0.01x} \\ \text{P(It lives for more than five days)} \quad P(T > 5) &= S(5) \\ \text{Survival function} \quad S(x) &= e^{-\int_0^x \lambda(u)du} \\ \text{Doing step by step} \quad 0.1\int_0^5 e^{0.01u}du &= 10(e^{0.05} - 1) \\ \text{Doing step by step} \quad 0.3\int_0^5 0.3du &= 1.5 \\ \text{Hence,} \quad S(5) &= \exp[-(1.5 + 10e^{0.05} - 10)] \end{aligned}$$

Thus the chances of the organism living for 5 days and more is 13.363 percent.

(b) In this part, it is required to find the probability that the organism survives for a period of between 7 and 10 days.

$$\begin{aligned} \text{Hazard rate function (data)} \quad \lambda(x) &= 0.3 + 0.1e^{0.01x} \\ \text{Survival function} \quad S(x) &= e^{-\int_0^x \lambda(u)du} \end{aligned}$$

$$\begin{aligned} P(7 < T < 10) &= P(T > 7) - P(T > 10) = S(7) - S(10) \\ &= \exp[-(2.1 + 10e^{0.07} - 10)] - \exp[-(3 + 10e^{0.1} - 10)] \end{aligned}$$

Prob. 59. Age is defined as the number of years at which the probability of the device/organism not having failed is 0.5. It will be seen below that the relevant equation cannot be solved in a closed form and hence an iterative numerical solution is needed.

$$\begin{aligned}
\text{Hazard rate function is} \quad \lambda(x) &= 1.2 + 0.3e^{0.5x} \\
\text{Survival function is} \quad S(x) &= e^{-\int_0^x \lambda(u)du} \\
\text{Transcendental equation in xm} \quad Sx_m &= e^{-\int_0^{x_m} \lambda(u)du} = 0.5 \\
\text{exp part of integral} \quad 0.3\int_0^{x_m} e^{0.5u}du &= 0.6(e^{0.5x_m} - 1) \\
\text{simpler part of integral} \quad 1.2\int_0^{x_m} du &= 1.2x_m \\
\text{For simplicity, let} \quad y(x_m) &= 1.2x_m + 0.6(e^{0.5x_m} - 1) \\
\text{equation to be solved} \quad e^{-y(x_m)} &= 0.5 \\
\text{Taking Ln on both sides} \quad -y(x_m) &= \ln(0.5) = -0.6931 \\
\text{equation for age} \quad 1.2x_m + 0.6(e^{0.5x_m} - 1) &= 0.6931 \\
\text{equation simplified} \quad x_m + 0.5e^{0.5x_m} &= 1.0776 \\
\text{exp upto 2 terms} \quad x_m + 0.5(1 + 0.5x_m + 0.5{x_m}^2/2 &= 1.0776 \\
\text{simplifying} \quad x_m^2 + 20x_m - 9.248 &= 0 \\
\text{solution} \quad x_m &= 0.45 \\
\text{verifying} \quad 0.45 + 0.5e^{0.50.45} &= -0.0014
\end{aligned}$$

Thus, the mean life (age) is 0.45

Prob. 61. This and the next few problems are on the use of Weibull law by which the hazard rate function increases according to a power law. Such law applies in general to technical devices.

(a) It is required to find the probability that the organism with known hazard rate function will survive for longer than 50 days.

$$\begin{aligned}
\text{Hazard rate function} \quad \lambda(x) &= (2)(10^{-5})x^{1.5} \\
P(T > 50days) &= S(50) \\
\text{Survival function} \quad S(50) &= e^{-\int_0^{50} \lambda(u)du} \\
\text{Substituting} \quad S(50) &= \exp[\int_0^{50} (2)(10^{-5})u^{1.5}du] \\
\text{Exponent} \quad &= (0.8)10^{-5}50^{2.5}
\end{aligned}$$

$$
\begin{aligned}
& = 0.14142 \\
\text{Hence} \quad S(50) & = e^{-0.14142} = 0.86812
\end{aligned}
$$

Thus, the probability of the organism surviving for longer than 50 days is 86.812 percent.

(b) It is required to find the probability that an organism will survive for a period between 50 and 70 days. On the same lines as before, we shall find the value of S(70)

$$
\begin{aligned}
\text{Hazard rate function} \quad \lambda(x) & = (2)10^{-5}x^{1.5} \\
P(T > 70days) & = S(70) \\
\text{Survival function} \quad S(70) & = e^{-\int_0^{70}\lambda(u)du} \\
\text{Substituting} \quad S(70) & = \int_0^{70}(2)10^{-5}u^{1.5}du \\
Exponent & = (0.8)10^{-5}70^{2.5} = 0.32797 \\
\text{Hence} \quad S(50) & = e^{-0.32797} = 0.7204 \\
\text{P(50 to 70 days)} \quad S(50) - S(70) & = 0.86812 - 0.7204 \\
& = 0.1477
\end{aligned}
$$

The chance of an organism being alive during the period 50 to 70 days is 14.77 percent.

Prob. 63. In this and the next problem we find the median life for organisms governed by Weibull law.

$$
\begin{aligned}
\text{Hazard rate function is} \quad \lambda(x) & = (4)10^{-5}x^{2.2} \\
\text{Survival function is} \quad S(x) & = e^{-\int_0^{x}\lambda(u)du} \\
\text{equation in xm} \quad S(x_m) & = e^{-\int_0^{xm}\lambda(u)du} = 0.5 \\
\text{Integral in the exponent} \quad & (4)(10^{-5})X^{3.2}/(3.2) \\
\text{To solve} \quad & \frac{4\times 10^{-5}}{3.2}X_m^{3.2} = -ln\frac{1}{2} \\
X_m & = \left(\frac{(3.2)ln2}{4}\times 10^{5}\right)^{1/3.2} \approx 30.4
\end{aligned}
$$

Prob. 65. Note, $N_x = N_0 S(x)$. Hence $ln\frac{N_{x+1}}{N_x} = ln\frac{S(X+1)}{S(X)} = lnS(X+1) - lnS(X)$

$$\begin{aligned}
\text{Hazard rate function is} \quad & = \lambda(x) \\
\text{Survival function is} \quad S(x) & = e^{-\int_0^x \lambda(u)du} \\
\text{Survival function is} \quad S(x+1) & = e^{-\int_0^{x+1} \lambda(u)du} \\
\text{Shorter Survival} \quad S(x) & = e^{-\int_0^x \lambda(u)du} \\
\text{Ln} \quad LnS(x+1) & = -\int_0^{x+1} \lambda(u)du \\
\text{Ln} \quad LnS(x) & = -\int_0^{x} \lambda(u)du \\
\text{Subtracting} \quad LnS(x+1) - LnS(x) & = -\int_x^{x+1} \lambda(u)du \\
\text{Hence, } -ln\frac{N_{X+1}}{N_X} & = \int_x^{x+1} \lambda(u)du
\end{aligned}$$

12.6 Limit Theorems

Prob. 1. As X is exponentially distributed with parameter $\lambda = \frac{1}{2}$, the density function is:

$$f(x) = \frac{1}{2}e^{\frac{-x}{2}}$$

So we can find Probability P as:

$$P(X \geq 3) = \int_3^\infty \frac{1}{2}e^{\frac{-x}{2}} = -e^{\frac{-x}{2}}\Big|_3^\infty = e^{\frac{-3}{2}} \approx 0.2231$$

Now computing EX we have:

$$EX = \int_0^\infty xf(x)dx = \int_0^\infty \frac{x}{2}e^{\frac{-x}{2}}dx = 2$$

Thus, according to Markov's inequality:

$$P(X \geq 3) \leq \frac{EX}{3} = \frac{2}{3} \approx 0.667$$

and the exact value is much lower.

Prob. 3. We will split the summation in two parts, before and after a and then estimate one of them:

$$EX = \sum_{x} xP(X = x) = \sum_{x<a} xP(X = x) + \sum_{x \geq a} xP(X = x)$$

we can only split the sums because they are both nonnegative and convergent, and since each of the summands is a nonnegative number,

$$EX \geq \sum_{x \geq a} xP(X = x)$$

and observe that the sum extends over values of $x \geq a$, so

$$EX \geq \sum_{x \geq a} xP(X = x) \geq \sum_{x \geq a} aP(X = x) = a \sum_{x \geq a} P(X = x) = aP(X \geq a)$$

dividing by a, the Markov inequality follows.

Prob. 5. Exact: $P(|X| \geq 1) = 1/2$; Chebyshev's inequality: $P(|X| \geq 1) \leq \frac{4}{3}$

Prob. 7. Using Chebyshev's inequality

$$P(|X - \mu| \geq c) \leq \frac{\sigma^2}{c^2}$$

Thus,

$$P(|X - 10| \geq 5) \leq \frac{9}{5^2} = 0.36$$

Prob. 9. $\frac{1}{n} \sum_{i=1}^{n} X_i$ converges to 0.9 as $n \to \infty$

Prob. 11. Since $E|X_i| = \infty$, we cannot apply the law of large numbers as stated in Section 12.6.

Prob. 13. The sample size should be at least 380.

Prob. 15. 0.1587 (without histogram correction); 0.1711 (with histogram correction)

Prob. 17. (a) 0.0023 (without histogram correction); 0.0029 (without histogram correction) (b) 0.83

Prob. 19. With $n = 200$ and $p = 1/2$, we obtain the following
(a) 0.1114 (b) 0.1664 (c) 0.1679

Prob. 21. (a) -11.2 (b) 0.579

Prob. 23. 69

Prob. 25. 385

Prob. 27. (a) 0.3660 (b) 0.3679 (c) 0.243

Prob. 29. (a) 0.1849 (b) 0.1755 (c) 0.1896

Prob. 31. Likely not.

Prob. 33. (a) 0.6065, 0.3033, 0.0758 (b) 0.8391

Prob. 35. 0.1429

Prob. 37. 0.9515

12.7 Statistical Tools

Prob. 1. Re-arranging the sample in increasing order to find median, we will get:

$$0, 3, 6, 12, 13, 17, 18, 21, 25, 47$$

So the median is

$$\frac{13+17}{2} = 15$$

$$\sum_{k=1}^{10} x_k = 162 \qquad \text{and} \qquad \sum_{k=1}^{10} {x_k}^2 = 4246$$

Hence,

$$\bar{X}_n = \frac{162}{10} \approx 16.2$$

and

$$S_n^2 = \frac{1}{n-1}\left[\sum_{k-1}^{n} {x_k}^2 - \frac{1}{n}(\sum_{k-1}^{n} x_k)^2\right] = \frac{1}{9}(4246 - \frac{162^2}{10}) \approx 180.2$$

Prob. 3.

$$\sum_{k=1}^{7} x_k f_k = 3830 \qquad \text{and} \qquad \sum_{k=1}^{7} {x_k}^2 f_k = 46782$$

Hence in order to find the mean,

$$\bar{X}_n = \frac{3830}{321} \approx 11.93$$

and

$$S_n^2 = \frac{1}{n-1}\left[\sum_{k=1}^{7} {x_k}^2 f_k - \frac{1}{n}\left(\sum_{k=1}^{7} x_k f_k\right)^2\right] = \frac{1}{320}(46782 - \frac{3830^2}{321}) \approx 3.389$$

Prob. 5. Finding out the frequency of each clutch size:

Relative Frequency	Frequency
0.05	15
0.09	27
0.12	36
0.19	57
0.23	69
0.12	36
0.13	39
0.07	21

$$\sum_{k=1}^{8} x_k f_k = 1707 \qquad \text{and} \qquad \sum_{k=1}^{8} {x_k}^2 f_k = 10749$$

Hence in order to find the mean,

$$\bar{X}_n = \frac{1707}{300} = 5.69$$

and

$$S_n^2 = \frac{1}{n-1}\left[\sum_{k-1}^{l} {x_k}^2 f_k - \frac{1}{n}(\sum_{k-1}^{l} x_k f_k)^2\right] = \frac{1}{299}(10749 - \frac{1707^2}{300}) \approx 3.465$$

Prob. 7.

$$\sum_{k=1}^{n}(X_k - \bar{X}) = \sum_{k=1}^{n}\left(X_k - \frac{1}{n}\sum_{k=1}^{n} X_k\right) = \sum_{k=1}^{n} X_k - \sum_{k=1}^{n}\left(\frac{1}{n}\sum_{k=1}^{n} X_k\right) = \sum_{k=1}^{n} X_k - n\frac{1}{n}\sum_{k=1}^{n} X_k = 0$$

Prob. 9. Grouping the summation by the elements of the same value we have:

$$\bar{X} = \frac{1}{n}\sum_{k=1}^{n} X_k = \frac{1}{n}\sum_{k=1}^{l} X_k f_k$$

Where l is number of X_k, i.e. the total number of distinct samples.

Prob. 11. The three samples are given below in the table:

Number of sample	$Sample_1$	*Number of sample*	$Sample_1$
1	0.01	6	.46
2	.1	7	.55
3	.19	8	.64
4	.28	9	.73
5	.37	10	.82

Number of sample	$Sample_2$	*Number of sample*	$Sample_2$
1	.03	6	.48
2	.12	7	.57
3	.21	8	.66
4	.30	9	.75
5	.39	10	.84

Number of sample	$Sample_3$	*Number of sample*	$Sample_3$
1	.05	6	.45
2	.13	7	.53
3	.21	8	.61
4	.29	9	.69
5	.37	10	.77

(a) For $Sample_1$:

$$\sum_{k=1}^{10} x_k = 4.15 \qquad \text{and} \qquad \sum_{k=1}^{10} {x_k}^2 = 2.2309$$

Hence,

$$\bar{X}_n = \frac{4.15}{10} \approx 0.415$$

and

$$S_n^2 = \frac{1}{n-1}\left[\sum_{k-1}^{n} {x_k}^2 - \frac{1}{n}(\sum_{k-1}^{n} x_k)^2\right] = \frac{1}{9}(2.3095 - \frac{4.15^2}{10}) \approx 0.06525$$

For $Sample_2$:

$$\sum_{k=1}^{10} x_k = 4.35 \qquad \text{and} \qquad \sum_{k=1}^{10} {x_k}^2 = 2.5605$$

Hence,

$$\bar{X}_n = \frac{4.35}{10} \approx 0.435$$

and

$$S_n^2 = \frac{1}{n-1}\left[\sum_{k-1}^{n} {x_k}^2 - \frac{1}{n}(\sum_{k-1}^{n} x_k)^2\right] = \frac{1}{9}(2.5605 - \frac{4.35^2}{10}) \approx 0.07425$$

For $Sample_3$:

$$\sum_{k=1}^{10} x_k = 3.81 \qquad \text{and} \qquad \sum_{k=1}^{10} {x_k}^2 = 2.209$$

Hence,

$$\bar{X}_n = \frac{3.81}{10} \approx 0.381$$

and

$$S_n^2 = \frac{1}{n-1}\left[\sum_{k-1}^{n} {x_k}^2 - \frac{1}{n}(\sum_{k-1}^{n} x_k)^2\right] = \frac{1}{9}(2.209 - \frac{3.81^2}{10}) \approx 0.08415$$

(b) Combining the three samples:

$$\sum_{k=1}^{30} x_k = 12.31 \qquad \text{and} \qquad \sum_{k=1}^{30} {x_k}^2 = 7.0004$$

Hence,

$$\bar{X}_n = \frac{12.31}{30} \approx 0.41033$$

and

$$S_n^2 = \frac{1}{n-1}\left[\sum_{k-1}^{n} {x_k}^2 - \frac{1}{n}(\sum_{k-1}^{n} x_k)^2\right] = \frac{1}{29}(7.0004 - \frac{12.31^2}{30}) \approx 0.06721$$

(c) True value of mean will be $\mu = 0.5$ and variance can be computed to $\sigma^2 = 1/12$, as we expect the combined sample, because of being larger, provides a smaller variance, since $var(\bar{X_n}) = \sigma^2/n$.

Prob. 13.

$$\sum_{k=1}^{10} x_k = 162 \qquad \text{and} \qquad \sum_{k=1}^{10} {x_k}^2 = 4246$$

Hence,

$$\bar{X_n} = \frac{162}{10} \approx 16.2$$

and

$$S_n^2 = \frac{1}{n-1}\left[\sum_{k-1}^{n} {x_k}^2 - \frac{1}{n}(\sum_{k-1}^{n} x_k)^2\right] = \frac{1}{9}(4246 - \frac{162^2}{10}) \approx 180.18$$

For sample error:

$$S_{\bar{X}} = \frac{S_n}{\sqrt{n}} \approx 4.245$$

Prob. 15. Since X_k is from a normal distribution then

$$P(-1.96 \leq Z \leq 1.96) = 0.95$$

and the confidence interval is given by:

$$[\bar{X_n} - 1.96\frac{\sigma}{\sqrt{n}}, \bar{X_n} + 1.96\frac{\sigma}{\sqrt{n}}]$$

Now using the fact that the $\bar{X_n} = 0.061$ and the $\sigma = 0.7427$ for this sample, the confidence interval is given by $[-0.3993, 0.5213]$.

Prob. 17. Use your calculator and prepare a data set for in-class discussion.

Prob. 19. The sample mean is

$$\bar{X_n} = \frac{117}{162} \approx 0.722$$

The estimate for the probability of germination success of seeds is therefore $\hat{p} = 0.722$. The standard error is

$$S.E. = \sqrt{\frac{\hat{p}(1-\hat{p})}{n}} = \sqrt{\frac{0.722 \times 0.288}{162}} \approx 0.03583$$

We can thus report the result as 0.722 ± 0.03583. Since n is large, we find a confidence interval as

$$[0.722 - (1.96)(0.03583), 0.722 + (1.96)(0.03583)] = [0.6518, 0.7922]$$

Prob. 21. To facilitate the computation, we construct the following table:

x_k	y_k	$x_k - \bar{x}$	$y_k - \bar{y}$	$(y_k - \bar{x})(x_k - \bar{x})$
-3	-6.3	-2.5	-4.4167	11.04175
-2	-5.6	-1.5	-3.7167	5.57505
-1	-3.3	-0.5	-1.4167	0.70835
0	0.1	0.5	1.9833	0.99165
1	1.7	1.5	3.58333	5.37495
2	2.1	2.5	3.9833	9.95825
$\bar{x} = -0.5$	$\bar{y} = -1.8833$	$\sum(x_k - \bar{x})^2$ $= 17.5$	$\sum(y_k - \bar{y})^2$ $= 67.96833$	$\sum(x_k - \bar{x})(y_k - \bar{y})$ $=33.65$

Now,

$$\begin{aligned} \hat{b} &= \frac{33.65}{17.5} = 1.92286 \\ \hat{a} &= -1.8833 - (1.92286)(-0.5) = -0.92187 \end{aligned}$$

Hence, the linear regresion line is given by

$$y = 1.92x - 0.92$$

The coefficient of determination r^2 is given by

$$r^2 = \frac{[\sum(x_k - \bar{x})(y_k - \bar{y})]^2}{\sum(x_k - \bar{x})^2 \sum(y_k - \bar{y})^2}$$

Substituting the values we will find

$$r^2 = \frac{(33.65)^2}{(17.5)(67.96)} = 0.9521$$

Prob. 23. We know that

$$\bar{y} - a - b\bar{x} = 0$$

Where

$$\bar{y} = \frac{1}{n} \sum y_k \qquad \text{and} \qquad \bar{x} = \frac{1}{n} \sum x_k$$

So the sum of the residuals about the linear regression line is equal to zero.

Prob. 25. To facilitate the computation, we construct the following table:

x_k	y_k	$x_k - \bar{x}$	$y_k - \bar{y}$	$(y_k - \bar{x})(x_k - \bar{x})$
69	15	−10.8	−2	21.6
70	15	−9.8	−2	19.6
72	16	−7.8	−1	7.8
75	16	−4.8	−1	4.8
81	17	1.2	0	0
82	17	2.2	0	0
83	16	3.2	−1	−3.2
84	18	4.2	1	4.2
89	20	9.2	3	27.6
93	20	13.2	3	39.6
$\bar{x} = 79.8$	$\bar{y} = 17$	$\sum(x_k - \bar{x})^2 = 589.6$	$\sum(y_k - \bar{y})^2 = 30$	$\sum(x_k - \bar{x})(y_k - \bar{y}) = 122$

Now,

$$\begin{aligned} \hat{b} &= \frac{122}{589.6} = 0.20692 \\ \hat{a} &= 17 - (0.20692)(79.8) = 0.487788 \end{aligned}$$

Hence, the linear regresion line is given by

$$y = 0.20x + 0.48$$

The coefficient of determination r^2 is given by

$$r^2 = \frac{[\sum(x_k - \bar{x})(y_k - \bar{y})]^2}{\sum(x_k - \bar{x})^2 \sum(y_k - \bar{y})^2}$$

Substituting the values we will find

$$r^2 = \frac{(122)^2}{(589.6)(30)} = 0.841$$

12.9 Review Problems

Prob. 1.

(a) There are 25 students and the total number of days in a year are 365. So the first person have a choice of having birthday on any of the days. Therefore,

$$p_1 = \frac{365}{365}$$

The second person has a choice of only 364 days to have a birhtday, so the probability for 2 people is,

$$p_2 = \frac{365}{365}\frac{364}{365}$$

similarly, for the 25 people,

$$p_{25} = \frac{365}{365}\frac{364}{365}\frac{363}{365}\cdots\frac{341}{365} = \frac{365!}{340!\,365^{25}} = 0.431$$

(b) We can write the general equation as:

$$p_n = \frac{365!}{(365-n)!\,365^n}$$

then it is easy to see that

$$\begin{aligned} p_{n+1} &= \frac{365!}{(365-(n+1))!\,365^{n+1}} \\ &= \frac{365!\,(365-n)}{(365-n)(365-(n+1))!\,365^n 365} \\ &= \frac{365!\,(365-n)}{(365-n)!\,365^n 365} \\ &= p_n\frac{365-n}{365} \end{aligned}$$

Prob. 3. There are $15 \times 14 \times 13$ ways to choose three plants (considering the order) for the first plot, and considering that choices of order (ABC, ACB, BAC, ...) does not matter, we can divide the total by 6, and the choices for the first plot are $15 \times 14 \times 13/6$,and proceed in this way for the rest of the plots, the total will be $15!/6^5 = 168,168,000$.

Prob. 5.

(a) We know that 0.42 of the seeds of a certain plant germinates, so the expected number of germinating seeds in a sample of ten seeds is:

$$\# \text{ of germinating seeds} = 0.42 \times \# \text{ of seeds} = 0.42 \times 10 = 4.2 \approx 4$$

(b) The probability that none of the seeds will grow in a plant are:

$$p = (1 - 0.42)^{10} = (0.58)^{10} \approx 0.0043$$

(c) On 5 pots, it will be: $5 \times 0.0043 = 0.0215$.

(d) The probability of number of pots having no seeds is: $P = 1 - 0.0043^5$
.

Prob. 7.

(a The chance of passing in the first trial is 0.2, and chance of failing is 0.8, So the the probability of passing in the second trial after failing in the first trial is:

$$p = 0.8 \times 0.2 = 0.16$$

(b) The chance of passing in the first trial is 0.2, and simmilarly is the chance of passing in the second trial. So the probability of passing in the second trial is: 0.2.

Prob. 9. (a) $\mu = 170$, $\sigma = 6.098$ (b) 0.4013

Prob. 11. A white flowering plant is of genotype cc. If the red-flowering parent plant is of genotype CC is crossed with a white-flowering plant, then all offspring are of genotype Cc and therefore produce red flowers. If the red-flowering parent plants is of genotype Cc is crossed with a white-flowering plant, then with probability 0.5 an offspring is of genotype Cc (and therefore red flowering) and with a probability 0.5 of genotype cc (and therefore white flowering). Assume that percentage of red-flowering parent plants of genotype Cc is x, so the percentage of red flowering parent plants of genotype CC would be $1 - x$. Therefore, the percentage red-flowering parent plants would be:

$$P(\text{red offspring}) = 0.9 = x(0.5) + (1 - x)(1) = 1 - 0.5x$$

which solves to $x = 0.2$, so the percentage of red-flowering parent plants would be 20 percent.

Prob. 13.

(a) $V = \frac{1}{n}\sum_{k=1}^{n}(X_k - \bar{X_n})^2 = \frac{n-1}{n}\frac{1}{n-1}\sum_{k=1}^{n}(X_k - \bar{X_n})^2 = \frac{n-1}{n}S^2$

(b) $EV = E(\frac{n-1}{n}S^2) = \frac{n-1}{n}ES^2 = \frac{n-1}{n}\sigma^2$